Climate Change and Africa

At the beginning of the twenty-first century, no environmental
issue is of such truly global magnitude as the issue of climate change. The poorer, developing countries are the least
equipped to adapt to the potential effects of climate change,
although most of them have played an insignificant role in causing it; African countries are amongst the poorest of the developing countries. This book presents the issues of most relevance
to Africa, such as past and present climate, desertification,
biomass burning and its implications for atmospheric chemistry and climate, energy generation, sea-level rise, ENSO-
induced drought and flood, adaptation, disaster risk reduction,
the UNFCCC and Kyoto Protocol (especially the Clean Development Mechanism), capacity-building, and sustainable development. It provides a comprehensive and up-to-date review of
these and many other issues, with chapters by the leading experts from a range of disciplines. *Climate Change and Africa*
will prove an invaluable reference for all researchers and policy
makers with an interest in climate change and Africa.

CLIMATE CHANGE AND AFRICA

PAK SUM LOW

CAMBRIDGE
UNIVERSITY PRESS

CAMBRIDGE UNIVERSITY PRESS
Cambridge, New York, Melbourne, Madrid, Cape Town, Singapore, São Paulo

Cambridge University Press
The Edinburgh Building, Cambridge CB2 2RU, UK

Published in the United States of America by Cambridge University Press, New York

www.cambridge.org
Information on this title: www.cambridge.org/9780521836340

First published 2005
This digitally printed first paperback version 2006

A catalogue record for this publication is available from the British Library

ISBN-13 978-0-521-83634-0 hardback
ISBN-10 0-521-83634-4 hardback

ISBN-13 978-0-521-02995-7 paperback
ISBN-10 0-521-02995-3 paperback

This book is dedicated to the late Professor David O. Hall and the late Dr Ruben O. Agwanda, and to all those who love Africa and its people.

Contents

Notes on contributors

EDITOR

Dr Pak Sum Low has been the Regional Adviser on Environment and Sustainable Development at United Nations Economic and Social Commission for Asia and the Pacific (UNESCAP) since June 2001. He has previously worked in the United Nations Environment Programme (1991–1999); Climatic Research Unit, School of Environmental Sciences, University of East Anglia, Norwich, UK (1988–1991); Institute of Applied Physical Chemistry (1987–1988) and Institute for Atmospheric Chemistry (1986–1987), Jülich Nuclear Research Centre, Germany. He was also a consultant for the secretariats of the United Nations Framework Convention on Climate Change (1999–2000) and the United Nations Convention to Combat Desertification (2000). First trained as a chemical engineer at the University of Canterbury, New Zealand, in the early 1970s, Dr Low later specialized in environmental studies, focusing on atmospheric research, and completed a Masters at the University of Adelaide, Australia and a Ph.D. at the University of Tasmania, Australia. He has travelled extensively within Africa and Asia and the Pacific.

CONTRIBUTORS OF FOREWORDS

Dr Mostafa Kamal Tolba, Former Executive Director, United Nations Environment Programme (UNEP) (1976–1992), and currently President, International Centre for Environment and Development (Geneva and Cairo). Professor of Microbiology, Cairo University. Dr Tolba is the recipient of honorary degrees from several universities around the world. He holds a special B.Sc. degree in Botany and a Ph.D. in Plant Pathology. He is the author of almost 100 papers (1950–1973) on plant diseases, anti-fungal substances and the physiology of micro-organisms, and of more than 600 statements and numerous books and articles on environment and sustainable development.

Ambassador Michael Zammit Cutajar was Executive Secretary of the secretariat of the United Nations Framework Convention on Climate Change (UNFCCC) based in Bonn, Germany until his retirement in January 2002. The UNFCCC secretariat provides substantive and technical support and advice to the intergovernmental negotiations within the scope of the Convention, including those on its Kyoto Protocol. A citizen of Malta, born 1940, Michael Zammit Cutajar devoted a large part of his career to work in and around the United Nations on international cooperation for development and environment. He undertook several assignments with UNCTAD and was part of the founding secretariat of UNEP. He worked with NGOs in his early and middle career. He currently contributes to discussions on the evolving global strategy to combat climate change, chairs the Advisory Group of the World Bank's Community Development Carbon Fund and advises the Government of Malta as its Ambassador for International Environmental Affairs.

Ambassador Mikko Pyhälä (b. 1945) has been in the diplomatic service of Finland since 1972. His posts include Peru, Pakistan, India, Italy, Mexico and Czechoslovakia. He was with the United Nations Environment Programme (1991–1995) as Chief of Clearing-house Unit, Chief of GEF-Unit and Secretary to the Scientific and Technical Advisory Panel to the GEF. He has published books and reports on environment and development. Ambassador Pyhälä received the Finlandia Award for best non-fiction book 1992 for *Amazonia* (co-authored with Dr Jukka Salo), and Peru's National Biodiversity Award 2002. He was conferred Dr.h.c. by the National Amazonian University of Peru.

Professor Dieter H. Ehhalt was Director of the Institute for Atmospheric Chemistry at the Research Centre in Jülich and Professor of Geophysics at the University of Cologne, Germany, until his retirement in August 2000. He is a Fellow of the American Geophysical Union and a member of the Academia Europaea. His current research interests are atmospheric chemistry and physics. On these and related subjects he has published over 250 articles. He was Coordinating Lead Author for the chapter 'Atmospheric chemistry and greenhouse gases' in the *Third Assessment Report of the IPCC* (2001), as well as Lead Author for the earlier Assessment Report (1995). From 1995 to 1997 he served as

Vice Chairman of IPCC Working Group I. From 1983 to 2002 he was co-editor of the *Journal of Atmospheric Chemistry*.

CONTRIBUTORS OF PREFACES

Professor Godwin Olu Patrick Obasi, a Nigerian, was Secretary-General of the World Meteorological Organization from January 1984 to December 2003. He holds a B.Sc. Honours in Mathematics and Physics (1959) from McGill University, Montreal, Canada, an M.Sc. (1960) and a D.Sc. (1963) in Meteorology both from the Massachusetts Institute of Technology, USA, and won the Carl Rossby Award for the best doctoral thesis. He was Dean of the Faculty of Science, Professor of Meteorology and Chairman of the Department of Meteorology at the University of Nairobi, Kenya (1967–1976). Professor Obasi has published several scientific articles and has been Consulting Editor to many journals. He is currently Vice-President of the Third World Academy of Sciences, and a fellow or honorary member of many meteorological/hydrological professional societies. He has been awarded university honorary degrees and honoured by several academies of sciences and governments. He has been at the forefront to draw the world's attention on the issue of climate change and contributed to initiation of several multilateral environmental agreements. He initiated the establishment of the African Centre of Meteorological Applications for Development in Niamey, Niger; the Drought Monitoring Centres in Nairobi, Kenya and Harare, Zimbabwe; and the ASEAN Specialized Meteorological Centre in Singapore.

Dr Wulf Killmann holds an M.Sc. and a Ph.D. in Wood Technology, both from the University of Hamburg, Germany. After having started his professional career in Germany, he worked for 20 years in Development Cooperation in Africa, Asia and Latin America, mostly for the German Agency of Technical Cooperation. Following assignments in Liberia, the Philippines and Pakistan, he built up Forestry Cooperation Programmes in Malaysia, and later on in Honduras. Since 1999 Dr Killmann has held the position of Director, Forest Products and Economics Division with the Food and Agriculture Organization of the United Nations. Since 2001 he has been the Chair of FAO's Interdepartmental Working Group on Climate Change.

Dr Sálvano Briceño is the Director of the Geneva-based Secretariat of the United Nations International Strategy for Disaster Reduction (ISDR). A Venezuelan, he has a Doctorate in Public Law from the University of Paris II (Panthéon-Sorbonne) and a Master of Public Management from Harvard University. His professional career has focused on public policy and management of international programmes in fields related to environment, climate, disaster reduction, and sustainable development. His roles have included Director General responsible for environmental education, professional development and international relations in the Ministry of Environment and Renewable Natural Resources, Venezuela; Executive Officer of IUCN's Commission on Education, where he worked on environmental education programmes and coordinated a worldwide network of experts; Coordinator of UNEP's Caribbean Environment Programme at Kingston, Jamaica, where he collaborated closely with the Pan-Caribbean Disaster Preparedness and Prevention Programme (PCDPPP); Coordinator of the BIOTRADE and GHG Emissions Trading Initiatives at UNCTAD; Deputy Executive Secretary of the UN Convention to Combat Desertification (UNCCD); and Coordinator of Intergovernmental and Institutional Support at the Secretariat of the UN Framework Convention on Climate Change (UNFCCC) where he was responsible for assisting the Executive Secretary with management, legal, information support, external relations and interagency relations.

Dr Hassan Virji is Deputy Director for Global Change System for Analysis, Research and Training (START), an organization that is co-sponsored by the Earth System Science partnership comprising the major international research programmes on global environmental change (IHDP, IGBP, WCRP and DIVERSITAS). Prior to his current position, he served as the Executive Secretary of the US Interagency Subcommittee on Global Change Research while based at the US National Science Foundation, as the Deputy Executive Director of the International Geosphere-Biosphere Programme based at the Royal Swedish Academy of Sciences in Stockholm, and as the Associate Programme Director for the Climate Dynamics Programme of the US National Science Foundation. He has also held academic and research faculty positions at the University of Nairobi and the University of Wisconsin. Dr Virji holds a Ph.D. degree in Meteorology from the University of Wisconsin-Madison, USA.

CONTRIBUTORS OF CHAPTERS

Dr Ibrahim Abdel Gelil is currently a professor of the academic chair of H H Sheikh Zayed Bin Sultan Al Nahayan, and Director, Environmental Management Programme, Arabian Gulf University, Bahrain. He was the CEO (1997–2002) of the Egyptian Environmental Affairs Agency (EEAA), and the chairman (1994–1997) of the Egyptian Organization for Energy Planning (OEP), the Government of Egypt's agency responsible for energy policy planning and analysis and promotion of energy efficiency. In addition, he is a board member of many national Committees, including the World Energy Council, the National Specialized Councils, the National Council of Natural Resources and Environment, the Egyptian Academy for Science and Technology. He has authored and co-authored more than 50 publications that were published nationally and internationally.

Dr Reuben Agwanda (1956–2001) obtained his Ph.D. in Statistics from the London School of Hygiene and Tropical Medicine, UK, in 1998. He joined the Kenya Medical Research Institute (KEMRI) in 1981, where he rose through the ranks to become a Senior Research Officer. He facilitated several training courses on statistics and the use of microcomputers for health research and data analysis organized by WHO and the International Development Research Centre (IDRC) of Canada. He published over 15 articles in medical journals.

Dr Stephen O Andersen is currently the Director of Strategic Projects in the USEPA Climate Protection Partnerships Division and Co-Chair of the Montreal Protocol Technology and Economic Assessment Panel. He has been Deputy Director of the USEPA Stratospheric Protection Division where he specialized in industry partnerships, international cooperation, and market incentives. Earlier he was Professor of Environmental Economics at the College of the Atlantic and the University of Hawaii and a visiting scholar in Kyoto University. He also worked for consumer, environmental and legal NGOs. He received his Ph.D. from the University of California, Berkeley. He co-authored with K. Madhava Sarma a book on *Protecting the Ozone Layer: The United Nations History (2002)*, and with Durwood Zaelke a book on *Industry Genius: Inventions and People Protecting the Climate and Fragile Ozone Layer (2003)*.

Professor Meinrat O. Andreae leads the Biogeochemistry Department of the Max Planck Institute for Chemistry in Mainz, Germany. His scientific interests include biogeochemistry, atmospheric chemistry, chemical oceanography, biomass burning, chemistry/climate interactions and the sulphur cycle. Much of his group's research work is being undertaken in the course of field campaigns in Amazonia, Africa, the boreal region (e.g., Siberia), and South-East Asia. His research group is investigating the role of fire in ecology; climate and atmospheric chemistry; and the exchange of trace gases in the soil–plant–atmosphere system. These research activities are closely tied to the IGBP, and they involve extensive international collaboration.

Ayaa Kojo Armah holds Masters degrees in Marine Biology from the University of Ghana and in Coastal Management from the University of Newcastle. Currently he is a Senior Lecturer in the Department of Oceanography and Fisheries in the University of Ghana. He has published over 25 articles and three books: *Coastal Zone Profile of Ghana* (1998); *Save Sea Turtles – A Primer for Sea Turtle Conservation for Coastal Communities in Ghana* (1997); and *The Coastal Zone of West Africa: Problems and Management* (1997). He has served as a consultant for UNIDO, IOC-UNESCO and UNEP.

Dr Robert C. Balling, Jr. is currently the Director of the Office of Climatology and Professor in the Department of Geography at Arizona State University. He has published over 100 articles in the professional scientific literature, and served as a consultant to the United Nations Environment Programme and the World Meteorological Organization. He has published three books: *The Heated Debate: Greenhouse Predictions Versus Climate Reality* (1992); *Interactions of Desertification and Climate* (1996); and *The Satanic Gases: Clearing the Air about Global Warming* (2000).

Dr Reid Basher is a senior officer at the UN Secretariat for the International Strategy for Disaster Reduction (UN/ISDR), and heads its Platform for the Promotion of Early Warning (PPEW) in Bonn, Germany, and Director Applications Research at the International Research Institute for Climate Prediction (IRI), New York. His interests lie in the interplay between the science, policy and applications practice of climate risk, encompassing climate change, the management of seasonal variability, and the issues of extremes of climate and disasters. As Director of Applications Research at the International Research Institute (IRI) for Climate Prediction, New York, from 1999 to 2003, he managed substantial USAID-funded projects on climate impacts reduction in Africa. A New Zealand citizen, he has been responsible for a broad range of climate research and applications work in New Zealand, including public and commercial climate services and the direction of the national cooperative climate network and national climate database. Dr Basher's qualifications include a Diploma of Business Administration in addition to science degrees. He worked in Fiji for two years, developed a seasonal outlook scheme for tropical cyclones in the South Pacific islands, and was an active participant in and consultant to activities of the South Pacific Regional Environmental Programme (SPREP). He has been a Lead Author in impacts reports of the Intergovernmental Panel on Climate Change (IPCC), and Chair of the Working Group on Climate Data of the World Meteorological Organization (WMO).

Dr John C. Beier is a tenured Professor in the Department of Epidemiology and Public Health at the University of Miami School of Medicine. He is the Director of the Global Public Health Programme at the University of Miami. He was formerly a Professor in the Department of Tropical Medicine, School of Public Health and Tropical Medicine, Tulane University. Dr Beier received training in medical entomology from the Johns Hopkins University and the University of Notre Dame. He has published over 150 papers and book chapters in the areas of mosquito ecology and malaria transmission dynamics. He directs programmes on mosquito biology and malaria control in several countries in Africa.

Dr Guy Brasseur is Director of the Department of Biogeochemical Systems at the Max Planck Institute for Meteorology in Hamburg, Germany; Scientific Director of the German Climate Computer Centre; Chair of the Scientific Committee of IGBP; as well as a member of the International Academy for Astronautics and Aeronautics, and the Academia Europaea. He was previously Senior Scientist and Director of the Atmospheric Chemistry Division, NCAR at Boulder, Colorado (1988–99). His major areas of scientific interest are atmospheric chemistry, aeronomy, ozone depletion and climate change, with specific interest in the development of Earth system models. He has authored over 130 peer-reviewed publications, and numerous book chapters and other reports, and co-authored (with Susan Solomon) a book entitled *Aeronomy of the Middle Atmosphere* (Kluwer, 1986). Dr Brasseur served (with J. Orlando and G. Tyndall) as editor of *Atmospheric Chemistry and Global Change* (Oxford University Press, 1999). He was Editor-in-Chief for *Journal of Geophysical Research – Atmospheres* (1992–1996). He won the NCAR Outstanding Publication Award (1996), and the European Physical Society Award for outstanding contributions in environmental physics (2002).

Dr Sálvano Briceño (see above)

Dr Robert Brinkman retired from FAO in 1998. He served FAO as a field expert 1961–1971, and from 1987 as Chief, Soils Service and Director, Land and Water Development Division. He obtained a Ph.D. in soil science at Wageningen Agricultural University, where he taught from 1971 to 1987, interrupted by Advisory work for the World Bank, FAO and the Netherlands Ministry of Development Cooperation. He has published over 100 scientific papers and handbook chapters.

Jake Brunner is Senior Director for Mainland Asia at Conservation International (CI), a biodiversity conservation group based in Washington, DC. Prior to joining CI, he was a senior associate at the World Resources Institute, where he managed research programmes in South-East Asia and West and Central Africa on forest and water resources management. His research interests include the impact of governance on natural resources management, and the application of GIS and remote sensing for environmental policy analysis. Jake Brunner was educated at Oxford University and Imperial College of Science and Technology, London, UK.

Dr Paul V. Desanker is Associate Professor of Geography and Associate Director for Research for the Alliance for Earth Sciences, Engineering and Development in Africa (AESEDA) at Pennsylvania State University, USA. He was formerly Research Assistant Professor in the Department of Environmental Science, University of Virginia. He was a member of the IPCC Task Group on Climate Impact Assessment (TGCIA) and now serves on the UNFCCC Least-Developed Country Expert Group and on the Afforestation-Reforestation Working Group of the Clean Development Mechanism (CDM) Executive Board. He was the Convening Lead Author for the chapter on Africa in the *IPCC Third Assessment Report,* and the editor of a special issue on African climate change in *Climate Research* (Volume 18, Special 8, August 2001). His other research covers the Miombo ecosystem and the region of southern Africa, including the study of land-use change, carbon accounting, and impacts of, and adaptation to, climate change. He is the coordinator of the IGBP/START/LUCC Miombo Network, as well as the Miombo GOFC/GOLD pilot project.

Dr Ruth Doherty is currently a Research Fellow at the Institute of Atmospheric and Environmental Science, School of Geo-Sciences at the University of Edinburgh, UK. She has previously worked in the Environmental and Societal Impacts Group at the National Centre for Atmospheric Research, Colorado, USA, and in the Climatic Research Unit at the University of East Anglia, UK. Her main research interests are in climate change and climate variability, both over the twentieth century and in future scenarios as simulated by Global Climate Models. She has a number of publications in this area. She is also interested in climate/land-surface/chemistry interactions.

Dr Richard D. Duke is a consultant with McKinsey & Company, USA. His dissertation developed an economic rationale for government 'buydowns' of emerging clean energy technologies based on spillover of learning-by-doing. His publications include various empirical and theoretical analyses of solar home systems markets, an examination of the economics of market transformation programmes, and a techno-economic assessment of the government policies necessary to catalyse markets for grid-connected residential photovoltaics. He also produced a documentary (viewable at www.princeton.edu/duke) on rural energy services delivery in Kenya and South Africa.

Dr Siri H. Eriksen is currently a Senior Research Fellow at the Centre for International Climate and Environmental Research in Oslo, Norway, and a Visiting Fellow at the Climatic Research Unit, University of East Anglia, UK. She has published on topics regarding environmental change and local vulnerability in Africa, including Transport and GHG emissions (1995); Land Tenure and Wildlife Management (1996); and Linkages between Climate Change and Desertification in East Africa (2001). She completed a Ph.D. thesis on *Responding to Global Change: Vulnerability and Management of Local Agro-ecosystems in Kenya and Tanzania* in 2000.

Dr John I. Githure received his Ph.D. in parasitology from the University of Nairobi in 1989. He is the Head of the Human Health Division at the International Centre of Insect Physiology and Ecology (ICIPE) and Chief Research Officer at Kenya Medical Research Institute (KEMRI), both in Nairobi. He has many years of experience in coordinating and managing research projects and organizing meetings and training courses. He has had a long and distinguished career in malaria and leishmaniasis research and has performed assignments as a malaria expert with many organizations, including WHO, the World Bank, the European Union and USAID. He has over 70 publications in peer reviewed journals.

Dr René Gommes coordinates the activities of the Agrometeorology Group in the Environment and Natural Resources Service of FAO. Although his basic training is in biogeochemistry and plant ecology, he spent most of his career in WMO and FAO as an agricultural climatologist. His main professional interests include agro-climatic risk assessments, the impact of extreme geophysical factors on food security, and the development of operational tools for agrometeorologists in developing countries. He has worked in about 40 countries.

Dr Alex Guenther is a senior scientist and the head of the Biogeochemistry Section of the Atmospheric Chemistry Division of the National Center for Atmospheric Research, Boulder, Colorado, USA. He has been investigating the biosphere–atmosphere exchange of trace gases for nearly 20 years and has led field investigations in Africa and four other continents using enclosures, above-canopy towers, tethered balloons and aircraft. Volatile organic compound emissions from vegetation have been a topic of particular interest for him and his algorithms describing the response of these emissions to changing environmental conditions are widely used in regional air quality models and global chemistry simulations.

Jacques du Guerny is a demographer. He started his career in public health research, then joined the United Nations population research and programmes specializing on Asian and African problems. As the former chief of the FAO Population Programme Service, he focused on inter-relations between demographic factors, food requirements and environmental issues, as well as on linkages between agriculture and the HIV/AIDS epidemics. He is presently retired in Provence, France.

Professor David Oakley Hall (1935–1999) studied at the Universities of Natal and California, before completing his Ph.D. on photosynthesis at UC Berkeley. In 1964 David began 35 years of research and teaching work at King's College London, where he developed the undergraduate course in Cell Biology, and supervised the formation of the School of Biological Sciences. His research and organizational skills ranged from laboratory work to initiating international committees, and to training a worldwide network of researchers with skills in field measurement and analysis of photosynthesis for UNEP. In later life he focused on photosynthesis as a source of sustainable energy and food for the world, and became an enthusiastic ambassador for the merits of biomass energy. He founded and edited the journal *Biomass and Bioenergy*, published more than 400 articles, was the author of 12 books, and 19 students obtained doctorates under his supervision. He is remembered with warmth, admiration and respect by all who worked and studied with him.

Dr David A. Hastings has been Scientific Affairs Officer for the United Nations Economic and Social Commission for Asia and the Pacific, in Bangkok, since April 2002. Prior to that he lectured at Kwame Nkrumah University of Science and Technology (Kumasi, Ghana) and Michigan Technological University (USA); worked for the Ghana Geological Survey as Senior Geophysicist and Assistant Director; the US Geological Survey's Earth Resources Data Center as Senior Applications Scientist; and the National Oceanic and Atmospheric Administration's National Geophysical Data Center as Chief of Data Integration and Remote Sensing, later Chief of Environmental Indicators. He studied physics and international relations at Tufts University (USA) and the Università di Bologna (Italy), geosciences and engineering at Brown University and the University of Arizona (both USA). He has worked, lectured, and traveled in over 100 countries on six continents, and has published over 100 papers. His latest books (co-edited with Ryutaro Tateishi) are *Global Environmental Databases: Present Situation – Future Direction Volume I* (2000) and *Volume II* (2002). Scientific websites developed by David receive several thousand hits and over 10 gigabytes of downloads weekly.

Dr Günter Helas is a senior scientist at the Biogeochemistry Department of the Max Planck Institute for Chemistry in Mainz, Germany. He has been with the Institute since 1977. Presently, he is interested in biogeochemical cycling of atmospheric trace compounds relevant to climate, such as those from vegetation fires, among others. He is also involved in research on biofuels and their emissions in Africa, and on aerosol morphology and composition.

Dr Larry W. Horowitz is a physical scientist at the NOAA Geophysical Fluid Dynamics Laboratory (GFDL) in Princeton, New Jersey. His research focuses on tropospheric chemistry and air quality. He has developed global chemical transport models (CTMs) to simulate the atmospheric distribution of ozone and its precursors, including reactive nitrogen, carbon monoxide, and volatile organic compounds. He has used these models to study the impact of natural and anthropogenic emissions on the abundance of trace species in the atmosphere, and to project the future concentrations of these species.

Professor Mike Hulme is Executive Director of the Tyndall Centre for Climate Change Research and is based in the School of Environmental Sciences at University of East Anglia, Norwich, UK, where he has worked for the last 17 years. His general research interest is global climate change, the construction and application of climate change scenarios, and African climate. He has published over 120 scientific journal articles and book chapters on these and other topics, together with over 220 reports and popular articles. He was a Coordinating Lead Author for the chapter on 'Climate scenario development' for the *Third Assessment Report of the IPCC* (2001), as well as a contributing author for several other chapters.

Dr Chris Justice is a professor in the Geography Department, University of Maryland, USA. He leads the Land Discipline Group for the NASA Moderate Imaging Spectroradiometer (MODIS) at NASA/GSFC. He is responsible for the MODIS Fire Product. Dr Justice is the Project Scientist for the NASA Land-Cover and Land-Use Change (LCLUC) Program, a member of the Science Advisory Panel for the NOAA Office of Global Programs, and on the Strategic Objective Team for USAID's Central Africa Regional Project for the Environment. He has current projects on forest and land cover, monitoring and modelling in central Africa. He played a key role in IGBP Data and Information System and in the development of NASA EOS Science Working Group on Data, and is currently the Fire Implementation Team Leader for the GOFC Project, which is part of the Global Terrestrial Observing System.

Dr Daniel M. Kammen holds the Class of 1935 Distinguished Chair in Energy, and is professor in the Energy and Resources Group, and the Goldman School of Public Policy at the University of California, Berkeley, USA. He is also the founding director of the Renewable and Appropriate Energy Laboratory (RAEL). He received his education in physics from Cornell and Harvard Universities. He was Assistant Professor in the Woodrow Wilson School of Public and International Affairs at Princeton University (1993–1998), where he also chaired the Science, Technology and Environmental Policy Program (STEP). Dr Kammen is the author of over 150 publications and four books, on energy science and policy, development, and risk, including *Should We Risk It?* (Princeton University Press, 1999).

Dr Evans Kituyi holds a Ph.D. in Environmental Chemistry and is currently a lecturer at the Department of Chemistry of the University of Nairobi, Kenya. He also founded and leads the Industrial Ecology Research Group at the same department. Between 2000 and 2004, he served as a Research Fellow on energy and climate change policy issues at the Nairobi-based African Centre for Technology Studies (ACTS), which he joined upon completing his doctoral research work with the Biomass Burning Group of the Max Planck Institute for Chemistry in Mainz, Germany. His research then focused on improving the understanding of the role of household woodfuel burning in Africa on local, regional and global climate and tropospheric chemistry. He has written several publications in the field of biomass energy technology and policy in which part of his research interests currently lie.

Dr Bart G. J. Knols has a Ph.D. in medical entomology (Wageningen University, the Netherlands) specializing in vectors of tropical diseases. He has worked in East and Southern Africa for the last 11 years and has published over 50 scientific articles and several book chapters on the behavioural and chemical ecology of African malaria mosquitoes. He has undertaken several

international consultancies in disease vector control and served as an external expert to the Stockholm Convention on Persistent Organic Pollutants (POPs) for Southern Africa. He is the chairman of the vector control and repellence group of the Research Initiative on Traditional Anti-malarial Methods (RITAM).

Dr Michael Komenda studied Chemistry at the University of Cologne (1992–1998). He finished his diploma thesis on *Emissions of Volatile Organic Compounds from Domestic Fires in Tropical Africa: A Comparison of Laboratory and Field Studies* in 1998, and received his Ph.D. in atmospheric chemistry in 2001. He was a Postdoctoral Fellow (2001–2003) at the Institute of Chemistry and Dynamics of the Geosphere: ICG-II: Troposphere at the Jülich Research Centre, Germany. His research was focused on the emission of volatile organic compounds from vegetation. Since 2003 he has been working for Schwarz BioSciences GmbH as a research scientist.

Dr Ralf Koppmann is Deputy Director of the Institute of Chemistry and Dynamics of the Geosphere: ICG-II: Troposphere at the Jülich Research Centre, Germany, and Head of the research group Organic Trace Gases. He has been working in the field of atmospheric chemistry of volatile organic compounds (VOC) for 15 years. His main research topics have been the sources of VOC (oceanic, biomass burning and anthropogenic emissions), the global distribution and photochemistry of VOC. His research group is currently focusing on the emission and photochemistry of biogenic VOC.

David Kpelle holds a B.Sc. degree in Natural Resource Management from the Kwame Nkrumah University of Science and Technology, Kumasi, Ghana, Postgraduate Diploma in Wildlife Management from the College of African Wildlife Management in Tanzania and an M.Sc. in Tropical Coastal Management Studies from the University of Newcastle, UK. From 1986 to 2000, he led the Zoological Survey Team under the Protected Areas Management Programme of Ghana Wildlife Division. Later he was appointed the National Coordinator of the GEF-funded Coastal Wetlands Management Project. He is currently the Director of Programmes at Conservation International, Ghana.

Dr Gordon A. Mackenzie is a Senior Energy Planner at the UNEP Risø Centre on Energy, Climate and Sustainable Development at Risø National Laboratory, Denmark. Following his Ph.D. and postdoctoral research in physics, he has focused on energy and environmental issues in developing countries. He was in the team that developed the approach to climate change mitigation analysis now used in most studies of this kind. He was Energy Adviser at the Department of Energy in Zambia (1984–1987), and Chief Technical Adviser at the Department of Energy in Lesotho (1999–2002).

Dr George Manful is a Programme Officer in the GEF Support Unit of the secretariat of the United Nations Framework Convention on Climate Change (UNFCCC) based in Bonn, Germany. He served as a senior Adviser to the Executive Secretary from August 2002 to March 2003. Prior to joining the secretariat in January 1999, he coordinated the preparation of the national communication of Ghana to the UNFCCC. He has extensive experience in coordinating the implementation of environmental management and climate change capacity-building projects. He holds a Ph.D. in Environmental Science from the University of Gent, Belgium. He was a consultant for the World Bank on the implementation of the Ghana Environmental Resource Management Project from 1995 to 1997, and Director of Operations at the Environmental Protection Agency of Ghana in 1998.

Dr Kennedy Masamvu is a Senior Agrometeorologist in charge of the SADC Secretariat in Gaborone, Botswana. He holds a Ph.D. in Remote Sensing from Bristol University, UK, and an M.Sc. in Agrometeorology from Reading University, UK. He has been working for the SADC Regional Early Warning System for Food Security since 1994. His main interests are in agrometeorology, remote sensing and GIS for use in early warning systems for food security. He has participated in many national and international projects on the impacts of climate variability on agricultural production.

Dr Mohammed U. Mohammed is currently an Associate Professor at the department of Geology and Geophysics of Addis Ababa University, Ethiopia. He obtained his Ph.D. in 1992 in France from the University of Aix-Marseille III, after conducting research in a CNRS laboratory on the environment and climate of the last few thousand years in Ethiopia. He also received grants for research visits from the French Ministry of Foreign affairs, from the CNRS (France), from the Royal Society (UK) and from the START office (Washington DC, USA). He has authored and co-authored several articles on the environmental and climatic history of Ethiopia and the surrounding regions. He is currently a member of the scientific steering committee of the IGBP-PAGES-PEP III project.

Gray Munthali is an agrometeorologist in the Meteorological Department of the Ministry of Transport and Civil Aviation in Malawi. He monitors the impact of weather on food production and provides critical information to the National Early Warning System in order to maintain agricultural and food security. He has participated in various national and international conferences on agricultural and climate issues. Gray Munthali has a masters degree in agricultural meteorology from the University of Reading, UK, and a bachelor's degree in meteorology from the University of Nairobi, Kenya. He received additional training in management studies from the University of Malawi.

Dr Freddy Nachtergaele is an agronomist working for the FAO as a Technical Officer on Soil Resources and Land Classification since 1989. Prior to that he was a land resources expert for FAO in field projects in East Africa and in South-East Asia. He coordinates the update of the Soil Map of the World and FAO's Global Agro-ecological Zones work. He is the author of numerous scientific articles in the field of agro-ecological zoning, land evaluation, land-use planning and soil classification.

Dr Mark New is a Lecturer at the School of Geography, Oxford University, UK. He has published widely in the area of observed climate change, climate prediction and climate impacts on hydrological systems and water resources. He was a contributing author to the chapter on 'Observed climate variability and change' to the *IPCC Third Assessment Report* (2001). His current research includes mechanisms of climate variability under global warming, probabilistic climate prediction and interactions between climate and soil moisture.

Dr Todd Ngara was in charge of the Advisory Services of the Zimbabwe Department of Meteorological Services for 15 years. He later joined the University of Zimbabwe as a lecturer in the Department of Geography and Environmental Studies. He supervised the preparation of the Zimbabwe Initial National Communication to the UNFCCC. In that capacity, he had worked on several occasions as a consultant for the UNFCCC secretariat based in Bonn, Germany. He is currently working at the Technical Support Unit (TSU) for the IPCC National Greenhouse Gas Inventory Programme, which is based at the Institute for Global Environmental Strategies (IGES) in Japan.

Professor Godwin Olu Patrick Obasi (see above)

Dr Eric O. Odada is the Director of the Pan-African START Secretariat (PASS) and a professor at the University of Nairobi, Kenya. His research interests include sedimentary processes in large lakes and oceans and the derivation of palaeoclimate records from lake sediments. He has conducted research on the large lakes of the East African Rift Valley for many years and published two books: *The Limnology, Climatology and Palaeoclimatology of the East African Lakes* (Gordon and Breach, 1996); and *The East African Great Lakes: Limnology, Palaeolimnology and Biodiversity* (Kluwer, 2002).

Dr Daniel Ochieng' Olago is a lecturer in Geology, and has a background in Quaternary Science. His research interests include palaeoenvironments and palaeoclimate and their implications and relevance to the present and future; human impact on the environment; past and present; and the physical and chemical dynamics of lacustrine systems. Dr Olago is currently a member of the East African Natural History Society (EANHS), the IGBP-PAGES Scientific Committee, and the CLIVAR-VACS Panel for Africa. He has authored and co-authored several papers on Quaternary palaeoclimate and palaeoenvironment of East Africa.

Keith Openshaw is an energy, environment and forestry economist and he has degrees in Forestry and Economics from Aberdeen University, Scotland. He has worked in sub-Saharan Africa since 1968, with 14 years' residency in Kenya and Tanzania. He has undertaken forestry assessments, environmental analyses and energy surveys in over 20 African countries. He introduced to Kenya the ceramic charcoal stove, which is now manufactured in several African countries. He has been employed by FAO, Norwegian Aid, the World Bank and the Swedish Academy of Sciences. Until his retirement in January 2004, Keith Openshaw was with the International Resources Group of Washington, DC. He has written over 150 articles, one book and several book chapters on natural resources and renewable energy.

Dr Luanne Otter is currently a Senior Research Scientist in the Climatology Research Group at the University of the Witwatersrand in South Africa. Trained as an ecophysiologist, she has over the last few years been involved in the field of biogeochemistry. Her main research interest is biogenic trace gas fluxes (NO, CH_4, VOCs) and their controlling factors. Dr Otter is the regional coordinator of the Southern African Regional Science Initiative (SAFARI-2000), and a member of the Pan-African Committee for the System for Analysis, Research and Training (START) programme.

Rolph A. Payet is currently the Principal Secretary of the Ministry of Environment and Natural Resources of the Republic of Seychelles. He was Director-General of the Policy, Planning and Services Division, Ministry of Environment, from 1999 to 2003, and directly responsible for the operation of the National Meteorological Services and the Climate Centre. He is also the Interim Coordinator of the UNEP Regional Seas Programme of Eastern Africa. Rolph Payet was review editor of Chapter 17 of the *IPCC Third Assessment Report*, Working Group 2, and is a lead author of the IPCC Fourth Assessment Report, Working Group 2. He was editor/co-author of the Seychelles Initial National Communication to the UNFCCC. He is the chief negotiator of the Seychelles delegation at the UNFCCC meetings.

Dr Francis E. 'Jack' Putz is a Professor of Botany and Forestry at the University of Florida and a Senior Research Associate with the Centre for International Forestry Research (CIFOR) in Indonesia. His research focus is on the ecological basis of tropical forestry and on market-based incentives for forest management as a conservation strategy. His principal study sites are currently in Bolivia, but he has also conducted research in much of South-East Asia and in southern Africa.

Robert A. Reinstein is currently an international consultant, specializing in energy, environment and international trade. He was Deputy Assistant Secretary for Environment, Health and Natural Resources at the US Department of State from 1990 to 1993 and the chief US negotiator for the UNFCCC. He was also chairman of Working Group III (on response strategies, 1991–1992) and of Working Group II (on impacts, adaptation and mitigation, 1992–1993) of the IPCC. He was alternate chief US negotiator for the Montreal Protocol on ozone-depleting substances in 1987 and for the London Amendment to the Protocol (1990) and chief negotiator for the Copenhagen Amendment (1992).

Dr Ian H. Rowlands is an Associate Professor in the Department of Environment and Resource Studies at the University of Waterloo, Canada. He is the editor of *Climate Change Co-operation in Southern Africa* (Earthscan, 1998), the author of *The Politics of Global Atmospheric Change* (Manchester University Press, 1995) and the co-editor of *Global Environmental Change and International Relations* (Macmillan, 1992). Dr Rowlands has also authored numerous articles, book chapters and consultancy reports on various subjects, including international environmental policy, energy management and policy issues, global climate change, business and the environment, and energy/environment issues in southern Africa.

K. Madhava Sarma is a consultant on environment currently based in Chennai, India. He worked for UNEP for more than nine years as the Executive Secretary of the Secretariat for the Vienna Convention and the Montreal Protocol. Earlier, he held senior positions in the Government of India and helped articulate the developing-country view on environmental issues. He has co-authored (with Stephen Andersen) a book on *Protecting the Ozone Layer: The United Nations History* (Earthscan, 2002).

Dr Mary Scholes is an associate professor in the Department of Animal, Plant and Environmental Sciences at the University of the Witwatersrand, Johannesburg, South Africa. She lectures in plant physiology, savanna ecology, environmental biology and biogeochemistry. Her research activities focus on soil fertility and biogeochemistry in savannas, forests and croplands. She has authored over 60 articles and eight book chapters. Dr Scholes is a member of the editorial boards of *Applied Soil Ecology* and *Ecosystems and Biogeochemistry*, and a foreign member of the Royal Swedish Academy of Natural Sciences. She also serves as a board member for the International Centre for Tropical Agriculture (CIAT), and on the Science Steering Committee of the IGBP.

Ivan Scrase studied Geography at Oxford University (1987–1991) and Environmental Technology at Imperial College,

London (1991–1992). Since then he has worked in energy research for the late Professor David Hall at King's College and at the Association for the Conservation of Energy in London. He has also worked as a coordinator of a rural development programme in Peru. He has published papers on biomass energy, energy conservation and integrated approaches to policy assessment. He returned in 2000 to Imperial College's Centre for Environmental Technology to study for a Ph.D. on approaches to environmental assessment, funded by the Environment Agency of England and Wales, and the Economic and Social Research Council.

Gillian Simmonds is currently a Senior Policy Advisor at the CBI, UK. She was formerly a senior researcher at the Centre for the study of Regulated Industries, University of Bath, UK, where her areas of specialization included energy, environment, and social and consumer policy and regulation. Her specific research interests include climate change policy and practice, the relationship between energy-related environmental policy and social development concerns, and competition policy and regulation with a specific focus on universal and public service obligations and consumer participation, redress and representation.

Randall Spalding-Fecher is Director of ECON Analysis (Oslo) office in South Africa. He was formerly a senior researcher at the Energy and Development Research Centre (EDRC), University of Cape Town, South Africa. His work focuses on CDM project development, climate change policy and mitigation analysis, energy and environmental economics, and the links between energy policy and sustainable development, including indicators for energy and sustainable development. He has published widely, including *The CDM Guidebook: A Resource for the Clean Development Mechanism Project Developers in Southern Africa* (EDRC, 2002).

Dr John Todd is an environmental educator and consultant. He lectured and conducted research in environmental management and environmental technology at the University of Tasmania, Australia, from 1978 to 2002. He now works as a consultant in air quality, renewable energy and biomass, and remains an honorary research associate in the Centre for Environmental Studies, University of Tasmania. He has a strong interest in Environmental Education and influencing the community through effective education practices. He has extensive experience with government committees and policy formulation relating to technical aspects of environmental management. He has worked in all Australian States and Territories and several Pacific and SE Asian countries. Dr Todd was Chairman of the Solid Fuel Burning Appliances Committee of Standards Australia (1982–1998), receiving the Standards Australia Award in 1994, and Chairman of Solar

Energy Society of Australia and New Zealand (1998–2000). He has published over 120 conference and journal papers, mainly in the fields of renewable energy and pollution, including *Wood-Smoke Handbook: Woodheaters, Firewood and Operator Practice,* Environment Australia, Canberra (2002).

Ingeborg M. C. J. van Schayk is a social scientist with strong interest in public health and gender in developing countries. She has worked and lived in Africa for more than nine years, and worked for UNESCO in Costa Rica for two years. Currently she is a consultant with the National Library of Medicine (USA) and serves as the communications coordinator for the Multilateral Initiative for Malaria in Africa (MIM), through the MIMCom network. She is specializing in communication enhancement for malaria research in the African context.

Dr Kristin von Czapiewski completed her Diploma thesis on *Emission of volatile organic compounds from small domestic fires in tropical and subtropical Africa* in 1995 after studying chemistry at the Universities of Greifswald and Oldenburg in Germany. She completed her Ph.D. thesis on *Investigations of volatile organic compounds emitted from biomass burning.* She was Postdoctoral Fellow (1999–2001) at the Centre for Atmospheric Chemistry, York University, Toronto, Canada. Her research was focused on kinetic isotope effects for the reactions of non-methane hydrocarbons with hydroxyl (OH) radicals.

Dr Shem O. Wandiga is a Professor in the Department of Chemistry, University of Nairobi, Kenya. He also served as Deputy Vice-Chancellor (Administration and Finances) of the University (1988–1994). He is interested in research on the sources and sinks of biogenic gases, persistent organochlorine pesticides in the tropics, trace metal concentrations in various environmental media, and complexes of Group VB metals with sulphur and oxygen binding ligands, and he has published several papers in these areas. Professor Wandiga has been Chair of the Pan-African Committee of START since 2001, and a member of the Advisory Committee on Environment of the ICSU since 1999. He was also a member of ICSU's General Committee (1996–1998), Chair of the Kenya National Academy of Sciences (1992–2002), and past Chair of the national IGBP Committee (1992–2002). He was elected, as a Kenya representative, to the Executive Board of UNESCO (1995), and President of the External Relations and Programme Commission of the Board (1997).

Dr George Wiafe is an Oceanography Lecturer at the Department of Oceanography and Fisheries, University of Ghana, and a visiting Oceanography Lecturer at the Ghana Armed Forces Command and Staff College. He holds a Masters degree in Marine Biology from the University of Newcastle and a Ph.D. in Oceanography from the University of Ghana. He has published 19 articles on issues relating to plankton ecology in the Gulf of Guinea and the North Sea, nutrient fluxes, and conservation of marine resources, as well as a manual (with CD-ROM) on marine zooplankton of West Africa.

Dr David S. Wilkie is a wildlife ecologist with a post-doctoral anthropology specialization in human behavioural ecology. He has over 18 years of research experience in the socio-economic aspects of household level natural resource use in Central and West Africa, and in Central and South America. His research in the Congo Basin has focused on: determining the local and regional impact of forager and farmer subsistence practices on forest plant and animal composition, distribution and abundance; and the household economic determinants of Efe hunter-gatherer adoption of agriculture into their suite of subsistence activities. His other research interests include examining the impacts of trade and the commercialization of non-timber forest products on forest animal populations; the role that logging plays in promoting bushmeat markets; the income and price elasticities of demand for bushmeat; and the use of satellite imagery and aerial photo-

graphy to model the location, extent and rate of land transformation within rainforests.

Dr Peter P. Zhou is an applied geophysicist and is currently the Director of Energy, Environment, Computer and Geophysical Applications (EECG) Consultants Pty in Gaborone, Botswana. He has been involved in climate change studies since 1992 and has participated in various multilateral, bilateral and national projects. Dr Zhou has been working on issues in energy sector in southern Africa since 1984. He has been contributing to the IPCC Assessment Reports since 1995, first as a reviewer, later as a lead author. He has also contributed to the UNFCCC process on issues related to technology transfer in 1999–2001 and as a CDM Meth panel member in 2002–2004. Dr Zhou has many publications, including Sustainable Mobility: Perspectives from the Transport Sector in East and Southern Africa (see *Industry and Environment*, Vol. 23, No. 4, UNEP (2000); Climate Change Mitigation Options and Strategies for Africa (see *Climate Change in Africa*, a GLOBE Southern Africa Publication);. Integrated Assessment and Mitigation (see *TIEMPO*, Issue 34, December 1999, IIED, London and University of East Anglia, Norwich, UK).

Peer reviewers

Professor Dilip Ahuja, National Institute of Advanced Studies, India.

Dr Stephen Andersen, United States Environment Protection Agency, USA.

Dr Robert C. Balling, Jr., Arizona State University, USA.

Dr Abdelkrim Ben Mohamed, Institute for Radioisotopes, Niger.

Professor Peter Brimblecombe, University of East Anglia, UK.

Dr Neil de Wet, University of Waikato, New Zealand.

Dr Paul V. Desanker, Pennsylvania State University, USA.

Dr Robert K. Dixon, Department of Energy, USA.

Dr Richard D. Duke, McKinsey & Company, USA.

Dr David Duthie, United Nations Environment Programme, Kenya.

Dr Graham Farmer, FAO–Regional Inter-Agency Coordination Support Office (FAO–RIACSO), South Africa.

Dr Jes Fenger, National Environmental Research Institute, Denmark.

Dr Michael H. Glantz, National Center for Atmospheric Research, USA.

Dr Hiremagalur N. B. Gopalan, United Nations Environment Programme, Kenya.

Dr Michael Graber, United Nations Environment Programme, Kenya.

Dr Mark Griffith, UNEP Regional Office for Latin America and the Caribbean, Mexico.

Dr Sujata Gupta, Asian Development Bank, the Philippines.

Tom Hamlin, United Nations Environment Programme (Paris Office), France.

Dr David A. Hastings, United Nations Economic and Social Commission for Asia and the Pacific, Thailand.

Professor John E. Hay, University of Waikato, New Zealand.

Dr Günter Helas, Max Planck Institute for Chemistry, Germany.

Professor Mike Hulme, University of East Anglia, UK.

Professor Phil Jones, University of East Anglia, UK.

Professor Christopher O. Justice, University of Maryland, USA.

Professor Daniel M. Kammen, University of California, Berkeley, USA.

Dr Stjepan Keckes, Marine science consultant, Croatia.

Dr Mick Kelly, University of East Anglia, UK.

Dr Lambert Kuijpers, Technical University, The Netherlands.

Dr Henry Lamb, University of Wales, UK.

Dr Mark G. Lawrence, Max Planck Institute for Chemistry, Germany.

Dr Ti Le-Huu, United Nations Economic and Social Commission for Asia and the Pacific, Thailand.

Dr Gordon A. Mackenzie, UNEP Risoe Centre on Energy, Climate Change and Sustainable Development, Denmark.

Dr George Manful, United Nations Framework Convention on Climate Change Secretariat, Germany.

Dr Alan Miller, International Finance Corporation, USA.

Dr Isabelle Niang-Diop, Technology University C.A. Diop, Senegal.

Keith Openshaw, International Resources Group, USA (retired in 2004).

Seth Osafo, United Nations Framework Convention on Climate Change Secretariat, Germany.

Dr Jean Palutikof, Meteorological Office, UK.

Vivian Raksakulthai, Environment consultant, Thailand.

Robert A. Reinstein, Reinstein & Associates International, Inc., USA.

Professor Henning Rodhe, Stockholm University, Sweden.

Dr Ian Rowlands, University of Waterloo, Canada.

Professor Jochen Rudolph, York University, Canada.

Dr Hesphina Rukato, The New Partnership for Africa's Development (NEPAD) Secretariat, South Africa.

Dr Mohd. Nor Salleh, TroBio Forest Sdn Bhd, Malaysia.

Dr Pedro A. Sanchez, Columbia University, USA.

K. Madhava Sarma, Environment consultant, India.

Professor Cherla B Sastry, C & R Associates Canada, Canada.

Randall Spalding-Fecher, ECON Analysis, South Africa.

Dr Anna Tengberg, United Nations Environment Programme, Kenya.

Dr Wassila M. Thiaw, National Oceanic and Atmospheric Administration, USA.

Dr Alexandre Timoshenko, United Nations consultant, Russian Federation.

Dr John Todd, Eco-Energy Options, Tasmania, Australia.

Gregory Tosen, Eskom Research, Development and Demonstration, South Africa.

Emeritus Professor Jan C. van der Leun, Utrecht University, The Netherlands.

Dr Robert van Slooten, Economics consultant, UK.

Dr Hassan Virji, International START Secretariat, USA.

Xueman Wang, Convention on Biological Diversity Secretariat, Canada.

Professor Donald A. Wilhite, University of Nebraska, USA.

Dr Andrew Yager, United Nations Development Programme, USA.

Professor Francis D. Yamba, University of Zambia, Zambia.

Dr Peter P. Zhou, EECG Consultants, Botswana.

Editor's note

This book project was first initiated in mid-1999 before I left the United Nations Environment Programme (UNEP) in Nairobi, Kenya. Originally it was intended as a special journal issue to be published by the African Centre for Technology Studies (ACTS), but the enthusiastic response from potential contributors has made the publication of a book possible. However, only a few manuscripts were submitted by the end of that year. The project then followed me to Bonn, Germany, where I was working as a consultant for the United Nations Framework Convention on Climate Change (UNFCCC) secretariat from July to December 1999. Following that, it went with me to Mainz, Germany, where I spent three months in the Max-Planck-Institute for Chemistry, which was then headed by Professor Paul Crutzen. Then, back to Bonn again, during my brief engagements as a consultant for the UNFCCC secretariat and the United Nations Convention to Combat Desertification (UNCCD) secretariat. Then on to Kuala Lumpur, Malaysia, as I worked as a freelance consultant, and finally to Bangkok, Thailand, when in June 2001 I joined the United Nations Economic and Social Commission for Asia and the Pacific (UNESCAP) as its Regional Adviser on Environment and Sustainable Development. Since then, the project has started to pick up momentum, with most chapters submitted, peer reviewed, revised, edited and proofread by the end of 2002. Four new chapters (5, 20, 25 and 27) were added in 2003.

In many ways, the slow progress of this book has, to a large extent, reflected the winding path of my professional career over the past few years. And the fact that I have been in full time employment with UNESCAP since June 2001 has meant countless late nights between numerous official missions dedicated to completing this book. I really owe the authors, especially those who submitted their papers before 2002, a sincere apology for not having moved the project on faster. Due to the lengthy processes of soliciting papers, peer reviewing, formatting, revising, editing and proofreading (some papers have undergone a few rounds), as well as a number of early technical difficulties, such as transforming all chapters into a standardized format, the process of producing this book has been far more time-consuming than I could have ever anticipated when I began the project.

This book is about *climate change* with special focus on *Africa*. It contains a total of 31 chapters, written by 62 authors and co-authors. The chapters are broadly divided into five parts: *Science*; *Sustainable Energy Development, Mitigation and Policy*; *Vulnerability and Adaptation*; *Capacity-Building*; *and Lessons From the Montreal Protocol*, covering a wide spectrum of topics. All chapters are independent from one another and yet they are interlinked. Together, they provide a coherent picture of the challenges and opportunities that African countries are facing amid the growing evidence and concerns of climate change and its impacts, as well as their response efforts and specific needs under the UNFCCC and its Kyoto Protocol. The chapters were all written by leading experts in various fields, including both well-known authors, with a long list of publications, and very promising, up and coming young authors. About 40% of the lead authors and co-authors are from Africa. Originally, I was hoping to have at least 50% African authors from both Anglophone and Francophone African countries. However, despite persistent efforts, I was unfortunately unsuccessful in including any papers from Francophone African experts, though a few have been peer reviewers.

The first nine chapters cover scientific issues. Odada and Olago (Chapter 1) provide a very comprehensive review of the climatic, hydrological and environmental oscillations in the tropics during the Holocene, with special reference to Africa, and suggest that they are linked to a number of factors, such as changes in earth surface temperatures, sea surface temperatures (SSTs), ocean and atmospheric circulation patterns, regional topography, land surface albedo, etc. 'The relative importance of these forcing factors and the extent of the linkages between them are still unclear, but the data suggest that the climate and hydrology of the tropical regions may be adversely affected by the anthropogenically driven rise in global temperatures and land use.' Mohammed (Chapter 2) assesses the relative importance of the different forcings on the dynamic nature of the environment in Ethiopia during the Holocene, and highlights the widespread human impacts on the environment during the twentieth century, especially in its second half. Hulme *et al.* (Chapter 3) provide an assessment of future climate change scenarios for Africa and discuss these possible future climates in the light of modelling uncertainties and in the context of other causes of African climate variability and change. They suggest that in view of the uncertainties, it is important to place emphasis 'on reducing vulnerability to adverse climate-events and increasing capacity to adapt to short-term and seasonal weather conditions and climatic variability'. The interactions of climate with desertification are clearly demonstrated by Balling (Chapter 4), who has previously written extensively on this topic. Hastings (Chapter 5), using satellite imagery and in situ data processed in a geographic information system, offers us some perspectives on Africa's climate. Some approaches described in the paper may be useful for operational climate-related decision-making. The important role of atmospheric chemistry in climate is extensively discussed in Chapters 6 to 9. Brasseur *et al.* (Chapter 6) provide an excellent overview of the chemical processes that determine the budget of tropospheric ozone (O_3) and the formation of the hydroxyl radical (OH), with emphasis on their role in the tropics. Koppmann *et al.* (Chapter 7) highlight the importance of natural and human-induced biomass burning in Africa as a source of volatile organic compounds, which play an active role in atmospheric chemistry. Kituyi *et al.* (Chapter 8) point out the need to control biomass burning, and this could provide opportunities for emission mitigation through the Clean Development Mechanism (CDM), one of the three Kyoto Protocol mechanisms. Otter and Scholes (Chapter 9) provide a comprehensive account of the importance of soil micro-organisms as sources and sinks of many trace gases (e.g. methane, nitrous oxide and nitric oxide) and the various approaches to measure the fluxes of these trace gases between soil and atmosphere. They also discuss the impacts of the biogenic trace gases on the atmospheric chemistry of the southern African region.

The next eight chapters deal with issues relating to sustainable energy development, including both technological and policy measures that have implications for climate change mitigation. These include biomass energy, enthusiastically promoted by Hall and Scrase (Chapter 10); the examination of the nexus between population growth and the demand for agricultural land in the different regions of Africa (Chapter 11 by Openshaw); and the opportunities offered by CDM for sustainable energy development (Chapter 12 by Spalding-Fecher and Simmonds). The proposed regional approaches in sub-Saharan Africa to global climate change policy merit serious consideration and discussion (Chapter 13 by Zhou and Chapter 14 by Rowlands). Duke and Kammen (Chapter 15) discuss the promotion of solar home systems in Africa and its implications for global carbon emission reduction. Justice *et al.* (Chapter 16) discuss the significance of climate change to African nations and the related needs and opportunities. They

highlight the importance of scientific equity and the urgent need for investment in African scientific infrastructure to help African scientists inform and advise African governments and decision makers on the likely impacts of climate change on their nations' economy and resource base. Abdel Gelil (Chapter 17) discusses the climate energy policy in Egypt and 'demonstrates a developing country's success story of better management of indigenous energy resources while striving to meet domestic energy demand and secure sufficient oil exports earnings that are needed to finance economic development'. It may serve as a model for other African countries of similar condition.

To many developing and least-developed countries, including African countries, *vulnerability and adaptation* to climate change are their major concerns. Seven chapters are devoted to these issues, ranging from the potential impacts of sea-level rise on populations and agriculture (Chapter 18 by Gommes *et al.*) and on coastal biodiversity in Ghana (Chapter 19 by Armah *et al.*), to the impacts of El Niño Southern Oscillation (ENSO) on the socioeconomic activities in Africa (Chapter 20 by Obasi). Two specific case studies further illustrate the impacts of ENSO events: one by Payet (Chapter 21) on the Seychelles during the 1997–1999 ENSO events, the other by van Schayk *et al.* (Chapter 22) on the outbreak of *Paederus* dermatitis in East Africa during the 1997–1998 El Niño event. Payet also emphasizes the need to use the response to ENSO events as a policy window for developing adaptive capacity to the long-term climate change. The linkage between climate change and ENSO is a focus for current research. It has been observed that the frequency, magnitude and persistence of El Niño have increased over the past two decades and this trend is projected to increase in the future. Eriksen (Chapter 23) discusses the importance of indigenous plants as a livelihood resource in household coping mechanisms during the 1996 drought in a dryland agricultural area in Kenya, and concludes that 'policies aimed at enhancing local indigenous plants and household capacity to cope with climatic variability can improve local welfare'. Desanker *et al.* (Chapter 24) discuss the requirements needed to use the tool of integrated assessment modelling for climate change impact and adaptation options assessment at the regional and national levels in Africa. Basher and Briceño (Chapter 25) provide the only chapter that discusses climate and disaster risk reduction in Africa. They suggest that the increased numbers of climate-related disasters over the last few decades appear to be mostly due to growing vulnerability and closer awareness and reporting of events, and hence attention must remain focused on the vulnerabilities and risks associated with existing climate variability, even though disaster risk reduction provides a potent means to advance the adaptation agenda.

Three chapters cover the important issue of *capacity-building*. While Mackenzie (Chapter 26) provides his own experiences in capacity enhancement activities related to climate change mitigation analysis – by no means a simple exercise – in Botswana, Tanzania and Zambia, Manful (Chapter 27) with his deep involvement in capacity-building activities in the UNFCCC secretariat, provides an informative account of the existing capacity-building activities in African countries, and further highlights their specific needs. Todd (Chapter 28), as an environmental educationist and energy efficiency expert, emphasizes the importance of education (both school and community) and public awareness in promoting energy efficiency and hence slowing the emissions of greenhouse gases in Africa. This is a 'no-regret' policy that has 'direct social benefits irrespective of their impact on climate change'. The imbalance of number of chapters between the previous two parts and this one is an acknowledged shortcoming of the book. Indeed, I would have liked to see a few more chapters in this section, including some case studies and good practices in African countries, especially with regard to the lessons learned in enabling activities during the preparation of their initial national communications, which are far more than just a reporting requirement under the UNFCCC. It is important for the developing and least-developed countries to use this national communication process to build up or strengthen their human, scientific, technical, technological and institutional capacity, so as to ensure the sustainability of their effective participation in the UNFCCC and its Kyoto Protocol processes.

The three chapters on *Lessons From the Montreal Protocol* are linked to climate change issues. They are included with a view to sharing some of the experiences and lessons learned from the implementation of the Montreal Protocol on Substances that Deplete the Ozone Layer adopted in 1987. While Sarma (Chapter 29) and Andersen (Chapter 31) believe that the success of the Montreal Protocol has much to offer to the UNFCCC and its Kyoto Protocol processes, Reinstein (Chapter 31) questions whether the Montreal Protocol is a good model for responding to climate change. Reinstein's view is thought-provoking and merits further discussion.

I would like to express my heartfelt thanks to all the authors. Indeed, without their great patience and persistent support, this book would have never been realized. I have been most privileged to have known and worked with many of them. I was saddened that two of the authors, Professor David O. Hall, who was born in South Africa, and Dr Ruben O. Agwanda, a Kenyan, passed away during the preparation of their manuscripts. This book is dedicated to them. I corresponded with the late Professor Hall, in particular, during the process of his manuscript preparation, though we never met. I was profoundly moved when I was told by one of his colleagues during his memorial service, held on 26 November 1999 in London, that he was very concerned about the completion of the paper even when he was hospitalized. This truly reflects the dedicated character of the late Professor Hall as a prominent scientist. As revealed by Dr K. Krishna Rao in his obituary, less than two

days before the late Professor Hall peacefully passed away, 'he spent his afternoons discussing research work and future projects, arranging examiners for his Ph.D. students and watching cricket on TV.'

The book is most honoured to have the forewords kindly written by Dr Mostafa K. Tolba, Ambassador Michael Zammit Cutajar, Ambassador Mikko Pyhälä and Professor Dieter H. Ehhalt, and the prefaces by Professor Godwin O. P. Obasi, Dr Wulf Killmann, Dr Sálvano Briceño and Dr Hassan Virji, who are all distinguished scientists and experts in their own fields. Their forewords and prefaces provide a good overview of the contents and focus of the book, and, in many ways, complement the chapters well, and hence have added great value to the book.

All chapters in this book have gone through a very robust peer-review process. My special thanks also to the 62 peer reviewers, who have kindly contributed their time and effort to review the chapters, sometimes at very short notice, and sometimes for more than one paper. They have ensured that the papers selected for publication in the book are of undisputed high quality.

I would also like to gratefully acknowledge the financial support of the sponsors: FAO, UN/ISDR, WMO, the START Secretariat and *Tiempo* – a quarterly information bulletin and associated website (http://www.cru.uea.ac.uk/tiempo/) on global warming and the third world. Their sponsorships have contributed towards a reduced price for the book, which makes it more affordable to the majority of readers. Needless to say, the views expressed in this book are those entirely of the authors and they do not necessarily reflect those of the sponsors, the editor or the publisher.

Many individuals have also contributed to this book. Vivian Raksakulthai, Ian Grange, Will Keenan, Rick Whisenand, Kai Yin Low and Ling Si Low have kindly proofread some of the chapters, and Pieter Bakker, Khun Patnarin Sutthirak and Khun Patanapong Siriwatananukul have kindly assisted in formatting many chapters. I am most grateful for their assistance.

Throughout my academic and professional career, a number of people have had a profound influence on the path that I have chosen to take. Professor Emeritus Miles Kennedy, former Head of Department of Chemical and Process Engineering, University of Canterbury, New Zealand, greatly inspired me during my formative years; Dr John Hails first transformed me from a chemical engineer to an environmental scientist in the late 1970s; Dr Terry Smith and Dr John Todd guided me during my early days in atmospheric research; Chris Purton, Bill Moriarty and Dr John Green informally taught me many things about meteorology; the late Professor Harry Bloom, Professor Dieter Ehhalt, Dr Urlich Schmidt, Professor Peter Brimblecombe, Professor Peter Liss, Professor Stuart Penkett and Professor Paul Crutzen have had a great influence on me with regard to the field of atmospheric chemistry. Professor Trevor Davies and Dr Mick Kelly first got me interested in ozone and climate change issues when I joined the Climatic

Research Unit (CRU) of the University of East Anglia, UK, in 1988; and it was a privilege to work with them. Other colleagues at CRU, including Professor Tom Wigley, Dr Astrid Ogilvie, Dr Graham Farmer, Dr Phil Jones, Dr Jean Palutikof, Dr Tom Holt, Dr Keith Briffa, Dr Sarah Raper and Dr Mike Hulme, were most helpful during my three exciting and fruitful years with the unit, when global warming and climate change were still hotly debated in the scientific community. These colleagues have been the prominent pioneers in climatic research. Dr Mostafa K. Tolba was my mentor on sustainable environmental policy and sustainable development, which he has been tirelessly promoting through his numerous speeches and writings, even after his retirement in December 1992 from UNEP as its Executive Director. Dr Wo-Yen Lee greatly broadened my outlook on various global environmental issues during our time as colleagues in UNEP before he retired in 1993. And Ambassador Michael Zammit Cutajar, who provided the UNFCCC secretariat with great momentum when he was its first Executive Secretary, was always an inspiration for his diplomatic skills in steering the climate change negotiations.

Since leaving UNEP in mid-1999, many friends and colleagues, including Dr Nay Htun, Ambassador Mikko Pyhälä and his wife, Pia, Dr Naigzy Gebremedhin, Dr Steve Andersen, Dr Suvit Yodmani, K. Madhava Sarma, Mahboob Elahi, Michelle Lee, Dr Mark Griffith, Hassane Bendahmane, Theodor Kapiga, Martha Perdomo, Dr George Manful, Dr Angela Wilkinson, Aidar Karatabanov, Dr Manab Chakraborty, Dr Anna Tengberg, Professor Alexandre Timoshenko, Dr Stejpan Keckes, Carmen Tavera, Song Li, Dr Mohd Nor Salleh, Lai Tan Fatt, Tan Meng Leng, Tan Hoo and Chow Kok Kee, have consistently provided me with their kind advice and encouragement. I am most grateful for their genuine and unfailing friendships.

I am also most grateful to my colleagues at UNESCAP, including Dr Kim Hak-Su, the Executive Secretary, and Ravi Sawhney, former Director of Environment and Sustainable Development Division, who have been very supportive of my work and hence provided the conducive working environment in which this book project could move forward. Dr Rezaul Karim and Phang Pin Suang were also very supportive of my work before they retired from UNESCAP.

The impressive artistic image selected for the book cover was kindly contributed by Emeritus Professor Yuan Li, a distinguished amateur photographer whom I met in November 1994 during his sabbatical leave at Kenyatta University, Kenya.

Last but not least, the authors and I are most delighted that Cambridge University Press has decided to publish the book. I am particularly grateful to Dr Matt Lloyd, Senior Commissioning Editor, Earth and Space Sciences; Emily Yossarian, Senior Publishing Assistant; Dr Sally Thomas, Physical Sciences Editor; Jayne Aldhouse, Production Editor; and their colleagues, as well as Margaret Patterson, copy-editor, for providing all the necessary guidance and assistance in ensuring that this book is published smoothly and in the most professional manner.

This book belongs to all authors, contributors and peer reviewers, and to all those who love Africa and its people.

Pak Sum Low
UNESCAP, Bangkok, Thailand

Foreword by Mostafa K. Tolba

No environmental issue has been of such truly global magnitude as the issue of climate change. And no other global environmental issue has been so controversial, not because of lack of scientific knowledge but rather because it is a result of every human action and will have a direct impact on all human endeavour everywhere, North and South, East and West.

Some hide behind the lack of scientific certainty, making it an excuse not to act to deal with a major potential catastrophe. As a scientist, I have never seen any scientific subject where scientists agreed on all its aspects one hundred per cent. We go by the majority – not just a simple majority, but a real, solid majority. And that is what we have now.

We now know enough to indicate that the poor developing countries are the least equipped to adapt, on their own, to climate change, although most of them played, and will certainly continue to play, an insignificant role in causing it.

African countries are among the poorest of the developing countries. Most of the least developed countries are in Africa.

So, this book is really coming at the right time, and it presents the issues of relevance to Africa – sea-level rise to a continent surrounded by two oceans and two seas; energy in a continent where the most used source of energy is firewood (destroying a carbon sink and an oxygen generator and a soil stabilizer for a continent with large areas of marginal soil); and desertification in a continent suffering from repeated droughts and hard-hitting desertification problems. These and several other issues in this book should catch the attention of both scientists and policy makers in a continent that is starting to put together for implementation an action programme to achieve sustainable development recently endorsed by the World Summit, 2002, in Johannesburg.

In Africa we need to believe in self-reliance; we need to get out of the shell of asking others to help us. Forty per cent of the articles in this book are written by Africans alone or with colleagues of other nationalities. We can build our own continent.

I have known the editor of this book since he joined UNEP some two years before I left it. I was impressed with his scientific knowledge and integrity.

I do recommend this book to everybody who is still hesitating about climate change. But, I recommend it more to my fellow Africans.

Mostafa K. Tolba
International Centre for Environment and Development
Cairo, Egypt

Foreword by Michael Zammit Cutajar

Global climate change is not just our greatest environmental challenge. It is also a symptom of the unequal and unbalanced development of the global economy. Generated by the consumption of the rich, it places an additional handicap on the survival of the poor. The 'creeping catastrophe' of climate change is thus an additional factor of inequity and stress in our global – and globalizing – community. Nowhere does this burden weigh more heavily than on Africa's fragile states and ecosystems, already under pressure from internal forces and from external shocks.

The papers in this rich collection, patiently assembled by a committed scientist, Pak Sum Low, range over the entire climate change *problématique*, with a special emphasis on Africa. One can spin from them many of the strands in the web of the climate change discussion:

- We need to continually improve our understanding of the process of climate change, so as to anchor our responses more soundly in science.
- We need to know what impact climate change will have on particular regions or countries, so as to strengthen the political case for these responses (by answering the question 'What does it mean for me?') and also to better design adaptation strategies.
- Strengthening the resilience of poor rural communities to external shocks is an obvious development strategy in developing countries and also a sensible first defence against climate change. This is a win–win synergy.
- The important interaction between climate change, drought and desertification – a key linkage for many African countries – can point to other win–win opportunities in the domain of soil and forest management.
- Great benefits can be reaped – for the economy, ecology and human health – from 'low-tech' improvements in using biomass fuels, so widely relied upon in Africa.
- At the same time, even poor countries should safeguard their long-term development by opening the door to cleaner energy futures, whether by preparing the switch from oil to gas or by experimenting with photovoltaics. Clean energy strategies will need to be backed by foreign investment, some of which can be

leveraged by the Global Environment Facility and projects under the Clean Development Mechanism, including those financed by the World Bank's carbon funds.
- While African countries are still low emitters on a global scale, improvement of their capacities to measure emissions and removals of greenhouse gases is a call upon external assistance that should not go unheeded.
- So too is the strengthening of their capacities to integrate climatic considerations in their national development strategies, in particular in their technological perspectives.

These topics feed into the ongoing formulation of a collective strategy to protect the global climate, founded on the United Nations Framework Convention on Climate Change and its first offspring, the Kyoto Protocol. They weigh in the balance between mitigation and adaptation, between North and South.

The intergovernmental negotiations on climate change mobilize an array of defensive forces and arguments from all sides, in which economic 'short-termism' clouds the long-term vision of the common good. Negotiators are engaged in the closed game of handing off the perceived burden of emission reduction, instead of working together to open up the technological and economic opportunities ahead. The powers that guide them must still be insufficiently convinced of the climatic threat. Until this sinks in, much can be done to underpin a dynamic and inclusive future global accord: deepen the science, improve measurement, prepare for adaptation, test-market instruments, explore new technologies and their applications, and build institutional capacities where they are needed. For all their limitations, by design and by default, the targets and mechanisms of the Kyoto Protocol will drive such actions in the right direction.

Michael Zammit Cutajar
Ambassador for International Environmental Affairs, Malta;
Formerly Executive Secretary,
UNFCCC Secretariat
Bonn, Germany

Foreword by Mikko Pyhälä

The reduction of the snowcap of the mighty Mount Kilimanjaro by four fifths over the last 90 years is one of the most immediately discernible impacts of climate change in Africa, and irrefutable evidence of what is happening to our planet. Under the present trends, scientists tell us, the remaining ice fields on the mountain are likely to disappear between 2015 and 2020 (see Thompson *et al.* (2002): *Science*, **298**, 589–593). If this were the case, there would be significant implications for the water resources of the African countries that are dependent on the melted water coming from the mountain. The integrity of already fragile ecosystems with their endemic species would suffer. In addition, a tropical peak without the snowcap would be far less attractive to the tourists who are the source of income for many local people.

Climate change will hit all nations directly or indirectly and its warming impact is likely to be strongest in areas near the poles. However, of all continents, human suffering as a consequence of climate change may well be most dramatic in Africa. This region has probably never faced in its history as formidable a challenge as adaptation to climate change, requiring the migration or transformation of not only natural (often already endangered) ecosystems, but also of agricultural production systems. The Africans can do little themselves to prevent the climate change, which has not, for the most part, been caused by them.

On the basis of equity, and in accordance with the principle of common but differentiated responsibilities and respective capabilities for all Parties to the United Nations Framework Convention on Climate Change (UNFCCC), we in the North should not only focus our efforts on mitigation of climate change through adherence to, and compliance with, the Kyoto Protocol, which enters into force on 16 February 2005, but also on assisting poorer, i.e. non-Annex I countries, in promoting sustainable development, and sharing their burden of adaptation.

The countries of the European Union have taken measures to mitigate climate change, such as energy conservation, and the development of renewable energy sources. They are also urging application of environmental taxes. The transfer of appropriate technologies to non-Annex I Parties on affordable terms is a particular question of survival for our humanity. Adequate financial resources must be provided to the least developed and developing countries in order to address their specific capacity-building needs in relation to the effective implementation of the Convention. This book provides important insights, for policy makers and development practitioners, into various aspects that are related to climate change and Africa.

I first got to know Dr Pak Sum Low in Nairobi, Kenya. We worked together in the Global Environment Facility (GEF) Unit of the United Nations Environment Programme (UNEP) from 1992 to 1995. He was responsible for climate change and ozone depletion issues and provided basic scientific contributions to the GEF where UNEP had a normative and scientific role, including support to the Scientific and Technical Advisory Panel (STAP) of the GEF. Dr Low made attempts to facilitate a far-sighted research plan on the assessment of tropospheric ozone as a greenhouse gas initiated by Professor Paul Crutzen and his colleagues in the International Global Atmospheric Chemistry (IGAC) for GEF funding in 1993–1995, and again in 1998, though unfortunately without success.

I can only laud Dr Low for keeping his deep commitment with Africa, even after his departure a few years ago, as evidenced by his conceiving and editing this volume of scientifically merited articles by leading experts, on the ways in which Africa relates to global climate change. This comprehensive volume will leave an outstanding mark on literature contributing to climate change science, technology and policy in Africa.

Mikko Pyhälä
Ambassador, Asia and Oceania
Helsinki, Finland

Foreword by Dieter H. Ehhalt

The African continent extends from about 35° N to about 35° S latitude straddling the equator. This particular geography determines its climate regimes – mostly tropical and subtropical with extensive arid and semiarid zones around 20° N and 20° S. In some parts of the continent the regional climate, in particular annual rainfall, is highly variable. Many countries are prone to recurrent droughts, others to flooding. As a consequence Africa is highly vulnerable to additional climate stresses. It is ironic that the continent which – owing to its low industrialization – has contributed least to the projected man-induced global warming should be suffering the most from it.

The problem is exacerbated by population pressure: Africa is the continent with the highest population growth rate in the world. It is further exacerbated by widespread poverty which limits the capability for adaptation.

It is therefore all the more important to take a close look at the environmental and economic problems facing the African continent in the context of global warming. This book is an important step in this direction. It addresses many of the pertinent issues in environmental science, such as the prediction of the regional climates, droughts and desertification, sea-level rise, biomass burning and its role in the emission of trace gases, and tropical photochemistry. In similar detail it also addresses the questions of sustainable energy development, and vulnerability and adaptation.

Equally important, many of the articles are authored or co-authored by African scientists. It underlines the progress that the African scientific community has made in environmental and climate research and its readiness to provide input and advice to the policy-making process. The editor, Pak Sum Low, who has a long-standing interest in environmental research, and his authors are to be commended for putting together this fine volume.

Dieter H. Ehhalt
Institute for Atmospheric Chemistry
Jülich, Germany

Preface by Godwin O. P. Obasi

The climatic variations of the past, except probably the recent glacier retreat, have been essentially natural, with little or no human influence. However, the present-day concern is that, for the first time in history, the human element has been added to the climatic equation. Thus, emissions of some of the greenhouse gases into the atmosphere from human activities have now modified the concentrations of these gases quite significantly compared to pre-industrial levels. The anthropogenic gases include carbon dioxide (CO_2), methane (CH_4), nitrous oxide (N_2O) and halocarbon compounds. In recent years, therefore, the major debate worldwide has centred on how real it is that the climate of the Earth is changing, what the climate change expectations are, and what measures humankind should take to avert the potential climate change and its impacts.

Evidence is there to support concern over the state of global climate. Indeed, since the middle to late nineteenth century, when a marked increase in CO_2 from anthropogenic activities has been measured, the observations show that over the last 100 years, the earth's atmosphere has warmed by about 0.6 °C, while the global sea level has risen by between 10 and 20 cm; spatial and temporal patterns of precipitation are changing; night-time temperatures over land have generally increased by double the increase of daytime temperatures; regional changes such as increased precipitation over land are evident; most of the world's glaciers have been retreating since 1850 while the Arctic ice is thinning, etc.

It is expected that climate change will lead to undesirable consequences in Africa. For example, with the projected sea-level rise, the coastal nations sharing low-lying lagoonal coasts, such as those of western, central and eastern Africa, will be susceptible to further erosion of the beaches and damage to coral reefs, with significant adverse impacts on the tourism industry. Sea-level rise will also lead to the flooding of rich agricultural fields, such as the rice paddies on the coastal plains in Gambia and Guinea. In some parts of West Africa, coastal erosion is already reported to have reached 30 m annually. A few studies indicate, for instance, that for a 1-metre rise in sea level, 2,000 km^2 of land may be lost to inundation in the lower Nile Delta, 6,000 km^2 mostly in

the wetlands of Senegal, 1,800 km^2 in Côte d'Ivoire, 2,600 km^2 in Nigeria and 2,117 km^2 in Tanzania. Some of the coastal areas and low-lying islands could be rendered uninhabitable. In the absence of any protective barriers, the estimation is that 1-metre sea-level rise will lead to the displacement of the entire population of Alexandria, 4 million people in the Nile Delta, 3.7 million in Nigeria and up to 180,000 people in Senegal. The projected rise in sea level will also adversely affect freshwater availability in the coastal areas due to the intrusion of saline water upstream of the rivers and into the freshwater aquifers. Some recent observations indicate that salt-water contamination has been observed 80 km upstream of the Zambezi River and up to 120 km upstream of the Gambia River during the dry season. Another phenomenon that will affect freshwater availability in the coastal regions is the decrease in the run-off of a number of rivers, such as the Nile, the Zambezi and most of the large rivers in the Sahel.

Already, there are serious concerns that climate change will spell more serious problems for Africa. Catastrophic droughts of the Sahel in the late 1960s and early 1970s heightened the concern for potential climate change, while recent weather- and climate-related natural disasters impacting on the lives and property of people are grave enough to focus the attention of scientists and policy makers alike on various issues concerning climate change. A few examples of such events in the region include the devastating floods in Mozambique in 2000, which reduced its gross national product by 11.6 per cent. In the West African subregion, devastating floods have been reported in major cities. A severe drought has been ravaging many parts of Africa in recent decades. Drought resulting from rain failure led to food shortage in the Greater Horn of Africa in 2002, which has in turn caused a serious humanitarian crisis involving about 30 million people. Severe tropical cyclones are a constant threat to the Indian Ocean Island countries.

Climate change is also expected to lead to a more vigorous hydrological cycle, with the likely results that more severe droughts will occur in some places and floods in others. Small changes in the mean climate can produce relatively large changes in the frequency of extreme weather events, such as floods and droughts. Over the coastal zone, a higher sea level combined with increased storm surges and a high tide could lead to extensive coastal flooding as regularly occurs in Lagos, the latest being in May 2003. Over the south-west Indian Ocean, tropical cyclones remain a constant threat to the island countries and to those of the south-eastern coast of the African continent.

Among the other most pressing challenges facing Africa is the alleviation of poverty, hunger and food insecurity and the sustainable management of agriculture and natural resources. Indeed, agriculture is the sector most sensitive to variability in the weather and climate. It is, therefore, necessary that further work be done to ensure that these concerns are dealt with through improved research and adequate planning, in the interests of present and future generations. This publication will go a long way in raising awareness on climate change in Africa, thus constituting a worthwhile effort in addressing the problems of climate, climate variability and change that have implications for social and economic development for Africa.

Godwin O. P. Obasi
Formerly Secretary-General
World Meteorological Organization
Geneva, Switzerland

Preface by Wulf Killmann

The key to food security is regular access to food, in a context where many factors seem to be competing to make its supply scarce and irregular. In Africa, as in most of the world, the variability of climate over seasons, years and decades, has been a dominant factor. This was dramatically illustrated by the latest drought episode in the West African Sahel between the 1960s and the late 1980s. Some papers recently published suggest a link between the drought and atmospheric pollution in the developed world, a reminder that human activities can affect climate, locally and globally, at many scales.

Over the last two decades, war has overtaken climate fluctuation as the dominant factor in food insecurity, particularly in Africa where civil unrest and war have killed people, driven them off their land and led to the creation of large refugee settlements. However, even humanitarian crises must be seen in their climatic context, as many tense and unstable situations have been created by high food prices due to drought. Many farmers are forced off the land at harvest or planting times, and many choose to grow cassava, safely concealed in the ground, rather than the conspicuous maize cobs that are so attractive to refugees and soldiers. Compared with cereals, cassava is drought resistant but poor in proteins, so that some farmers' attempts to ensure food production actually contribute to malnutrition.

Africa remains the continent where climate variability most directly affects people's well-being. This is why Africa is also the continent where climate change will add most to the uncertainties and risks associated with farming and ranching, because large land areas are currently semi-arid and exposed to desertification. Deforestation spreads fast, few farmers have access to markets – except their local market – and to alternative sources of income, and populations are growing faster than resources in semi-arid and highland areas.

Ever since 1988, when FAO established an Interdepartmental Working Group on Climate Change, the Organization has actively contributed to the climate change debate by assisting its members and collaborating with the IPCC and the UNFCCC in areas of its mandate, such as the definition of forest-related terms. FAO also tries to ensure that the voice of the crop farmers, pastoralists and those living off the forest, is heard in international circles.

This publication is one of the very few which is specifically dedicated to climate change and Africa. It provides a comprehensive coverage of climate change science as applied to the continent, the role of the continent as a contributor to climate change but also as a likely victim of the changes. It further covers the role of energy policies and the potential of energy from biomass, as well as the primordial role of education in increasing awareness and, in the longer term, the response capacity of individuals and national institutions.

By lending its support to this book, FAO hopes to contribute to building greater climate change awareness, not only on the African continent, but also among scientists and decision makers in the developed world, and in particular, hopes to attract more support for activities dealing with the climate change related risks and opportunities faced by Africa.

Wulf Killmann
Food and Agriculture Organization
Rome, Italy

Preface by Sálvano Briceño

Droughts, floods, storms and other climatic phenomena are natural features of planet Earth and have been occurring for millions of years. The earliest humans, cradled in Africa, learned over time how the seasonal rains came and went. They observed how in some years the rains could be excessive, deficient or untimely, sometimes with disastrous consequences, and gradually they developed ways to better cope with the uncertainties. Episodes of significant climate shifts stretching for tens and hundreds of years also occurred and forced major shifts in the locations and activities of societies.

Today, our understanding of the rains is scientific and extensive, yet the continent suffers more and more from the vagaries of the climate. Drought in particular affects millions of people each year – people who are often already suffering from poverty and disease and who are least able to resist or cope with the added stress of food and water shortages. And on the horizon there is the looming threat that climate change may make matters even worse.

For many, there seems no escape from the repeated impacts of climatic hazards. But closer examination shows that disasters occur *only* when the hazard is coupled to human vulnerability. Surprisingly, it is the human situation that mainly causes the problem – especially populations in risky and degraded locations without the means to understand and avoid or manage the risks. Disasters are thus a manifestation of poverty, inadequate governance, meagre public services, and unsustainable development. This insight is a powerful one, for it shows that we must address a complex set of socio-economic and environmental factors in order to reduce the risk of disasters.

It is also now clear that the substantial increase in disasters over recent decades has been mostly due to changes in socio-economic factors, such as population increases in high-risk areas and land degradation, rather than to changes in climate, though some observed changes in climate, such as the relatively greater and stronger El Niño episodes over the last 30 years, will have exacerbated disaster situations in affected countries. The prospects of increased weather hazards for the future reported by the IPCC are of very great concern, and add even more urgency to the pressing immediate need to reduce vulnerabilities and to better manage existing climate variability and extremes.

Experience shows that informed communities, backed up with strong government action, can identify and substantially reduce their risks. African institutions are evolving to systematically and regionally deal with disaster reduction and disaster management, but the widespread famines of recent years show that much more remains to be done.

The United Nations General Assembly has called for urgent action to stem the rising impacts of disasters, especially on the fragile economies of its developing country members. While climatic hazards will be with us for the foreseeable future, the human vulnerability that leads to disaster need not be. The International Strategy for Disaster Reduction (ISDR) Secretariat and its partners stand ready to assist the efforts of African governments, regional bodies, development organizations, humanitarian organizations, and others to more effectively grapple with the growing problem of risk and disasters in the region. This book provides a valuable contribution of knowledge relevant to this important task.

Sálvano Briceño
United Nations Inter-Agency Secretariat of the International Strategy for Disaster Reduction
Geneva, Switzerland

Preface by Hassan Virji

There has been a rapid growth in the number of programmes and activities addressing climate change in relation to Africa. Many of these programmes aim to improve the understanding of the nature of climate variability and the potential for human systems to adapt to climatic changes that might result from global warming. The Earth System Science Partnership, comprising the major international global change research programmes, DIVERSITAS, International Human Dimensions Programme (IHDP), International Geosphere–Biosphere Programme (IGBP) and World Climate Research Programme (WCRP), sponsors the global change SysTem for Analysis, Research and Training (START). START focuses on regional implementation of global change research and related capacity-building. START views capacity-building as research-driven, with the objective of entraining researchers in collaborative regional research networks.

START activities in Africa are designed to:

- Engage the African research community in global change research of the Earth System Science Partners;
- Promote collaboration within the African research community to provide sound, scientific assessments of impacts of global changes, including climate variability and change, and coping strategies relevant to critical socio-economic sectors;
- Provide targeted capacity-building opportunities for African researchers, largely within the region;
- Enhance regional research capability at individual and institutional levels, and
- Provide input to policy/decision-making communities, at various levels, on critical sectors.

Programmatically, the START effort on Africa emphasizes collaborative regional research on (i) climate change and climate variability, including regional climate scenarios and extreme climate events, and assessments of impacts of, adaptation and vulnerability to climatic variability and change; (ii) global environmental change and water resources, food systems and coastal zone resources; (iii) land-use/land-cover change, ecosystems/biogeochemical changes, and biodiversity.

Given its strategic focus on research-driven capacity-building, START conducts a comprehensive capacity-building programme that includes awards for short- and longer-term fellowships, Visiting Scientist and Young Scientist Awards, Advanced Institutes and Young Scientists' Conferences. Promoting collaboration with the policy/decision-making community, and fostering regional research and institutional networks is an inherent aspect of these activities. Currently, there are over 40 regional and national projects involving over 200 scientists and at least 30 institutions in Africa. Strategic partnerships with IPCC, WMO and others have enabled START to address issues at the interface of science and policy. More information on START is available on the Internet at http://www.start.org.

A number of contributions in this volume are a product of START-supported capacity-building projects. It is our hope that the ideas developed in this volume will further strengthen the goals and activities of those working at the science-policy interface in Africa.

Hassan Virji
International START Secretariat,
Washington, DC, USA

Abbreviations

ACMAD	Africa Centre for Meteorological Applications for Development
ACTS	African Centre for Technology Studies
AESEDA	Alliance for Earth Sciences, Engineering and Development in Africa
AGRHYMET	Regional Centre for Training and Application in Agrometeorology and Operational Hydrology
AIACC	Assessments of Impacts and Adaptation to Climate Change
AIDS	Acquired Immunodeficiency (or Immune Deficiency) Syndrome
AIJ	Activities Implemented Jointly
AMCEN	African Ministerial Conference on the Environment
AOSIS	Alliance of Small Island States
APF	Adaptation Policy Framework
AVHRR	Advanced Very High Resolution Radiometer
BEA	Bureau of Environmental Analysis
BIBEX	Biomass Burning Experiment
CARPE	Central African Regional Programme for the Environment
CBD	Convention on Biological Diversity
CCGT	Combined Cycle Gas Turbine
CDM	Clean Development Mechanism
CEAO	West African Economic Community
CEEEZ	Centre for Energy, Environment and Engineering (Zambia) Ltd
CEEST	Centre for Energy, Environment, Science and Technology (Tanzania)
CEO	Chief Executive Officer
CER	Certified Emission Reduction
CFC	Chlorofluorocarbon
CFL	Compact Fluorescent Lamp
CIAT	International Centre for Tropical Agriculture
CIFOR	Center for International Forestry Research
CILSS	Permanent Interstate Committee for Drought Control in the Sahel
CLIPS	Climate Information and Prediction Services
CLIVAR	Climate Variability and Predictability Programme
CNG	Compressed Natural Gas
COMAP	Comprehensive Mitigation Assessment Process
COP	Conference of the Parties
CORDIO	Coral Reef Degradation in the Indian Ocean
CRED	Center for Research on the Epidemiology of Disasters
CREC	Center for Regional Environmental Change
CRU	Climatic Research Unit
CSD	Commission on Sustainable Development
CTI	Climate Technology Initiative

DAAD	German Academic Exchange Service
DECAFE	Dynamique et Chimie Atmosphérique es Forêt Equatoriale
DMC	Drought Monitoring Centre
DMS	Dimethyl Sulphide
DRC	Democratic Republic of Congo
DSM	Demand-Side Management
EANHS	East African Natural History Society
EBRD	European Bank of Reconstruction and Development
ECDPM	European Centre for Development Policy Management (Maastricht)
ECMWF	European Centre for Medium-Range Weather Forecasts
ECOWAS	Economic Community of West African States
EECG	Energy, Environment, Computer and Geophysical Applications Pty
EEEDA	Eskom Energy Effective Design Awards
EEZ	Exclusive Economic Zone
ELA	Equilibrium Line Altitude
ENSO	El Niño Southern Oscillation
ERDC	Engineer Research and Development Center
ESCO	Energy Service Company
ESKOM	Electricity Supply Kommission of South Africa
ESMAP	Energy Management Programme
ET	Emission Trading
EXPRESSO	Experiment for Regional Sources and Sinks of Oxidants
FANR	Food, Agriculture and Natural Resources
FAO	Food and Agriculture Organization of the United Nations
FDI	Foreign Development Investment
FEWS	Famine Early Warning System
FEWSNET	Famine Early Warning System Network
FID	Flame Ionization Detector
FINESSE	Financing Energy Services for Small Enterprises
FOS	Fire of Savannahs
GACMO	Greenhouse Gas Abatement Costing Model
GC	Gas Chromatography
GCM	Global Climate Model
GCOS	Global Climate Observing System
GDP	Gross Domestic Product
GEF	Global Environment Facility
GFDL	Geophysical Fluid Dynamics Laboratory
GHA	Greater Horn of Africa
GHG	Greenhouse Gas
GIEWS	Global International Early Warning System
GIS	Geographic Information System
GLOBE	Global Legislators Organization for a Balanced Environment
GNP	Gross National Product
GOFC	Global Observation of Forest Cover
GOLD	Global Observation of Land Dynamics
GRASS	Geographic Resources Analysis Support System
GTOS	Global Terrestrial Observing System
GTZ	German Technical Cooperation
GWP	Global Warming Potential
HC	Hydrocarbon

HCFC	Hydrochlorofluorocarbon
HFC	Hydrofluorocarbon
HIV	Human Immunodeficiency Virus
IAM	Integrated Assessment Modelling
IATFDR	Inter-Agency Task Force on Disaster Reduction
IBRD	International Bank for Reconstruction and Development
ICIPE	International Centre of Insect Physiology and Ecology
ICRAN	International Coral Reef Action Network
ICSU	International Council for Scientific Unions
ICZM	Integrated Coastal Zone Management
IDNDR	International Decade for Natural Disaster Reduction
IDRC	International Development Research Centre
IEA	International Energy Agency
IFAD	International Fund for Agricultural Development
IFPRI	International Food Policy Research Institute
IFRC	International Federation of Red Cross and Red Crescent Societies
IGAC	International Global Atmospheric Chemistry
IGAD	Intergovernmental Authority on Development
IGADD	Intergovernmental Authority on Drought and Development
IGBP	International Geosphere–Biosphere Programme
IGES	Institute for Global Environmental Strategies
IHDP	International Human Dimensions Programme
IIED	International Institute for Environment and Development
IMAGE	Integrated Model to Assess the Greenhouse Effect
IMAGES	Intermediate Model for the Annual and Global Evolution of Chemical Species
IMF	International Monetary Fund
INC	Intergovernmental Negotiating Committee
IOC	Intergovernmental Oceanographic Commission
IOGOOS	Indian Ocean Global Ocean Observing System
IPCC	Intergovernmental Panel on Climate Change
IRI	International Research Institute for Climate Prediction
ISCCP	International Satellite Cloud Climatology Project
ISCCS	Integrated Solar Combined Cycle Systems
ISDR	International Strategy for Disaster Reduction
ISO	International Organization for Standardization
ITCZ	Inter-Tropical Convergence Zone
ITD	Inter-Tropical Discontinuity
IUCC	Information Unit on Climate Change
IUCN	International Union for Conservation of Nature and Natural Resources
JI	Joint Implementation
KEMRI	Kenya Medical Research Institute
LCLUC	Land-Cover and Land-Use Change
LDCs	Least-Developed Countries
LEAP	Long-range Energy Alternatives Planning system
LED	Light Emitting Diodes
LEG	LDC Experts Group
LGM	Last Glacial Maximum
LPG	Liquefied Petroleum Gas
LUCC	Land-use and Land-Cover Change programme
MIM	Multilateral Initiative for Malaria in Africa

MODIS	Moderate Resolution Imaging Spectroradiometer
MS	Mass Spectrometer
NAPA	National Adaptation Programme of Action
NASA	National Aeronautics and Space Administration
NCAR	National Center for Atmospheric Research
NCEP	National Center for Environmental Prediction
NDVI	Normalized Difference Vegetation Index
NEPAD	New Partnership for Africa's Development
NGC	Non-Governmental Centre
NGO	Non-Governmental Organization
NMHC	Non-Methane Hydrocarbon
NMHS	National Meteorological and Hydrological Services
NOAA	National Oceanic and Atmospheric Administration
NPP	Net Primary Production
OCHA	UN Office for the Coordination of Humanitarian Affairs
ODS	Ozone-Depleting Substance
OECD	Organization for Economic Cooperation and Development
OEP	Organization for Energy Planning
OSFAC	Observatoire Satellital des Forêts d'Afrique Centrale
PAGES-PEP III	Past Global Changes Pole–Equator–Pole III (Euro-African Transect)
PCDPPP	Pan-Caribbean Disaster Preparedness and Prevention Programme
PCF	Prototype Carbon Fund
PFC	Perfluorocarbons
POP	Persistent Organic Pollutant
ppb	parts per billion
PPEW	Platform for the Promotion of Early Warning
ppmv	parts per million by volume
PSMSL	Permanent Service for Mean Sea Level
PV	Photovoltaic
PVMTI	Photovoltaic Market Transformation Initiative
PWI	Precipitable Water Index
QBO	Quasi-Biennial Oscillation
R&D	Research and Development
RADM	Regional Acid Deposition Model
RAEL	Renewable and Appropriate Energy Laboratory
RCOF	Regional Climate Outlook Forum
REIO	Regional Economic Integration Organization
RETS	Renewable Energy Technologies
RITAM	Research Initiative on Traditional Anti-malarial Methods
SADC	Southern African Development Community
SADCC	Southern African Development Coordination Conference
SAFARI	Southern African Fires and Atmosphere Research Initiative
SAPP	Southern African Power Pooling
SBSTA	Subsidiary Body for Scientific and Technological Advice
SCEE	Southern Centre for Energy and Environment
SDFZ	Silti Debrezait Fault Zone
SEI	Stockholm Environment Institute
SIDS	Small Island Developing States
SLR	Sea-Level Rise
SOI	Southern Oscillation Index

SPREP	South Pacific Regional Environmental Programme
SSA	Sub-Saharan Africa
SST	Sea Surface Temperature
START	System for Analysis, Research and Training
STEP	Science, Technology and Environmental Policy Programme
TCI	Temperature Condition Index
TEAP	Technology and Economic Assessment Panel
TEM	Terrestrial Ecosystem Model
TFAP	Tanzania Forest Action Plan
TGCIA	Task Group on Climate Impact Assessment
TSU	Technical Support Unit
TWAS	Third World Academy of Sciences
UCCEE	UNEP Collaborating Centre on Energy and Environment (now URC)
UEA	University of East Anglia
UNCCD	United Nations Convention to Combat Desertification
UNCED	United Nations Commission on Environment and Development
UNCTAD	United Nations Conference on Trade and Development
UNDP	United Nations Development Programme
UNEP	United Nations Environment Programme
UNESCAP	United Nations Economic and Social Commission for Asia and the Pacific
UNESCO	United Nations Educational, Scientific and Cultural Organization
UNFCCC	United Nations Framework Convention on Climate Change
UNIDO	United Nations Industrial Development Organization
UN/ISDR	United Nations International Strategy for Disaster Reduction
UNITAR	United Nations Institute for Training and Research
UPDEA	Union des Producteurs et Distributeurs d'Energie en Afrique
URC	UNEP Risoe Centre on Energy, Climate and Sustainable Development (since May 2003)
USAID	United States Agency for International Development
USCSP	United States Country Studies Programme
USEPA	United States Environmental Protection Agency
USGS	United States Geological Survey
VACS	Variability of the African Climate System
VAM	Vulnerability Assessment and Monitoring
VCI	Vegetation Condition Index
VOC	Volatile Organic Compound
WAGP	West African Gas Pipeline
WCRP	World Climate Research Programme
WFB	Wonji Fault Belt
WFP	World Food Programme
WFPS	Water-Filled-Pore-Space
WHO	World Health Organization
WMO	World Meteorological Organization
WSSD	World Summit on Sustainable Development
YD	Younger Dryas

SI prefixes

Prefix	Abbreviation	Factor
deca-	da	10
hecto-	h	10^2
kilo-	k	10^3
mega-	M	10^6
giga-	G	10^9
tera-	T	10^{12}
peta-	P	10^{15}
exa-	E	10^{18}
deci-	d	10^{-1}
centi-	c	10^{-2}
milli-	m	10^{-3}
micro-	μ	10^{-6}
nano-	n	10^{-9}
pico-	p	10^{-12}
femto-	f	10^{-15}
atto-	a	10^{-18}

Unit abbreviations

Linear measure
- millimetre mm
- centimetre cm
- decimetre dm
- metre m
- kilometre km

Square measure
- square metre m^2
- square kilometre km^2
- hectare ha

Cubic measure
- cubic centimetre cm^3
- cubic metre m^3

Capacity measure
- millilitre ml
- litre l

Weight
- milligram mg
- gram g
- kilogram kg
- gigagram Gg
- tonne t
- dry tonne dt

Other units
- pascal Pa
- joule J
- watt W
- kilowatt kW
- kilowatts of electricity kWe
- megawatt MW
- kilowatthour kWh
- terawatthour TWh
- second s
- year yr

Chemical formulae

CH_4	methane
C_2H_4	ethylene (ethene)
C_2H_6	ethane
C_3H_6	propylene
C_5H_8	isoprene
CH_3O_2	methyl peroxy radical
CO	carbon monoxide
CO_2	carbon dioxide
H^+	hydrogen ion
H_2	hydrogen
HNO_3	nitric acid
HO_2	peroxy radical
K	potassium
N	nitrogen
N_2O	nitrous oxide
NH_3	ammonia
NH_4^+	ammonium ion
NO	nitric oxide
NO_2	nitrogen dioxide
NO_3^-	nitrate ion
NO_x	nitrogen oxides
N_2O_2	peroxide
N_2O_3	nitrous anhydride
N_2O_5	nitric anhydride
O_2	oxygen
O_3	ozone
OH	hydroxyl radical
P	phosphorus

PART I

Science

1 Holocene climatic, hydrological and environmental oscillations in the tropics with special reference to Africa

ERIC ODADA AND DANIEL OLAGO

University of Nairobi, Kenya

Keywords

Tropical Africa; Holocene climate; hydrological fluctuations; environmental oscillations

Abstract

The tropics have experienced large and sometimes abrupt fluctuations in the water balance since the beginning of the Holocene period. Water levels were generally high in the equatorial region and northern hemisphere at the beginning of the Holocene, a trend that was asynchronous with many southern hemisphere records. Apart from a desiccation event in many African lakes between 8,000 and 7,500 yr BP (Before Present), water levels continued to be high until c. 5,000 yr BP. Southern hemisphere sites experienced intermediate to high lake levels at c. 6,000 yr BP. The tropical lakes experienced a drying phase between 5,000 and 3,000 yr BP, and these arid conditions have continued to the present day. Tropical glaciers have, on the other hand, been gradually receding during the Holocene period, but there have been several minor advances. After about 8,000 yr BP, glacial events show much less spatial consistency. The environmental response has been less marked, but the major trend is from wet/moist vegetation in the early Holocene to drier vegetation from the middle Holocene to the present. The climatic, hydrological and environmental oscillations of the low-latitude regions during the Holocene are linked to changes in earth surface temperatures, sea surface temperatures (SSTs), ocean and atmospheric circulation patterns, regional topography, land surface albedo, etc. The relative importance of these forcing factors and the extent of the linkages between them are still unclear, but the data suggest that the climate and hydrology of the tropical regions may be adversely affected by the anthropogenically driven rise in global temperatures and land use.

1.1 INTRODUCTION

Africa is a rich repository of palaeoenvironmental and palaeoclimatic change. It is unique in that it is the only continent that, almost symmetrically, straddles the equator, and hence experiences both northern and southern hemispheric climatic influences. This, coupled with the influence of the oceans that surround it, results in an intriguing palaeo-record that offers the possibility of understanding the links in climate between the high latitudes and the tropics, and inter-hemispheric teleconnections. The sediments from both the large and small lakes, swamps and mires, ranging in altitude from sea level to over 5,000 m above sea level, and extending from the northern mid-latitudes to the southern mid-latitudes, provide an array of palaeo-proxies and range of sensitivities to the regional and global climate system that are essential in the elucidation of natural climate and environmental variability in the past. They also provide the prospect of delinking present natural variability from anthropogenically induced variability, in order to be able to better assess the extent of the current anthropogenic impact on the climate and environment, and to better predict the possible future trends in the dynamic earth system as a whole.

1.1.1 The Pre-Holocene

The Holocene period (10,000 yr BP to present) is set against (i) the backdrop of the Last Glacial Maximum (LGM) from 22,000 to 14,000 yr BP, the time during which climate was most markedly contrasted with the Holocene over the whole of the last glacial–interglacial cycle; and (ii) the rapid and somewhat unstable transition from LGM to Holocene climate, from 14,000 to 10,000 yr BP.

The LGM

Lake levels

During the arid interval spanning the LGM (21,000 to 14,000 yr BP and, in some areas, up to 12,500 yr BP), low water levels were recorded across a broad belt from near the Tropic of Cancer to

Climate Change and Africa, ed. Pak Sum Low. Published by Cambridge University Press. © Cambridge University Press 2005.

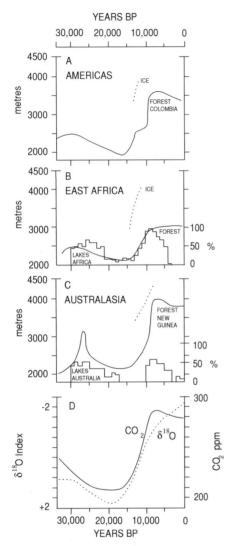

Figure 1.1. Synopsis of glacier, vegetation and lake-level changes in (A) Americas, (B) Africa, and (C) Australasian sector. The broken line denotes deglaciation age and elevation, solid curve upper forest limit, and block-histogram percentage of lakes with high water level status by millennium intervals, with percentage scale given to the right. In part (A) the deglaciation curve is for the Northern Andes, forest limit for Colombia. In part (B) the deglaciation curve includes one Ethiopian site in addition to East Africa, the forest limit is East Africa, and the lake histogram refers to tropical Africa at large. In part (C) the deglaciation curve includes one Borneo site in addition to New Guinea, the forest limit is one of New Guinea, and the lake histograms refer to Australia. Part (D) shows the long-term variation of atmospheric CO_2 levels (in ppm) reconstruction from polar ice cores (Neftel *et al.*, 1982; Crowley, 1983), solid line; and an index of $\delta^{18}O$ of seawater in the low-latitude ocean (Imbrie *et al.*, 1984), broken line. (from Hastenrath, 1991).

about 3–4° N, while intermediate and high water levels prevailed from near the equator to around 10° S (Figure 1.1) (Street-Perrott and Harrison, 1985; Street-Perrott *et al.*, 1985, 1989). In South America and southern Africa, some sites between 30 and 35° S recorded high lake levels (Street-Perrott *et al.*, 1989), and high

groundwater discharge characterized the western part of southern Africa (e.g. Heine, 1978; Lancaster, 1979; Butzer, 1984; Vogel and Visser, 1981; Deacon and Lancaster, 1988). Meanwhile, the lakes in southern Australia were drying up, and most were completely desiccated between 17,000 and 15,000 yr BP (Bowler, 1975; Bowler *et al.*, 1976). The greatest aridity in equatorial and northern tropical Africa occurred between 16,000 and 14,000 yr BP, while in southern hemispheric South America (30–35° S) this occurred between 15,000 and 12,000 yr BP (Table 1.1) (Street-Perrott *et al.*, 1989).

Glacier fluctuations

Maximum glaciation in South America occurred between 16,000 and 14,000 yr BP (Figure 1.1), and was accompanied by an equilibrium line altitude (ELA) lowering of 400–1,500 m (Clapperton, 1983; Mercer, 1983). Adopting a mean annual temperature lapse rate of 0.6 °C per 100 m (Andressen and Ponte, 1973; Bradley, 1985), and assuming no change in precipitation, the 1,500 m descent in snowline in South America indicates a decrease in mean annual temperatures of 7–9 °C (Table 1.2) (Weingarten *et al.*, 1991).

Similar ELA depressions are observed on the East African mountains (Figure 1.1). Estimates of the firnline on Mount Elgon during the last glaciation (using the morainic method) indicate a temperature depression of 3.5 °C, as compared to that of the present day (Hamilton and Perrott, 1979). On the Ethiopian Mountains, the periglacial belt of the last cold episode lies 700–800 m lower than the present height (3,500 m compared to 4,200–4,300 m today), and the temperature is estimated to have decreased by about 7 °C (Hurni, 1981). The ^{14}C dates for the Lake Mahoma basal sediments (Ruwenzori Range) indicate a maximum of the main glacial stage before 14,700 yr BP (Livingstone, 1962).

On the mountains of Australia and New Guinea, the ELA lowerings were between 800 and 1,200 m (Galloway, 1965; Galloway *et al.*, 1973), and maximum glaciation in New Guinea occurred shortly before 15,000 yr BP (Figure 1.1) (Hope *et al.*, 1976; Mercer, 1983). Evidence of frost action has been found on the higher mountains of the Eastern Escarpment of South Africa (the Lesotho and Drakensberg mountains), and the Inyanga and Chimanimani mountains of Zimbabwe (Harper, 1969). These stadials have not been dated, but are associated with temperature depressions of about 9 °C and 5 °C for the older and younger glaciations, respectively. In the cascade ranges of North America, Porter *et al.* (1986) observed an 800–1,040-m ELA lowering associated with lower accumulation season precipitation than present, and estimate a mean annual temperature depression of about 4 °C. In Mexico, late Pleistocene glaciation is associated with an ELA lowering of 675 m on Mount Ajusco (White, 1986). Within Central and Southern Europe, ELAs were lower by 1,000–1,200 m (Andersen, 1981).

Table 1.1. Tropical precipitation at the LGM (relative to present).

Area	Time period (yr BP)	% Rainfall	Source
Ziway-Shala Basin, Ethiopia (7° to 8° (30' N)	LGM	−9 to −32	Street, 1979
Southern Africa	LGM	+150 to +200	Lancaster, 1979; Shaw, 1986
East and Central Africa (between 4° S to 12° N and 28° E to 42° E)	LGM	−30	Bonnefille et al., 1990
Lake Tanganyika, Tanzania	LGM	−15	Vincens et al., 1993

Table 1.2. Tropical temperature lowering at the LGM (relative to present).

Area	Time period (yr BP)	Temperature (°C)	Proxy	Source
Eastern Colombian Andes, South America	LGM	−6 to −7	Pollen	Van Der Hammen, 1974
Wonderkrater, South Africa	LGM	−5 to −6	Pollen	Scott, 1990
Sacred Lake, Mount Kenya	LGM	−5 to −8	Pollen	Coetzee, 1967
New Guinea	18,000 to 16,000	−7 to −11	Pollen	Flenley, 1979a
Muchoya Swamp, Uganda	LGM	−5 to −8	Pollen	Morrison, 1968
Lake Tanganyika, north basin, Tanzania	LGM	−5 to −6	Pollen	Vincens, 1989a
East and Central Africa (between 4° S to 12° N and 28° E to 42° E)	LGM	−4 ± 2	Pollen	Bonnefille et al., 1990
Lake Tanganyika, Tanzania	LGM	−4.2 ± 3.6	Pollen	Vincens et al., 1993
Mount Elgon, Kenya	LGM	−3.5	ELA	Hamilton and Perrott, 1979
High Semyen, Ethiopia	LGM	−7	ELA	Hurni, 1981
Mount Kenya, Kenya	LGM	−5	ELA	Osmaston, 1975
Ethiopian Mountains	LGM	−7	ELA	Hurni, 1981
South America	LGM	−7 to −9	ELA	Weingarten et al., 1991
Cascade Ranges, North America	LGM	−4	ELA	Porter et al., 1986
Global average (glaciers)	LGM	−4.2 to −6.5	ELA	Broecker and Denton, 1990

Broecker and Denton (1990) proposed that the broad similarity of mountain snowline depression across virtually all of the earth's climatic zones was caused by high mountain temperature lowering of nearly equal magnitude (4.2–6.5 °C) in both polar hemispheres and the tropics. This phenomenon cannot be explained by the Milankovitch mechanisms, as these produced opposite insolation signals in both hemispheres (Dawson, 1992). Mercer (1983) tentatively proposed that the later advance of the tropical glaciers was somehow related to the final major readvances of the northern hemisphere ice sheets prior to their rapid downwasting. Under such circumstances, the global warming that led to the melting of the last ice sheets may have begun worldwide between 15,000 and 14,000 yr BP (Broecker and Denton, 1990).

Vegetation
Vegetation changes were similarly affected in the global tropics. High-altitude vegetation zones were depressed to lower altitudes, with little loss of the integrity of the ecozones (although on local scales this may have been quite significant) (Figure 1.1). Tropical low-altitude vegetation ecozones expanded/contracted latitudinally. Pollen evidence from South America, e.g. the Eastern Colombian Andes (Laguna de Fuquene, 2,580 m) indicates that scrub vegetation similar to the modern subpáramo above the forest limit surrounded the site, suggesting colder, drier conditions particularly between 20,000 and 14,000 yr BP (van der Hammen, 1974). For a vegetation depression of 1,000 m, and using the usual values for lapse rates, a mean annual temperature value of 6–7 °C lower than today is derived for high-altitude regions in South America during the last glaciation (Table 1.2) (van der Hammen, 1974). Pollen from coastal central Brazil (19° S, 46° 46′ W) indicates that conditions were drier between 17,000 and 14,000 yr BP (Ledru, 1993). During this period much of the Amazon area was occupied by savannah (Brown and Ab'Saber, 1979) and the tropical rainforests retreated into remnant enclaves or *refugia* (e.g. Spencer and Douglas, 1985; Livingstone, 1975, 1980).

In the New Guinea Highlands, the forest limit was lower than today from about 25,000 to 15,000 yr BP (Flenley, 1979a). The montane forest vegetation in New Guinea and the Sumatran Highlands was also depressed in a broadly synchronous manner as that in equatorial Africa and South America; maximum amounts of forest depression limits were between 800 and 1,500 m (Flenley, 1979a; Newsome and Flenley, 1988). Temperatures in New Guinea were possibly below present-day values by 2–8 °C between 28,000 and 18,000 yr BP, and between 7 and 11 °C from 18,000 to 16,000 yr BP (Flenley, 1979a).

The LGM in West Africa was marked by drastic reductions of forest, freshwater swamps and moist savannah communities in the Niger Delta (Sowunmi, 1981). During this period, forests occurred in three isolated *refugia*: Upper Guinea, Cameroon–Gabon and eastern Zaire (Maley, 1987). Pooid grasses (identified by grass cuticle analysis), and *Olea hochstetteri* pollen imply palaeotemperatures several degrees lower than today in southern Ghana (Talbot *et al.*, 1984). The δ^{13}C values in Lake Bosumtwi (Talbot and Johannessen, 1992) and Lake Barombi Mbo (Giresse *et al.*, 1994) support pollen evidence of grassland expansion during the LGM. In East Africa, highland ecozones were depressed to lower altitudes during the LGM: grasslands expanded and forests became fragmented, reflecting a cooler, drier climate than today (Figure 1.1). Sites include Mount Elgon (Hamilton, 1987; Street-Perrott *et al.*, 1997), Cherangani (Coetzee, 1967), Mount Kenya (Coetzee, 1967; Olago, 1995; Street-Perrott *et al.*, 1997; Olago *et al.*, 1999), Aberdares (Perrott and Street-Perrott, 1982), Lake Naivasha (Maitima, 1991), Lake Victoria (Kendall, 1969), Lake Bogoria (Vincens, 1986), Muchoya Swamp (Morrison, 1968; Taylor, 1990), Burundi Highlands (Aucour and Hillaire-Marcel, 1993; Aucour *et al.*, 1994; Bonnefille and Riollet, 1988; Bonnefille *et al.*, 1990; Roche and Bikwemu, 1989), Lake Tanganyika (Vincens, 1989a, 1989b). Early estimates of pollen-inferred temperature depression, relative to present, were as follows: Muchoya Swamp, 5–8 °C (Morrison, 1968); Mount Kenya, 7 °C (Coetzee, 1967); and north basin of Lake Tanganyika, 5–6 °C (Vincens, 1989a). More recently, Bonnefille *et al.* (1990) used multivariate statistical analysis (based on an extensive modern data set (356 sites) from East and Central Africa between latitudes 4° S and 12° N and longitudes 28° to 42° E, covering desert, subalpine grassland and all forest types) on a 40,000-year (radiocarbon dated) pollen profile from Kashiru Swamp, Burundi (32° 8′ S, 29° 34′ E) to derive quantitative estimates of past temperatures in tropical Africa (Table 1.2). They derived, for the last glacial period, a temperature decrease of 4 °C ± 2 °C, which is slightly lower than previously inferred values, and a simultaneous 30% decrease in mean annual rainfall, which is in broad agreement with the concomitant lake level declines in the East African region. Using a similar method, Vincens *et al.* (1993) gave an estimated temperature lowering of 4.2 ± 3.6 °C,

and a mean precipitation lowering of 15% (with a large deviation) for Lake Tanganyika. Further south in the Wonderkrater sequence of northern Transvaal, a cold phase with an estimated temperature drop of 5–6 °C is inferred to have occurred sometime between 25,000 and 11,000 yr BP (Scott, 1990). The cold phase is associated with relatively humid climates, except for the coldest period, which corresponds with a dry spell (Scott, 1990).

1.1.2 The LGM to Holocene transition

Lake levels

Following the arid LGM, water levels began to rise again at about 12,500 yr BP across a vast area extending from at least 8° S in East Africa to 25° N in the eastern Sahara (Figure 1.1). Conversely, in Australia and New Zealand, low water levels at 12,000 yr BP are indicative of widespread aridity (Street-Perrott *et al.*, 1989). In southern hemispheric South America (30–35° S), lake levels fell significantly between 15,000 and 12,000 yr BP (Street-Perrott *et al.*, 1989). Upper limits of last glacial climates are indicated at: 13,000 ± 1,400 yr BP in northern Africa, with means centring around 14,000 and 12,000 yr BP; 15,000 yr BP in southern Africa; and 13,000 ± 1,500 yr BP with a significant median at 12,000 yr BP in intertropical Africa (Table 1.3) (Littmann, 1989).

Glacier fluctuations

In the Ruwenzori, the ice began to retreat at about 14,700 yr BP (Livingstone, 1962), and is broadly concordant with the disappearance of ice from the summit region of the Aberdares and from cirques on Mounts Elgon and Badda at >12,200 yr BP (Hamilton and Perrott, 1979), >11,000 yr BP (Perrott, 1982), and >11,500 yr BP (Street, 1979), respectively. Hastenrath (1991) notes that deglaciation in the various equatorial and tropical mountains progressed from around 3,000 m altitude after about 15,000 yr BP towards the 4,000 m level at 8,000 yr BP, thus showing a lag of about 1,000 years per 200 m elevation (Figure 1.1). Differences in deglaciation dates as recorded on the various sites of these mountains are partly a factor of altitude differences; lower sites would experience earlier deglaciation than higher altitude sites. After this major deglaciation, later glacial events show much less spatial consistency (Hastenrath, 1991).

Vegetation

At 13,000 yr BP, *Quercus* forest invaded the Fuquene plateau in the Eastern Colombian Andes and the surrounding mountains, and indicates a warming phase in the Eastern Colombian Andes (van der Hammen, 1974), and generally there is evidence of a wetter

Table 1.3. Tropical precipitation during the deglacial period (relative to present).

Area	Time period (yr BP)	Rainfall mm/yr	Rainfall %	Author
Victoria basin	>12,500, 10,000	−125 to −180	−10 to −15	Hastenrath and Kutzbach, 1983
Nakuru−Elmenteita basin, Kenya	12,000 to 10,000		+35	Vincent *et al.,* 1989

Table 1.4. Tropical temperatures during the deglacial period (relative to present).

Area	Time period (yr BP)	Temperature (°C)	Proxy	Author
Sumatra Highlands, Australasia	11,000 to 8,000	−2 to −4	Pollen	Flenley, 1979a
Lynch's Crater, North-East Australia	11,000 to 7,000	−2 to −4	Pollen	Flenley, 1979a
Bateke Plateau, Congo (3° 31′ S, 15° 21′ E, 700 m)	10,850 (YD)	−6	Pollen	Elenga *et al.,* 1991
Muchoya Swamp, Uganda	Pre−11,000	−5 to −8	Pollen	Morrison, 1968
Lake Tanganyika, north basin, Tanzania	Pre−12,000	−5 to −6	Pollen	Vincens, 1989a
Pilkington Bay, Lake Victoria	Just before 10,000 (YD)	Not inferred	Pollen	Kendall, 1969

climate in the South American Andes between 12,800 and 9,800 yr BP (e.g. Cienaga del Visitador) (Flenley, 1979a, 1979b). The forest limit in the Eastern Colombian Andes rose until around 9,500 yr BP (van der Hammen, 1974). In coastal central Brazil, the increase in *Araucaria* forest elements in the pollen record, interpreted as being related to shifts in the confluence zone of the Antarctic polar front and warm tropical air (a zone characterized by heavy precipitation), shows that moister cooler conditions set in at about 12,000 yr BP (Ledru, 1993). Disappearance of *Araucaria* forest between 11,000 and 10,000 yr BP is related to the Younger Dryas (YD) event (Ledru, 1993).

In northern Africa, there was a spread of both deciduous and evergreen oaks and pine forests from 12,000 yr BP (Brun, 1991). Pooid grasses (identified by grass cuticle analysis), and *Olea hochstetteri* pollen at the Pleistocene–early Holocene boundary imply palaeotemperatures several degrees lower than today in southern Ghana (Talbot *et al.,* 1984). From 13,000 to 10,000 yr BP, there was forest extension in the Lake Barombi Mbo area (Giresse *et al.,* 1994). Two short dry phases, marked by positive $\delta^{13}C$ spikes, are dated by interpolation at 11,200 and 10,300 yr BP (Giresse *et al.,* 1994).

In East Africa, pollen records indicate that the cool, dry conditions of the LGM persisted up to about 12,000 yr BP (Tables 1.3 and 1.4) (Coetzee, 1967; Kendall, 1969; Vincens, 1989a; Maitima, 1991) and, in fewer cases, up to about 11,000 yr BP (Morrison, 1968; Perrott and Street-Perrott, 1982). The records during this period are, however, not entirely similar; a temperature increase is inferred in the Burundi Highlands between 15,000

and 13,000 yr BP (Roche and Bikwemu, 1989; Bonnefille *et al.,* 1990). A slight climatic amelioration, marked by some expansion of montane forest at about 14,000 yr BP, is also observed around Muchoya and Ahakagyezi Swamps in the Rukiga Highlands (Taylor, 1990). The period 12,000 to 10,000 yr BP marked a time of climatic transition, with rising temperatures and an increasingly moist climate; details remain unclear, and there were probably large climatic fluctuations, the most prominent of which is a brief return to arid conditions that is correlated with the YD event between 11,000 and 10,000 yr BP (Coetzee, 1967; Kendall, 1969; Flenley, 1979a, 1979b; Hamilton, 1987; Bonnefille *et al.,* 1995; Olago, 1995; Olago *et al.,* 1999). Concurrent dry episodes are recorded in pollen and lake records from West Asia, Ethiopia and West Africa (Gasse and Van Campo, 1994). Due to rising temperatures and increased precipitation, forests began to spread and diversify in the following areas of tropical Africa: Mount Elgon area (Hamilton, 1987), Pilkington Bay, Lake Victoria (Kendall, 1969), Ahakagyezi (Taylor, 1990), north basin of Lake Tanganyika (Vincens, 1989a) and other sites. The $\delta^{13}C$ data from Muchoya Swamp support the pollen evidence, as they show an increase in C_3 plants (Aucour and Hillaire-Marcel, 1993; Aucour *et al.,* 1994). In the region of Congo, forest underwent major expansion from about 12,000 yr BP (Giresse and Lanfranchi, 1984). Further south, around the south basin of Lake Tanganyika, montane forest communities abruptly retreated at 12,000 yr BP, whereas the Zambezi woodlands greatly expanded and diversified; these changes occurred within the context of a climatic amelioration, primarily related to temperature increases (Vincens, 1989c).

1.2 THE HOLOCENE PALAEO-RECORD

1.2.1 Lake level changes

A rise is noted in water levels at c. 10,000 yr BP in Africa, following a short, dry interlude which is broadly correlated with the YD event: data on the lake level trends indicate that they responded first near the equator and rose progressively later towards the central Sahara, and, by 9,000 yr BP, a belt of high lake levels extended from 4° S to 33° N, suggesting that large areas now arid were regularly receiving substantial tropical rainfall (Figures 1.1 and 1.2; Table 1.5) (Street-Perrott et al., 1989).

Freshwater conditions also existed in the lakes of Rajasthan, India, between 10,000 and 9,500 yr BP (Singh et al., 1974; Swain et al., 1983), corresponding to the rise in lake levels observed in Africa. Initiation of overflow in Lake Victoria, for example, occurred at 11,200 cal. (calendar) yr BP (Johnson et al., 2000). In southern Africa, low lake levels and dry conditions were characteristic of the early Holocene (Shaw, 1986; Finney and Johnson, 1991; Ricketts and Johnson, 1996); there were, however, some exceptions, e.g. high levels occurred around 9,130 yr BP, as evidenced by spring deposits at Namutoni (Rust, 1984). More generally, the lake level pattern of southern Africa does not appear to be as spatially and temporally homogenous as the coincidental wet period in equatorial and northern Africa (Patridge, 1993; Scott, 1993). By 9,000 yr BP, intermediate to high water levels had developed in some of the basins in Australasia; low lake levels prevailed in southern hemispheric South America; and intermediate to high lake levels in the equatorial areas of South America (Figures 1.1 and 1.2) (Street-Perrott et al., 1989).

Apart from a global cooling event recorded at about 8,000 yr BP (cf. Begét, 1983), which generally correlates with a large and abrupt desiccation event in many African lakes between 8,000 and 7,500 yr BP (Street-Perrott et al., 1985) or 8,000 to 7,000 yr BP across West Asia, East and West Africa (Gasse and Van Campo, 1994), water levels continued to be high until c.5,000 yr BP, when arid conditions returned, and intermediate-level lakes were restricted to the narrow latitudinal range (2° S to 13° N) that they occupy today (Figures 1.1 and 1.2). Southern Africa also experienced high lake levels at about 6,000 yr BP (Shaw, 1986). In Australasia, water levels generally continued to rise to intermediate- and high-level status until about 6,000 yr BP, as in several areas of Central and South America (Street-Perrott et al., 1989). A drying phase began soon after 5,000 yr BP in Africa and, by 3,000 yr BP, almost all areas north of 16° S had lower water levels than at 6,000 yr BP (Street-Perrott et al., 1989). For example, Lake Turkana achieved permanent closed basin status at 3,900 yr BP, and the lake levels approach modern conditions about 1,250 yr BP (Halfman et al., 1992). Arid intervals are also recorded between 4,000 and 3,000 yr BP in West Asia (Gasse

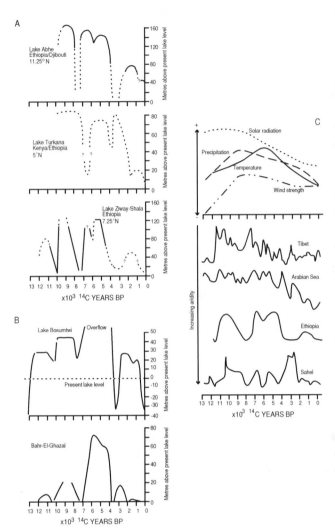

Figure 1.2. A. Water-level fluctuations of some African intertropical lakes, 13,000 to 0 ^{14}C yr BP (Sources: Lake Abhe – Gasse and Street, 1978; Lake Ziway-Shala – Gillespie et al., 1983; Lake Turkana – Owen et al., 1982). Dashed lines indicate uncertainty (from Street-Perrott and Harrison, 1984). B. Water-level fluctuation in Lake Bosumtwi (Talbot et al., 1984) and Bahr-El-Ghazal (Servant and Servant-Vildary, 1980). C. Climate change and droughts throughout tropical–subtropical Asia and Africa during the past 13,000 years. The variation of solar insolation (Berger, 1979) and environmental parameters (temperature, precipitation and southerly wind strength) are inferred from a numerical climate model (Prell and Kutzbach, 1992). There are four different environmental records, each indicating relative aridity: Tibet, a pollen record from Lake Sumxi in the western Tibetan highlands (Gasse and Van Campo, 1994); Arabian Sea, variations in carbonate concentration along a sediment core drilled off the coast of Oman (Sirocko et al., 1993); Ethiopia, water-level variation in Lake Abiyata in the Ethiopian rift (Gasse and Van Campo, 1994); Sahel, an oxygen isotope record of a former lake in the Chad basin (Gasse and Van Campo, 1994). Major dry spells occurred at 11,000–13,000, 8,000–9,000 and 3,000–4,000 years ago (all ages are given here in calendar years) (from Zahn, 1994, refs. above included).

Table 1.5. Tropical precipitation during the Holocene (relative to present)

Area	Time period (yr BP)	Rainfall		Source
		mm/yr	%	
Ziway–Shala basin, Ethiopia (7° to 8° 30′ N)	9,400 to 8,000		+25	Street, 1979; Gillespie *et al.*, 1983
Turkana basin	10,000 to 7,000	+80 to +140	+10 to +19	Hastenrath and Kutzbach, 1983
Lake Turkana, Kenya	10,000 to 4,000	+200	+27	Vincens, 1989b
Nakuru–Elmenteita basin	10,000 to 8,000	+260 to +300	+29 to +33	Hastenrath and Kutzbach, 1983
Naivasha basin	9,200 to 5,650	+90 to +155	+10 to +17	Hastenrath and Kutzbach, 1983
Victoria basin	AD 1880	+170 to +220	+14 to +18	Hastenrath and Kutzbach, 1983
Lake Naivasha	AD 1890s	+150		Vincent *et al.*, 1989

and Van Campo, 1994). These arid conditions have continued, with a few significant and short-lived lake level changes which cannot be readily correlated between regions, to the present day, with a pattern similar to that occurring today being established by about 2,000 yr BP (Street-Perrott *et al.*, 1989). The lakes in Rajasthan, India, began to dry up at about 5,000 yr BP as well, and were probably completely desiccated between 3,500 and 1,500 yr BP (Swain *et al.*, 1983). In Australasia, conditions were approaching those of the present day by 3,000 yr BP (Street-Perrott *et al.*, 1989). High lake levels are, however, recorded at about 3,130 yr BP and between 1,700 and 1,600 yr BP in southern Africa (Deacon and Lancaster, 1988), with lower levels being generally characteristic (Shaw, 1986).

In southern Africa, Lake Chilwa (622 m a.s.l.) stood 9 m above present levels between AD 1650 and 1760 (Crossley *et al.*, 1984), and during historical times it has stood at modern levels or lower (Owen and Crossley, 1989). Lake Malawi, on the other hand, had a lowstand between AD 1500 and 1850 (^{210}Pb dates), overlapping with the Little Ice Age (Owen *et al.*, 1990), and in 1915 when outflow via the Shire River ceased; outflow resumed in 1935 after the lake level had risen 6 m (Beadle, 1981).

Several lake level variations occurred in recent times in eastern Africa. In the Ziway-Shala basin, low lake levels occurred in 1926, 1933, 1956, 1967 and 1978, and high levels in 1938 and 1970 to 1972: lake levels in Lake Abiyata and Lake Shala have thus varied by at least 6 m and 9 m, respectively (Street-Perrott, 1981). The cyclicity of the rainfall record of Addis Ababa is reflected, at least in part, by the lake level changes, but their significance on a secular timescale is hard to determine since the available records are too short (op. cit.). Between 1895 and 1968, Lake Turkana's levels fluctuated over a range of 20 m (+20 m above the 1968 level in the 1890s), and lower levels were recorded between 1950 and 1960, as compared to the 1968 level (Butzer, 1971).

High lake levels were recorded during the early and mid-1960s. A sharp rise in the level of Lake Victoria, Lake Turkana, Lake Naivasha, Lake Elementeita, Lake Nakuru, Lakes Edward

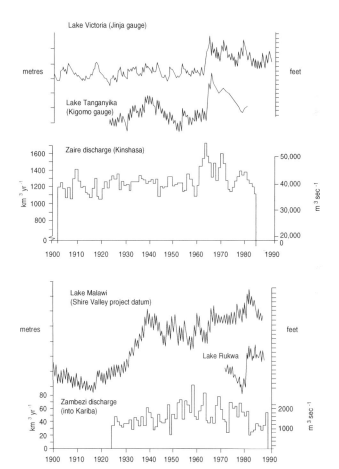

Figure 1.3. Fluctuations in African lake levels and river discharges in the twentieth century (from Grove, 1996).

and Albert, and Lake Tanganyika occurred in the early 1960s (Figure 1.3) (Richardson and Richardson, 1972; Beadle, 1981; Vincent *et al.*, 1989; Grove, 1996). The excessive rainy season of 1961–2 over East Africa, causing a discontinuous rise of Lake Victoria and the discharge of the White Nile lasting nearly 20 years, coincided with a large anomaly of SST, surface winds and convective cloudiness at the western equatorial Indian Ocean (Flohn, 1987). Since about 1965, there has been a decline in river discharges and lake levels, except in southern hemispheric Africa, where lakes such as Malawi and Rukwa reached their highest levels around 1980 (Figure 1.3) (Grove, 1996).

High levels in Lake Turkana in the 1890s are correlated with high lake levels in Lake Naivasha, where Vincent *et al.* (1989) estimated that precipitation was 0.15 m yr^{-1} greater than in the period 1935–75 (based on a water balance model) (Table 1.5).

Lake Naivasha was nearly twice as extensive in the early 1920s as in 1960 (Richardson and Richardson, 1972). The declining lake levels reflect in part a decreasing rainfall trend, averaging about 5 mm yr^{-1} over the basin between 1920 and 1949 (Sansome, 1952), as well as perhaps increasing human consumption from river influents and borehole pumping (Richardson and Richardson, 1972).

1.2.2 Glacier fluctuations

It has been noted above that tropical mountain glaciers show much less spatial consistency in the Holocene compared to the LGM and deglacial periods. Johannessen and Holmgren (1985) dated a moraine from Teleki Valley, Mount Kenya, at 6,070 ± 225 to 4,135 ± 70 yr BP, suggesting an ELA at least 200 m lower than present and a reduction of mean temperature of about 1.2 °C (or less if there was an increase in precipitation). This correlates with the moraine in Hobley Valley, dated at 6,277 to 5,425 yr BP (Perrott, 1982).

Evidence of systematic Holocene glacier fluctuations in equatorial Africa has been largely derived from work by Karlén *et al.* (1999) on Mount Kenya. They record six major periods of glacier advances from *c.* 6,000 cal. yr BP (5,300 ^{14}C yr BP) to the present (Figure 1.4). These are in cal. yr BP (^{14}C yr BP): shortly before 5,700 (4,800), 4,500–3,900 (4,050–3,750), 3,500–3,300 (3,300–3,150), 3,200–2,300 (3,330–2,250), 1,300–1,200 (1,380–1,320), and 600–400 (650–320). Many of the glacier expansions seem to coincide with periods of low precipitation and low temperature (Karlén *et al.*, 1999). Because the glacier advances occur at times when other palaeo-proxies from East Africa and other areas affected by the Indian monsoon indicate a relatively dry climate, low temperature is possibly the primary cause of the glacier expansions. Precipitation is likely to have been effective in some cases, and may have been a limiting factor to the extent of the advances (Karlén *et al.*, 1999).

Evidence of a cold period in East Africa at 2,000 yr BP, concurrent with that inferred from historical records of snow cover on the Semyen Mountains, is provided by similar estimated ages for moraines on Mount Kilimanjaro (Wood, 1976). Historical records state that twice within the last 2,000 years (*c.*AD 100 and 1800–50), permanent snowfields existed on the presently snow-free

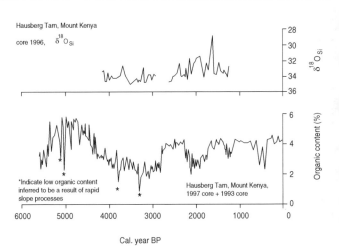

Figure 1.4. Organic carbon content of sediments from Hausberg Tarn, Mount Kenya, reflecting variations in rock flour influx from Josef and Cesar glaciers; and variations in $\delta^{18}O_{Si}$ of biogenic opal which reflects variations in short-term water discharge from the glaciers and long-term variations in lake water temperature (from Karlén *et al.*, 1999).

Semyen Mountains (4,543 m) of Ethiopia (Wood, 1976). A ^{14}C date of 1,960 ± 140 yr BP for a formerly vast lake in the now dry Dobi Graben of Ethiopia indicates that the early cold period was also wet (Wood, 1976). Certain moraine complexes in the Americas and across the mountains of the equatorial tropics can be ascribed to the period 1500–1800 (Little Ice Age). The glacial period of the nineteenth century in Ethiopia was cool and dry, as indicated by consistently low Nile floods from 1790 to 1839, and by three major drought famines in Ethiopia (1800, 1826–7, and 1835) (Wood, 1976). Recent glacier recessions are documented for, e.g., the Lewis Glacier on Mount Kenya from 1880 (as inferred from numerical modelling), the Carstensz Glacier on Mount Java, New Guinea, around the middle of the nineteenth century (Hastenrath, 1991) and the Venezuelan Andes, where several small glaciers above 4,700 m (the present day snow line), confined to the highest peaks, have been receding rapidly in recent years (Schubert, 1984). The present glacier recessions are attributed to a precipitation decrease of the order of 150 mm yr^{-1}, concomitant with a small decrease of cloudiness and surface albedo during the last two decades of the nineteenth century, and to a temperature increase of a few tenths of a degree during the twentieth century, concentrated in the 1920s (Hastenrath, 1991). The drastic decrease in precipitation in the latter part of the nineteenth century is also reflected in a drastic drop of the water level of East African lakes.

1.2.3 Vegetation

In the region of central coastal Brazil, South America, *Araucaria* forest reappeared at 10,000 yr BP (Figure 1.1) (Ledru, 1993). From pollen and spore analysis of sediments from Lake Mucubaji

Table 1.6. Tropical temperatures during the Holocene (relative to present)

Area	Time period (yr BP)	Temperature (relative to present) (°C)	Proxy	Source
Sumatra Highlands, Australasia	11,000 to 8,000	−2 to −4	Pollen	Flenley, 1979a
Sumatra Highlands, Australasia	Pre-9,000	−1.6	Pollen	Newsome and Flenley, 1988
Lynch's Crater, North-East Australia	11,000 to 7,000	−2 to −4	Pollen	Flenley, 1979a

(8° 47' N, 70° 49' W; alt. 3,540 m) in the Venezuelan Andes, it is inferred that vegetation has been fairly stable over the last 8,300 years (Salgado-Labouriau et al., 1992). An arid interval is recorded at c. 5,000 yr BP in coastal central Brazil (Ledru, 1993). After 3,000 yr BP, forest elements declined and grass pollen increased in the Eastern Colombian Andes (van der Hammen, 1974), presumably reflecting destruction of forest by man (Flenley, 1979a, b).

Cooler temperatures (2–4 °C lower than present) may have occurred from 11,000 to 8,000 yr BP in Sumatra, and possibly between 11,000 and 7,000 yr BP in Queensland, Australia (Table 1.6) (Flenley, 1979a). Slight temperature depressions (c. 1.6 °C) are recorded in the Sumatra Highlands for the period prior to 9,000 yr BP, after which forest limits began to rise (Newsome and Flenley, 1988). Palynological evidence from Lynch's Crater in northeastern Australia indicates a marked increase in precipitation in the early Holocene (Kershaw, 1974; 1978). A rapid rise in forest limits is indicated in the New Guinea Highlands, with highest elevations being achieved around 9,500 yr BP (Figure 1.1) (Flenley, 1979a,b). There is evidence of possible anthropogenic disturbance of rainforests in Sumatra after c. 8,200 yr BP (Newsome and Flenley, 1988).

In western Africa, pollen analysis of sediments from Senegal and Mauritania, from areas which are, at present, of Sahelian and Saharan nature, shows rapid extension of humid vegetation towards the north from about 9,000 yr BP with maximum intensity at about 8,500 yr BP, and correlated with the intensification of the Atlantic monsoon (Lézine, 1989). The maximum intensity is dated at 8,500 yr BP and corresponds with a large body of evidence on lacustrine extensions in the Sahelian and Saharan subtropical latitudes (Lézine, 1989). Tropical savannahs shifted 500–700 km northwards of their present range between 7,000 and 6,500 yr BP (this appears to have occurred simultaneously in the eastern and central Sahara), receding slightly after 6,000 yr BP to 300–400 km north of their present range (Neumann, 1991). From 5,200 yr BP onwards, aridity increased and the savannah formations retreated to the south until the present status was reached by 3,300 yr BP (Neumann, 1991). In western Africa, grass pollen assemblages dominate before 9,000 yr BP, and a dramatic rise in the abundance of arboreal pollen suggests that forest rapidly replaced grassland

after 9,000 yr BP (Talbot et al., 1984). Continuous forest became established between 7,500 and 5,000 yr BP in Lake Bosumtwi (Talbot and Hall, 1981); the early stages are concomitant with the more humid conditions recorded in the Niger delta between c.7,600 and 6,960 yr BP when freshwater swamp and rainforest components increase in the pollen assemblages (Sowunmi, 1981). In north-west Africa there was a rapid extension of humid vegetation to the north at 9,000 yr BP, correlated with the intensification of the Atlantic monsoon (Lézine, 1989). The maximum intensity is dated at 8,500 yr BP and corresponds with a large body of evidence on lacustrine extensions in the Sahelian and Saharan subtropical latitudes (Lézine, 1989). The interval 4,000 to 2,000 yr BP was more humid in Senegal and Mauritania (Lézine, 1989). Palynological and sedimentological records from Lake Barombi Mbo, West Cameroon, show climatic deterioration between 4,000 and 3,000 yr BP, and between 2,500 and 2,000 yr BP (Maley, 1992). Reduction in rainforest communities after 3,000 yr BP in the Niger Delta is attributed to human disturbance through agricultural practices (Sowunmi, 1981).

Between 10,000 and 8,000 yr BP, pollen data from various sites in East Africa indicate important montane forest extensions at high-altitude sites previously occupied by Ericaceous Belt taxa during the last glacial period and suggest warmer climate and significantly increased precipitation. A pollen spectrum from the low-altitude Galana Boi Beds of the north-east Lake Turkana basin is characterized by montane forest and thicket taxa at 9,880 yr BP, not found in modern pollen, and these are thought to have been transported to the lake by rivers originating in the Ethiopian Highlands, suggesting both increased run-off and an extension of the highland forests (Owen et al., 1982). The occurrence of more humid vegetation types near the lake (compared to present vegetation) suggests an increase in rainfall measuring at least 200 mm yr^{-1} above the modern average (Vincens, 1989b). During the early Holocene at Cherangani, the lowest part of the Ericaceous Belt, with Stoebe dominant, became established. Major forest persisted in Lake Victoria up to 7,000 yr BP (Kendall, 1969). In Kashiru, an important extension of montane forest occurred between about 10,000 and 5,000 yr BP (Roche and Bikwemu, 1989). Around the north basin of Lake Tanganyika, an increase in rainfall is inferred at 10,000 yr BP from the pollen spectrum, and is supported by a great increase in fern pollen, transported also by rivers and streams (Vincens, 1989a). This wet and warm phase characterized a major part of the Holocene around the northern basin of Lake Tanganyika (Vincens, 1989a). Possible vegetation change corollaries to the lake regression phase between 8,000 and 6,000 yr BP in East Africa are: (i) the onset of cool conditions, depression of the treeline and the establishment of Afroalpine grassland at Sacred Lake (Street-Perrott and Perrott, 1993); (ii) a shift from evergreen to semi-deciduous forest between 7,000 and 6,000 yr BP in the Pilkington Bay area (Kendall, 1969); (iii) an increase in dry tree taxa and *Chenoams* in the Naivasha area at *c.* 6,500 yr BP (Maitima, 1991); and (iv) a shift from moist montane *Hagenia* forest to a mixed montane or bamboo forest in Muchoya Swamp at *c.* 6,000 yr BP, marked by a sharp decline in *Hagenia* and a sharp increase in Ericaceae, which was explained as being possibly due to human influence (Morrison, 1968). Conditions in Lake Turkana were similar to present by the middle Holocene (Owen *et al.*, 1982). Similarly, the $\delta^{13}C$ of palaeosoils in a transect from Naivasha to the Mau Escarpment shows a mid-Holocene rise in the altitude of the savannah–forest ecotone and reflects the onset of drier conditions in the Naivasha basin (Ambrose and Sikes, 1991).

Aucour *et al.* (1994) observed an increase in C_4 plant types from about 4,500 yr BP in Kashiru, Burundi. A similar observation was made in the Nilgiri Hills, southern India, by Sukumar *et al.* (1993) for the period 4,700 to 3,000 yr BP. In both cases these changes are attributed to an increase in aridity.

The initiation of drier conditions on Mount Kenya at about 4,000 yr BP, indicated by an increase in *Podocarpus* pollen at Sacred Lake and the Hobley valley mire, and at about 3,720 yr BP at Lake Kimilili, Mount Elgon (alt. 4,265 m) (Perrott, 1982), and by a shift towards more positive $\delta^{13}C$ values in the Sacred Lake cores and C_4 plants in Kiluli Swamp, is also evident at other sites from pollen and stable carbon isotope records (Olago *et al.*, 1999). Dry montane forest taxa replaced wet montane forest taxa in the high-altitude regions, while at lower altitudes, trees gave way to herbaceous elements (e.g. Vincens, 1986; Maitima, 1991). Significant reduction of forest occurred at Lake Victoria after 3,000 yr BP and this has been attributed to increased human activity (Kendall, 1969), although the initial impact was probably climatic. The expansion of dry forest taxa around Ahakagyezi and Muchoya swamps occurred after 3,900 and 3,400 yr BP, respectively (Hamilton *et al.*, 1986; Taylor, 1990). In Kashiru Swamp in the Burundi Highlands, a significant extension of Gramineae and Ericaceae, accompanied by a decline of all forest elements, except *Podocarpus, Maytenus* and *Hypericum* occurred (Roche and Bikwemu, 1989). In the region of Congo, a slight trend towards aridity is evident from 3,000 yr BP south of the equator (Giresse and Lanfranchi, 1984). A progressive degradation of the arboreal cover and a concomitant development of Gramineae are observed

from 2,500 yr BP in the north basin of Lake Tanganyika, and are related to an increasingly dry climate and possibly to human interference (Vincens, 1989a). The dry conditions initiated at 4,000 yr BP generally characterized the rest of the Holocene, although slightly more humid conditions are superimposed on this trend. Roche and Bikwemu (1989) note a cold and dry period centred around 2,500 yr BP in the Kashiru area.

Efforts are now under way to have high-resolution records for the past millennium in Africa. For example, a 1,100-yr record describes the hydrological response of Lake Naivasha to a succession of decade-scale fluctuations in the regional balance of rainfall and evaporation, providing an excellent record of rainfall and drought in equatorial East Africa (Verschuren *et al.*, 2000). The data indicate that, over the past millennium, equatorial East Africa has alternated between contrasting climatic conditions, with significantly drier climate than today during the Medieval Warm Period (*c.* AD 1000–1270) and a relatively wet climate during the Little Ice Age (AD 1270–1850), which was interrupted by three prolonged dry episodes. The arid periods or drought events were broadly coeval with phases of high solar radiation, and intervening periods of increased moisture were coeval with phases of low solar radiation (Verschuren *et al.*, 2000). The drought periods matched oral historical records of famine, political unrest and large-scale migration of indigenous peoples, while the wet periods were prosperity years.

1.3 MECHANISMS OF CHANGE

1.3.1 LGM

Most Global Circulation Models (GCMs) have concentrated mainly on the contrasting climatic extremes of the Last Glacial Maximum (LGM) and the early Holocene periods (Street-Perrott, 1991). During the LGM (at 18,000 yr BP), the seasonal cycle and the annual total of solar radiation reaching the earth were similar to the present (Berger, 1979; Kutzbach and Street-Perrott, 1985; Kutzbach and Wright, 1985). Global circulation models show that tropical Africa, for example, was considerably more arid than at present (e.g. Manabe and Hahn, 1977; Kutzbach and Guetter, 1986), a trend consistent with the southward displacement of the Inter-Tropical Convergence Zone (ITCZ) and related monsoon precipitation. The simulations are strongly corroborated by Nicholson and Flohn's (1980) inferences on the major circulation features over Africa and South America during the main arid phase of the Late Glacial (20,000 to 12,000 yr BP), based on geological data. Most areas of Africa appear to have been dominated by dry north-east and northerly winds. This southern shift of the north-east trade winds led to a corresponding southward displacement of the summer south-west monsoon of West Africa and the south-west monsoon of East Africa, resulting in monsoon precipitation over a much more restricted land area (Nicholson and Flohn, 1980). However, displaced westerly flow (due to a strong baroclinic zone along the ice-sheet margin over northern Europe) would have brought relatively frequent depressions, and hence relatively moist conditions, to North Africa (op. cit.). In northern areas of South America, the displacement of the ITCZ (over an area presently occupied by equatorial rainforest) led to a more restricted winter north-east monsoon and pronounced aridity. By contrast, southerly airstreams (presently characteristic of South America) may have been more vigorous due to enhanced upwelling in the area of the Benguela current, along the eastern margin of a strengthened South Pacific anticyclone. Similar trends are observed in the equatorial belt and the northern hemisphere tropical summer-rainfall region of South and Central America (Street-Perrott et al., 1989).

Land-based studies in India (e.g. Sukumar et al., 1993) and studies of deep-sea cores in the Arabian Sea (e.g. Prell, 1984; Clemens and Prell, 1990; Sirocko et al., 1991; Rostek et al., 1993) suggest that the south-west monsoon winds were weaker during the glacial than in the early Holocene. GCMs also simulate weaker southern hemisphere monsoons during the LGM (Kutzbach and Guetter, 1986). Sirocko et al. (1991) suggest that the glacial seasonal timespan of the south-west monsoon season was also greatly reduced. This is supported by the model of Barnett et al. (1989), who show that the build-up of the continental heat low over South Asia (which produces south-west monsoons) is a function of snow and ice cover in high and mid-Asian latitudes, and indicates that the reduced glacial intensity of the heat low was a function of a shorter time span between snow melting in spring and the onset of renewed autumn snow precipitation.

Studies of gas inclusions in ice cores from Greenland and Antarctica, for instance, indicate that during the last glaciation the carbon dioxide (CO_2) level was only 56% of the level today (200 ppmv versus 355 ppmv), and about one third lower than typical Holocene values (200 ppmv versus 280 ppmv) (Berner et al., 1980; Delmas et al., 1980; Neftel et al., 1982, 1985; Lorius et al., 1984; Barnola et al., 1987). Methane (CH_4) concentrations were also about 50% lower during the glacials than in interglacials (350 ppbv compared to 650 ppbv in the early Holocene), as indicated by measurements of ancient CH_4 locked within the polar ice cores from both the northern and southern hemispheres (Stauffer et al., 1988; Chappellaz et al., 1993; Jouzel et al., 1993). The synchronicity of vegetation changes across the global low-latitude belts suggests that these environmental changes were driven by globally pervasive climatic factors, including temperature, precipitation and atmospheric trace gas changes. Pollen-derived estimates of the temperature depression during the LGM for various tropical sites average about 6 °C. More recent temperature estimates derived from multivariate statistical analysis of pollen assemblages from Kashiru Swamp (Bonnefille et al., 1990) and

Lake Tanganyika (Vincens et al., 1993) indicate a decrease of 4 °C ± 2 °C and 4.2 °C ± 3.6 °C, respectively. These estimates are generally lower than the earlier inferred values for East Africa, but are more plausible in view of the effect of the large changes in atmospheric CO_2 concentrations on treeline fluctuation (Street-Perrott, 1994; Olago, 1995; Street-Perrott et al., 1997; Olago et al., 1999). Stable carbon isotope analysis of organic lake sediments has revealed that this period was marked by the expansion of C_4 plants across the tropics, for example, in East Africa (Aucour et al., 1994; Street-Perrott et al., 1997; Olago et al., 1999), West Africa (Giresse et al., 1994) and southern India (Sukumar et al., 1993). Sukumar et al. (1993) attributed the C_4 plant dominance during the LGM in the Nilgiri Hills (alt. c. 2,000 m), southern India, to a weak south-west summer monsoon and thus reduced precipitation. The altitude of their site indicates that vegetation ecosystems would have responded more dramatically to pCO_2 stress rather than to precipitation decreases (Street-Perrott, 1994). Thus, although the south-west monsoon was weaker during the LGM, it was probably not the primary factor in governing C_3/C_4 plant distribution at their site.

1.3.2 The LGM to Holocene transition

Initial deglaciation following the LGM was accompanied by two sharp increases in CO_2 and CH_4 concentrations at 14,450 and 11,500 yr BP (calendar years) (Barnola et al., 1987; Chappellaz et al., 1993). Barnola et al. (1987) noted that the increases in temperature and in CO_2 were almost simultaneous, and proposed that this coherence was due to temperature-coupling of the radiative effect of CO_2 and its associated feedback mechanisms. Considerable amplification of the seasonal cycle in the northern hemisphere occurred between 15,000 and 6,000 yr BP due to changes in both perihelion and axial tilt (Kutzbach and Street-Perrott, 1985). The period 11,000 to 10,000 yr BP (Kutzbach and Street-Perrott, 1985) or c. 9,000 yr BP (Hecht, 1985), when perihelion occurred near the summer solstice, marks the time during which the maximum seasonal amplification occurred (Kutzbach and Otto-Bliesner, 1982). Seasonal contrasts were about 7% greater during the summer and 7% less during the winter as compared to today across the low and middle latitudes of both hemispheres, and there was increased heating of the land surface (Kutzbach and Otto-Bleisner, 1982; Hecht, 1985). The GCMs indicate that the amplified seasonal cycle of solar radiation, coupled with SSTs close to those of the modern ocean, created an intensified monsoon cycle over Africa and India, and predict increased rainfall over Africa (Kutzbach, 1981; Kutzbach and Otto-Bleisner, 1982; Kutzbach and Street-Perrott, 1985). Using a coupled atmosphere/mixed-layer ocean model, Kutzbach and Gallimore (1988) observed a slightly greater intensification of the monsoon (as compared to non-coupled atmospheric GCMs) resulting from small positive ocean feedbacks.

These data are compatible with the rise of lake levels in many lakes in Africa, and the warming was accompanied by high CO_2 and CH_4 concentrations (e.g. Chappellaz et al., 1993; Greenland Ice Core Project (GRIP) Members, 1993; Dansgaard et al., 1993; Taylor et al., 1993). They demonstrate the links between extrinsic and intrinsic climate-influencing mechanisms, and their consequent globally pervasive effects. The YD event in the tropics, just prior to 10,000 yr BP and accompanied by sharp decreases in CO_2 and CH_4 in polar ice cores (Chappellaz et al., 1993), has been attributed largely to precipitation decrease rather than lower temperatures from both lake level sensitivity experiments (Hastenrath and Kutzbach, 1983), and vegetation (Bonnefille et al., 1995). The decrease in CH_4 was observed to coincide with severe droughts in northern, western and eastern Africa, Tibet, and northern South America, and was related to reduced CH_4 emissions from dry wetlands in these areas (Street-Perrott, 1993).

1.3.3 The Holocene

The YD–early Holocene warming was accompanied by high CO_2 and CH_4 concentrations (e.g. Chappellaz et al.; 1993; Greenland Ice Core Project (GRIP) Members, 1993; Dansgaard et al., 1993; Taylor et al., 1993) and a large increase in precipitation in the tropical regions. Kutzbach and Otto-Bleisner (1982) suggest that the dry phase in the tropics at about 5,000 yr BP may be partly due to orbital variations as the solar radiation peak at 9,000 yr BP returned to near modern values by 5,000 yr BP. The post-glacial short-term anomalies in tropical and subtropical records occur during a state of apparent stability of external boundary conditions, which change only little and indicate that tropical climate is very sensitive to subtle changes of its forcing mechanisms (Zahn, 1994). It is suggested that variation in cross-equatorial heat transport by ocean surface currents in the thermohaline conveyor belt circulation may well have contributed to the short-term instabilities in monsoonal climate domains (Street-Perrott and Perrott, 1990). In addition, comparison of the rainfall pattern in the Sahel region (western Africa) with the global ocean SST patterns shows that during the past 80 years, dry episodes in Sahel (reduction of moisture flux) have been correlated with warming of surface ocean in the southern hemisphere and the northern Indian Ocean, and cooling of the North Atlantic and North Pacific (Folland et al., 1996). Duplessy et al. (1992) have reconstructed the past variations in salinity from sedimentary cores taken from two sites in the north-east Atlantic and noted low salinity events during 14,500–13,000 yr BP, YD, 8000–7000 and 5000–3000 yr BP, corresponding to known low lake levels in East Africa. The reduction in precipitation over land is associated with a southward shift in the equatorial rain belt (Street-Perrott, 1994). Guilderson et al. (1994) further postulate that variable tropical SSTs may explain the interhemispheric synchronicity of global climate changes

recorded in ice cores, snowline reconstructions and vegetation records, and that radiative changes due to cloud type and cover are plausible mechanisms for maintaining cooler tropical SSTs in part. However, despite apparent covariation between fluctuations in sea surface conditions in the North Atlantic and the monsoon record, there is no identifiable, direct mechanism that relates the intensity of the oceanic thermohaline conveyor belt to the monsoon strength (Gasse and Van Campo, 1994).

Early Holocene atmospheric CO_2 concentrations were 268 ppmv at 10,500 yr BP to 260 ppmv at 8,200 yr BP, and these millennial scale variations are attributed to changes in terrestrial biomass and SST (Indermühle et al., 1999). Although the summer insolation curve accounts for the general envelope of the post-glacial monsoon circulation, the pre-10,000 and 8,000 to 7,000 yr BP arid events occurred when seasonal contrasts in solar radiation in the northern tropics were at their maximum, and thus cannot solely explain these events (Gasse and Van Campo, 1994). The following mechanism has been put forward as a possible explanation for the abrupt dry events (including the 4,000 to 3,000 yr BP): (i) increases in tropical land cover, wetland area, soil moisture and lake filling, resulting from high monsoon rains, reduce surface albedo and increase methane production, enhancing the direct effects of insolation forcing; (ii) methane production falls when lakes reach their maximum extent, and high evaporation/evapotranspiration rates cool the continents, switching off the monsoon; and (iii) the land dries and warms again, increasing the ocean–land pressure gradient and resulting in inland penetration of moist oceanic air and precipitation (Gasse and Van Campo, 1994).

Nesje and Johannessen (1992) made a compilation of radiocarbon-dated records of Holocene glacier fluctuations in North and South America, Antarctica, Africa, Scandinavia, the Alps, the Himalayas and New Zealand. Based on this compilation, the number of glacier advances on a 250-year radiocarbon timescale over the last 10,000 years was used to make a frequency diagram of Holocene glacier advances (Figure 1.5). The calendar-dated record of global acid fallout (H_2SO_4 + HX) from volcanic eruptions north of 20° S, estimated from the acidity signal of annual layers in the Crête and Camp Century ice cores from Greenland (Hammer et al., 1980), was summarized on a 250-year timescale and transformed to radiocarbon years for comparison. The results suggest that volcanic aerosols represent one of the primary forcing mechanisms of worldwide Holocene glacier and climate fluctuations on a decadal to century timescale–local and regional topographic and climatic effects notwithstanding, as the latter cannot explain the irregular, step-wise character of the frequency curve of Holocene glacial advances (Nesje and Johannessen, 1992).

During the early Holocene, the effect of enhanced summer insolation to the northern hemisphere apparently reduced the climatic impact from volcanic eruptions. On the other hand, volcanic eruptions combined with reduced summer insolation during the late Holocene apparently led to greater glacier expansions than during the early Holocene (Nesje and Johannessen, 1992). Periods of glacier advances, most obvious after 5,000 years, occurred during intervals of global acid fallout exceeding 100×10^6 t. A climatic forcing curve (synthesis of summer solar insolation to the northern hemisphere and volcanic eruptions weighted 1:1) yielded a correlation coefficient of $r = 0.90$, as compared to the correlation coefficient of $r = 0.71$ for Holocene glacier advances versus volcanic eruptions (Nesje and Johannessen, 1992). They do not exclude 100-year timescale variations in solar irradiance, as well as local, regional topographic and climatic effects as contributing to modulation of climate and glacier fluctuations. Karlén and Kuylenstierna (1996), however, note that several periods with many volcanic eruptions and large atmospheric loads of aerosols do not coincide with any major cold period in the Scandinavian record. They instead attribute the Holocene record of glacier fluctuations in Scandinavia to a combination of land isostatic rebound and orbitally forced changes in irradiation.

This does not, however, obviate the conclusions of Nesje and Johannessen (1992), because they integrated a large data set including glaciers from tropical sites, where land isostatic rebound effects were negligible, and which probably responded more to integrated changes in solar irradiation, precipitation and volcanic aerosol loadings, rather than to direct temperature changes alone. Comparison of the global data set of glacier advances (Nesje and Johannessen, 1992) and that from Mount Kenya (Karlén et al., 1999) suggests that the advances of the Mount Kenya glaciers are influenced in part by changes in volcanic aerosol loading and correlate well with inferred cold periods for the Holocene in general (Figure 1.5).

Solar variation is becoming an increasingly exciting prospect as a major driver of Holocene climatic change. Several palaeo-proxy records from the high-latitude northern hemisphere, and to a lesser extent, Antarctica, show the influence of the Suess solar cycle of c. 211 years, on Holocene climate and environment (Chambers et al., 1999). However, owing to the short-term nature of ^{14}C excursions, attempts to detect solar forcing in sedimentary proxy records have often been constrained by the general lack of precision in dating (Chambers et al., 1999). It is now reasonably established that periods of lower solar activity lead to less effective screening of the earth by the solar wind from cosmic ray bombardment (Chambers et al., 1999). Increased cosmic ray bombardment leads not only to increased production of ^{14}C in the atmosphere, but would also lead to increased cloudiness (Svensmark and Friis-Christensen, 1997); the cooling effect created by increased cloud formation would be greatest during solar minima, the timing of these being embodied in the ^{14}C record throughout the Holocene (Figure 1.6).

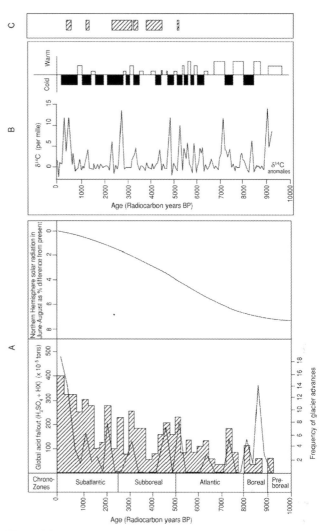

Figure 1.5. (A). Shaded scale: frequency of Holocene glacier advances based on a compilation of radiocarbon-dated records of Holocene glacier variations in North and South America, Antarctica, Africa, Scandinavia, the Alps, the Himalayas and New Zealand. Solid line: record of global acid fallout (H_2SO_4 + HX) ($\times 10^6$ t) from volcanic eruptions north of $20°$ S estimated from the acidity signal of annual layers in the Crête and Camp Century ice cores from Greenland. Upper: northern hemisphere solar radiation in June–August as percentage difference from present (from Nesje and Johannessen, 1992). (B) Histogram: generalized view of variations in the Holocene tree limit in Scandinavia. The inferred tree-limit altitude is reduced because of glacio-eustatic recovery, which is estimated to be about 150 m since 9000 yr BP. The height of the boxes above the mid-line indicates relative summer warmth, based on tree-lines in northern Scandinavia. Boxes below the mid-line only indicate glacier advances and do not indicate the severity of cold events. The lower panel shows variations in solar irradiation, calculated as the difference between dendro-age and the ^{14}C age of tree rings (δ^{14}C anomalies; Wigley and Kelly, 1990). Accepting a lag of 150 years (the timescale of the solar index is displaced by 150 years to the right) for the tree-limit, since pine trees survive even when summer temperature is too low to permit germination, there is a good correspondence between times with large ^{14}C anomalies (low solar irradiation) and low tree-limit (from Karlén and Kuylenstierna, 1996). (C) Glacier advances on Mount Kenya (from Karlén et al., 1999).

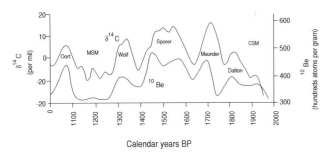

Figure 1.6. Trends in $\delta^{14}C$ and ^{10}Be from AD 1000 to present, together with the solar minima (named) and solar maxima (CSM – Contemporary Solar activity Maximum; MSM – Medieval Solar activity Maximum), showing that the concentrations of these cosmogenic isotopes in south polar ice are apparently related to historic solar activity (from Chambers *et al.*, 1999).

The high-resolution, 1000-yr hydrological record for Lake Naivasha shows a relationship with solar radiation: arid periods or drought events were broadly coeval with phases of high solar radiation, and intervening periods of increased moisture were coeval with phases of low solar radiation (Verschuren *et al.*, 2000).

Tropical climatic and environmental changes in the Holocene, therefore, appear to be controlled by changes in (i) air temperature (up to 2 °C), due to variations in solar irradiation and volcanic aerosol loadings; and (ii) precipitation, influenced by changes in sea surface temperatures and changing land-surface conditions.

1.4 DISCUSSION

1.4.1 Precipitation variability

Eastern Africa experiences precipitation variability on a five- to six-year cycle, which is particularly influenced by El Niño Southern Oscillation (ENSO)-induced changes in the 'short' rains period of October–November (Nicholson, 1996). Time series analysis of mid-Holocene to recent sediment cores from Lake Turkana reveals several cycles, including 11 and 22 years, which may be more likely due to multiples of the 5–6 year recurrence of ENSO, than due to sunspot activity (single and double sunspot cycles, respectively) (Johnson, 1996). However, new statistical analysis indicates that periodicities shorter than a century cannot be confirmed in Lake Turkana, given the present uncertainty in the calibration of the radiocarbon timescale (Johnson, 1996).

Time series analysis of a varved core from northern Lake Malawi (Pilskaln and Johnson, 1991; Johnson, 1996) shows periodicities at 2.6 and 3.5 years, very similar to periodicities in rainfall anomalies for this part of Africa (cf. Nicholson, 1996). High-resolution, well-dated cores are particularly required from the tropics to better understand the long- and short-term periodicities that characterize the climate of the region. The mechanisms

underlying these cycles need to be elucidated by use of high-performance regional and global models with reliable palaeo-data sets for validation.

1.4.2 Atmospheric trace gas concentrations

The doubling of atmospheric trace gas concentrations may produce a 1.5 to 4.5 °C global mean warming by the end of this century (Houghton *et al.*, 1990). This expected increase in CO_2 is unprecedented in the 400,000 yr record of CO_2 concentrations measured from the ice cores (Petit *et al.*, 1999). Meehl and Washington (1993) used a global coupled ocean–atmosphere model to simulate the effects of doubled CO_2. They observed an increase in SSTs and evaporation, and a mean increase in precipitation in the South Asian monsoon region; as a partial consequence, interannual variability of the area-averaged monsoon rainfall was enhanced. The increased variability in precipitation was a consequence of higher mean SSTs (due to increased CO_2) that produced a non-linear increase in evaporation. This is consistent with observations, which showed an increase in interannual variability of the Indian monsoon precipitation, associated with warmer land and ocean temperatures in the monsoon region. Melillo *et al.* (1993) used a process-based terrestrial ecosystem model (TEM) with doubled CO_2 (from an ambient concentration of 312.5 ppmv), which describes how important ecosystem processes such as photosynthesis, respiration, decomposition and nutrient cycling interact to affect net primary production (NPP). The model suggests that the NPP in tropical and dry temperate regions is dominated by effects of elevated CO_2, while in northern and moist temperate ecosystems, the NPP increases reflect, primarily, the effects of elevated temperature in enhancing nitrogen availability. In addition, Bonan *et al.* (1992) studied the effect of boreal forest vegetation on climate change, and their results showed that the boreal forests and climate were an interactively coupled system and could initiate important climate feedback, which can extend to low latitudes.

Vegetation plays an important role in the regulation of other greenhouse gases, e.g. hydrogen peroxide (H_2O_2), CO_2, CH_4, nitrous oxide (N_2O), non-methane hydrocarbons (NMHCs), and via influences of NMHCs on ozone (O_3) (Overpeck, 1993). The exact role of the biosphere in changing the global carbon cycle remains to be worked out. Deglacial trace gas changes may have been partly responsible for the apparent global synchronicity of large abrupt climatic change about 13,000 yr BP, and a complete understanding of how vegetation can influence trace gases and, consequently, climate will require a careful study of past global change. This is particularly important because the response of natural vegetation to present-day environmental change is difficult to decipher, due to large human impact on natural vegetation in many parts of the world (Overpeck, 1993). It is therefore crucial to initiate sensitivity studies on large-scale response of tropical forests to elevated

CO_2, and to establish linkages with boreal forests to better guide policies on trace gas emissions.

1.4.3 Land-use/land-cover change

Presently, existing precipitation in the Sahel savannah, Western Africa, could support a much richer flora and fauna, were it not for human activities (Cloudsley-Thompson, 1974). Increased albedo, atmospheric dust, and reduced soil moisture resulting from anthropogenic deforestation enhance drought in the region (Cloudsley-Thompson, 1993). However, it has been noted that, from the 1950s to the present, the Sahel has undergone a significant decrease in rainfall (average 650 mm per annum in the 1950s and 1960s, to 500 mm per annum in the 1970s and 1980s) (Fosberg and Gash, 1998). The rainy season, from May to October, did not change, and the decrease in rainfall (convective type) is attributed exclusively to a decrease in the number of rainfall events: there is speculation that the reduced rainfall is associated with deforestation, i.e. reduced evaporation from forest area and greater run-off back to the sea (Fosberg and Gash, 1998). However, while high rates of anthropogenic deforestation have been reported in a few West and Central African countries, a slow and natural recolonization of secondary forests on previously cultivated areas is also observed in other places (Lambin, 1998). Episodic conflicts have generated large movements of refugees that have led to margin land cover changes in some places. Except in Gabon and Cameroon, most of the humid forest regions in Central Africa are land-locked. In the surrounding biomes, savannah fires are widespread and have significant ecological impacts on a wider region, given the long-range transport of atmospheric emissions. Climate–land cover interactions, therefore, require much more investigation, as the impacts of anthropogenically driven land cover change can, potentially, significantly alter regional climate boundary characteristics in the short term, and may feedback positively into the global climate system as a whole in the long term.

1.4.4 Recent ecosystem changes in African Great Lakes

Disturbance factors affecting the African lakes (Cohen et al., 1996) are: (i) fishing activities, including overfishing and lack of effective management practices (Aloo, 1996; Kudhongania et al., 1996); (ii) discharge of pollutants; (iii) damage to watersheds, leading to cultural eutrophication and excess sedimentation problems; (iv) introduction of exotic species and translocations of species within lakes, leading to extinctions and drastically altered community structures (Kudhongania et al., 1996; Twongo, 1996); and (v) regional climate change. The development of effective conservation strategies will require substantial changes in how the lakes are treated. For example, there is evidence of increased productivity in Lake Victoria since 1900, reflected by increases

in the cyanobacteria population; and good preservation of organic matter and organic indicators suggest anoxic conditions have existed over the last 200 years (Lipiatou et al., 1996). Evidence of increased human population of the basin and subsequent deforestation is supported by the large increase in phenanthrene accumulation rate derived from combustion of fossil fuels at low temperature and the increase in retene accumulation attributed to either wood combustion or soil erosion (Lipiatou et al., 1996).

1.5 CONCLUSION

Long-term records of change are required to assess the significance of historically documented and modern-day change in the tropics. More high-resolution proxy records are required from these regions. Given the rapidity of human impact on the environment and the associated uncertainties regarding the impact on global climate and environment, as well as the modes of feedback and degree of interaction between the various components/factors of change, there is a need to better understand the natural climate and environmental variability. Rapid progress is critical. The mechanisms underlying the abrupt, large-scale climatic events in the Holocene need to be understood as they occurred during a period with climatic boundaries similar to those of today.

ACKNOWLEDGEMENTS

The authors wish to thank Pak Sum Low for his kind invitation to contribute to this volume. Special thanks are due to Professor Phil Jones of the Climatic Research Unit, University of East Anglia, United Kingdom, and an anonymous reviewer for very constructive comments and help in editing the text.

REFERENCES

Aloo, P. A. (1996) Anthropogenic impact on fisheries resources of Lake Naivasha. In T. C. Johnson and E. O. Odada (Eds.), *The Limnology, Climatology and Palaeoclimatology of the East African Lakes*, pp. 325–335. Australia, Gordon and Breach.

Ambrose, S. H. and Sikes, N. E. (1991) Soil carbon isotope evidence for Holocene habitat change in the Kenya Rift Valley. *Science*, **253**, 1402–1405.

Andersen, B. G. (1981) Late Weichselian ice sheets in Eurasia and Greenland. In G. H. Denton and T. J. Hughes (Eds.), *The Last Great Ice Sheets*, pp. 3–66. New York, John Wiley and Sons.

Andressen, R. L. and Ponte, R. R. (1973) *Climatología e Hidrología; Estudio Integral de las Cuencas de los Rios Chama y Capazon, Sub-Proyecto No. II*. Merida, Venezuela, Universidad de Los Andes, 135pp.

Aucour, A. and Hillaire-Marcel, C. (1993) A 30,000 yr record of ^{13}C and ^{18}O changes in organic matter from an equatorial peatbog. In P. K. Swart, K. C. Lohmann, J. McKenzie and S. Savin (Eds.), *Climate Change in Continental Isotopic Records*. Geophysical Monograph, **78**, 343–351. American Geophysical Union.

Aucour, A., Hillaire-Marcel, C. and Bonnefille, R. (1994) Late Quaternary biomass changes from ^{13}C measurements in a highland peatbog from equatorial Africa (Burundi). *Quat. Res.*, **41**, 225–233.

Barnett, T. P., Dumenil, L., Schlese, U., Roeckner, E. and Latif, M. (1989) The effect of Eurasian snow cover on regional and global climate variations. *J. Atm. Sci.*, **46**, 661–685.

Barnola, J. M., Raynaud, D., Korotkevich, Y. S. and Louis, C. (1987) Vostok ice core provides 160,000 year record of atmospheric CO_2. *Nature*, **329**, 408–414.

Beadle, L. C. (1981) *The Inland Waters of Tropical Africa: An Introduction to Tropical Limnology*. London and New York, Longman, 479pp.

Begét, J. E. (1983) Radiocarbon-dated evidence of worldwide early Holocene change. *Geology*, **11**, 389–393.

Berger, A. L. (1979) Long-term variations of daily insolation and Quaternary climatic changes. *J. Atm. Sci.*, **35**, 2362–2367.

Berner, W., Oeschger, H. and Stauffer, B. (1980) Information on the CO_2 cycle from ice core studies. *Radiocarbon*, **22**, 227–234.

Bonan, G. B., Pollard, D. and Thompson, S. L. (1992) Effect of boreal forest vegetation on global climate. *Nature*, **359**, 716–718.

Bonnefille, R. and Riollet, G. (1988) The Kashiru pollen sequence (Burundi) palaeoclimatic implications for the last 40,000 yr BP in Tropical Africa. *Quat. Res.*, **30**, 19–35.

Bonnefille, R., Roeland, J. C. and Guiot, J. (1990) Temperature-rainfall estimates for the past 40,000 years in equatorial Africa. *Nature*, **346**, 347–349.

Bonnefille, R., Riollet, G., Buchet *et al.* (1995) Glacial/interglacial record from intertropical Africa, high resolution pollen and carbon data at Rusaka, Burundi. *Quat. Sci. Rev.*, **14**, 917–936.

Bowler, J. M. (1975) Deglacial events in southern Australia: their nature, age and palaeoclimatic significance. In R. P. Suggate and M. M. Cresswell (Eds.), *Quaternary Studies*, pp. 75–82. Wellington, NZ, Royal Society.

  (1976) Aridity in Australia: origins and expression in aeolian landforms and sediments. *Earth Sci. Rev.*, **12**, 279–310.

Bowler, J. M., Hope, G. S., Jennings, J. N., Singh, G. and Walker, D. (1976) Late Quaternary climates of Australia and New Guinea. *Quat. Res.*, **6**, 359–394.

Bradley, R. S. (1985) *Quaternary Paleoclimatology: Methods of Paleoclimatic Reconstruction*. Boston, London, Unwin Hyman. 472pp.

Broecker, W. S. and Denton, G. H. (1990) What drives glacial cycles? *Sci. Am.*, **262**, 49–56.

Brown, K. S. Jr. and Ab'Saber, A. N. (1979) Ice-age forest refuges and evolution in the neotropics: correlation of palaeoclimatological, geomorphological and pedological data with modern biological endemism. *Paleoclimas*, **5**, 1–30.

Brun, A. (1991) Reflexions sur les pluviaux et arides au Pléistocène

supérieur et à l'Holocène en Tunisie. *Palaeoecol. Afr.*, **22**, 157–170.

Butzer, K. W. (1971) *Recent History of an Ethiopian Delta*. Univ. Chicago, Dept. of Geography, Res. Paper 136, 184pp.

  (1984) Late Quaternary environments in South Africa. In J. C. Vogel (Ed.), *Late Cainozoic Palaeoclimates of the Southern Hemisphere*, pp. 235–264. Rotterdam, Balkema.

Chambers, F. M., Ogle, M. I. and Blackford, J. J. (1999) Palaeoenvironmental evidence for solar forcing of Holocene climate: linkages to solar science. *Prog. Phys. Geog.*, **23**, 181–204.

Chappellaz, J., Blunier, T., Raynaud, D. *et al.* (1993) Synchronous changes in atmospheric CH_4 and Greenland climate between 40 and 8 kyr BP. *Nature*, **366**, 443–445.

Clapperton, C. M. (1983) The glaciation of the Andes. *Quat. Sci. Rev.*, **2**, 83–155.

Clemens, S. C. and Prell, W. L. (1990) Late Pleistocene variability of Arabian Sea summer monsoon winds and continental aridity: aeolian records from the lithogenic component of deep-sea sediments. *Palaeoceanography*, **5**, 109–146.

Cloudsley-Thompson, J. L. (1974) The expanding Sahara. *Envir. Conserv.*, **1**, 5–13.

  (1993) The future of the Sahara. *Envir. Conserv.*, **20**, 335–338.

Coetzee, J. A. (1967) Pollen analytical studies in East and Southern Africa. *Palaeoecol. Afr.* **3**, 146pp.

Cohen, A. S., Kaufman, L. and Ogutu-Ohwayo, R. (1996) Anthropogenic threats, impacts and conservation strategies in the African Great Lakes: a review. In T. C. Johnson and E. O. Odada (Eds.), *The Limnology, Climatology and Palaeoclimatology of the East African Lakes*, pp. 575–624. Australia, Gordon and Breach.

Crossley, R., Davison-Hirchsmann, S., Owen, R. B. and Shaw, P. (1984) Lake level fluctuations during the last 2000 years in Malawi. In J. C. Vogel (Ed.), *Late Cainozoic Palaeoclimates of the Southern Hemisphere*, pp. 305–316. Rotterdam, Balkema.

Crowley, T. J. (1983) The geologic record of climatic change. *Rev. Geophys. Space Phys.*, **21**, 827–877.

Dansgaard, W., Johnsen, S. J., Clausen *et al.* (1993) Evidence for general instability of past climate from a 250 kyr ice core record. *Nature*, **364**, 218–220.

Dawson, A. G. (1992) *Ice Age Earth: Late Quaternary Geology and Climate*. London, Routledge, 293pp.

Deacon, J. and Lancaster, N. (1988) *Late Quaternary Palaeoenvironments of Southern Africa*. Oxford, Clarendon Press, 225pp.

Delmas, R. J., Ascencio, J. M. and Legrand, M. (1980) Polar ice evidence that atmospheric CO_2 20,000 years BP was 50% of present. *Nature*, **284**, 155–157.

Duplessy, J. C., Labeyrie, L., Arnold, M. *et al.* (1992) Changes in surface salinity of the North Atlantic ocean during the last deglaciation. *Nature*, **358**, 485–488.

Elenga, H., Vincens, A. and Schwartz, D. (1991) Presence d'elements forestiers montagnards sur les Plateaux Batéké (Congo) au Pléistocène supérieur: nouvelles données palynologiques. *Palaeoecol. Afr.*, **22**, 239–252.

Finney, B. P. and Johnson, T. C. (1991) Sedimentation in Lake Malawi, Eastern Africa, during the past 10,000 years: a continuous palaeoclimatic record for the southern tropics. *Palaeogeogr. Palaeoclimatol. Palaeoecol.*, **85**, 351–366.

Flenley, J. R. (1979a) The late Quaternary vegetational history of the equatorial mountains. *Prog. Phys. Geogr.*, **3**, 488–509.

(1979b) *The Equatorial Rain Forest: A Geological History.* London, Butterworth and Co. Ltd., 162pp.

Flohn, H. (1987) East African rains of 1961/62 and the abrupt change of the White Nile discharge. *Palaeoecol. Afr.*, **18**, 3–18.

Folland, C. K., Palmer, T. N. and Parker, D. E. (1996) Sahel rainfall and worldwide sea temperatures, 1901–85. *Nature*, **320**, 602–607.

Fosberg, M. and Gash, J. (1998) Land Management implications of the HAPEX – Sahel Study. *Global Change Newsletter* No. 35, 8–9. A newsletter of the International Geosphere Biosphere Programme (IGBP).

Galloway, R. W. (1965) Late Quaternary climates in Australia. *J. Geol.*, **73**, 603–618.

Galloway, R. W., Hope, G. S., Löffler, E. and Peterson, J. A. (1973) Late Quaternary glaciation and periglacial phenomena in Australia and New Guinea. *Palaeoecol. Afr.*, **8**, 125–138.

Gasse, F. and Van Campo, E. (1994) Abrupt post-glacial climatic events in West Asia and North Africa monsoon domains. *Earth Planet. Sci. Lett.*, **126**, 435–456.

Gillespie, R., Street-Perrott, F. A., Switsur, R. (1983) Post-glacial arid episodes in Ethiopia have implications for climate prediction. *Nature*, **306**, 680–683.

Giresse, P. and Lanfranchi, R. (1984) Les climats et les oceans de la region Congolaise pendant l'Holocene-Bilans selons les echelles et les methodes de l'observation. *Palaeoecol. Afr.*, **16**, 77–88.

Giresse, P., Maley, J. and Brenac, P. (1994) Late Quaternary palaeoenvironments in the lake Barombi Mbo (West Cameroon) deduced from pollen and carbon isotopes of organic matter. *Palaeogeogr. Palaeoclimatol. Palaeoecol.*, **107**, 65–78.

Greenland Ice Core Project (GRIP) Members (1993) Climate instability during the last interglacial period recorded in the GRIP ice core. *Nature*, **364**, 203–207.

Grove, A. T. (1996) African river discharges and lake levels in the Twentieth Century. In T. C. Johnson and E. O. Odada (Eds.), *The Limnology, Climatology and Palaeoclimatology of the East African Lakes*, pp. 95–100. Australia, Gordon and Breach Publishers.

Guilderson, T. P., Fairbanks, R. G. and Rubenstone, J. L. (1994) Tropical temperature variations since 20000 years ago: modulating inter-hemispheric climate change. *Science*, **263**, 663–665.

Halfman, J. D., Jacobson, D. F., Cannella, C. M., Haberyan, K. A. and Finney, B. P. (1992) Fossil diatoms and the mid to late Holocene palaeolimnology of Lake Turkana, Kenya: a reconnaissance study. *J. Palaeolimnol.*, **7**, 23–35.

Hamilton, A. and Perrott, R. A. (1978) Date of deglaciation of Mount Elgon. *Nature,* **273**, 49.

(1979) Aspects of the glaciation of Mount Elgon, East Africa. *Palaeoecol. Afr.*, **11**, 153–161.

Hamilton, A., Taylor, D. and Vogel, J. C. (1986) Early forest clearance and environmental degradation in south-west Uganda. *Nature*, **320**, 164–167.

Hamilton, A. C. (1982) *Environmental History of East Africa: A Study of Quaternary.* London, Academic Press.

(1987) Vegetation and climate of Mt. Elgon during late Pleistocene and Holocene. *Palaeoecol. Afr.*, **18**, 283–304.

Hammer, C. U., Clausen, H. B. and Dansgaard, W. (1980) Greenland ice sheet evidence of post-glacial volcanism and its climate impact. *Nature*, **288**, 230–235.

Harper, G. (1969) Periglacial evidence in southern Africa during the Pleistocene epoch. *Palaeoecol. Afr.*, **4**, 71–101.

Hastenrath, S. (1991) *Climate Dynamics of the Tropics.* Dordrecht, Boston, London, Kluwer Academic Publishers, 488pp.

Hastenrath, S. and Kutzbach, J. E. (1983) Palaeoclimate estimates from water and energy budgets of East African Lakes. *Quat. Res.*, **19**, 141–153.

Hecht, A. D. (1985) A retrospective of the past 20 years. In A. D. Hecht (Ed.), *Palaeoclimate Analysis and Modelling*, pp. 1–26. New York, Wiley Interscience.

Heine, K. (1978) Radiocarbon chronology of late Quaternary lakes in the Kalahari, southern Africa. *Catena*, **5**, 145–149.

Hope, G. S., Peterson, J. A., Allison, I. and Radock U. (Eds.) (1976) *The Equatorial Glaciers of New Guinea.* Rotterdam, Balkema, 244pp.

Houghton, J. T., Jenkins, G. T. and Ephraums, J. J (Eds.) (1990) *Climate Change: The IPCC Scientific Assessment.* Cambridge, Cambridge University Press.

Hurni, H. (1981) Simen mountains – Ethiopia: palaeoclimate of the last cold period (late Würm). *Palaeoecol. Afr.*, **13**, 127–137.

Imbrie, J., Hays, J. D., Martinson, D. G. *et al.* (1984) The orbital theory of Pleistocene climate: support from a revised chronology of the marine $U^{18}O$ record. In D. A. Berger, J. Imbrie, J. Hays, G. Kukla, and B. Saltzman (Eds.), *Milankovitch and Climate, Part 1*, NATO ASI Series C: Mathematical and Physical Sciences Vol. 126, pp. 269–305. Dordrecht, Reidel Publishers.

Indermühle, A., Stocker, J. F., Joos, F. *et al.* (1999) Holocene carbon-cycle dynamics based on CO_2 trapped in ice at Taylor Dome, Antarctica. *Nature*, **398**, 121–126.

Johannessen, L. and Holmgren, K. (1985) Dating of a moraine on Mount Kenya. *Geografiska Annaler*, **67** A 1–2, 123–128.

Johnson, T. C. (1996) Sedimentary processes and signals of past climatic changes in the large lakes of the East African rift valley. In T. C. Johnson and E. O. Odada (Eds.), *The Limnology, Climatology and Palaeoclimatology of the East African Lakes*, pp. 367–412. Australia, Gordon and Breach Publishers.

Johnson, T. C., Kelts, K. and Odada, E. O. (2000) The Holocene history of Lake Victoria. *Ambio*, **29** (1), 2–11.

Jouzel, J., Barkov, N. I., Barnola, J. M. *et al.* (1993) Extending the Vostok ice-core record of palaeoclimate to the penultimate glacial period. *Nature*, **364**, 407–412.

Karlén, W. and Kuylenstierna, J. (1996) On solar forcing of Holocene climate: evidence from Scandinavia. *The Holocene*, **3**, 359–365.

Karlén, W., Fastook, J. L., Holmgren, K. *et al.* (1999) Glacier fluctuations on Mount Kenya since *c.* 6000 cal. years BP: Implications for Holocene climatic change in Africa. *Ambio*, **28**, 409–418.

Kendall, R. L. (1969) An ecological history of the Lake Victoria Basin. *Ecol. Monogr.* **39**, 121–176.

Kershaw, A. P. (1974) A long continuous pollen sequence from Northeastern Australia. *Nature*, **251**, 222–223.

(1978) Record of the last interglacial-glacial cycle from northeastern Queensland. *Nature*, **272**, 159–161.

Kudhongania, A. W., Ocenodongo, D. L. and Okaronon, J. O. (1996) Anthropogenic perturbations on the Lake Victoria ecosystem. In T. C. Johnson and E. O. Odada (Eds.), *The Limnology, Climatology and Palaeoclimatology of the East African Lakes*, pp. 625–632. Australia, Gordon and Breach.

Kutzbach, J. E. (1981) Monsoon climate of the early Holocene: climate experiment with the Earth's orbital parameters for 9000 years. *Science*, **214,** 59–61.

Kutzbach, J. E. and Gallimore, R. G. (1988) Sensitivity of a coupled atmosphere/mixed-layer ocean model to changes in orbital forcing at 9000 yr BP. *J. Geophys. Res.*, **93** (D1), 803–821.

Kutzbach, J. E. and Guetter, P. J. (1986) The influence of changing orbital parameters and surface boundary conditions on climate simulations for the past 18000 years. *J. Atm. Sci.*, **43**, 1726–1759.

Kutzbach, J. E. and Otto-Bleisner, B. L. (1982) The sensitivity of the African-Asian monsoonal climate to orbital parameter changes for 9000 years BP in a low-resolution general circulation model. *J. Atm. Sci.*, **39**, 1177–1188.

Kutzbach, J. E. and Street-Perrott, F. A. (1985) Milankovitch forcing of fluctuations in the level of tropical lakes from 18 to 0 kyr BP. *Nature*, **315**, 130–134.

Kutzbach, J. E. and Wright, H. E. (1985) Simulations of the climate of 18,000 yr BP: results for the North American/North Atlantic/European sector and comparison with the geologic record. *Quat. Sci. Rev.*, **4**, 147–187.

Lambin, E. (1998) An emerging land-use/land cover change transect in West and Central Africa. *Global Change Newsletter* No. **35**, 4–5. A newsletter of the International Geosphere Biosphere Programme (IGBP).

Lancaster, N. (1979) Evidence for a widespread late Pleistocene humid period in the Kalahari. *Nature*, **279**, 145–146.

Ledru, M. P. (1993) Late Quaternary environmental and climatic changes in Central Brazil. *Quat. Res.*, **39**, 90–98.

Lézine, A. M. (1989) Vegetational palaeoenvironments of northwest tropical Africa since 12,000 yr BP: pollen analysis of continental sedimentary sequences (Senegal-Mauritania). *Palaeoecol. Afr.*, **20**, 187–188.

Lipiatou, E., Hecky, R. E., Eisenreich *et al.* (1996) Recent ecosystem changes in Lake Victoria reflected in sedimentary natural and anthropogenic organic compounds. In T. C. Johnson and E. O. Odada (Eds.), *The Limnology, Climatology and Palaeoclimatology of the East African Lakes*, pp. 523–541. Australia, Gordon and Breach.

Littmann, T. (1989) Climatic change in Africa during the Last Glacial: facts and problems. *Palaeoecol. Afr.*, **20**, 163–179.

Livingstone D. A. (1962) Age of deglaciation in the Ruwenzori Range, Uganda. *Nature*, **194**, 859–860.

(1975) Late Quaternary climatic change in Africa. *Ann. Rev. Ecol. Syst.*, **6**, 249–280.

(1980) Environmental changes in the Nile headwaters. In M. A. J. Williams and H. Faure (Eds.), *The Sahara and the Nile,* pp. 339–359. Rotterdam, Balkema.

Lorius, C., Raynaud, D., Petit, J. R., Jouzel, J. and Merlivat, L. (1984) Late glacial maximum-Holocene atmospheric and ice thickness changes from Antarctic ice core studies. *Ann. Glaciol.*, **5**, 88–94.

Maitima, J. M. (1991) Vegetation response to climatic change in Central Rift Valley, Kenya. *Quat. Res.*, **35**, 234–245.

Maley, J. (1987) Fragmentation de la forêt dense humide Africaine et extension des biotopes montagnards au Quaternaire recent: nouvelles données polliniques et chronologiques. Implication palaéoclimatiques et biogéographiques. *Palaeoecol. Afr.*, **18**, 307–336.

(1992) Mise en evidence d'une pejoration climatique entre ca. 2,500 et 2,000 ans B.P. en Afrique tropicale humide. *Bull. Geol. Soc. Fr.*, **3**, 363–365.

Manabe, S. and Hahn, D. B. (1977) Simulation of the tropical climate of an ice age. *J. Geophys. Res.*, **82**, 3,889–3,911.

Meehl, G. A. and Washington, W. M. (1993) South Asian summer monsoon variability in a model with doubled atmospheric carbon dioxide concentration. *Science*, **260**, 1101–1104.

Melillo, J. M., McGuire, A. D., Kicklighter, D. W. *et al.* (1993) Global climate change and terrestrial net primary production. *Nature*, **363**, 234–240.

Mercer, J. H. (1983) Cenozoic glaciation in the Southern Hemisphere. *Ann. Rev. Earth Planet. Sci.*, **11**, 99–132.

Morrison, M. E. S. (1968) Vegetation and climate in the uplands of south-western Uganda during the later Pleistocene Period, I. Muchoya Swamp, Kigezi District. *J. Ecol.*, **56**, 363–384.

Neftel, A., Oeschger, H., Schwander, J., Stauffer, B. and Zumbrunn, R. (1982) Ice core sample measurements give atmospheric CO_2 content during the past 4,000 yr. *Nature*, **295**, 220–223.

Neftel, A., Morr, E., Oeschger, H. and Stauffer, B. (1985) Evidence from polar ice cores for the increase in atmospheric CO_2 in the past two centuries. *Nature*, **315**, 45–47.

Nesje, A. and Johannessen, T. (1992) What were the primary forcing mechanisms of high-frequency Holocene Climate and glacier variations? *The Holocene* **2**, 79–84.

Neumann, K. (1991) In search for the green Sahara: palynology and botanical remains. *Palaeoecol. Afr.*, **22**, 203–212.

Newsome, J. and Flenley, J. R. (1988) Late Quaternary vegetational history of the Central Highlands of Sumatra. II. Palaeopalynology and vegetational history. *J. Biogeogr.*, **15**, 555–578.

Nicholson, S. and Flohn, H. (1980) African environmental and climatic changes and the general circulation in late Pleistocene and Holocene. *Climatic Change*, **2**, 313–348.

Nicholson, S. E. (1996) A review of climate dynamics and climate variability in eastern Africa. In T. C. Johnson and E. O. Odada (Eds.), *The Limnology, Climatology and Palaeoclimatology of the East African Lakes*, pp. 25–56. Australia, Gordon and Breach.

Olago, D. O. (1995) *Late Quaternary Lake Sediments of Mount Kenya, Kenya.* D.Phil. Thesis, University of Oxford.

Olago, D. O., Street-Perrott, F. A., Perrott, R. A., Ivanovich, M. and Harkness, D. D. (1999) Late Quaternary glacial-interglacial cycle of climatic and environmental change on Mount Kenya, Kenya. *J. Afr. Earth Sci.*, **29**, 593–618.

Osmaston, H. A. (1975) *Models for the Estimation of Firnlines of Present and Pleistocene Glaciers.* London, Heinemann.

Overpeck, J. T. (1993) The role and response of continental vegetation in the global climate system. In J. A. Eddy and H. Oeschger (Eds.), *Global Changes in the Perspective of the Past*, pp. 221–237. New York, John Wiley and Sons.

Owen, R. B. and Crossley, R. (1989) Recent sedimentation in Lakes Chilwa and Ciuta, Malawi. *Palaeoecol. Afr.,* **20**, 109–117.

Owen, R. B., Barthelemew, J. W., Renaut, R. W. and Vincens, A. (1982) Palaeolimnology and archaeology of Holocene deposits north-east of Lake Turkana, Kenya. *Nature*, **298**, 523–529.

Owen, R. B., Crossley, R., Johnson, T. C. *et al.* (1990) Major low levels of Lake Malawi and implications for speciation rates in cichlid fishes. *Proc. Roy. Soc. Lond.*, **240** (1299), 519.

Patridge, T. C. (1993) Warming phases in southern Africa during the last 150,000 years: an overview. *Palaeogeogr. Palaeoclimatol. Palaeoecol.*, **101**, 237–244.

Perrott, R. A. (1982) A high altitude pollen diagram from Mount Kenya: its implications for the history of glaciation. *Palaeoecol. Afr.*, **14**, 57–75.

Perrott, R. A. and Street-Perrott, F. A. (1982) New evidence for a late Pleistocene wet phase in northern intertropical Africa. *Palaeoecol. Afr.*, **14**, 57–75.

Petit, J. R., Jouzel, J., Raynaud, D. *et al.* (1999) Climate and atmospheric history of the past 420,000 years from the Vostok ice core, Antarctica. *Nature*, **399**, 429–435.

Pilskaln, C. and Johnson, T. C. (1991) Seasonal signals in Lake Malawi sediments. *Limnol. Oceanogr.*, **36**, 544–557.

Porter, S. C., Pierce, K. L. and Hamilton, T. D. (1986) Late Wisconsin mountain glaciation in the Western United States. In S. C. Porter (Ed.), *Late Quaternary Environments of the United States*, pp. 71–114. London, Longman.

Prell, W. L. (1984) Monsoonal climate of the Arabian sea during the late Quaternary: a response to changing solar radiation. In A. Berger, J. Imbrie, J. Hays, G. Kukla and B. Saltzman (Eds.), *Milankovitch and Climate, Part 1.* NATO ASI Series C: Mathematical and Physical Sciences 126, pp. 349–366. Dordrecht, Reidel.

Richardson, J. L. and Richardson, A. E. (1972) History of an African Rift lake and its climatic implications. *Ecol. Monogr.*, **72**, 499–534.

Ricketts, R. D. and Johnson, T. C. (1996) Early Holocene changes in lake level and productivity in Lake Malawi as interpreted from oxygen and carbon isotopic measurements of authigenic carbonates. In T. C. Johnson and E. O. Odada (Eds.), *The Limnology, Climatology and Palaeoclimatology of the East African Lakes*, pp. 475–508. Australia, Gordon and Breach.

Roche, E. and Bikwemu, G. (1989) Palaeoenvironmental change on the Zaire-Nile ridge in Burundi; the last 20,000 years: an interpretation of palynological data from the Kashiru Core, Ijenda, Burundi. In W. C. Mahaney (Ed.), *Quaternary and Environmental Research on East African Mountains*, pp. 231–244. Rotterdam, Balkema.

Rostek, F., Ruhland, G., Bassinot, F. C. *et al.* (1993) Reconstructing sea surface temperature and salinity using $\delta^{18}O$ and alkenone records. *Nature*, **364**, 319–321.

Rust, U. (1984) Geomorphic evidence of late Quaternary climatic changes in Etosha, South West Africa/Namibia. In J. C. Vogel (Ed.), *Late Cainozoic Palaeoclimates of the Southern Hemisphere*, pp. 279–286. Rotterdam, Balkema.

Salgado-Labouriau, M. L., Bradley, R. S., Yuretich, R. and Weingarten, B. (1992) Palaeoecological analysis of the sediments of Lake Mucubaji, Venezuelan Andes. *J. Biogeogr.*, **19**, 317–327.

Sansome, H. W. (1952) The trend of rainfall in East Africa. *E. Afr. Meteorol. Dep. Tech. Mem.*, **1**, 14pp.

Schubert, C. (1984) The Pleistocene and recent extent of the glaciers of the Sierra Nevada de Merida, Venezuela. In W. Lauer (Ed.), *Natural Environment and Man in Tropical Mountain Ecosystems*, pp. 66–69. Stuttgart, Erdwissenschaftliche Forschung, Band XVIII, Frantz Steiner Verlag, Wiesbaden GMBH.

Scott, L. (1990) Palynological evidence for late Quaternary environmental change in southern Africa. *Palaeoecol. Afr.* **21**, 259–268.

—— (1993) Palynological evidence for late Quaternary warming episodes in southern Africa. *Palaeogeogr. Palaeoclimatol. Palaeoecol.*, **101**, 229–235.

Servant, M. and Servant-Vildary, S. (1980) L'environnement Quaternaire du bassin du Tchad. In M. A. J. Williams and H. Faure (Eds.), *The Sahara and the Nile,* pp. 133–162. Rotterdam, Balkema.

Shaw, P. (1986) The palaeohydrology of the Okavango delta – some preliminary results. *Palaeoecol. Afr.*, **17**, 51–58.

Singh, G., Joshi, R. D., Chopra, S. K. and Singh, A. B. (1974) Late Quaternary history of vegetation and climate of the Rajasthan desert, India. *Phil. Trans. Roy. Soc. Lond.*, **B267**, 467–501.

Sirocko, F., Sarnthein, M., Lange, H. and Erlenkeuser, H. (1991) Atmospheric summer circulation and coastal upwelling in the Arabian Sea during the Holocene and the last glaciation. *Quat. Res.*, **36**, 72–93.

Sowunmi, M. A. (1981) Nigerian vegetational history from the late Quaternary to the present day. *Palaeoecol. Afr.*, **13**, 217–234.

Spencer, T. and Douglas, I. (1985) The significance of environmental change: diversity, disturbance and tropical ecosystems. In I. Douglas and T. Spencer (Eds.), *Environmental Change and Tropical Geomorphology*, pp. 13–33. London, George Allen and Unwin.

Stauffer, B., Lochbronner, E., Oeschger, H. and Schwander, J. (1988) Methane concentration in the glacial atmosphere was only half that of the pre-industrial Holocene. *Nature*, **332**, 812–814.

Street, F. A. (1979) Late Quaternary precipitation estimates for the Ziway-Shala basin, Southern Ethiopia. *Palaeoecol. Afr.*, **12**, 137–158.

Street-Perrott, F. A. (1991) General circulation (GCM) modelling of palaeoclimates: a critique. *The Holocene*, **1**(1), 74–80.

(1993) Ancient tropical methane. *Nature*, **366**, 411–412.

(1994) Palaeoperspectives: changes in terrestrial ecosystems. *Ambio*, **43**, 37–43.

Street-Perrott, F. A. and Harrison, S. P. (1985) Lake levels and climate reconstruction. In A. D. Hecht, (Ed.), *Palaeoclimate Analysis and Modelling*, pp. 291–340, New York, Wiley.

Street-Perrott, F. A. and Perrott, R. A. (1990) Abrupt climatic fluctuations in the tropics – the influence of Atlantic Ocean circulation. *Nature*, **343**, 607–612.

(1993) Holocene vegetation, lake levels and climate of Africa. In H. E. Wright *et al.* (Eds.), *Global Climates Since the Last Glacial Maximum*, pp. 318–356. Minneapolis, University of Minnesota Press.

Street-Perrott, F. A., Roberts, N. and Metcalfe, S. (1985) Geomorphic implications of late Quaternary hydrological and climatic changes in the Northern Hemispheric tropics. In I. Douglas and T. Spencer (Eds.), *Environmental Change and Tropical Geomorphology*, pp. 165–183. London, George Allen and Unwin.

Street-Perrott, F. A., Marchand, D. S., Roberts, N. and Harrison, S. P. (1989) *Global Lake-Level Variations from 18,000 to 0 Years Ago: A Palaeoclimatic Analysis*. Washington, DC, United States Department of Energy, 213pp.

Street-Perrott, F.A., Huang, Y., Perrott, R. A. *et al.* (1997) Impact of lower atmospheric carbon dioxide on tropical mountain ecosystems. *Science*, **278**, 1,422–1,426.

Sukumar, R., Ramesh, R., Pant, R. K. and Rajagopalan, G. (1993) A $\delta^{13}C$ record of late Quaternary climate change from tropical peats in southern India. *Nature*, **364**, 703–706.

Svensmark, H. and Friis-Christensen, E. (1997) Variation of cosmic ray flux and global cloud coverage – a missing link in solar-climate relationships. *J. Atmos. Terr. Phys.*, **59**, 1225–1232.

Swain, A. M., Kutzbach, J. E., and Hastenrath, S. (1983) Estimates of Holocene precipitation for Rajasthan, India, based on pollen and lake-level data. *Quat. Res.*, **19**, 1–17.

Talbot, M. R. and Hall, J. B. (1981) Further late Quaternary leaf fossils from Lake Bosumtwi, Ghana. *Palaeoecol. Afr.*, **13**, 83–92.

Talbot, M. R., and Johannessen, T. (1992) A high resolution palaeoclimatic record for the last 27,500 years in tropical west Africa from the carbon and nitrogen isotopic composition of lacustrine organic matter. *Earth Planet. Sci. Lett.*, **100**, 23–37.

Talbot, M. R., Livingstone, D. A., Palmer, P. G. *et al.* (1984) Preliminary results from sediment cores from Lake Bosumtwi, Ghana. *Palaeoecol. Afr.*, **16**, 173–192.

Taylor, D. M. (1990) Late Quaternary pollen records from two Ugandan mires: evidence for environmental change in the Rukiga Highlands of south-west Uganda. *Palaeogeogr. Palaeoclimatol. Palaeoecol.*, **80**, 283–300.

Taylor, K. C., Hammer, C. U., Alley, R. B. *et al.* (1993) Electrical conductivity measurements from the GISP2 and GRIP Greenland ice cores. *Nature*, **366**, 549–552.

Twongo, T. (1996) Growing impact of water hyacinth on nearshore environments on Lakes Victoria and Kyoga (East Africa). In T. C. Johnson and E. O. Odada (Eds.), *The Limnology, Climatology and Palaeoclimatology of the East African Lakes*, pp. 633–655. Australia, Gordon and Breach.

van der Hammen, T. (1974) The Pleistocene changes of vegetation and climate in the tropical South America. *J. Biogeogr.*, **1**, 3–26.

Verschuren, D., Kathleen, R. L. and Cumming, B. F. (2000) Rainfall and drought in equatorial east Africa during the past 1,100 years. *Nature*, **403**, 410–413.

Vincens, A. (1986) Diagramme pollinique d'un sondage Pleistocene supérieur-Holocene du Lac Bogoria (Kenya). *Review of Palaeobotany and Palynology*, **47**, 162–192.

(1989a) Palaeoenvironments du bassin nord-Tanganyika (Zaire, Burundi, Tanzanie) au cours des 13 derniers milles ans: apport de la palynologie. *Rev. Palaeobot. Palynol.*, **61**, 69–88.

(1989b) Early Holocene pollen data from an arid East African region, Lake Turkana, Kenya: botanical and climatic implications. *Palaeoecol. Afr.*, **20**, 87–97.

(1989c) Les forêts zambéziennes du bassin Sud-Tanganyika. Evolution entre 25,000 et 6,000 ans BP. *C. R. Acad. Sci. Paris*, **308**, Série II, 809–814.

Vincens, A., Chalié, F., Bonnefille, R., Guiot, J. and Tiercelin, J. J. (1993) Pollen-derived rainfall and temperature estimates for Lake Tanganyika and their implication for late Pleistocene water levels. *Quat. Res.*, **40**, 343–350.

Vincent, C. E., Davies, T. D. and Beresford, A. K. C. (1979) Recent changes in the level of Lake Naivasha, Kenya, as an indicator of equatorial westerlies over East Africa. *Climate Change*, **2**, 175–189.

Vincent, C. E., Davies, T. D., Brimblecombe, P. and Beresford, A. K. C. (1989) Lake levels and glaciers: indicators of changing rainfall in the mountains of East Africa. In W. C. Mahaney (Ed.), *Quaternary and Environmental Research on East African Mountains*, pp. 199–216. Rotterdam, Balkema.

Vogel, J. C. and Visser, E. (1981) Pretoria radiocarbon dates II. *Radiocarbon*, **23**, 43–80.

Weingarten, B., Salgado-Labouriau, M. L., Yuretich, R. and Bradley, R. (1991) Late Quaternary environmental history of the Venezuelan Andes. In R. Yuretich (Ed.), *Late Quaternary Climatic Fluctuations of the Venezuelan Andes*. Department of Geology and Geography, University of Massachusetts, Amherst, MA, Contribution No. 65: 63–94.

White, S. E. (1986) Quaternary glacial stratigraphy and chronology of Mexico. In V. Šibrava, D. Q. Bowen and G. M. Richmond (Eds.), *Quaternary Glaciations in the Northern Hemisphere*, pp. 210–205. *Quat. Sci. Rev.*, **15**. Oxford, Pergamon Press.

Wigley, T. M. L. and Kelly, P. M. (1990) Holocene climatic change, ^{14}C wiggles and variations in solar irradiance. *Philos. Trans. R. Soc. London, Series A*, **330**, 547–560.

Wood, C. A. (1976) Climate change in Ethiopia within the last 2,000 years. *Abstract volume of 4th AMQUA Conference (Tempe, Arizona, Oct. 1976)*.

Zahn, R. (1994) Fast flickers in the tropics. *Nature*, **372**, 621–622.

2

The relative importance of the different forcings on the environment in Ethiopia during the Holocene

MOHAMMED U. MOHAMMED

Addis Ababa University, Ethiopia

Keywords

Ethiopia; Main Ethiopian Rift; Late Quaternary; Holocene environment; climate; volcanism; tectonics; human impact; crater lakes; palaeosoils; sediments; pollen; diatoms; radiocarbon dates; drought; rainfall variability; ENSO; deforestation; Little Ice Age

Abstract

The environment within the Ethiopian region has been sensitive to different natural and anthropogenic factors on various timescales and during different times of the Quaternary. During the Holocene (the last 10,000 years), the intensity of tectonics and volcanism in the Main Ethiopian Rift Valley has reduced, and their effects have been localized to marginal zones. However, climate change and variability on different timescales have been affecting the region within and outside the Rift Valley. During the twentieth century, the accelerated human impact has combined with highly variable rainfall, both on the inter-annual and inter-decadal timescales, to play a major role in limiting the availability of resources.

2.1 INTRODUCTION

Ethiopia is located in a geodynamically active zone and geographically sensitive area, Consequently, volcano-tectonism, as well as climate change and variability on different spatial and temporal scales, have affected its environment. Superimposed on these, increasing human impact is modifying the land cover, land surface and hydrology. This chapter presents a review of the relative importance of the different environmental forcings during the Holocene (the last 10,000 years) in Ethiopia. This is a time when the major regional features of the environment, as we see them now, have been shaped.

2.2 THE MAIN ETHIOPIAN RIFT: A MODEL AREA FOR THE STUDY OF THE VARIOUS ENVIRONMENTAL FORCINGS

The most geodynamic zone in Ethiopia is located in the Rift Valley, which is the northern branch of the East African Rift system. In particular, the Main Ethiopian Rift (MER), with its intermediate environment between high- and low-altitude zones, is a favourable site for the presence of lacustrine deposits in present-day lake systems. These contain climatic signals that continue to the present day. The region is also being affected by increasing human impacts, mainly due to subsistence agriculture.

Steep border fault escarpments bound this area 70–80 km apart, limiting the Ethiopian Plateau to the west and the south-eastern plateau to the east. The latter exceeds 2,500 m in elevation over wide areas and some peaks can attain over 4,000 m.

The modern climate of the region varies markedly over short distances, mainly characterized by alternating wet and dry seasons following the annual movement of the Intertropical Convergence Zone (ITCZ). This separates the air streams of the north-east from those of the south-east monsoons (Nicholson, 1996). Mean annual rainfall varies from 600 mm at the Rift floor to about 1,200 mm around 2,500–3,000 m. Temperature varies from 15 °C to 20 °C, respectively. Natural vegetation in the Rift floor is *Acacia* woodland, which changes to dry montane forest with *Podocarpus* and *Juniperus* dominating above 2,000 m. First ericaceous scrub and then afroalpine moorland cover the land above 3,200–3,500 m (Makin *et al.*, 1976; Friis, 1992). The natural vegetation has undergone severe degradation mainly since the twentieth century.

The MER is a continental rift with slow crustal attenuation since the Late Oligocene to Miocene (Di Paola, 1972; Woldegabriel *et al.*, 1990; Ebinger, *et al.*, 1993; Boccaletti *et al.*, 1992). The present symmetrical rift was fully defined by 3–5 million years (Woldegabriel, 1987).

Recent tectonic deformation, however, is restricted to two belts running NNE: the Silti Debrezait Fault Zone (SDFZ) to

Climate Change and Africa, ed. Pak Sum Low. Published by Cambridge University Press. © Cambridge University Press 2005.

the west and the Wonji Fault Belt (WFB) to the east. These two belts of recent tectonic deformation consist of hundreds of steep normal faults with throws of about 50 m (Mohr, 1960, 1962; Di Paola, 1972; Mohr *et al.,* 1980; Woldegabriel *et al.,* 1990).

The SDFZ is 5–10 km wide and 80 km long, with very recent basalt effusion, and the vertical movement along the steep normal faults seems to be very recent (Iasio, 1997).

The WFB is a 6–10 km wide belt of normal and dilatational faulting characterized by important volcanic activity from the Early Pleistocene, and active until historical times. Recent tectonism related to the SDFZ and the WFB is evidenced by the very young and well-preserved fault escarpments often affecting recent formations, and by the intense seismicity of the region. This indicates that tectonic deformation could still be active in the central part of the MER, at least intermittently.

2.3 ENVIRONMENTAL IMPACT OF VOLCANO TECTONIC ACTIVITIES

The effects of recent deformations are seen where offsets have produced morphologic features such as inconspicuous hydrological thresholds and displacement of river courses. At times, hydrologic systems have been interrupted or disintegrated (Coltorti *et al.,* 2002). Moreover, changes in base levels have created frequent geomorphic thresholds, leading to catastrophic erosion. There is also a gradient increase and rejuvenation on slopes, producing highly convex slope segments or even steps that are highly unstable under an aggressive climate. The erosive action of rejuvenated streams is forced to switch from incision to aerial stripping over hard rocks, from which extended badlands have resulted (Carnicelli *et al.,* 1998).

Active volcanism is relatively in quiescence in the MER, although widespread hydrothermal sources show the recent volcano tectonic activity in the area, as well as its control on the quality of surface water and groundwater (Cherenet, 1998). The presence of several recent volcanic events is inferred from volcanic centres aligned along active fault systems (Di Paola, 1972), as well as the ash and tephra layers in outcrop sediments (Street, 1979) and in sediment cores collected from the Rift Valley lakes (Le Turdu *et al.,* 1999; Telford *et al.,* 1999; Telford and Lamb, 1999). Some of the tephra layers have given radiocarbon dates on adjacent sediments, which vary from the last few thousand to few hundred years.

Diatom records have shown that some of the Holocene changes in the hydrochemistry of Lake Awassa could not be explained by climate change alone. Rather, it was hypothesized that a pulsed inflow of groundwater, made saline due to carbon dioxide degassing from magma and reacting with silicate minerals, may have been responsible (Telford *et al.,* 1999). In crater lake Tilo, hydrothermalism is supposed to have influenced the mineral sediment (silica and calcite) accumulation rate during the Holocene (Telford and Lamb, 1999). These findings suggest the necessity of differentiating climatic events in crater lake sediment records from volcanic influences, thus refuting the idea that crater lakes act only as big rain gauges (Lamb *et al.,* 2000). On the other hand, there are no data in Ethiopia establishing a cause–effect relationship between volcanism and climate during the Holocene.

Soils developed on tephra and ash are found to be susceptible to erosion, particularly under the present day human impact on vegetation cover (Carnicelli *et al.,* 1998).

2.4 EFFECTS OF CLIMATE CHANGES

During the Late Quaternary the effects of climate changes on land surface and water resources have generated famines, migrations, civilization foundations and collapses. In Africa, climatic events have induced cultural changes through their impacts on the quality, amount, distribution, inter-annual variability and spatial unpredictability of food and water resources (Hassan, 2002). The recent events of drought and famine in some African countries have attracted international attention towards the importance of this phenomenon. Studies have revealed that the different environments in Ethiopia have reacted to climate changes and variability on different timescales (Mohammed *et al.,* 2001).

2.4.1 Climate during the transition from the last glacial to the Holocene

Evidence of pre-Holocene extensive glaciation of some of the Ethiopian mountains comes from geomorphic data on glacial moraines (Hastenrath, 1977; Hurni, 1989; Messerli *et al.,* 1977). These studies have suggested the lowering of the present-day solifluxion limits (4,300–4,400 m) to an elevation of 3,600 m, with a corresponding lowering of vegetation belts due to the decrease in temperature and rainfall amounts. In mid-altitude sites, low lake levels have recorded dry conditions, contemporaneous to the glacial extension on the high mountains (Street, 1979).

The minimum age for deglaciation on the Ethiopian mountains was between around 13,000 and 14,000 years BP (Mohammed and Bonnefille, 1998), and perhaps the modern vegetation belts began to be established during the beginning of the Holocene, both in response to wetter and warmer climate conditions (Hamilton, 1982; Lézine and Bonnefille, 1982). Strong evidence on the re-establishment of vegetation belts after deglaciation is, however, lacking.

2.4.2 Climate change during the Holocene

The early Holocene

In Ethiopia, regionally synchronous evidence of climate change has been obtained from studies such as geomorphology and stratigraphy (Street, 1979; Gasse and Street, 1978; Benvenuti et al., 2002; Sagri et al., 1999), from multi-proxy studies of lacustrine and other cores (sedimentology, geochemistry, pollen, diatoms, charcoal, magnetic susceptibility) (Hamilton, 1982; Lézine and Bonnefille, 1982; Bonnefille et al., 1986; Mohammed, 1992; Bonnefille and Mohammed, 1994; Mohammed et al., 1996; Mohammed and Bonnefille, 1998; Tamrat, 1997; Telford et al., 1999; Telford and Lamb, 1999) and from studies of Palaeosoil records (Barakhi et al., 1998; Machando et al., 1998).

Lake level records come from the Ethiopian Rift Valley and the Afar located at low altitudes, although their main water supply is from the surrounding highlands. This implies a wider spatial significance of the climate change record, which they contain.

Evidence of high lake stands was found between 10,000 and 5,000 years BP, with intervals of abrupt aridity. Lake levels dropped after 4,000–5,000 years BP, with a minor rise centred at about 2,000 years BP (Gillespie et al., 1983). Palaeolimnological data of continuous Holocene cores from crater lakes in Ethiopia have also shown the major Holocene climate shifts, although the abrupt intervals need to be precise (Telford and Lamb, 1999). The stable isotopic record of a sedimentary core in the same crater lake was, however, able to show the presence of an arid interval in the Early Holocene between 7,900–7,600 years BP.

Correspondingly, during the Early Holocene, major soil formation, including peat deposition, took place both in southern (Sagri et al., 1999) and in northern Ethiopia. Stratigraphy of the latter has also shown intervals of soil degradation phases, perhaps in response to the Early Holocene arid intervals (Dramis and Mohammed, 1999). A major phase of soil degradation began after about 4,000 years BP (Barakhi et al., 1998; Belay, 1997) in relation to Late Holocene aridity. There were, however, minor intervals of soil formation related to Late Holocene low amplitude wet conditions (Machando et al., 1998).

The late Holocene climate fluctuations

This time interval, after 4,000–5,000 years BP, is characterized by a general trend towards a dry climate (Gillespie et al., 1983). From the geomorphic evidence of lake-level changes, it was not possible to resolve the low-amplitude, small-scale climate signals of the last few thousand years and to differentiate climate signals from human impacts. This became possible when short cores were collected from highland swamps and from lakes.

Table 2.1. Climate signals reconstructed from pollen and lake level records in Ethiopia

Age ^{14}C Yr BP	Climatic condition	
	Wetness	Temperature
? 3,000	Very dry	Very Cold?
3,000–2,500	Dry	Cold
2,500–1,000	Wet	Cold
1,000–500	Dry[a]	Warm
500–100	Wet[a]	Cold
100–present	From wet to dry and high inter-annual variability (Fekadu and Bauwens, 1998)	Warming trend

[a]Recent works indicated 'Medieval Warm Epoch' dry and 'Little Ice Age' wet conditions (see text).
Mohammed and Bonnefille, 2002.

The results from pollen analysis on a site from the eastern Plateau are summarized in Table 2.1 (Bonnefille and Mohammed, 1994).

Palaeosoil studies of the last 4,000 years in the highlands of northern Ethiopia (Tigray) show phases of soil formation interpreted in terms of wetter conditions at 4,000–3,500, 2,500–1,500, 1,000–960 years BP, and two degradational phases at 3,500–2,500 and 1,500–1,000 years BP (Machando et al., 1998).

Historical records of climate events are rare in Ethiopia. They come mainly from royal chronicles (Pankhurst, 1985), travellers' accounts and indirectly from the Nile flow records at Aswan, Egypt (Tousson, 1925; Hassan, 1981). The Nile record (Hassan, 1981) shows several decadal-scale Nile flood variations linked mainly to the rainfall on the Ethiopian highlands.

Based on the records from 640–1921 AD, Hassan (1981) argued for the presence of a possible link between low Nile discharge and cold climate in Europe. The high frequency of famine events from the sixteenth to nineteenth centuries (Pankhurst, 1985; Machando et al., 1998) happened during the high-latitude cold period of the Little Ice Age (Lamb, 1982). Pollen data on the Ethiopian highlands during approximately the same period have revealed a much colder climate than that at present (Bonnefille and Mohammed, 1994). A phase of soil erosion and reduced stream power was deduced from geomorphological studies in the south Ethiopian Rift Valley and this could be related to climate aridity during the northern hemisphere cold climate of the 'Little Ice Age' (Sagri et al., 1999). However, recent information from Kenya (Verschuren et al., 2000) and from northern Ethiopia (Darbyshire et al., 2003) has shown a transition from dry to wet conditions during the change from the 'Medieval Warm epoch' to the 'Little Ice Age'. Although the impact of climate changes during the above time intervals is evident and shows a very widespread effect, the nature and timing of the events need better understanding and precision for Africa.

Travellers' records show that lake levels were generally higher from the end of the nineteenth century to about the first half of the twentieth century (Donaldson, 1896; Harrison, 1901; Erlanger, 1901), and continued to shrink during the second half of the twentieth century. This is also observed in sedimentological records of other lakes in East Africa (Verschuren, 1999). It is also consistent with a general pattern of the long rainfall record in Ethiopia (Fekadu and Bauwens, 1998). During the latter period, some lakes in the Rift Valley were affected by human impacts (e.g. extraction of trona from Lake Abiyata). This is accentuated by a high inter-annual variability. Some others have shown a tendency to rise (e.g. Lake Beseka, Lake Awassa and the Debrezait lakes) in spite of the rainfall pattern. The implication for this is that different factors operate in an out-of-phase order, at least locally. The rise in the Debrezait lakes, for instance, has been shown to be related to the construction of an artificial reservoir that raised the groundwater table (Seifu, 1999).

High inter-annual variability of rainfall during the twentieth century, and particularly the events of famine during the beginning of the 1970s and 1980s have been attributed to the El Niño Southern Oscillation (ENSO) event (WMO, 1987). However, the role ENSO plays on Ethiopian rainfall is not yet clearly defined, as there are various factors influencing rainfall variability in Ethiopia and because the moisture sources are different (Conway, 2000). There are some preliminary meteorological data showing a recent increase in temperature trends (Billi, 1998), although relating this to global warming is premature.

The meteorological time series is too short to reveal any significant cyclicity or a major trend. In the future it will have to be supported by palaeo data from tree rings, stalagmites and the sedimentary record. Recent work on high-resolution ana-lysis of lake sediments from the Ethiopian Rift valley lakes has shown environmental changes since the late nineteenth century and re-vealed a drought stage around 1890 (Dagnachew et al., 2002), corresponding to a major famine which occurred from 1888 to 1892 in the area (Pankhurst, 1966).

2.5 THE HUMAN IMPACT

Ethiopia has one of the most ancient histories of human settlement and cultural activities. However, human impacts should be measured by changes perceived as significant deviations of the environment from its natural status. Moreover, generalized human impacts must be separated from localized effects. Looking for past human impacts in natural archives could be difficult, because of the problem in finding appropriate indicators of human impacts, incapacity to isolate human impacts from natural variability, weak power of resolution of the record, inability to find appropriate sites and material, and lack of precise dating methods for the events.

Although previous works and speculations tried to give a very old age for a generalized human impact in Ethiopia (Hamilton, 1982), new findings have shown that the scale has been variable in space and time (Mohammed, 1992). Signals of widespread intensive land degradation in the form of massive deforestation and the resulting accelerated soil erosion seem to be the phenomena of the twentieth century, particularly during the second half of the twentieth century (Mohammed and Bonnefille, 1991; Mohammed, 1992; Machando et al., 1998; Darbyshire et al., 2003). In the northern part of the country, where agricultural history seems to be more ancient, pollen indications have shown a much earlier forest clearance (Darbyshire et al., 2003). The intensive human impact during the twentieth century could be related to population explosion and political instability. Climate might have acted as an environmental stressor with high inter-annual variability.

2.6 CONCLUSIONS

This paper has tried to assess the dynamic nature of the environment in Ethiopia resulting from the presence of various natural factors and human impacts. The activities of tectonics and volcanism have declined through time during the Quaternary (Le Turdu et al., 1999) and moved to the marginal areas in the MER. During the Holocene, climate has played a major role in influencing the environment on different timescales and with different amplitudes, but with more or less regional effects. However, the nature, timing, duration and extent of the high-frequency climate changes and variabilities will have to be defined in order to stretch the climatic data beyond those obtained from instrumental records. During the twentieth century, particularly in its second half, widespread human impacts put pressure on the environment, under conditions of high decadal and inter-annual natural variabilities.

ACKNOWLEDGEMENTS

This review is mainly based on studies conducted by French, British and Italian researchers in collaboration with the Department of Geology and Geophysics, Addis Ababa University, under Ethio-French, EU/NERC and Ethio-Italian funded projects. The comments from the anonymous reviewers and Dr Pak Sum Low have improved the quality of this paper.

REFERENCES

Barakhi, O., Brancaccio, L., Calderoni, G. *et al.* (1998) The Mai sedimentary sequence – A reference point for the environmental evolution of the highlands of Northern Ethiopia. *Geomorphol.,* **23**, 127–128.

Belay, T. (1997) Variabilities of catena on degraded hill slopes of watiya catchment, Wollo, Ethiopia. *SINET: Ethiop. J. Sci.* **20**, 151–175.

Benvenuti, M., Carniccelli, S., Belluomini, G. *et al.* (2002) The Ziway-Shalla lake basin (main Ethiopian rift, Ethiopia): a revision of basin evolution with special reference to Late Quaternary. *J. Afri. Sci.,* **35**, 247–269.

Billi, P. (1998) Climatic change. In M. Sagri (Coordinator), *Land Resources Inventory and Environmental Changes Analysis and their Application to Agriculture in the Lake Region (Ethiopia).* Report submitted to EC. Project STD 3 contract No. Ts 3–CT 92–0076 (1993–98), 108–122.

Boccaletti, M., Getaneh, H. and Tortorici, L. (1992) The Main Ethiopian Rift: an example of oblique rifting. *Ann. Tectonicae,* **6**, 20–25.

Bonnefille, R. and Mohammed, U. (1994) Pollen inferred climatic fluctuations in Ethiopia during the last 3,000 yrs. *Palaeogeog. Palaeoclimatol. Palaeoecol.,* **109**, 331–343.

Bonnefille, R., Robert, C., Delibrias, G. *et al.* (1986) Paleo environment of Lake Abijata, Ethiopia during the past 2,000 yrs. In L. Forstick *et al.* (Eds.), *Sedimentation in the African Rift. London, Geol. Soc., Spec. Pap.* **25**, 253–265.

Carnicelli, S., Ferarri, G. A., Wolf, U. *et al.* (1998) Geomorphic history and soils. In M. Sagri (Coordinator), *Land Resources Inventory and Environmental Changes Analysis and their Application to Agriculture in the Lake Region (Ethiopia).* Report submitted to EC. Project STD 3 contract No. Ts 3-CT 92-0076 (1993–98), 90–102.

Cherenet, T. (1998). *Etude des mechanismes de minéralisation en fluorure et éléments associés de la région des lacs du Rift Ethiopien.* Thése doctorat de l'Université d'Avignon et des pays du Vaucluse, 210pp.

Coltorti, M., Pizzi, A., Corbo, L. and Sacchi, G. (2002) Fault activity, river capture and delta-growth on the eastern side of the Ziway-Shalla lake (Ethiopia). In F. Dramis, P. Molin, C. Cipolloni, and G. Fubelli, (Eds.), *Climate Changes, Active Tectonics and Related Geomorphic Effects in High Mountain Belts and Plateaux.* IAG International Symposium Proceedings, Addis Ababa, Ethiopia.

Conway, D. (2000) Some aspects of climate variability in the Northeast Ethiopian Highlands – Wollo and Tigray. *SINET. Ethiop. J. Sci.,* **23**, 139–161.

Dagnachew, L., Gasse, F., Radakovitch, O. *et al.* (2002) Environmental changes in a tropical lake (Lake Abiyata, Ethiopia), during recent centuries. *Palaeogeogr. Palaeoclimatol. Palaeocol.,* **187**, 233–258.

Darbyshire, I., Lamb, H. F. and Mohammed, M. U. (2003) Forest clearance and regrowth in northern Ethiopia during the last 3000 years. *The Holocene.* **13**, 537–546.

Di Paola, G. M. (1972) The Ethiopian Rift Valley (between 7°00′ and 8°40′ lat. North). *Bull. Volcanol.,* **36**, 517–560.

Donaldson, S. A. (1896) Expedition through Somali–land to lake Rudolf. *Geogr. J.,* **8**, 221–239.

Dramis, F. and Mohammed, M. U. (1999) Holocene climate phases from buried soils in Northern Ethiopia: Comparison with lake level in the Main Ethiopian Rift. In *Environmental Background to Hominid Evolution in Africa.* Book of Abstracts. XV INQUA Congress, Durban, South Africa, 54pp.

Ebinger, G. J., Yemane, T., Woldegabriel, G., Aronson, J. L. and Walter, R. C. (1993) Late Eocene – recent volcanism and faulting in the southern Main Ethiopian Rift. *J. Geol. Soc. London,* **105**, 99–108.

Erlanger, C. (1901) Baron Erlanger and Herr Neumann in Southern Abyssinia. *Geogr. J.,* **18**, 214–215.

Fekadu, M. and Bauwens, W. (1998) Water resources variability in Africa during the twentieth century. *Proceedings of the Abidjan (1998) Conference held at Abidjan, Côte d'Ivoire, November 1998.* IAHS Publ. No. 252, 297–305.

Friis I. (1992) *Forests and Forest Trees of Northeast Tropical Africa.* HMSO Kew Bulletin Additional Series XV, London, UK.

Gasse, F. and Street, F. A. (1978) Late Quaternary lake level fluctuations and environments of the Northern Rift valley and Afar Region (Ethiopia and Djibouti). *Palaeogeogr. Palaeoclimatol. Palaeoecol.,* **24**, 279–325.

Gillespie, R., Street-Perrot, A. F. and Switsor, R. (1983) Postglacial arid episodes in Ethiopia have implications for climatic prediction. *Nature,* **306**, 680–683.

Hamilton, A. C. (1982) *Environmental History of East Africa. A Study of the Quaternary.* London, Academic Press, 328pp.

Harrison, J. J. (1901) A journey from Zeila to Lake Rudolf. *Geogr. J.,* **18**, 258–275.

Hassan, F. (1981) Historical Nile floods and their implication for climatic change. *Science,* **212**, 1,142–1,145

(2002) Paleoclimate food and culture changes in africa: an overview. In F. Hassan (Ed.), *Droughts, Food and Culture, Ecological Change and Food Security in Africa's Later Prehistory.* New York, Kluwer Academic/Plenum Publishers, 347pp.

Hastenrath, S. (1977) Pleistocene mountain glaciation in Ethiopia. *J. Glacial.,* **28**, 309–313.

Hurni, H. (1989) Late Quaternary of Simien and other mountains in Ethiopia. In W. C. Mahaney (Ed.), *Quaternary and Environmental Research on East African Mountain,* pp.105–120. Rotterdam, Balkema.

Iasio, C. (1997) *Dipositi et suoli del tardi Quaternario nell area di Tora: Regione dei Laghi (rift Ethiopico).* Laurea Dissertation, Universita di Frienze, 159pp.

Lamb, H. H. (1982) *Climate History and the Modern World.* London, Methuen, 316pp.

Lamb, L. A., Leng, M. J., Lamb, H. F. and Mohammed, M. U. (2000) A 9,000 yr oxygen and carbon isotope record of hydrological change in a small crater lake. *The Holocene.* **10**, 167–177.

Le Turdu, C., Tiercelin, J. J., Gibert, E. *et al.* (1999) The Ziway Shalla basin system, Main Ethiopian Rift: influence of volcanism, tectonics and climatic forcing on basin formation and sedimentation. *Palaeogeogr. Palaeoclimatol. Palaeoecol.*, **150**, 135–177.

Lézine, A. M. and Bonnefille, R. (1982) Diagramme pollinique Holocene d'un sondage du lac Abiyata (Ethiopie, 7°42′ Nord). *Pollen et Spore*, **XXIV**, 463–480.

Machando, J. M., Perez–Gonzalez, A. and Benito, G. (1998) Paleoenvironmental changes during the last 4000 yrs in Tigray, Northern Ethiopia. *Quat. Res.*, **49**, 312–321.

Makin, M. J., Kingham, T. J., Addams, A. E., Brichall, C. J. and Eavis, B. W. (1976) *Prospects for Irrigation Development Around Lake Ziway, Ethiopia.* Land Resources Study 26. Land Resources Division, Ministry of Overseas Development, England, 316pp.

Messerli, B., Hurni, H., Kienholz, H. and Winiger, M. (1977) *Bale Mountains: the largest Pleistocene mountain glacier system of Ethiopia.* X INQUA Congr. Birmingham. Abstract.

Mohammed, M. U. (1992) *Paleoenvironnement et Paleolimatologie des dernieres millenaires en Ethiopie, contribution palynologique.* Thèse Doctorat, Université d'Aix-Marseille III. 209pp.

Mohammed, M. U. and Bonnefille, R. (1991) The recent history of vegetation and climate around Lake Langano (Ethiopia). *Palaeoecol. Afr.*, **22**, 275–285.

—— (1998) A Late Glacial to Late Holocene pollen record from a high land peat at Tamsaa, Bale Mountains, South Ethiopia. *Global and Planetary Change*, **16–17**, 121–129.

—— (2002) Late Holocene climatic fluctuations and historical records of famine in Ethiopia. In F. Hassan (Ed.), *Droughts, Food and Culture, Ecological Change and Food Security in Africa's Later Prehistory.* New York, Kluwer Academic/Plenum Publishers, 347pp.

Mohammed, M. U., Bonnefille, R. and Johnson, T. C. (1996) Pollen and isotopic records of Late Holocene sediments from lake Turkana, N. Kenya. *Palaeogeogr. Palaeoclimatol. Palaeoecol.*, **119**, 371–383.

Mohammed, M. U., Dagnachew, L., Gasse, F. *et al.* (2001) Late Quaternary climate changes in the Horn of Africa. In *Past Climate Variability Through Europe And Africa.* PAGES-PEPIII Int. Conf. 27–31 August, 2001, Aix-en Provence, France.

Mohr, P. A. (1960) Report on a geological excursion through Southern Ethiopia. *Bull. Geophys. Observ. Addis Ababa Univ.*, **2**, 9–19.

—— (1962) The Ethiopian Rift system. *Bull. Geophys. Observ. Addis Ababa Univ.*, **5**, 33–62.

Mohr, P. A., Mitchell, J. G. and Reynolds, R. G. H. (1980) Quaternary volcanism and faulting at O'A Caldera, Central Ethiopia. *Bull. Volcanol.*, **43**, 173–189.

Nicholson, S. E. (1996) A review of climate dynamics and climate variability in Eastern Africa. In T. C. Johnson, and E. O. Odada (Eds.), *The Limnology, Climatology and Paleoclimatology of the Eastern African Lakes.* Australia, Gordon and Breach. pp. 25–56.

Pankhurst, R. (1966) The great Ethiopian famine of 1888–92: a new assessment. *J. Hist. Med. Appl. Sci.*, **21**, 39–92.

—— (1985) *The History of Famine and Epidemics in Ethiopia Prior to the Twentieth Century.* Report to Relief and Rehabilitation Commission. Addis Ababa University Press.

Sagri, M., Atnafu, B., Benvenutti, M. *et al.* (1999) Geo-morphological evolution of the Lake Region (Ethiopia) and climatic changes. In *Environmental Background to Hominid Evolution in Africa.* Book of Abstracts. pp. 154–155. XV INQUA Congress, Durban, South Africa.

Seifu, K. (1999) *Hydrology and Hydrochemistry of the Bishoftu Crater Lakes (Ethiopia). Hydrological, Hydrochemical and Oxygen Isotope Modelling.* M.Sc. Thesis, Addis Ababa University, 128pp.

Street, F. A. (1979) *Late Quaternary Lakes in the Ziway-Shalla Basin, South Ethiopia.* Unpublished Ph.D. Thesis, University of Cambridge, UK, 457pp.

Tamrat, E. (1997) *Magnétostratigraphie et magnétisme de séquence lacustres Plio-Pleistocene et Holocene d'Afrique Oriental.* Thèse Doctorat, Université d'Aix-Marseille III, 302pp.

Telford, R. J. and Lamb, H. F. (1999) Ground water mediated response to Holocene climate change recorded by diatom stratigraphy of an Ethiopian crater lake. *Quat. Res.*, **52**, 63–75.

Telford, R. J., Lamb, H. F. and Mohammed, M. U. (1999) Diatom derived paleoconductivity estimates for Lake Awassa, Ethiopia: evidence for pulsed inflow of saline ground water. *J. Paleolimnol.*, **21**, 409–421.

Tousson, O. (1925) L'Histoire du Nile. *Mem. Inst. Egypte.*, **9** Part 2.

Verschuren, D. (1999) Sedimentation controls on the preservation and time resolution of climate proxy records from shallow fluctuating lakes. *Quat. Sci. Rev.*, **18**, 821–837.

Verschuren, D., Laird, K. R. and Cumming, B. F. (2000) Rainfall and drought in equatorial East Africa during the past 1,100 years. *Nature*, **403**, 410–414.

WMO (1987) *Climate System Monitoring.* Monthly Bulletins, World Meteorological Organization, Geneva, Switzerland.

Woldegabriel, G. (1987) *Volcano-tectonic History of the Central Sector of the Main Ethiopian Rift: a Geochronological, Geochemical, and petrological approach.* Ph.D. Thesis. Case Western Reserve University, Cleveland, OH, USA, 410pp.

Woldegabriel, G., Aronson, J. and Walter, R. C. (1990) Geology, geochemistry and rift basin development in the central sector of the Main Ethiopian Rift (MER). *Geol. Soc. Am. Bull.*, **102**, 439–458.

3   Global warming and African climate change: a reassessment

MIKE HULME[1], RUTH DOHERTY[2,*], TODD NGARA[3,†], AND MARK NEW[4]

[1]*Tyndall Centre for Climate Change Research, UK, and University of East Anglia, Norwich, UK*
[2]*University of Edinburgh, UK*
[3]*Ministry of Mines, Environment and Tourism, Harare, Zimbabwe*
[4]*University of Oxford, UK*

Keywords

African climate; the Sahel; climate change; scenarios; climate variability

Abstract

Understanding and predicting temporal variations in African climate has become the major challenge facing African and African-specialist climate scientists in recent years. Whilst seasonal climate forecasts have taken great strides forward, we remain unsure of the ultimate causes of the lower frequency decadal and multi-decadal rainfall variability that affects some African climate regimes, especially in the Sahel region. This work examining the variability of African climate, especially rainfall, is set in the wider context of our emerging understanding of human influences on the larger, global-scale climate. Africa will be no exception to experiencing these human-induced changes in climate, although much work remains to be done in trying to isolate those aspects of African climate variability that are 'natural' from those that are related to human influences. African climate scientists face a further challenge in that it is in this continent that the role of land-cover changes in modifying regional climates is perhaps most marked. It is for these reasons – large internal climate variability and the confounding role of land-cover change – that climate change 'predictions' (or scenarios) for Africa based on greenhouse gas warming remain at a low level of confidence. Nevertheless, it is of considerable interest to try and scope the magnitude of the problem that the enhanced greenhouse effect may pose for African climate and for African resource managers. Are the changes that are simulated by global climate models (GCMs) for the next century large or small in relation to our best estimates of 'natural' climate variability in Africa? How well do GCM simulations agree for the African continent? And what are the limitations/uncertainties of these model predictions? This chapter makes a contribution to this debate by providing an assessment of future climate change scenarios for Africa and by discussing these possible future climates in the light of modelling uncertainties and in the context of others causes of African climate variability and change.

3.1 INTRODUCTION

The climates of Africa are both varied and varying; *varied*, because they range from humid equatorial regimes, through seasonally arid tropical regimes, to sub-tropical Mediterranean-type climates, and *varying* because all these climates exhibit differing degrees of temporal variability, particularly with regard to rainfall. Understanding and predicting these inter-annual, inter-decadal and multi-decadal variations in climate has become the major challenge facing African and African-specialist climate scientists in recent years. Whilst seasonal climate forecasting has taken great strides forward, both its development and application (Stockdale *et al.*, 1998; Washington and Downing, 1999), the ultimate causes of the lower frequency decadal and multi-decadal rainfall variability that affects some African climate regimes, especially in the Sahel region, remain uncertain (see Rowell *et al.*, 1995 versus Sud and Lau, 1996; also Xue and Shukla, 1998). It is clear that the oceans, especially the Pacific Ocean, are important modulators of both inter-annual and inter-decadal climate variability in Africa. But the oceans alone do not hold all of the clues. This work examining the variability of African climate, especially rainfall, is set in the wider context of our emerging understanding of human influences on the larger, global-scale climate. Increasing greenhouse gas accumulation in the global atmosphere and increasing regional concentrations of aerosol particulates are now understood to have detectable effects on the global climate system (Santer *et al.*, 1996). These effects will be manifest at regional scales, although perhaps in more uncertain terms (Mitchell and Hulme, 1999).

*Currently at University of Edinburgh, UK.
†Currently at Institute of Global Environmental Strategies, Japan.

Climate Change and Africa, ed. Pak Sum Low. Published by Cambridge University Press. © Cambridge University Press 2005.

Africa will be no exception to experiencing these human-induced changes in climate. Much work remains to be done, however, in trying to isolate those aspects of African climate variability that are 'natural' from those that are related to human influences. African climate scientists face a further challenge in that in this continent the role of land-cover changes – some natural and some human-related – in modifying regional climates is perhaps most marked (Xue, 1997). This role of land-cover change in altering regional climate in Africa has been suggested for several decades now. As far back as the 1920s and 1930s theories about the encroachment of the Sahara and the desiccation of the climate of West Africa were put forward (Stebbing, 1935; Aubreville, 1949). These ideas have been explored over the last 25 years through modelling studies of tropical North African climate (e.g. Charney, 1975; Cunnington and Rowntree, 1986; Zheng and Eltahir, 1997). It is for these reasons – large internal climate variability as driven by the oceans and the confounding role of human-induced land-cover change – that climate change 'predictions' (or scenarios) for Africa based on greenhouse gas warming remain highly uncertain. While global climate models (GCMs) simulate changes to African climate as a result of increased greenhouse gas concentrations, these two potentially important drivers of African climate variability – El Niño Southern Oscillation (ENSO) (poorly) and land-cover change (not at all) – are not well-represented in the models.

Nevertheless, it is of considerable interest to try and explore the magnitude of the problem that the enhanced greenhouse effect may pose for African climate and for African resource managers. Are the changes that are simulated by GCMs for the next 100 years large or small in relation to our best estimates of 'natural' climate variability in Africa? How well do GCM simulations agree for the African continent? And what are the limitations/uncertainties of these model predictions? This chapter therefore makes a contribution to this exploration by providing an overview of future climate change scenarios for Africa, particularly with regard to simulations of greenhouse gas warming over the next 100 years. In Section 3.2 we consider a few salient features of African climate change and variability over the last 100 years, based on the observational record of African climate. Such a historical perspective is essential if the simulated climates of the next 100 years are to be put into their proper context. Section 3.3 reviews a number of climate change scenarios and analyses for regions within Africa, although such studies remain far from comprehensive. Section 3.4 then discusses these future climate simulations in the light of modelling uncertainties and in the context of other causes of African climate variability and change. We consider how much useful and reliable information these types of studies yield and how they can be incorporated into climate change impacts assessments. Our key conclusions are presented in Section 3.5.

3.2 AFRICAN CLIMATE VARIABILITY: A CONTEXT FOR FUTURE CLIMATE CHANGE

The continent of Africa is warmer than it was 100 years ago. Warming through the twentieth century has been at the rate of about 0.5 °C/century (Figure 3.1), with slightly larger warming in the June–August and September–November seasons than in December–February and March–May. The six warmest years in Africa have all occurred since 1987, with 1998 being the warmest year. This rate of warming is not dissimilar to that experienced globally and the periods of most rapid warming – the 1910–1930s and the post-1970s – occur simultaneously in Africa and the world.

Inter-annual rainfall variability is large over most of Africa, and for some regions, most notably the Sahel, multi-decadal variability in rainfall has also been substantial. Reviews of twentieth century African rainfall variability have been provided by, among others, Janowiak(1988), Hulme(1992) and Nicholson(1994). To illustrate something of this variability, we present an analysis for the three regions of Africa used by Hulme(1996b) – the Sahel, East Africa and south-east Africa. These three regions exhibit contrasting rainfall variability characteristics (Figure 3.2): the Sahel displays large multi-decadal variability with recent drying, East Africa a relatively stable regime with some evidence of long-term wetting, and south-east Africa also a basically stable regime, but with marked inter-decadal variability.

In recent years Sahel rainfall has been quite stable around the 1961–1990 annual average of 371 mm, although this 30-year period is substantially drier (about 25%) than earlier decades in the twentieth century. In East Africa, 1997 was a very wet year and, as in 1961 and 1963, led to a surge in the level of Lake Victoria (Birkett *et al.*, 1999). In south-east Africa, the dry years of the early 1990s were followed by two very wet years in 1995/96 and 1996/97. Mason *et al.* (1999) report an increase in recent decades in the frequency of the most intense daily precipitation over South Africa, even though there is little long-term trend in total annual rainfall amount.

Figure 3.2 also displays the trends in annual temperature for these same three regions. Temperatures for all three regions during the 1990s were higher than they have been during the twentieth century (except for a period at the end of the 1930s in the Sahel) and were between 0.2 °C and 0.3 °C warmer than the 1961–1990 average. This result is confirmed for South Africa using independent geothermal profiles (Tyson *et al.*, 1998). There is no simple correlation between temperature and rainfall in these three regions, although Hulme (1996c) noted that drying in the Sahel was associated with a moderate warming trend.

With regard to inter-annual rainfall variability in Africa, the ENSO is one of the more important controlling factors, at least

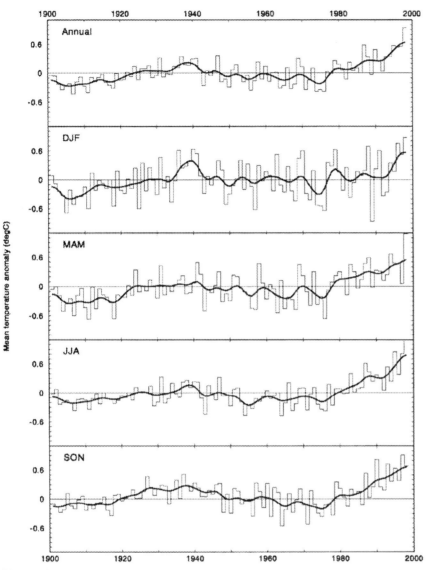

Figure 3.1. Mean surface air temperature anomalies for the African continent, 1901–1998, expressed with respect to the 1961–1990 average; annual and four seasons – DJF, MAM, JJA, SON. The smooth curves result from applying a 10-year Gaussian filter.

for some regions (Janowiak, 1988; Ropelewski and Halpert, 1987; 1989; 1996). These studies have established that the two regions in Africa with the most dominant ENSO influences are in eastern equatorial Africa during the short October–November rainy season and in south-eastern Africa during the main November–February wet season. Ropelewski and Halpert (1989) also examined Southern Oscillation and rainfall relationships during La Niña or high index years.

Most recently, Hulme *et al.* (2001) conducted an analysis of Southern Oscillation–rainfall variability for the African region over the period 1901–1998 using an updated and more comprehensive data set (Hulme, 1994b) than was used by these earlier studies. They used the Southern Oscillation Index (SOI) as

a continuous measure of Southern Oscillation behaviour rather than designating discrete 'warm' (El Niño; low index) and 'cold' (La Niña; high index) Southern Oscillation events as was done by Ropelewski and Halpert (1996). Their analysis confirmed the strength of the previously identified relationships for equatorial East Africa (high rainfall during a warm ENSO event) and southern Africa (low rainfall during a warm ENSO event). The former relationship is strongest during the September–November rainy season (the 'short' rains), with an almost complete absence of ENSO sensitivity in this region during the February–April season ('long' rains), as found by Ropelewski and Halpert (1996). The southern African sensitivity is strongest over South Africa during December–February before migrating northwards over

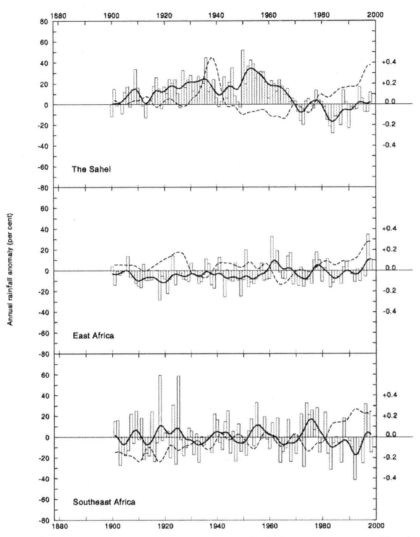

Figure 3.2. Annual rainfall (1900–1998; histograms and bold line; percentage anomaly) and mean temperature anomalies (1901–1998; dashed line; degree C anomaly) for three African regions, expressed with respect to the 1961–1990 average: Sahel, East Africa and south-eastern Africa. Note: For south-east Africa the year is July–June. The smooth curves for rainfall and temperature result from applying a 10-year Gaussian filter.

Zimbabwe and Mozambique during the March–May season. There is little rainfall sensitivity to ENSO behaviour elsewhere in Africa, although weak tendencies for Sahelian June–August drying (Janicot *et al.*, 1996) and north-west African March–May drying (El Hamly *et al.*, 1998) during El Niño warm phases can also be found.

3.3 REVIEW OF AFRICAN CLIMATE CHANGE SCENARIOS

There has been relatively little work published on future climate change scenarios for Africa. The various Intergovernmental Panel on Climate Change (IPCC) assessments have of course included global maps of climate change within which Africa has featured,

and in Mitchell *et al.* (1990) the African Sahel was identified as one of five regions for which a more detailed analysis was conducted. Kittel *et al.* (1998) and Giorgi and Francisco (2000) also identify African regions within their global analysis of inter-model differences in climate predictions, but no detailed African scenarios are presented.

Tyson (1991) published one of the first scenario analyses specifically focused on an African region. In this case some climate change scenarios for southern Africa were constructed using results from the first-generation GCM equilibrium $2 \times CO_2$ experiments. In a further development, Hulme (1994a) presented a method for creating regional climate change scenarios combining GCM results with the IPCC IS92 emissions scenarios and demonstrated the application of the method for Africa. In this study, mean annual temperature and precipitation changes from 1990 to 2050 under the IS92a emission scenario were presented.

Some more recent examples of climate scenarios for Africa use results from transient GCM climate change experiments. Hernes *et al.* (1995) and Ringius *et al.* (1996) constructed climate change scenarios for the African continent that showed land areas over the Sahara and semi-arid parts of southern Africa warming by the 2050s by as much as 1.6 °C and the equatorial African countries warming at a slightly slower rate of about 1.4 °C. These studies, together with Joubert *et al.,* (1996), also suggested a rise in mean sea level around the African coastline of about 25 cm by 2050. A more selective approach to the use of GCM experiments was taken by Hulme *et al.* (1996a). They describe three future climate change scenarios for the Southern African Development Community (SADC) region for the 2050s on the basis of three different GCM experiments. These experiments were selected to deliberately span the range of precipitation changes for the SADC region as simulated by GCMs. Using these scenarios, the study then describes some potential impacts and implications of climate change for agriculture, hydrology, health, biodiversity, wildlife and rangelands. A similar approach was adopted by Conway *et al.* (1996) for a study of the impacts of climate change on the Nile Basin. The Africa chapter (Zinyowera *et al.*, 1998) in the IPCC Assessment of Regional Impacts of Climate Change (IPCC 1998) also reported on some GCM studies that related to the African continent.

Considerable uncertainty exists in relation to large-scale precipitation changes simulated by GCMs for Africa (Hudson, 1997; Joubert and Hewitson, 1997; Feddema, 1999). Joubert and Hewitson (1997) nevertheless concluded that, in general, precipitation is simulated to increase over much of the African continent by the year 2050. These GCM studies show, for example, that parts of the Sahel could experience precipitation increases of as much as 15% over the 1961–1990 average by 2050. A note of caution is needed, however, concerning such a conclusion. Hulme (1998) studied the present-day and future simulated inter-decadal precipitation variability in the Sahel using the HadCM2 GCM. These model results were compared with observations during the twentieth century. Two problems emerge. First, the GCM does not generate the same magnitude of inter-decadal precipitation variability that has been observed over the last 100 years, casting doubt on the extent to which the most important controlling mechanisms are being simulated in the GCM. Second, the magnitude of the future simulated precipitation changes for the Sahel is not large in relation to 'natural' precipitation variability for this region. This low signal-to-noise ratio suggests that the greenhouse gas-induced climate change signals are not well defined in the model, at least for this region.

Although there have been studies of GCM-simulated climate change for several regions in Africa, the downscaling of GCM outputs to finer spatial and temporal scales has received relatively little attention in Africa. Hewitson and Crane (1998) and Hewitson and Joubert (1998) have applied empirical downscaling methods to generate climate change scenarios for South Africa using Artificial Neural Networks and predictors relating to upper air circulation and tropospheric humidity. The usual caveats, however, apply to these downscaled scenarios – they are still dependent on the large-scale forcing from the GCMs and they still only sample one realization of the possible range of future possible climates, albeit with higher resolution. The application of high-resolution regional climate models for climate change experiments in Africa is still in its infancy, although some studies have been completed or are now under way for southern Africa (e.g. Joubert *et al.*, 1999), East Africa, and West Africa.

The most recent set of climate change scenarios for Africa are those presented in Hulme *et al.* (2001). Their range of four future scenarios is the result of combining low, medium and high future greenhouse gas emissions scenarios, with low, medium and high estimates of the global climate sensitivity. At a global scale, this yields the four climate change scenarios shown in Table 3.1. The two middle cases were chosen deliberately because, even though the global warming is similar in each, the assumed geopolitical and techno-economic worlds that underlie the B2 and A1 emissions scenarios are quite different. The impacts on Africa of what may be rather similar global and regional climate changes could be quite different in these two cases. For example, global (and African) population is lower in the A1 world than in the B2 world, but carbon and sulphur emissions and CO_2 concentrations are higher (Table 3.1).

The regional patterns of climate change over Africa in the Hulme *et al.* (2001) scenarios were based on results from seven GCM experiments. Using a pattern scaling method, they combined the results of these different experiments to generate a median pattern-response for seasonal mean temperature and precipitation across Africa for periods centred on the 2020s, 2050s and 2080s. These scenarios were presented in both map form and as scatter plots for a selection of countries within Africa – Senegal, Tunisia, Ethiopia and Zimbabwe (Figure 3.3). Each country graph shows, for the 2050s, the distribution of the mean annual changes in mean temperature and precipitation for each GCM simulation and for each of the four scenarios. These changes are compared with the natural multi-decadal variability of annual mean temperature and precipitation extracted from one 1,400-year unforced model simulation. These graphs provide a quick assessment at a national scale of the likely range and significance of future climate change and again show the extent to which different GCMs agree in their regional response to a given magnitude of global warming.

For each country there is a spread of results relating to inter-model differences in climate response. For example, in Tunisia the change in annual rainfall is predominantly towards drying, although the magnitude of the drying under the A2-high scenario

Table 3.1. The four global climate change scenarios adopted by Hulme *et al.* (2001).

Scenario/ Climate sensitivity	Population (billions)	C emissions from energy (GtC)	Total S emissions (TgS)	Global ΔT (°C)	Global ΔSL (cm)	pCO$_2$ (ppmv)
B1-low/1.5 °C	8.76	9.7	51	0.9	13	479
B2-mid/2.5 °C	9.53	11.3	55	1.5	36	492
A1-mid/2.5 °C	8.54	16.0	58	1.8	39	555
A2-high/4.5 °C	11.67	17.3	96	2.6	68	559

Estimates are for the 2050s (i.e. 2055). Temperature and sea-level changes assume no aerosol effects and are calculated from a 1961–1990 baseline. C is annual carbon emissions (in gigatonnes or Gt) from fossil energy sources, S is annual sulphur emissions (in teragrams or Tg), ΔT is change in mean annual temperature, ΔSL is change in mean sea level and pCO$_2$ is the atmospheric carbon dioxide concentration (in parts per million by volume or ppmv).

is between 1% and 30%. Natural climate variability is estimated to lead to differences of ±10% between different 30-year mean climates, therefore the more extreme of these scenario outcomes would appear to be 'significant' for Tunisia. The picture would appear at first sight to be less clear for Zimbabwe where four of the GCMs suggest wetting and three suggest drying. However, the range of natural variability in annual rainfall when averaged over 30 years is shown to be about ±6% and most of the wetting scenarios fall within this limit. It is the drying responses under the more extreme A2, B2 and A1 scenarios that would appear to yield a more 'significant' result. We discuss the significance of these differences and similarities between different GCMs in Section 3.4, in the context of other uncertainties.

The Hulme *et al.* (2001) scenarios suggest a future annual warming across Africa ranging from below 0.2 °C per decade to over 0.5 °C per decade. This warming is greatest over the interior semi-arid tropical margins of the Sahara and central southern Africa, and least in equatorial latitudes and coastal environments. All of the estimated temperature changes exceed the one sigma level of natural temperature variability, even under the B1-low scenario. The inter-model range (an indicator of the extent of agreement between different GCMs) is smallest over northern Africa and the equator, and greatest over the interior of central southern Africa.

Future changes in mean seasonal rainfall in Africa are less well defined. Under the B1-low scenario, relatively few regions in Africa experience a change in either DJF or JJA rainfall that exceeds the one sigma level of natural rainfall variability. The exceptions are parts of equatorial East Africa where rainfall increases by 5% to 30% in DJF and decreases by 5% to 10% in JJA. With more rapid global warming (e.g. the B2, A1 and A2-high scenarios), increasing areas of Africa experience changes in DJF or JJA rainfall that *do* exceed the one sigma level of natural rainfall variability. For the A2-high scenario, large areas of equatorial Africa experience 'significant' increases in DJF rainfall of up to 50% or 100% over parts of East Africa, while rainfall decreases

'significantly' in JJA over parts of the Horn of Africa and north-west Africa. Some 'significant' JJA rainfall increases occur over the central Sahel region of Niger and Chad, while 'significant' decreases in DJF rainfall (15% to 25%) occur over much of South Africa and Namibia and along the Mediterranean coast. The inter-model range for these rainfall changes remains large, however, and with very few exceptions exceeds the magnitude of the median model response. Even for the seasonally wet JJA rainfall regime of the Sahel, inter-model ranges can exceed 100%, suggesting that different GCM simulations yield (sometimes) very different regional rainfall responses to a given greenhouse gas forcing. This large inter-model range in seasonal mean rainfall response is not unique to Africa and is also found over much of south and south-west Asia and parts of Central America (Carter *et al.*, 2000).

Given the important role that ENSO events exert on inter-annual African rainfall variability, at least in some regions, determining future changes in inter-annual rainfall variability in Africa can only be properly considered in the context of changes in ENSO behaviour. There is still ambiguity, however, about how ENSO events may respond to global warming. This is partly because global climate models only imperfectly simulate present ENSO behaviour. Tett *et al.* (1997) demonstrate that HadCM2 simulates ENSO-type features in the Pacific Ocean, but the model generates too large a warming across the tropics in response to El Niño events. Timmermann *et al.* (1999) however, have recently argued that their ECHAM4 model has sufficient resolution to simulate 'realistic' ENSO behaviour. They analyse their greenhouse gas forced simulations to suggest that in the future there are more frequent and more intense 'warm' and 'cold' ENSO events.

What effects would such changes have on inter-annual African rainfall variability? This not only depends on how ENSO behaviour changes in the future, but also on how realistically the models simulate the observed ENSO–rainfall relationships in Africa. Smith and Ropelewski (1997) looked at Southern

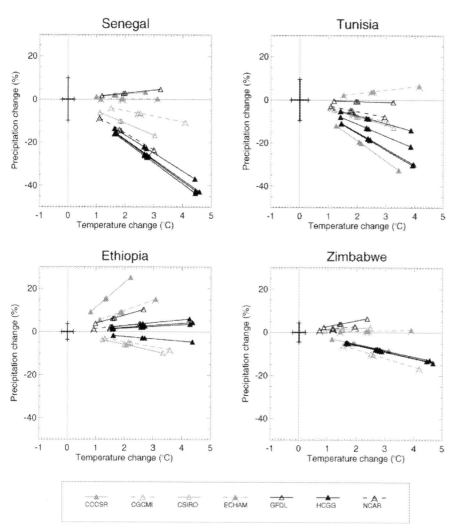

Figure 3.3. Change in mean annual temperature and precipitation for the 2050s (with respect to 1961–1990) for Senegal, Tunisia, Ethiopia and Zimbabwe. Results from seven GCMs are shown, scaled to reflect the four climate change scenarios adopted in this study: A2-high, A1-mid, B2-mid and B1-low. The bold lines centred on the origin indicate the two standard deviation limits of natural 30-year timescale climate variability defined by the 1,400-year HadCM2 control simulation (*Source*: Hulme *et al.*, 2001).

Oscillation–rainfall relationships in the National Centre for Environmental Prediction (NCEP) atmospheric GCM, where the model is used to re-create observed climate variability after being forced with observed sea surface temperatures (SSTs). Even in this most favourable of model experiments the model relationships do not always reproduce those observed. Over south-eastern Africa, the simulated rainfall percentiles are consistent with the observations reported by Ropelewski and Halpert (1996), but over eastern equatorial Africa the model simulates an opposite relationship to that observed.

Hulme *et al.* (2001) analysed 240 years of unforced simulated climate made using the HadCM2 GCM to see to what extent this model can reproduce observed ENSO–rainfall relationships in Africa. The two strongest ENSO signals in African rainfall

variability (cf. Section 3.2) were only imperfectly reproduced by the model. The East African negative correlation from November to April was rather too weak in the model and also too extensive, extending westwards across the whole African equatorial domain. The positive correlation over Southern Africa was too weak in HadCM2 and displaced northwards by some 10° latitude. The absence of any strong and coherent relationship during the June–October season *was* reproduced by this model.

On the basis of our assessment of the literature, we are not convinced that quantifying future changes to inter-annual rainfall variability in Africa due to greenhouse gas forcing is warranted. At the very least, this issue deserves a more thorough investigation of ENSO–rainfall relationships in the GCMs used here, and how these relationships change in the future (Doherty and Hulme, 2002). Such an analysis might also be useful in determining the extent to which seasonal rainfall forecasts in Africa that rely upon ENSO signatures may remain valid under scenarios of future greenhouse gas forcing.

3.4 UNCERTAINTIES AND LIMITATIONS TO KNOWLEDGE

In the introduction to this chapter we alluded to some of the limitations of climate change scenarios for Africa. The scenarios reviewed here are no exception. These limitations arise because of, *inter alia*, the problem of small signal-to-noise ratios in some scenarios for precipitation and other variables, the inability of climate model predictions to account for the influence of land-cover changes on future climate, and the relatively poor representation in many models of some aspects of climate variability that are important for Africa (e.g. ENSO).

We cannot place probability estimates on the climate change scenarios reviewed here with any confidence. While this conclusion may well apply for most, or all, world regions, it is particularly true for Africa where the roles of land-cover change and dust and biomass aerosols in inducing regional climate change are excluded from the climate change model experiments reported here. This concern is most evident in the Sahel region of Africa. None of the model-simulated present or future climates for this region display behaviour in rainfall regimes that is similar to that observed over recent decades. This is shown in Figure 3.4 where we plot the observed regional rainfall series for 1900–1998, as used in Figure 3.2, and then append 10 model-simulated evolutions of future rainfall for the period 2000–2100. These future curves are extracted from the same 10 GCM experiments used by Hulme *et al.* (2001). One can see that none of the model rainfall curves for the Sahel displays multi-decadal desiccation similar to what has been observed in recent decades. This conclusion also applies to the multi-century unforced integrations performed with the same GCMs (Brooks, 1999).

There are a number of possible reasons for this. It could be that the climate models are poorly replicating 'natural' rainfall variability for this region. In particular the possible role of ocean circulation changes in causing this desiccation (Street-Perrott and Perrott, 1990) may not be well simulated in the models. It could also be that the cause of the observed desiccation is some process that the models are not including. Two candidates for such processes would be the absence of a dynamic land cover/atmosphere feedback process and the absence of any representation of changing atmospheric dust aerosol concentration. The former of these feedback processes has been suggested as being very important in determining African climate change during the Holocene by amplifying orbitally induced African monsoon enhancement (Kutzbach *et al.*, 1996; Doherty *et al.*, 2000). This feedback may also have

contributed to the more recently observed desiccation of the Sahel (Xue, 1997). The latter process of elevated Saharan dust concentrations may also be implicated in the recent Sahelian desiccation (Brooks, 1999).

Without such a realistic simulation of observed rainfall variability, it is difficult to define with confidence the true magnitude of natural rainfall variability in these model simulations and also difficult to argue that these greenhouse gas-induced attributed rainfall changes for regions in Africa will actually be those that dominate the rainfall regimes of the twenty-first century. Notwithstanding these model limitations due to omitted or poorly represented processes, Figure 3.4 also illustrates the problem of small signal-to-noise ratios in precipitation scenarios. The 10 individual model simulations yield different signs of precipitation change for these three regions as well as different magnitudes. How much of these differences are due to model-generated natural variability is difficult to say.

One other concern about the applicability in Africa of climate change scenarios such as those reviewed here is the relationship between future climate change predictions and seasonal rainfall forecasts. There is increasing recognition (e.g. Downing *et al.*, 1997; Ringius, 1999; Washington and Downing, 1999) that for many areas in the tropics, one of the most pragmatic responses to the prospect of long-term climate change is to strengthen the scientific basis of seasonal rainfall forecasts. Where forecasts *are* feasible, this should be accompanied by improvements in the management infrastructure to facilitate timely responses. Such a research and adaptation strategy focuses on the short-term realizable goals of seasonal climate prediction and the near-term and quantifiable benefits that improved forecast applications will yield. At the same time, the strengthening of these institutional structures offers the possibility that the more slowly emerging signal of climate change in these regions can be better managed in the decades to come. It is therefore an appropriate form of climate change adaptation. This means that two of the objectives of climate change prediction should be to determine the effect global warming may have on seasonal predictability – will forecast skill levels increase or decrease or will different predictors be needed? – and to determine the extent to which predicted future climate change will impose additional strains of natural and managed systems over and above those that are caused by existing seasonal climate variability. For both of these reasons we need to improve our predictions of future climate change and in particular improve our quantification of the uncertainties.

3.5 CONCLUSIONS

The climate of Africa is warmer than it was 100 years ago. Although there is no evidence for widespread desiccation of the continent during the twentieth century, in some regions substantial

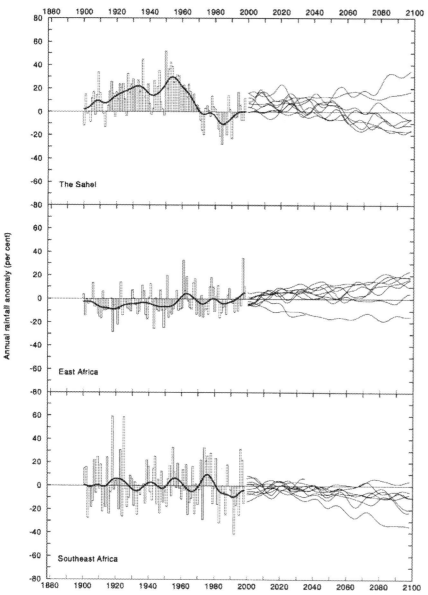

Figure 3.4. Observed annual rainfall anomalies for three African regions, 1900–1998 (cf. Figure 3.2), and model-simulated anomalies for 2000–2099. Model anomalies are for the 10 model simulations used in the Hulme *et al.* (2001) African scenarios. All anomalies are expressed with respect to either observed or model-simulated 1961–1990 average rainfall. The smooth curves result from applying a 20-year Gaussian filter (*Source*: Hulme *et al.*, 2001).

inter-annual and multi-decadal rainfall variations have been observed and near continent-wide droughts in 1983 and 1984 had some dramatic impacts on both environment and some economies (Benson and Clay, 1998). The extent to which these rainfall variations are related to greenhouse gas-induced global warming, however, remains undetermined. A warming climate will nevertheless place additional stresses on water resources, whether or not future rainfall is significantly altered.

Model-based predictions of future greenhouse gas-induced climate change for the continent clearly suggest that this warming will continue and, in most scenarios, accelerate so that the continent on average could be between 2 °C and 6 °C warmer in 100 years time. While these predictions of future *warming* may be relatively robust, there remain fundamental reasons why we are much less confident about the magnitude, and even direction, of regional *rainfall* changes in Africa. Two of these reasons relate to the rather ambiguous representation in most global climate models of ENSO-type climate variability in the tropics (a key determinant of African rainfall variability) and the omission in all current global climate models of any representation of dynamic land cover-atmosphere interactions and dust and biomass aerosols. Such interactions have been suggested to be important in determining African climate variability during the Holocene and may well have contributed to the more recently observed desiccation of the Sahel.

We suggest that climate change scenarios, such as those reviewed here, should nevertheless be used to explore the sensitivity of a range of African environmental and social systems, and economically valuable assets, to a range of future climate changes. Some examples of such exploration were presented by Dixon *et al.* (1996), although in these studies there was little coordinated and quantified use of a coherent set of climate futures. Further work can be done to elaborate on some of the higher order climate statistics associated with the changes in mean seasonal climate shown here – particularly daily temperature and precipitation extremes. It may also be worthwhile to explore the sensitivity of these model predictions to the spatial resolution of the models – i.e. explore the extent to which downscaled scenarios differ from GCM-scale scenarios – although such downscaling techniques do not remove the fundamental reasons why we are uncertain about future African rainfall changes.

The exploration of African sensitivity to climate change must also be undertaken, however, in conjunction with the more concrete examples we have of sensitivity to short-term (seasonal time-scale) climate variability. These estimates may be based on observed reconstruction of climate variability over the twentieth century, or on the newly emerging regional seasonal rainfall forecasts now routinely being generated for southern, eastern and western Africa (e.g. Jury *et al.*, 1999; IRI, 1999; NOAA, 1999; SAR-COF, 2003). Because of the uncertainties mentioned above about future regional climate predictions for Africa, initial steps to reduce vulnerability should focus on improved adaptation to existing climate variability (Downing *et al.*, 1997; Ringius, 1999). Thus, emphasis would be placed on reducing vulnerability to adverse climate-events and increasing capacity to adapt to short-term and seasonal weather conditions and climatic variability. The likelihood of significant economic and social benefits from adaptation to short-term climate variability in Africa justifies this activity (Desanker and Magadza, 2001). Additionally, and importantly, lessons from adaptation to short-term climate variability would build capacity to respond incrementally to longer-term changes in local and regional climates.

ACKNOWLEDGEMENTS

Todd Ngara was supported by an IGBP/START Fellowship. The comments of Dr Robert Balling, Jr., reviewer of an early draft of this chapter, are acknowledged.

REFERENCES

Aubreville, A. (1949) *Climats, forêts et désertification de l'Afrique tropicale.* Paris, Société d'Editions Géographiques, Maritimes et Coloniales.

Benson, C. and Clay, E. J. (1998) *The Impact of Drought on Sub-Saharan Economies.* World Bank Technical Paper No. 401, Washington, DC, World Bank, 80 pp.

Birkett, C., Murtugudde, R. and Allan, T. (1999) Indian Ocean climate event brings floods to East Africa's lakes and the Sudd Marsh. *Geophys. Res. Lett.*, **26**, 1,031–1,034.

Brooks, N. J. (1999) *Dust-climate Interactions in the Sahel-Sahara Zone of Northern Africa, with Particular Reference to Late Twentieth Century Sahelian Drought.* Unpublished Ph. D. thesis, University of East Anglia, Norwich, 325 pp.

Carter, T. R., Hulme, M., Malyshev, S. *et al.* (2000) *Interim Characterizations of Regional Climate and Related Changes up to 2100 Associated with the Provisional SRES Marker Emissions Scenarios.* Finnish Environment Institute Report, Helsinki, 30pp. plus 3 Appendices.

Charney, J. G. (1975) Dynamics of deserts and drought in the Sahel. *Q. J. Roy. Meteorol. Soc.*, **101**, 193–202.

Conway, D., Krol, M., Alcamo, J. and Hulme, M. (1996) Future water availability in Egypt: the interaction of global, regional and basin-scale driving forces in the Nile Basin. *Ambio*, **25**, 336–342.

Cunnington, W. M. and Rowntree, P. R. (1986) Simulations of the Saharan atmosphere: dependence on moisture and albedo. *Q. J. Roy. Meteorol. Soc.*, **112**, 971–999.

Desanker, P. and Magadza, C. (2001) Africa. In J. J. McCarthy O. F. Canziani N. A. Leary D. J. Dokken and K. S. White (Eds.), *Climate Change 2001: Impacts, Adaptation and Vulnerability*, pp. 487–531. Cambridge, Cambridge University Press.

Dixon, R. K., Guill, S., Mkanda, F. X. and Hlohowskyj, I. (Eds.) (1996) Vulnerability and adaptation of African ecosystems to global climate change. *Climate Research (Special Issue)*, **6** (2).

Doherty, R. M. and Hulme, M. (2002) The relationship between the SOI and extended tropical precipitation in simulations of future climate change. *Geophys. Res. Lett.*, **29**, 113–114.

Doherty, R. M., Kutzbach, J., Foley, J. and Pollard, D. (2000) Fully-coupled climate/dynamical vegetation model simulations over Northern Africa during the mid-Holocene. *Climate Dynamics*, **16**, 561–573.

Downing, T. E., Ringius, L., Hulme, M. and Waughray, D. (1997) Adapting to climate change in Africa. *Mitigation and Adaptation Strategies for Global Change*, **2**, 19–44.

El Hamly, M., Sebbari, R., Lamb, P. J., Ward, M. N. and Portis, D. H. (1998) Towards the seasonal prediction of Moroccan precipitation and its implications for water resources management. In E. Servat, D. Hughes J.M. Fritsch and M. Hulme (Eds.), *Water Resources Variability in Africa During the 20th Century.* pp. 79–88. Wallingford, UK, IAHS Publication No. 252.

Feddema, J. J. (1999) Future African water resources: interactions between soil degradation and global warming. *Climatic Change*, **42**, 561–596.

Giorgi, F. and Francisco, R. (2000) On the predictability of regional climate change. *Climate Dynamics*, **16**, 169–182.

Hernes, H., Dalfelt, A., Berntsen, T. *et al.* (1995) *Climate Strategy for Africa*. CICERO Report 1995: 3, University of Oslo, Norway, 83 pp.

Hewitson, B. C. and Crane, R. G. (1998) Regional scale daily precipitation from downscaling of data from the GENESIS and UKMO GCMs. J48–J50. In *Proceedings of 14th Conference on Probability and Statistics in The Atmospheric Sciences, Phoenix, Arizona, 11–16 of January 1998*. Boston, USA, American Meteorological Society.

Hewitson, B. C. and Joubert, A. (1998) *Climate Downscaling: Current South African Projections*. http://www.egs.uct.ac.za/fccc/

Hudson, D. A. (1997) Southern African climate change simulated by the GENESIS GCM. *South African J. Sci.*, **93**, 389–403.

Hulme, M. (1992) Rainfall changes in Africa: 1931–60 to 1961–90. *Int. J. Climatol.*, **12**, 685–699.

(1994a) Regional climate change scenarios based on IPCC emissions projections with some illustrations for Africa. *Area,* **26**, 33–44.

(1994b) Validation of large-scale precipitation fields in General Circulation Models. In M. Desbois.and F. Désalmand (Eds.), *Global Precipitations and Climate Change*, pp. 387–405. Berlin, Springer-Verlag.

(Ed.) (1996a) *Climate Change and Southern Africa: An Exploration of Some Potential Impacts and Implications in the SADC Region*. Norwich, UK, CRU/WWF, 104 pp.

(1996b) Climatic change within the period of meteorological records. In W. M. Adams A. S. Goudie and A. R. Orme (Eds.), *The Physical Geography of Africa*, pp. 88–102. Oxford, Oxford University Press.

(1996c) Recent climate change in the world's drylands. *Geophys. Res. Lett.*, **23**, 61–64.

(1998) The sensitivity of Sahel rainfall to global warming: implications for scenario analysis of future climate change impact. In E. Servat, D. Hughes J. M. Fritsch and M. Hulme (Eds.), *Water Resources Variability in Africa During the 20th Century*, pp. 429–436. Wallingford, UK, IAHS Publication No. 252.

Hulme, M., Doherty, R. M., Ngara, T., New, M. G. and Lister, D. (2001) African climate change: 1900–2100. *Climate Res.*, **17**, 145–168.

IPCC (1998) *The Regional Impacts of Climate Change: An Assessment of Vulnerability*. Cambridge, Cambridge University Press, 517 pp.

IRI (1999) *International Research Institute for Climate Prediction*. http://iri.ldeo.columbia.edu/

Janicot, S., Moron, V. and Fontaine, B. (1996) Sahel droughts and ENSO dynamics. *Geophys. Res. Lett.*, **23**, 515–518.

Janowiak, J. E. (1988) An investigation of inter-annual rainfall variability in Africa. *J. Climate*, **1**, 240–255.

Joubert, A. M. and Hewitson, B. C. (1997) Simulating present and future climates of southern Africa using general circulation models. *Prog. Phys. Geogr.*, **21**, 51–78.

Joubert, A. M., Mason, S. J. and Galpin, J. S. (1996) Droughts over southern Africa in a doubled-CO_2 climate. *Int. J. Climatol.*, **16**, 1149–1156.

Joubert, A. M., Katzfey, J. J., McGregor, J. L. and Nguyen, K. C. (1999) Simulating midsummer climate over southern Africa using a nested regional climate model. *J. Geophys. Res.*, **104**, 19,015–19,025.

Jury, M. R., Mulenga, H. M. and Mason, S. J. (1999) Exploratory long-range models to estimate summer climate variability over southern Africa. *J. Climate*, **12**, 1,892–1,899.

Kittel, T. G. F., Giorgi, F. and Meehl, G. A. (1998) Intercomparison of regional biases and doubled CO_2-sensitivity of coupled atmosphere-ocean general circulation model experiments. *Climate Dynamics*, **14**, 1–15.

Kutzbach, J., Bonan, G. , Foley, J. and Harrison, S.P. (1996) Vegetation and soil feedbacks on the response of the Africa monsoon to orbital forcing in the early to middle Holocene. *Nature*, **384**, 623–626.

Mason, S. J., Waylen, P. R., Mimmack, G. M., Rajaratnam, B. and Harrison, J. M. (1999) Changes in extreme rainfall events in South Africa. *Climatic Change*, **41**, 249–257.

Mitchell, J. F .B., Manabe, S., Meleshko, V. and Tokioka, T. (1990) Equilibrium climate change – and its implications for the future. In J. T. Houghton G. J. Jenkins, and J. J. Ephraums (Eds.), *Climate Change: The IPCC Scientific Assessment*, pp. 137–164. Cambridge, Cambridge University Press.

Mitchell, T. and Hulme, M. (1999) Predicting regional climate change: living with uncertainty. *Prog. Phys. Geogr.*, **23**, 57–78.

Nicholson, S. E. (1994) Recent rainfall fluctuations in Africa and their relationship to past conditions over the continent. *The Holocene*, **4**, 121–131.

NOAA, (1999) *An Experiment in the Application of Climate Forecasts: NOAA-OGP Activities Related to the 1997–98 El Niño Event*. Washington DC, OGP/NOAA/US Dept. of Commerce, 134pp.

Ringius, L. (1999) *The Climatic Variability and Climate Change Initiative for Africa*. Processed report for the World Bank, Washington DC.

Ringius, L., Downing, T. E., Hulme, M., Waughray, D. and Selrod, R. (1996) *Climate Change in Africa – Issues and Regional Strategy*. CICERO Report No. 1996: 8, Oslo, Norway, CICERO, 154pp.

Ropelewski, C. F. and Halpert, M. S. (1987) Global and regional scale precipitation patterns associated with the El Niño Southern Oscillation. *Mon. Wea. Rev.*, **115**, 1,606–1,626.

(1989) Precipitation patterns associated with the high phase of the Southern Oscillation. *J. Climate*, **2**, 268–284.

(1996) Quantifying Southern Oscillation–precipitation relationships. *J. Climate*, **9**, 1043–1059.

Rowell, D. P., Folland, C. K., Maskell, K. and Ward, M. N. (1995) Variability of summer rainfall over tropical North Africa

(1906–92): observations and modelling. *Q. J. Roy. Meteorol Soc.*, **121**, 669–704.

Santer, B. D., Taylor, K. E., Wigley, T. M. L. *et al.* (1996) Human effect on global climate. *Nature,* **384**, 523–524.

SARCOF (2003) *Southern Africa Regional Climate Outlook Forum.* http://www.dmc.co.zw/sarcof/sarcof.htm

Smith, T. M. and Ropelewski, C. F. (1997) Quantifying Southern Oscillation–precipitation relationships from an atmospheric GCM. *J. Climate*, **10**, 2,277–2,284.

Stebbing, E. P. (1935) The encroaching Sahara: the threat to the West African colonies. *Geogr. J.,* **85**, 506–524.

Stockdale, T. N., Anderson, D.L.T., Alves, J.O.S. and Balmaseda, M. A. (1998) Global seasonal rainfall forecasts using a coupled ocean-atmosphere model. *Nature*, **392**, 370–373.

Street-Perrott, F. A. and Perrott, R. A. (1990) Abrupt climate fluctuations in the tropics: the influence of Atlantic Ocean circulation. *Nature*, **343**, 607–612.

Sud, Y. C. and Lau, W. K.-M. (1996) Comments on paper variability of summer rainfall over tropical North Africa (1906–92): observations and modelling. *Q. J. Roy. Meteorol Soc.*, **122**, 1,001–1,006.

Tett, S .F .B., Johns, T. C. and Mitchell, J. F. B. (1997) Global and regional variability in a coupled AOGCM. *Climate Dynamics*, **13**, 303–323.

Timmermann, A., Oberhuber, J., Bacher, A., *et al.* (1999) Increased El Niño frequency in a climate model forced by future greenhouse warming. *Nature*, **398**, 694–697.

Tyson, P. D. (1991) Climatic change in southern Africa: past and present conditions and possible future scenarios. *Climatic Change*, **18**, 241–258.

Tyson, P. D., Mason, S. J., Jones, M. Q. W. and Cooper, G. R. J. (1998) Global warming and geothermal profiles: the surface rock–temperature response in South Africa. *Geophys. Res. Lett.*, **25**, 2,711–2,713.

Washington, R. and Downing, T. E. (1999) Seasonal forecasting of African rainfall: prediction, responses and household food security. *Geogr. J.*, **165**, 255–274.

Xue, Y. (1997) Biosphere feedback on regional climate in tropical North Africa. *Q. J. Roy. Meteorol. Soc.*, **123**, 1,483–1,515.

Xue, Y. and Shukla, J. (1998) Model simulation of the influence of global SST anomalies on Sahel rainfall. *Mon. Wea. Rev.*, **126**, 2,782–2,792.

Zheng, X. and Eltahir, A. B. (1997) The response to deforestation and desertification in a model of West African monsoon. *Geophys. Res. Lett.*, **24**, 155–158.

Zinyowera, M. C., Jallow, B. P., Maya, R. S. and Okoth-Ogendo, H. W. O. (1998) Africa. In *The Regional Impacts of Climate Change: An Assessment of Vulnerability*, pp. 29–84. IPCC, Cambridge, Cambridge University Press.

4   Interactions of desertification and climate in Africa

ROBERT C. BALLING, JR.

Arizona State University, Tempe, AZ, USA

Keywords

Desertification; climate change

Abstract

Much of the scientific literature dealing with interactions of desertification and climate is based upon hundreds of studies dealing with the African continent. Three decades ago, the severe drought in sub-Saharan Africa generated an extensive discussion of whether overgrazing and land degradation in the region had produced atmosphere/land surface feedbacks that exacerbated drought throughout the Sahel. Related studies focusing on the possibility of desertification producing warming trends in historical temperature records were also based on African data sets. Furthermore, evidence from southern Africa suggests that land degradation in drylands can lead to warmer afternoon temperatures and an increase in the diurnal temperature range; this finding is opposite of the decline in diurnal temperature range reported throughout most of the world. Many articles have appeared in the literature showing how variations in the climate system can impact drought conditions in Africa, and the linkages between sea surface conditions, atmospheric circulation, and precipitation patterns in Africa are reasonably well known thanks to this research. In most recent years, a substantial literature has developed regarding how anthropogenic greenhouse-induced climate changes could impact the African continent. Generally speaking, climate models suggest that a build-up of greenhouse gases in the atmosphere could lead to a warming of Africa, increased potential evapotranspiration rates, a reduction of soil moisture, and an increase in the frequency, intensity and magnitude of droughts. The countries of the world recognize the importance and potential threat of the desertification/climate change issue for Africa, and in response, the United Nations has developed an international convention to combat desertification, especially in Africa.

4.1 INTRODUCTION

Even before the emergence of the human species in Africa, drylands of the continent were degraded periodically due to natural causes, and interactions occurred between desertification and local and regional climates. However, with approximately 0.7 billion people now living in Africa, the effects of human-induced desertification are more pronounced today than at any time in the past. Africa's drylands are being seriously degraded at present, resulting in a complicated set of feedbacks between dryland surface conditions and local and regional climates. While a great deal of attention is being placed on how anthropogenic climate change (e.g. enhanced greenhouse effect) may alter the climate of Africa, the ongoing land degradation in the drylands is having an immediate and identifiable impact on local and regional climate conditions. The seriousness of the desertification problem in Africa is fully recognized by nations of the world, and their concern led to the adoption of an international convention to combat desertification.

The focus of the scientific community on desertification and climate can be traced, in a substantial part, to a fluctuation in climate that occurred in the twentieth century in Sahelian Africa. For nearly five decades in the twentieth century, precipitation levels in this region remained relatively high, allowing local inhabitants to sustain or even increase crop yields, cattle herds and human population levels. However, in the early 1960s, the precipitation levels fell in the region (Figure 4.1), and by the early 1970s, the human and ecological tragedy of the Sahel was a serious matter receiving considerable attention worldwide.

In one of the first attempts to explain the climate 'failure' in the Sahel, noted climatologist Reid Bryson (1973) proposed that pollution in the northern hemisphere was creating a 'human volcano' causing a cooling at a hemispheric level. In Bryson's view, that cooling was causing a southerly migration of the northern circumpolar vortex (the jet stream) as well as a southerly migration of the subtropical high pressure belt. This

Climate Change and Africa, ed. Pak Sum Low. Published by Cambridge University Press. © Cambridge University Press 2005.

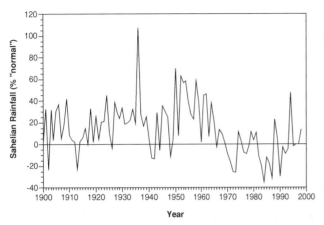

Figure 4.1. Sahel June–September rainfall, as percentage of the 1961–1990 mean, for the region 10°–20° N, 15° W–30° E based on station data.

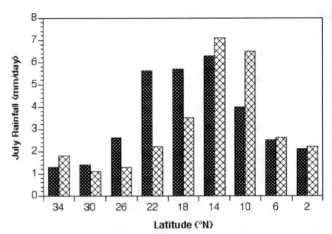

Figure 4.2. Simulated Sahelian area July rainfall rates for varying albedo conditions (from Charney, 1975); dark bars are for albedo = 0.14 and light bars are for albedo = 0.35.

pattern would restrict the ability of the rain-producing Intertropical Convergence Zone to move northward into the Sahel, and lower precipitation levels during the summertime rainy season would result. Bryson's argument placed the blame for the tragic Sahelian drought on industrial activity in the developed nations of the northern hemisphere.

Shortly afterwards, Cloudsley-Thompson (1974) presented the argument that overgrazing in the Sahel could increase both the surface albedo (reflectivity) and the soil compaction, leading to an increase in surface run-off and a reduction in available soil moisture. The 'expanding Sahara' could be the result of local activities and not the industrialization of the mid-latitudes of the northern hemisphere. Otterman (1974) similarly argued that overgrazing in dryland areas could increase albedo, reduce surface temperatures and near-surface air temperatures, stabilize the atmosphere, and reduce local rainfall levels.

A landmark article was published in 1975 by Jule Charney in which he presented a complex biogeophysical feedback model of desertification in the Sahel (Charney, 1975). The work received considerable attention in the popular press and scientific community, and the basic ideas put forth in this and related articles became known as the 'Charney hypothesis'. Charney and his associates developed a physically based numerical representation of a biogeophysical feedback mechanism that could initiate and/or reinforce drought in sub-Saharan Africa as a result of vegetation depletion. In their numerical model, the degradation of vegetation by natural or anthropogenic causes would increase the surface albedo, decrease the net short-wave radiation, decrease the surface temperature, and increase the relative emission of long-wave radiation due to a slight increase in the emissivity of the surface. These processes would reduce the net radiation at the surface, leaving less energy for warming the surface and the overlying

atmosphere. These changes in heat transfer would stabilize the lower layers of the atmosphere and suppress local convection. The reduction in local rainfall would further stress the remaining vegetation, thereby initiating a positive biogeophysical feedback (Charney, 1975; Charney *et al.*, 1977).

When Charney and his colleagues increased Sahelian albedo from 0.14 to 0.35, their model showed a southerly shift of the Intertropical Convergence Zone of several degrees of latitude in the Sahel during the normally rainy summer season. In some latitudinal bands, rainfall decreased by more than 50% due to this albedo effect (Figure 4.2). Within a few years, similar results were being reported by other investigators and for other dryland areas of the world (Ellsaesser *et al.*, 1976; Berkofsky, 1976; Walker and Rowntree, 1977).

However, despite the support from a variety of modelling studies, the Charney hypothesis was challenged immediately on theoretical and empirical grounds. Some scientists argued that removal of vegetation in drylands would lower soil moisture levels, decrease radiant energy used to evaporate and transpire water, and increase surface and near-surface temperatures (Jackson and Idso, 1975; Ripley, 1976). In areas where vegetation was reduced, the local surface and near-surface temperatures would increase, not decrease as proposed in the Charney hypothesis. Hydrological processes, not albedo effects, would dominate the surface energy balance changes associated with vegetation removal in most dryland environments.

The debate surrounding the Charney hypothesis continues to this day, but fundamentally, several important interrelated questions may be raised concerning climate change and desertification in Africa. First, there is the question of whether desertification in African drylands changes the local and regional climate. As we see in the Charney debate, desertification may produce feedbacks that lead to warming or cooling of the surface and the near-surface air, and either the warming or cooling effect could impact soil moisture levels. Second, there is a question regarding how variations in the sea surface temperatures and atmospheric circulation impact the climate variations in Africa's dryland areas. Finally, there is a growing body of literature on how anthropogenic climate change at the global scale (e.g. the greenhouse effect) may affect the dryland areas of Africa. The interactions of desertification and climate operate in both directions – desertification impacts local and regional climate and oppositely, variations in global and regional climate certainly impact desertification in Africa.

4.2 SIMULATING AFRICAN CLIMATE CHANGE CAUSED BY DESERTIFICATION

Approximately 100 articles have appeared in the professional literature in the 1980s and 1990s dealing directly with questions regarding the Charney hypothesis and its application to conditions in Africa (see Williams and Balling, 1996). Many of these articles describe numerical modelling efforts aimed at improving the simulation work first performed by Charney and his co-workers. In general, the modelling results have supported Charney's work for Africa; however, most of the research has shown that soil moisture is at least as important as any realistic changes to surface albedos in altering local climate. Furthermore, the albedo changes imposed on the models have usually been unrealistically large compared to albedo changes observed in areas undergoing desertification.

The 1990s has been a decade of substantial theoretical research on this issue. Scientists have linked a detailed local surface energy balance model to a global climate model and found that Sahelian-area temperature and precipitation patterns are very sensitive to soil moisture levels in the region (Bounova and Krishnamurti, 1993a, 1993b). Others have used a general circulation model and found that sub-Saharan desertification leads to (a) a reduction in moisture flux and rainfall in the Sahel; (b) an increase in moisture flux and precipitation to the south of the Sahel; (c) a reduction in the strength of the tropical easterly jet; (d) a strengthening of the African easterly jet; and (e) a decrease in the intensity of easterly waves in the region (Xue and Shukla, 1993).

The use of even more sophisticated general circulation models has revealed that in the Sahel, land degradation would cause both an increase in local temperatures and a decrease in rainfall levels (Xue, 1997). The results are linked to changes in local latent heat fluxes, moisture flux convergence, and subsidence in the atmosphere; the radiative effects of albedo changes seem secondary in causing a reduction in precipitation. In a somewhat opposite approach, researchers have used numerical models of climate to show that afforestation could increase local precipitation

in the Sahel, particularly in dry years and in the area that was afforested (Xue and Shukla, 1996).

Other recent research incorporates an atmospheric model coupled with a simple land surface scheme to investigate the sensitivity of West African monsoons to perturbations in the meridional distribution of vegetation (Zheng and Eltahir, 1998). The results of the numerical experiments demonstrate that West African monsoons and therefore rainfall distribution depend critically on the location of the vegetation perturbations. Changes in vegetation cover along the border between the Sahara desert and West Africa, which were assumed to represent the end result of desertification processes, would have only a minor impact on the simulated monsoon circulation. However, coastal deforestation may cause the collapse of the monsoon circulation and have a dramatic impact on the regional rainfall. Human impacts on vegetation influenced climate, but deforestation, and not desertification, dominated the feedbacks.

De Ridder (1998) used a one-dimensional version of a mesoscale atmospheric model coupled to a detailed land surface model to investigate the influence of the land surface in the Sahel on rain infiltration, surface evapotranspiration, and their impact on the convective available potential energy. The simulations show that the presence of a densely vegetated surface acts as a catalyst in the hydrological cycle, creating a positive feedback and enhancing precipitation recycling. De Ridder argued that this result is due to the relation between the characteristic drying-out time of the soil and the return frequency of the rain-triggering African easterly waves.

The role of vegetation disruption on local climate usually involves perturbations to the surface energy balance which in turn alter energy and moisture fluxes in the lower atmosphere. However, once the vegetation is depleted, the local atmosphere is further influenced by increased dust loads. The role of locally generated dust in altering the local climate in Africa remains a significant issue in establishing the linkage between land degradation in drylands and its impact on local climate. However, the literature on the subject is inconclusive, but most modelling efforts lead to a conclusion that increased mineral aerosol loads will cool the surface, warm the lower atmosphere, stabilize the atmosphere, and reduce local precipitation (Littman, 1991; Tegen et al., 1996; Moulin et al., 1997).

When compared to other dryland areas of the world, including those drylands of southern and eastern Africa, the Sahel appears to be in a particularly sensitive position allowing even small changes at the surface to create substantial changes in local temperature and precipitation (Bounova and Krishnamurti, 1993a, 1993b). Charney may have provided us with some very useful ideas about desertification, local climate change, and feedbacks to the desertification processes. The many other scientists who continued to run simulation models regarding this issue have shown that the desertification-climate linkage is far more complex than what Charney originally proposed.

4.3 MEASURING AFRICAN CLIMATE CHANGES CAUSED BY DESERTIFICATION

The findings of Charney and others in the mid-1970s stimulated a number of empirical efforts to measure and analyse the impact of desertification on local and regional climate in Africa. At least four articles have appeared in the literature in which historical temperature data from Africa are directly linked to land degradation in the dryland regions. Balling (1991) assembled gridded monthly temperature records for Africa and elsewhere, and he searched for matched pairs with one grid point in an area of severe desertification while the adjacent grid point 5° of latitude away would be in an area not impacted by desertification. He used the United Nations map prepared by Dregne (1977) to assign the desertification classification. Seven of the matched pairs came from Sahelian Africa, and the warming trend in the region that appeared to be desertification-induced was 0.057 °C per decade. The desertification warming signal was highly statistically significant, and the largest value found anywhere in the world came from the central Sahel.

Nasrallah and Balling (1993) used a spatial analytical approach to address the issue of desertification and temperature trends. Their analyses involved grid cells in North Africa, and once again, they found that warming and desertification were significantly and positively linked. Areas in North Africa with severe desertification were warming faster than nearby areas with little or no ongoing desertification. Furthermore, desertification appeared to have its greatest impact on temperatures from the high-sun months and its least impact during the low-sun months. Nasrallah and Balling (1994) repeated these analyses given an updated map of desertification classification, and the results remained the same – desertification in Africa appeared to be producing a recognizable local/regional warming signal in the historical temperature records.

Unganai (1997) examined temperature records from Zimbabwe from 1933 to 1993 and found another potentially important impact of African land degradation on local and regional climate. Throughout most of the world, scientists have noted an upward trend in minimum temperatures and no trend or a downward trend in maximum temperatures. The combination of the two trends has led to the widely recognized decrease in the diurnal temperature range. However, land degradation in drylands could have an opposite effect allowing afternoon temperatures to rise more quickly than early morning temperatures. Indeed, Unganai noted that unlike most other areas, the rural network of stations in Zimbabwe showed a rise in maximum temperatures, a decrease in minimum temperatures, and a substantial rise in the diurnal

temperature range. Local deforestation and degradation in Zimbabwe was suggested as a possible cause of this unusual pattern.

The effect of overgrazing on historical temperature trends in Africa is seen in the following relatively simple analysis. Fifty-seven 5° latitude by 5° longitude grid cells in Africa have nearly complete monthly temperature data from 1946 to 1998 – the data used in this example are the gridded near-surface air temperature data presented in the many reports of the Intergovernmental Panel on Climate Change (IPCC), and described in detail by Jones *et al.* (1999). When all grid cells are considered, the mean warming for the African continent is 0.104 °C per decade over the 53-year period. However, when the cells are stratified by overgrazing class presented in the *World Atlas of Desertification* (UNEP, 1992), the 31 cells most impacted by overgrazing have warmed at a rate of 0.135 °C per decade, while the 26 cells not impacted by overgrazing warmed at a rate of 0.066 °C per decade. A t-test reveals that the difference between the two trends is highly statistically significant (t = 3.03).

These rather simple calculations coupled with the existing literature suggest that overgrazing and land degradation in Africa are adding to the regional warming signal in historical temperature records. Whether this is caused by an increased mineral aerosol load in the atmosphere or alterations to the energy balance at the surface/atmosphere interface is a matter of debate. While any number of other explanations could be presented for the differential warming trend, including the possibility that the greenhouse thermal enhancement is greatest in the driest air masses, the disruption of the surface energy balance caused by overgrazing must be seriously considered as an important contributor.

4.4 AFRICAN DROUGHT CAUSED BY CLIMATIC VARIATIONS

While a substantial literature exists on how desertification in Africa may impact local and regional climate, a much larger literature exists on how variations and/or trends in climate impact desertification on the African continent. The primary focus of this research has been upon how the configura-tion of sea surface conditions and atmospheric circulation combine to enhance or depress rainfall in various parts of Africa. Once again, the focus has been on the Sahel, but drylands of East Africa and southern Africa are well represented in the literature.

At the local and regional levels, antecedent climate conditions can greatly impact current and near-future drought levels. For example, Walker and Rowntree (1977) used an 11-layer tropical model to examine the effect of Sahelian soil moisture on rainfall during August. In the case with dry initial conditions, only limited rainfall developed over 10 days and the rainfall was largely restricted to a latitudinal band between 5° N and 15° N. In the case with wet initial conditions, the August disturbances in the model produced extensive rainfall over a 20-day period that extended over a 5° N to 30° N latitudinal band. Later, Sud and Fennessey (1984) used a more complex numerical model for a 47-day simulation of June and July rainfall in the Sahel. When local evaporation in the model was held to zero, mean July precipitation actually increased in the Sahel. The increase in rainfall was related to the development of thermal lows, enhanced vertical motion, and the advection of low-level moisture. Nonetheless, Sud and Fennessey (1984) concluded that the local evaporation rates (which depend strongly on antecedent soil moisture conditions) influence local precipitation levels.

More recently, Bounova and Krishnamurti (1993a, 1993b) coupled a local energy balance model based on a statistical parameterization of soil moisture availability to a global spectral climate model, and they conducted numerical experiments on the onset and active phase of the West African monsoon. They found that soil moisture variations can control regional temperatures, the northward propagation of the maximum temperature zone, the mean streamlines of near-surface airflow, and ultimately, the spatial distribution and amount of precipitation. They too concluded that Sahelian-area climate is very sensitive to local soil moisture conditions, again confirming the linkage between soil moisture levels and precipitation levels.

The numerical modelling studies are strongly supported by a large literature regarding diagnostic research based on empirical data from the African continent. Winstanley (1974) used West African rainfall data to show a high correlation (+0.63) between June rainfall and the core rainy season (July–September) precipitation amounts, and he argued that a positive feedback associated with a soil moisture and local precipitation relationship could be contributing to the statistical relationship. Later, Lamb (1982, 1983, 1985) analysed rainfall and drought data for sub-Saharan Africa and concluded that the drought was showing persistence through the 1970s and early 1980s. Lare and Nicholson (1990) also concluded that (a) Sahelian drought definitely contains a significant persistence signal; and (b) surface processes are important in driving and maintaining this level of persistence. Similarly, Nicholson (1979) found an autocorrelation value of +0.30 for rainfall data collected in West Africa over the twentieth century, although Anyadike (1992, 1993) found no persistence in long-term (1916–1987) historical precipitation data from throughout the different climatic regions of Nigeria.

A large literature exists linking drought in Africa to variations throughout the climate system and, of this literature, many of the studies have linked sea surface conditions and related atmospheric circulation patterns to widespread drought patterns in various parts of Africa. Furthermore, the majority of studies on this issue have been empirical making use of statistical relationships between historical drought patterns and sea surface and atmospheric circulation data. For example, Lamb (1978) examined various tropical Atlantic circulation features for dry, wet, and average years in the Sahelian region. He found that drought characterized the July–September rainy season period in the West African Sahel when (a) the tropical Atlantic near-equatorial pressure trough, the kinematic axis separating the trade winds of the Northern and Southern Hemispheres, and the mid-Atlantic zone of maximum precipitation frequency and total cloudiness are located 200 to 500 km south of their mean positions; (b) the North Atlantic subtropical high extends further equatorward than average (although its centre may actually shift northward); and (c) the northern hemispheric trade winds are stronger than normal. In addition, the sea surface temperatures in Gulf of Guinea were higher than normal in these drought years.

Lamb (1983) and others (Newell and Kidson, 1984; Bah, 1987; Lamb and Peppler, 1992; Janicot, 1992a, 1992b; Fontaine and Bigot, 1993) continued this line of inquiry and were able to determine the impact of local sea surface temperatures on evaporation rates, the magnitude and direction of atmospheric moisture transfer, and thickness of various moisture-bearing layers of the atmosphere. Lamb and Peppler (1992) followed earlier work and showed that dry years in the Sahel are associated with relatively warm sea surface temperature conditions in the equatorial Atlantic (particularly during the July through September period) and somewhat cooler conditions to the west of West Africa and to north and east of north-eastern Brazil. Wet seasons in the Sahel are associated with a somewhat opposite pattern; relatively cool sea surface temperatures along the equatorial Atlantic are particularly prominent during the wetter years. Folland et al. (1986) linked drought in the Sahel to warm water in the equatorial Atlantic, the southern coast of West Africa, and the area off the coast of Ecuador, and cool water in the Caribbean, the mid-latitudes of the North Atlantic, and off the southern coast of Greenland.

Rowell et al. (1995) examined the variability of summer rainfall in tropical North Africa and found that a statistically significant linkage to sea surface temperatures existed at a global scale. They also noted the importance in this region of land-surface-moisture feedbacks in controlling amounts of precipitation. Fontaine et al. (1995) analysed August rainfall in West Africa over the period 1958 to 1989 and found that wet months have an increase in upper easterlies and lower south-westerlies, while droughts have less southward extension of the upper easterlies. Janicot et al. (1996) found that correlations between the El Niño Southern Oscillation index and rainfall in the Sahel have increased in recent years and that decreasing rainfall in the region is related to warm water in the east Pacific, the equatorial Pacific, and the Indian Ocean.

Obviously, a large amount of empirical research has been conducted on the linkage between sea surface temperature patterns, atmospheric circulation patterns, and climate conditions in dryland areas in Africa, particularly for the Sahelian region. Not surprisingly, many scientists have also studied these relationships using general circulation models, and in most cases, the results support the relationships found using the empirical approaches. For example, Druyan (1987, 1989) and Druyan and Hastenrath (1991, 1992) have been able to illustrate the sensitivity of Sahelian rainfall to changes in sea surface temperatures, and their results are consistent with established empirical relationships described earlier. Similarly, Palmer (1986), Semazzi et al. (1988), and Palmer et al. (1992) were able to correctly simulate dry/wet conditions in the Sahel under specified sea surface temperature patterns, including the El Niño Southern Oscillation phenomenon. Their results were quite consistent with the sea surface temperature linkages to Sahelian precipitation discussed earlier in this chapter.

Eltahir and Gong (1996) provided a theoretical argument suggesting that a cold pool of water in the region south of the West African coast should favour a strong monsoonal circulation favouring wet conditions in the Sahel. Druyan (1998) analysed two different June–September seasons using a general circulation model forced by sea surface temperatures and found that daily precipitation in the Sahel is highly correlated to mid-tropospheric vorticity, near-surface convergence, and 200 mb divergence. Zheng et al. (1999) used a numerical model and found that warm spring sea surface temperatures yield a wet summer in the Sahel. The rainfall anomaly begins over the sea and propagates over land and persists for several months due in part to a positive relationship between soil moisture and local rainfall.

Cook (1999) performed a series of simulations to show that the African easterly jet forms over West Africa in summer as a result of strong meridional soil moisture gradients. Positive temperature gradients associated with the summertime distributions of solar radiation, sea surface temperatures, or clouds were not large enough to produce the easterly jet in the absence of soil moisture gradients. Cook argued that moisture divergence in the Sahel is closely tied to the jet dynamics, and the jet's magnitude and position are sensitive to sea surface temperatures and land surface conditions. Yet another mechanism important to West African precipitation was found to be linked to surface conditions in the region.

The literature on African precipitation and its link to sea surface conditions and atmospheric circulation patterns is dominated by an emphasis on the Sahelian area, but research efforts have also focused on both southern and eastern Africa as well. For example, D'Abreton and Tyson (1995) showed that wet summers in southern Africa are related to an expansion of the Hadley Cell circulation. Mason (1995) showed that the sea surface temperatures in the South Atlantic Ocean areas explain greatest variance in South African rainfall, while Shinoda and Kawamura (1996) used a combination of atmospheric circulation indicators and sea surface temperatures in their analysis of rainfall over semi-arid southern Africa. Similarly, Mutai et al. (1998) revealed that July through September sea surface temperatures in the tropical Pacific (related to El Niño Southern Oscillation), tropical Indian and, to a lesser extent, tropical Atlantic Oceans contain moderately strong relationships (explained variance up to 60%) with the October through December rainfall totals averaged across the drylands of East Africa. This high level of explained variance is unusual for African dryland areas. Folland et al. (1991) commented that skill scores for seasonal forecasting of precipitation in Africa are not particularly high for either general circulation models or statistical models driven by variations in sea surface temperatures.

All of these studies provide significant insight into the underlying causes of drought in various parts of the African continent with special attention placed on how sea surface temperature conditions impact rainfall patterns in the Sahel and elsewhere. Because sea surface temperature patterns are relatively conservative through time (i.e. they do not change rapidly from week to week or even month to month), there is optimism that droughts in Africa can be forecast with moderate accuracy a month or season in advance. As forecasting skill scores continue to improve, early warning systems for drought will become more reliable, and this information should be useful in mitigating the negative human and ecological consequences of droughts in the future.

4.5 AFRICAN DESERTIFICATION AND GLOBAL WARMING

No issue in climatology has received more attention than the anthropogenic greenhouse effect, and the potential impacts of human-induced global warming on regions throughout the world have been investigated vigorously. Virtually every element of the greenhouse issue is a subject of considerable debate, and accordingly, the presentation that follows is based largely on the major reports of the IPCC (e.g. Houghton et al., 1996). The following are generalizations from the IPCC that shed light on future drought patterns in African drylands as driven by potential greenhouse-induced changes in climate:

(1) Numerical model experiments that include realistic increases in greenhouse gases and concentrations of sulphate aerosols generally show an increase in temperature throughout Africa between 1 °C and 3 °C by the middle of twenty-first century. However, in some numerical modelling experiments, parts of central Sahelian Africa are expected to cool slightly by 2050; the cooling is related to a projected increase in the intensity of convection and an increase in precipitation.

(2) These same experiments generally show a small increase in precipitation of approximately 0.5 mm per day over the next half century. Exceptions include a projected 1 to 3 mm per day increase in summer season rainfall in the central Sahel and in an area centred on Mozambique. These estimates for future rainfall changes are considered far more uncertain than the simulated increase in temperature over the next 50 years.

(3) Throughout most of Africa, the increase in temperature causes an increase in potential evapotranspiration that overwhelms any increases in precipitation and results in a reduction in soil moisture. Soil moisture levels throughout most of Saharan Africa are expected to decline by less than 1 cm. The northern portion of the Sahel is projected to see up to a 3-cm reduction. Similarly, most of the western half of southern Africa is expected to witness a decline in mean soil moisture of 1 to 2 cm, while most of eastern southern Africa is projected to have an increase in soil moisture of a few centimetres.

The uncertainties of these projections are related to limitations of the numerical climate models and the identification of the different 'forcings' that will drive future climate, including the climatic impact of future land-use changes. We know that climate in African drylands has varied considerably through time, and there is no reason to believe that today's climate will persist into the future. Climate change in African drylands is the rule, not the exception.

4.6 CONCLUSIONS

Dryland climates occur throughout the world in areas where the ratio of precipitation to potential evapotranspiration on an annual basis falls between 0.05 and 0.65. The causes of this imbalance may include many factors, including planetary-scale atmospheric circulation features and associated synoptic-scale meteorological systems that inhibit precipitation or raise potential evapotranspiration rates, geographic distance from moisture sources, the impact of mountain barriers in producing rain-shadow effects, and the lower atmospheric stabilization associated with cold ocean currents. Given the size of Africa, all of these factors contribute to the climate of the various dryland areas.

What has changed considerably in Africa's drylands is the exponential rise in human populations and their impacts upon the landscape. Many scientists believe that the end result of human activities is a degradation of the landscape, including changes in climate that exacerbate desertification processes. In the eyes of many scientists and policy makers, human activities in the dryland areas are ultimately decreasing available soil moisture levels. Yet, to be fair, a considerable debate surrounds each element of the desertification issue. For example, satellite records in the Sahel from 1980 to 1995 show us that while vegetation varies with rainfall, no evidence exists of any expansion or deterioration of the desert environment (Nicholson *et al.*, 1998).

Irrespective of these and other arguments, a sufficient number of scientists and policy makers are concerned about the desertification issue, particularly in Africa. By the middle of the 1990s, the United Nations was able to hammer out a convention to combat desertification. On 3 April 1995, Mexico became the first country to ratify the United Nations Convention to Combat Desertification in Countries Experiencing Serious Drought and/or Desertification, Particularly in Africa. The Convention deals largely with the areas of data collection, analysis and exchange of information, research, technology transfer, capacity-building and awareness-building, the promotion of an integrated approach in developing national strategies to combat desertification, and assistance in ensuring that adequate financial resources are available for programmes to combat desertification and mitigate the effects of drought. In substantive terms, status as a 'Party' would also allow a country to show solidarity with affected countries in facing an urgent and growing issue of global dimensions, benefit from cooperation with other affected countries, and with developed countries, in designing and implementing its own programmes to combat desertification and mitigate the effects of drought, and improve access to relevant technologies and data.

The international efforts to improve the vegetation in drylands may have an additional benefit to those concerned with global warming – the evidence is convincing that rehabilitation of drylands may cause significant cooling in some of the world hottest locations. Generally speaking, if the drylands 'green up', they should cool down as well. The carbon storage of expanding vegetation in drylands may not be substantial in a global sense, but on a country by country basis, the rehabilitation may help offset increased carbon dioxide emissions related to a dryland country's economic development.

Rehabilitating drylands will certainly take a lot more than nations signing an international agreement to combat desertification. Nonetheless, the issue is part of the brightening global awareness, and it will likely receive increased attention in this century. And while other environmental issues are grounded on concern about emerging and future consequences, the human and economic consequences of desertification and its related climate change have been evident for several decades, and will likely become evident in the decades to come.

ACKNOWLEDGEMENT

The author thanks Dr Mike Hulme for his valuable suggestions for improving this manuscript.

REFERENCES

Anyadike, R. N. C. (1992) Regional variations in fluctuations of seasonal rainfall over Nigeria. *Theor. Appl. Climatol*, **45,** 285–292.
    (1993) Seasonal and annual rainfall variations over Nigeria. *Int. J. Climatol.*, **13,** 567–580.
Bah, A. (1987) Towards the prediction of Sahelian rainfall from sea surface temperatures in the Gulf of Guinea. *Tellus*, **39A,** 39–48.

Balling, R. C. Jr. (1991) Impact of desertification on regional and global warming. *Bull. Am. Meteorol. Soc.*, **72**, 232–234.

Berkofsky, L. (1976) The effect of variable surface albedo on atmospheric circulation in desert regions. *J. Appl. Meteorol.*, **15**, 1,139–1,144.

Bounova, L. and Krishnamurti, T. N. (1993a) Influence of soil moisture on the Sahelian climate prediction I. *Meteorol. Atm. Phys.*, **52**, 183–203.

(1993b) Influence of soil moisture on the Sahelian climate prediction II. *Meteorol. Atm. Phys.*, **52**, 205–224.

Bryson, R. A. (1973) Drought in Sahelia: Who or what is to blame? *Ecologist.*, **3**, 366–377.

Charney, J. G. (1975) Dynamics of deserts and drought in the Sahel. *Q. J. Roy. Meteorol. Soc.*, **101**, 193–202.

Charney, J. G., Quirk, W. J., Chow, S-H. and Kornfield, J. (1977) A comparative study of the effects of albedo change on drought in semi-arid regions. *J. Atm. Sci.*, **34**, 1,366–1,385.

Cloudsley-Thompson, J. L. (1974) The expanding Sahara. *Environ. Conserv.*, **1**, 5–13.

Cook, K. H. (1999) Generation of the African Easterly Jet and its role in determining West African precipitation. *J. Climate*, **12**, 1,165–1,184.

D'Abreton, P. C. and Tyson, P. D. (1995) Divergent and non-divergent water vapour transport over southern Africa during wet and dry conditions. *Meteorol. Atm. Phys.*, **55**, 47–59.

de Ridder, K. (1998) The impact of vegetation cover on Sahelian drought persistence. *Boundary-Layer Meteorology*, **88**, 307–321.

Dregne, H. E. (1977) *Generalised Map of the Status of Desertification of Arid Lands.* Prepared by FAO, UNESCO and WMO for the 1977 United Nations Conference on Desertification.

Druyan, L. M. (1987) GCM studies of the African summer monsoon. *Climate Dynamics*, **2**, 117–126.

(1989) Advances in the study of sub-Saharan drought. *Int. J. Climatol.*, **9**, 77–90.

(1998) The role of synoptic systems in the inter-annual variability of Sahel rainfall. *Meteorol. Atm. Phys.*, **65**, 55–75.

Druyan, L. M. and Hastenrath, S. (1991) Modelling the differential impact of 1984 and 1950 sea-surface temperatures on Sahel rainfall. *Int. J. Climatol.*, **11**, 367–380.

(1992) GCM simulation of the Sahel 1984 drought with alternative specifications of observed SSTs. *Int. J. Climatol.*, **12**, 521–526.

Ellsaesser, H. W., MacCracken, M. C., Potter, G. L. and Luther, F. M. (1976) An additional model test of positive feedback from high desert albedo. *Q. J. Roy. Meteorol. Soc.*, **102**, 655–666.

Eltahir, E.A.B. and Gong, C. (1996) Dynamics of wet and dry years in West Africa. *J. Climate.*, **9**, 1,030–1,042.

Folland, C. K., Palmer, T. N. and Parker, D. E. (1986) Sahel rainfall and worldwide sea temperatures, 1901–85. *Nature*, **320**, 602–607.

Folland, C. K., Owen, J. A., Ward, M. N. and Colman, A. W. (1991) Prediction of seasonal rainfall in the Sahel region using empirical and dynamical methods. *J. Forecasting*, **10**, 21–56.

Fontaine, B. and Bigot, S. (1993) West African rainfall deficits and sea surface temperatures. *Int. J. Climatol.*, **13**, 271–285.

Fontaine, B., Janicot, S. and Moron, V. (1995) Rainfall anomaly patterns and wind field signals over West Africa in August (1958–89). *J. Climate*, **8**, 1,503–1,510.

Houghton, J. T., Meira Filho, L. G., Callander, B. A., Harris, N., Kattenberg, A. and Maskell, K. (Eds.) (1996) *Climate Change 1995: The Science of Climate Change.* Cambridge, Cambridge University Press.

Jackson, R. D. and Idso, S. B. (1975) Surface albedo and desertification. *Science*, **189**, 1,012–1,013.

Janicot, S. (1992a) Spatiotemporal variability of West African rainfall. Part I. Regionalizations and typings. *J. Climate*, **5**, 489–497.

(1992b) Spatiotemporal variability of West African rainfall. Part II: Associated surface and airmass characteristics. *J. Climate*, **5**, 499–511.

Janicot, S., Moron, V. and Fontaine, B. (1996) Sahel droughts and ENSO dynamics. *Geophys. Res. Lett.*, **23**, 515–518.

Jones, P. D., New, M., Parker, D. E., Martin, S. and Rigor, I. G. (1999) Surface air temperature and its changes over the past 150 years. *Rev. Geophys.*, **37**, 173–199.

Lamb, P. J. (1978) Case studies of tropical Atlantic surface circulation patterns during recent sub-Saharan weather anomalies: 1967 and 1968. *Mon. Weather Rev.*, **106**, 482–491.

(1982) Persistence of sub-Saharan drought. *Nature*, **299**, 46–47.

(1983) West African water vapour variations between recent contrasting sub-Saharan rainy seasons. *Tellus*, **35A**, 198–212.

(1985) Rainfall in sub-Saharan West Africa during 1941–83. *Z. Gletscherk. Glazialgeol.*, **21**, 131–139.

Lamb, P. J. and Peppler, R. A. (1992) Further case studies of tropical Atlantic surface atmospheric and oceanic patterns associated with sub-Saharan drought. *J. Climate.*, **5**, 476–488.

Lare, A. R. and Nicholson, S. E. (1990) A climatonomic description of the surface energy balance in the central Sahel. Part I: Shortwave radiation. *J. Appl. Meteorol.*, **29**, 123–137.

Littman, T. (1991) Rainfall, temperature and dust storm anomalies in the African Sahel. *Geogr. J.*, **157**, 136–160.

Mason, S. J. (1995) Sea-surface temperature – South African rainfall associations, 1910–89. *Int. J. Climatol.*, **15**, 119–135.

Moulin, C., Lambert, C. E., Dulac, F. and Dayan, U. (1997) Control of atmospheric export of dust from North Africa by the North Atlantic Oscillation. *Nature*, **387**, 691–695.

Mutai, C. C., Ward, M. N. and Colman, A. W. (1998) Towards the prediction of the East Africa short rains based on sea-surface temperature-atmosphere coupling. *Int. J. Climatol.*, **18**, 975–997.

Nasrallah, H. A. and Balling, R. C. Jr. (1993) Spatial and temporal analysis of Middle Eastern temperature changes. *Climatic Change*, **25**, 153–161.

(1994) The effect of overgrazing on historical temperature trends. *Agric. Forest Meteorol.*, **71**, 425–430.

Newell, R. E. and Kidson, J. W. (1984) African mean wind changes between Sahelian wet and dry periods. *J. Climatol.*, **4**, 27–33.

Nicholson, S. E. (1979) Revised rainfall series for the West African subtropics. *Mon Weather Rev.*, **107,** 620–623.

Nicholson, S. E., Tucker, C. J. and Ba, M. B. (1998) Desertification, drought, and surface vegetation: an example from the West African Sahel. *Bull. Am. Meteorol. Soc.*, **79,** 815–822.

Otterman, J. (1974) Baring high-albedo soils by overgrazing: a hypothesized desertification mechanism. *Science*, **186,** 531–533.

Palmer, T. N. (1986) Influence of the Atlantic, Pacific and Indian Oceans on Sahel rainfall. *Nature*, **322,** 251–253.

Palmer, T. N., Brankovic, C., Viterbo, P. and Miller, M. J. (1992) Modelling inter-annual variations of summer monsoons. *J. Climate.*, **5,** 399–417.

Ripley, E. A. (1976) Comments on the paper 'Dynamics of deserts and droughts in the Sahel' by J. G. Charney. *Q. J. Roy. Meteorol. Soc.*, **102,** 466–67.

Rowell, D. P., Folland, C. K., Maskell, K. and Ward, M. N. (1995) Variability of summer rainfall over tropical North Africa (1906–92): Observations and modelling. *Q. J. Roy. Meteorol. Soc.*, **121,** 669–704.

Semazzi, F. H. M., Mehta, Y. and Sud, Y. C. (1988) An investigation of the relationship between sub-Saharan rainfall and global sea surface temperatures. *Atmosphere-Ocean*, **26,** 118–138.

Shinoda, M. and Kawamura, R. (1996) Relationships between rainfall over semi-arid southern Africa, geopotential heights, and sea surface temperatures. *J. Meteorol. Soc. Jpn*, **74,** 21–36.

Sud, Y. C. and Fennessy, M. J. (1984) Influence of evaporation in semi-arid regions on the July circulation: a numerical study. *J. Climatol.*, **4,** 383–398.

Tegen, I., Lacis, A. A. and Fung, I. (1996) The influence on climate forcing of mineral aerosols from disturbed soils. *Nature*, **380,** 419–422.

UNEP (1992) *World Atlas of Desertification.* London, Edward Arnold.

Unganai, L. S. (1997) Surface temperature variation over Zimbabwe between 1897–1993. *Theor. Appl. Climatol.*, **56,** 89–101.

Walker, J. and Rowntree, P. R. (1977) The effect of soil moisture on circulation and rainfall in a tropical model. *Q. J. Roy. Meteorol. Soc.*, **103,** 29–46.

Williams, M. A. J. and Balling, R. C. Jr. (1996) *Interactions of Desertification and Climate.* London, Edward Arnold.

Winstanley, D. W. (1974) Seasonal rainfall forecasting in West Africa. *Nature*, **248,** 464.

Xue, Y. (1997) Biosphere feedback in regional climate in tropical North Africa. *Q. J. Roy. Meteorol. Soc.*, **123,** 1,483–1,515.

Xue, Y. and Shukla, J. (1993) The influence of land surface properties on Sahel climate. Part I: Desertification. *J. Climate*, **6,** 2,232–2,245.

(1996) The influence of land surface properties on Sahel climate. Part II: Afforestation. *J. Climate*, **9,** 3,260–3,275.

Zheng, X. and Eltahir, E. A. B. (1998) The role of vegetation in the dynamics of West African Monsoons. *J. Climate.*, **11,** 2,070–2,096.

Zheng, X., Elthahir, E. A. B. and Emanuel, K. A. (1999) A mechanism relating tropical Atlantic spring sea surface temperature and West African rainfall. *Q. J. Roy. Meteorol. Soc.*, **125,** 1,129–1,164.

5 Africa's climate observed: perspectives on monitoring and management of floods, drought and desertification

DAVID A. HASTINGS[*]

United Nations Economic and Social Commission for Asia and the Pacific Bangkok, Thailand

Keywords

Drought; desertification; floods; climate; ENSO; satellite observations; GIS; disaster management; weather modification; open-source; geomatics

Abstract

Satellite imagery and in situ data processed in a geographic information system offer perspectives on Africa's climate. Areas of relatively high precipitable water vapour extend more widely than areas of high precipitation, suggesting that improved weather modification (e.g. stimulating precipitation) might benefit desertification mitigation efforts. Processed Normalized Difference Vegetation Index data offer perspectives on the current state of desertification risk. The Vegetation Condition Index, used with compilations of environmental associations with ENSO, and with ENSO and weather forecasts, offer tools for agricultural, water resources and public health managers to respond to drought. Finally, evolving capabilities in satellite observations of flooding, global and regional digital elevation data, geomatics software, and regional cooperative mechanisms, offer tools for improved flood management.

5.1 INTRODUCTION

Africa's climate has been summarized many times in the past; it is not the purpose of this effort to repeat such exercises. Instead, this chapter uses scientific geographic information systems (GIS) to focus on aspects of climate parameters related to drought, desertification risk, and flooding. Such approaches deserve more widespread attention by those attempting to respond to climate change, including decision makers that must set national policy, or to build national or regional capacity responding to these issues. This chapter draws on significant progress over the last decade in compiling spatial environmental data, such as the Global Ecosystems Database (NOAA-EPA Global Ecosystems Database Project (1992)).

5.2 PRECIPITATION AND TEMPERATURE

Africa's precipitation ranges from essentially none, to about 5 metres annually (Figure 5.1). Lower latitudes in the southern hemisphere and the Mediterranean coast receive most precipitation early in the year, while lower latitudes in the northern hemisphere are wettest in mid-year (Figure 5.2).

Compilations of mean annual and monthly precipitation (computed by the author from source data of Legates and Willmott (1992)) have been analysed spatially in a GIS to determine the minimum, maximum, mean, median, and specific percentiles for the areas shown in the figures. Note that high values of annual precipitation (e.g. above 2,400 mm) occur over less than 2% of the sample area, and (as shown in Figures 5.1 and 5.2) most high precipitation occurs offshore. Overall, Figures 5.1 and 5.2, plus Table 5.1, show much of Africa to be arid or semi-arid. Fully 20% of the area shown in Figures 5.1 and 5.2 receive on average 222 mm or less of annual precipitation.

Similarly, significant geographical and seasonal temperature variations occur within Africa (Figure 5.3). Note that Figures 5.1 to 5.3 are windowed from global compilations (in this case for meteorological station data) of the Global Ecosystems Database (Kineman and Ohrenschall, 1992; Global Ecosystems Database Project, 2000). As they are not so simplified as traditional presentations, some of the details shown are artifacts of the uneven distribution (in time and space) of source data – especially those shown over oceanic areas.

5.3 NORMALIZED DIFFERENCE VEGETATION INDEX (NDVI)

For over two decades NDVI has been used as an indicator of the vigour of vegetative activity as represented by indirectly

---

[*]Formerly with the National Oceanic and Atmospheric dministration (USA); Kwame Nkrumah University of Science and Technology, and the Geological Survey Department (Ghana).

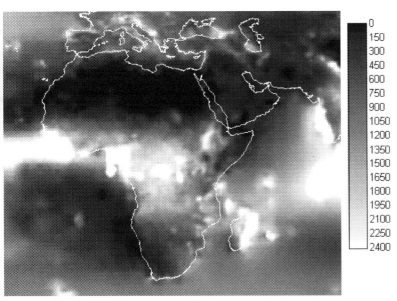

Figure 5.1. Mean annual precipitation for Africa, in millimetres; adapted from Legates and Willmott (1992) compilations from meteorological station data covering approximately 1920–1980.

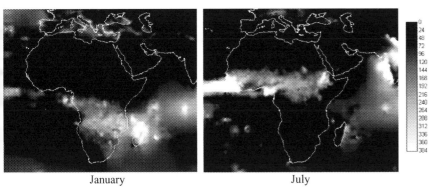

January   July

Figure 5.2. Mean monthly precipitation for January and July, in millimetres; adapted from Legates and Willmott (1992) data.

observable chlorophyll activity (Hastings and Emery, 1992). Low values of NDVI have been associated with lack of vegetation, dormant states of existing vegetation or stress caused by drought, over-irrigation, or disease.

Figure 5.4 depicts relative minimum and maximum annual NDVI values. These views were compiled by the author's computing minimum and maximum monthly mean climatological (typical) NDVI values using the Advanced Very High Resolution Radiometer (AVHRR) land climatologies of Gutman *et al.* (1997), using scientific (as opposed to ones used mostly for cartographic output) geographic information systems (GISs). Note that coastal West Africa, and northern Central Africa, plus a small area north of Lake Victoria, have the highest minimum NDVI, whereas areas generally slightly north of those areas have the highest maximum NDVI. It is not surprising that arid regions exhibit low NDVI year-round.

Figure 5.5 shows typical January and July NDVI values, adapted from Gutman *et al.* (1997). The correspondence between NDVI and precipitation is, not surprisingly, strong.

Spatial principal components analysis is sometimes used to depict subtle features of multispectral or multitemporal data. Figure 5.6 depicts the first two principal components computed by the author from the Gutman *et al.* (1997) NDVI climatologies.

Spatial/temporal principal components are difficult to explain to the non-specialist. Eastman and Fulk (1993) performed a similar analysis of an earlier compilation of NDVI data, equating the first principal component to the annual mean NDVI, and the second principal component to seasonality difference between the northern/southern hemispherical rainy/dry seasons. Though statistical principal components should not be literally equated with physical phenomena (one statistician graphically noted that 'the least-squares fit is always statistically better than the true fit!'), the

Table 5.1. *Areal statistics on precipitation in Africa, for annual, January and July mean precipitation values (and for the areas) shown in Figures 1 and 2.*

Percentile	Precipitation (mm)		
	Annual	January	July
Minimum	0	0	0
2%	20	0	0
10%	119	3	1
20%	222	10	4
40%	525	36	23
Median	695	54	34
Mean	787	54	39
60%	862	72	54
80%	1,220	119	106
90%	1,583	175	177
98%	2,390	270	383
Maximum	5,187	531	1,540

general association is strong. Light-coloured areas in the second principal component tend to correspond to areas receiving monsoonal precipitation around the middle of the calendar year, while dark-coloured areas tend to correspond to monsoonal rains early in the year.

5.4 PRECIPITABLE WATER INDEX

The precipitable water index (PWI, Gutman *et al.*, 1997) is a less common indicator of climate. Computed from differences in thermal response (caused by differential atmospheric absorption due to atmospheric water vapour at different wavelengths in the thermal spectrum) viewed by the National Oceanic and Atmospheric Administration's AVHRR, PWI is a useful indicator of water vapour available (under favourable conditions) to become precipitation.

Figure 5.7 shows annual minimum and maximum PWI, computed from Gutman *et al.* (1997) climatologies. Precipitable water vapour is generally low in desert areas, yet it does vary over time. Moreover, in Sahelian regions like Senegal and Chad, and in areas to the north and east of the Kalihari Desert, areas where desertification is considered to be occurring at a rapid rate, the maximum annual PWI is relatively high.

Figure 5.8 shows typical January and July PWI. Note that the patterns generally follow those for NDVI, but in some areas, especially at the edges of areas of high NDVI, areas of relatively high PWI are more extensive. As for Figure 5.7, this suggests that the water vapour may be available at times, if weather modification techniques can be improved to stimulate rain. Although weather modification techniques (notably cloud seeding or other techniques to induce precipitation) have been shelved for over a

decade[1], can an eventual revisit to the topic be made more successful, perhaps helping to stem desertification trends in Africa? This chapter does not advocate renewed efforts in this arena, but suggests that water vapour may exist at times, and that vapour may be allowed to become local rain.

Figure 5.9 depicts the first and second principal components of PWI. As for NDVI, the first principal component tends to correspond with the mean PWI, where the second component tends to depict seasonality between the two hemispheres' main rainy seasons. Note that the second principal component depicts some component of precipitable water vapour over parts of several desert areas during their respective hemisphere's rainy seasons.

5.5 FLOODING IN AFRICA

Though Asia is the continent most affected by flooding, floods frequently devastate parts of Africa. Figure 5.10 depicts areas affected by flooding in the years 1985–1989, 1993, and 1998–2002 (years for which the Dartmouth Flood Observatory (2003) provides such information). '5' signifies 5 years of the total 11 years surveyed in which floods affected the depicted areas (shaded in white), where '1' signifies, out of the 11 years of the compilation, a so-depicted area was affected by flooding in a single year. By comparison, some areas in Bangladesh were affected by the most flooding during the period of compilation, in 8 of the 11 years.

The Dartmouth Flood Observatory contains tabulated data on major flooding (including damage, dates, countries/districts/rivers affected), maps compiling areas affected by flooding, sample images depicting selected flood events, and other research-oriented data describing flooding. Not surprising, the areas most affected by flooding tend to correspond to areas that receive significant seasonal precipitation.

Computerized flood modelling can be initialized with some combination of precipitation information, river gauge data,satellite-based graphics depicting flooded areas, and digital elevation models. The software itself is sometimes optimized to different basin characteristics, including the size of the basin, the dendritic or other drainage patterns (e.g. divergently radial, deranged, braided, patterns of some deltas or artificial patterns of irrigated areas). In addition, if source data such as digital elevation models come from general-purpose sources, they may have to be modified to add features that influence the flow of flood waters. For example, raised road beds, buildings, vegetation, irrigation and terraced padi, which may have been smaller than the

---

[1] Such discontinuation has been attributed to perceived ineffectiveness, cost, and the legal and institutional ramifications of these activities. However, some companies appear to have some success in cloud seeding of specific areas by infusing careful information management.

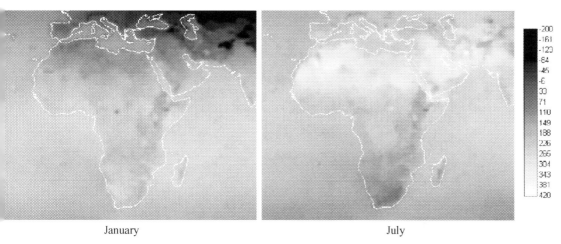

January   July

Figure 5.3. Mean temperature for January and July, in tenths of degrees Celsius; adapted from Legates and Willmott (1992) data.

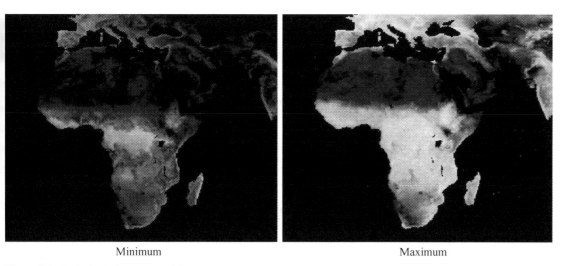

Minimum   Maximum

Figure 5.4. Typical relative annual minimum and maximum NDVI for Africa. Black signifies low values, white signifies high values; computed by the author from Gutman *et al.* (1997).

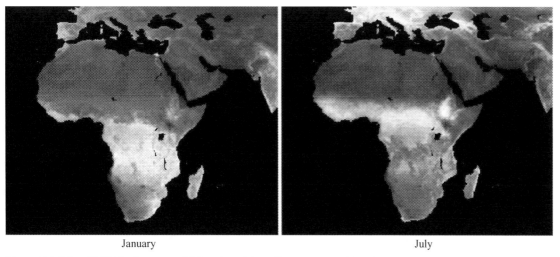

January   July

Figure 5.5. Mean NDVI for January and July, adapted from Gutman *et al.* (1997) climatologies. Lighter tones signify higher NDVI values.

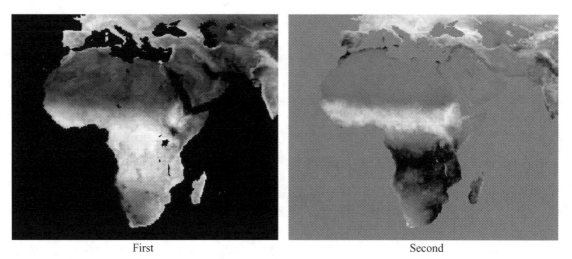

First   Second

Figure 5.6. First and second principal component of mean monthly NDVI; computed by the author from Gutman *et al.* (1997) climatologies.

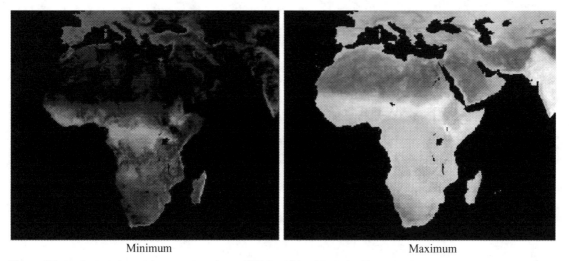

Minimum   Maximum

Figure 5.7. Typical relative minimum and maximum PWI for Africa. Black signifies low values; white, high values; computed by the author from Gutman *et al.* (1997) climatologies.

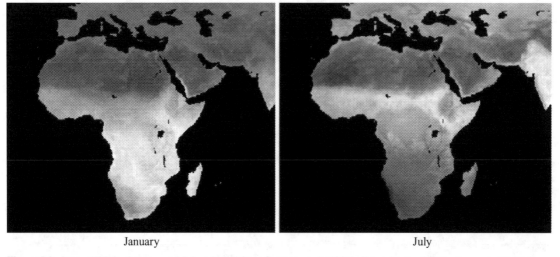

January   July

Figure 5.8. Mean PWI for January and July, adapted from Gutman *et al.* (1997) climatologies. Lighter tones signify higher PWI values.

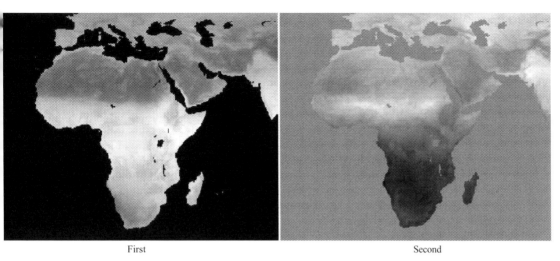

First   Second

Figure 5.9. First and second principal component of mean monthly PWI; computed by the author from Gutman *et al.* (1997) source data.

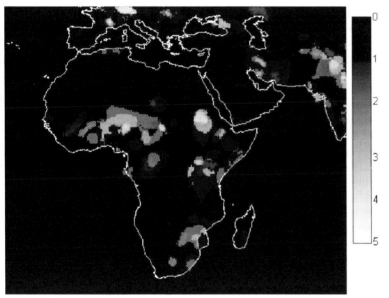

Figure 5.10. Number of years (out of 11 surveyed, as noted in the text) affected by flooding in Africa. Compiled by the author from Dartmouth Flood Observatory (2003) compendia.

minimum-mapping-unit of the original data, may have to be added to make the data more suitable for flood management purposes.

Data have become available since the mid-1990s, or are on the verge of availability in all these areas. Perhaps the most interesting ones are:

(a) Improved spatial and spectral resolution of experimental moderate resolution sensors, as the National Aeronautics and Space Administration's (NASA's) Moderate Resolution Imaging Spectroradiometer (MODIS), is currently available. Ideas from MODIS and other recent efforts are being used as design input for the operational successor to AVHRR.

(b) Availability of sub-metre-pixel (optionally stereoscopic) IKONOS and QuickBird data and derived detailed digital elevation models.

(c) Operational Radarsat imagery for cloud-penetrating views of floods in progress.

(d) Impending availability of global 3 arc-second (nominal 100 m) gridded digital elevation data (between 60° North and South Latitudes) from the Shuttle Radar Topography Mission.

(e) Improved robustness and integration of flood/inundation forecast models, especially of open-source models which offer rapid repair and enhancement, and of open-source geomatics software like the Geographic Resources Analysis Support System (GRASS) GIS (http://grass.itc.it/index.html). Increased attention paid to modelling and GIS (Benciolini *et al.*, 2002) is also an important factor.

(f) New capabilities for operational flash flood nowcasting.

(g) Improved availability and quality of web-based information.

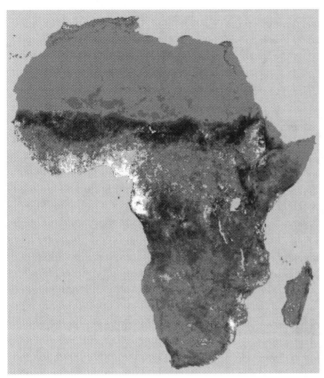

Figure 5.11. Sahelian drought in mid-1984. Blacks depict drought, white signifies anomalously high NDVI (from Tucker, 2002).

Such advances foretell improved flood management. Efforts to improve regional and global coordination, such as the International Charter: Space and Major Disasters, and the United Nations Economic and Social Commission for Asia and the Pacific (UN-ESCAP)'s efforts to promote regional cooperative mechanisms for using space-based and other data for disaster management, may dovetail with these increased technological capabilities to help flood managers worldwide.

5.6 DROUGHT AND DESERTIFICATION IN AFRICA

Depictions above of precipitation, NDVI (and generally low PWI though this is not necessarily determinate due to the circumstances that relate precipitable water vapour to actual precipitation) lend perspectives for drought managers. NDVI has been used for over two decades for depicting drought in Africa (e.g. the United States Agency for International Development (USAID's Famine Early Warning System of the early 1980s), and for planning possible food relief deployment. Figure 5.11 is a depiction of Sahelian drought in mid-1984 by Tucker (2002).

In addition, improved understanding of regional associations of environmental phenomena with El Niño and La Niña events may help the improvement of regional medium-term forecasts of certain climatic events or phenomena. These include drought, increased risk of desertification, and floods, as well as public health

problems associated with drought or increased precipitation (such as river blindness, malaria and other water-related illnesses), and may be valuable for water resource (e.g. hydroelectric power dam project) managers.

Kogan (1997) has developed an offshoot of NDVI, which he calls the Vegetation Condition Index (VCI), Temperature Condition Index (TCI), and Vegetation and Temperature Condition Index, combining data from AVHRR thermal channels with NDVI. His goal for VCI is an enhanced depiction of drought and other phenomena that affect vegetation. Figure 5.12 shows illustrations of VCI for southern Africa, toward the end of a typical January of an El Niño year vs. a La Niña year.

An improved drought management model might incorporate:

(a) **Medium-term forecasting**: Using improved predictability of El Niño and La Niña events, and improved understanding of environmental associations such as shown in Figure 5.11, agriculture (and water resources, and health-care) managers would strategize with their constituents (farmers, health clinics, and water managers) on the upcoming forecast, and its implications. For example, farmers, in conjunction with agriculture specialists, would develop strategies for planting (timing, strains to plant, anticipated fertilization, water issues, etc.).

(b) **Monitoring**: Using NDVI and VCI, monitor the progress of the growing season. Agriculture specialists would update farmers on current conditions, all concerned would refine plans for the season.

(c) **Planning near-term response activities**: With improved 10-day weather forecasts, and expectations that such forecasts will increase in accuracy, the upcoming week's activities can be tentatively planned – and revised as forecasts evolve.

An overall drought preparedness and management programme might proactively incorporate risk assessment and planning for various sectors and population groups, with a view to reducing the risk by identifying and adopting appropriate mitigation measures (Wilhite, 2000). Monitoring, prediction and early (regularly updated) warning are considered by Wilhite to be the foundation for appropriate response and recovery measures, which require the development of organization framework and institutional capacity. These aspects are beyond the scope of this chapter, but they have been comprehensively discussed in Wilhite (2000).

As for desertification risk, efforts have been made to use time-series compilations of NDVI to help monitor the risk (Hastings and Di, 1994). For example, an annual NDVI maximum (filtered to eliminate the spurious highs that occasionally occur from noise in the data, etc.) would show whether or not vegetation had responded at all during the year – as in an arid region 'blooming' from an unusual rainfall where dormant vegetation or seeds/spores existed that could be germinated.

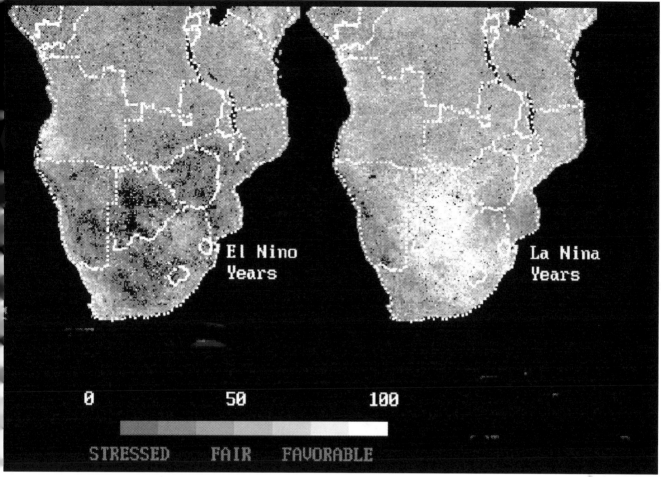

Figure 5.12. Vegetation and Temperature Condition Index typical of El Niño and La Niña conditions (from Kogan, 2001). Note the marked contrast in favourable vs. unfavourable growing conditions between El Niño and La Niña years.

Figure 5.13 shows the maximum climatological NDVI, computed by the author from Gutman *et al.* (1997) source data. Though the darkest greys generally associate with desert, and brighter whites with areas of healthy vegetation, this image has flagged relatively low values of NDVI, generally at the edges of areas of lower NDVI (e.g. very arid regions or deserts) also in near-whites. Comparing this image with the right side of Figure 5.4 should make the flagged values even clearer. These flagged areas might be considered as most susceptible to desertification – say, if land management practices (such as overgrazing) are not adapted to the environmental conditions. More cautious planners may wish to broaden the area of desertification risk into areas of higher NDVI (brighter, non-flagged, areas in Figure 5.13).

Figure 5.14 shows the results of a separate study reported by Hastings (1997) and Hastings and Tateishi (1998). A cluster analysis was made of 48 data layers of Gutman *et al.* (1997): 12 monthly values each of NDVI, PWI, and of the standard deviations of each for the years of the Gutman *et al.* (1997) compilation. An experimental clustering of all 48 layers into two classes produced a

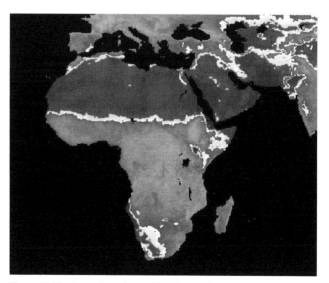

Figure 5.13. Annual maximum NDVI, computed by the author from Gutman *et al.* (1997) data, flagged (whitest shades) to depict areas that might be considered, from their marginal NDVI values, to be susceptible to desertification. Methodology from Hastings and Di (1994).

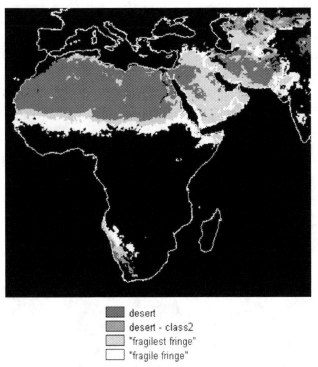

Figure 5.14. Four subclasses of 'deserts' and 'fragile fringes' (areas subject to desertification) from Hastings and Tateishi (1998), derived from clustering monthly mean NDVI, PWI, and standard deviations thereof from Gutman and others (1997).

clustering that could be described as 'desert and non-desert' (Hastings, 1997). Further clustering solely in that 'desert' cluster led to one experimental map depicting four classes as shown in Figure 5.14.

Note that Figures 5.13 and 5.14 show different aspects of desertification. Figure 5.13 uses only the maximum of 12 mean monthly Gutman *et al.* (1997) NDVI values, whereas Figure 5.14 shows the results of a hierarchical clustering experiment using NDVI, PWI, and their temporal standard deviations as described in Hastings and Tateishi (1998). There are other possible methods for attempting to monitor deserts, and risk of desertification, including methods using more spectrally sophisticated data, such as those from NASA's MODIS, and methods that might incorporate Kogan (1997)'s Vegetation and Temperature Condition Index, cited above.

Compiling annual maximum NDVI, and flagging such areas of marginal NDVI value, could provide for annual reports on the state of desertification risk. In a dry year, the flagged areas might increase in area. In a wet year, the flagged areas might get smaller. The areas flagged here are somewhat arbitrary, though they have been peer reviewed by desert specialists in Egypt, Kuwait, and elsewhere. The flagging could be modified to suit particular needs, such as the development and monitoring of desertification risk for the United Nations Convention to Combat

Desertification (UNCCD). Such information could also complement the United Nations Environment Programme's *Atlas of Desertification*, which was compiled mostly from traditional sources, not using much satellite imagery or GIS-based analysis.

5.7 CONCLUSIONS

Display and analysis of African climatic data in scientific GIS offer opportunities for enhanced assessment of relationships between climatic and other environmental phenomena. Africa has extensive desert areas, and drylands that are at risk of desertification. Though flooding is much less devastating in Africa than in Asia, nevertheless, floods impact parts of Africa in any given year.

Until recently, flood and drought management has been hampered by inadequate data and computer models for flood forecasting. However, advances over the past five years in data and information content, rigour, and delivery mechanisms, and in integrating open-source modelling and GIS software, are setting the stage for improved drought and flood management capabilities. Methodologies are outlined for enhanced flood and drought management, which incorporate long-term regional associations of drought and flood with El Niño and La Niña events, current monitoring of conditions, and one to two week weather forecasts to help guide near-future operations.

Data and software tools made available in the last five years, or soon to be available, offer promise of greatly improved flood management capabilities. Before they are effective, developers will have to optimize their use in enhanced management systems.

Contributions to drought management are shown from NDVI and VCI, for annual and growing season prognoses, and for shorter-term monitoring and planning of tactical responses to current and forecast conditions. Similar benefits appear possible in associating water and public health care phenomena with climate, then monitoring climatic trends with NDVI and VCI, and associated data.

NDVI and PWI have been shown to be useful for monitoring the state of desertification risk, supporting UNCCD National Action Plans and other efforts to combat desertification. Possible contributions of PWI for monitoring available atmospheric water vapour, and for supporting the usefulness of weather modification techniques (if technically and economically feasible) for combating desertification.

ACKNOWLEDGEMENTS

The author would like to thank Pak Sum Low for suggesting the subject of this chapter. Donald Wilhite, Robert Balling, Jr. and an anonymous reviewer provided valuable comments, in what must be a record for speed (within two days) by all three reviewers.

REFERENCES

Benciolini, B., Colli, M. and Zatelli, P. (2002) *Proceedings of the Open-Source Free Software GIS – GRASS Users Conference 2002*, University of Trento, Italy. http://www.ing.unitn.it/~grass/conferences/GRASS2002/home.html

Dartmouth Flood Observatory (2003) *2002 Global Register of Major Flood Events*. Dartmouth Flood Observatory, Hannover, NH, USA. http://www.dartmouth.edu/artsci/geog/floods/

Eastman, J. R. and Fulk, M. (1993) Long sequence time-series evaluation using standardized principal components. *Photogramm. Eng. Remote Sensing.* **59**, 1307–1312.

Global Ecosystems Database Project (2000) *Global Ecosystems Database Version II: Database, User's Guide, and Dataset Documentation*. US Department of Commerce, National Oceanic and Atmospheric Administration, National Geophysical Data Center, Boulder, CO, USA. Two CDROMs and publication on the World Wide Web.

Gutman, G., Tarpley, D., Ignatov, A. and Olson, S. (1997) *Global AVHRR-Derived Land Climatology*. National Oceanic and Atmospheric Administration, National Geophysical Data Center, Boulder, CO, USA.

Hastings, D. A. (1997) Land cover classification: new techniques, new source data. *Proceedings, 18th Asian Conference on Remote Sensing.* pp. JS-2–JS-6. Malaysian Remote Sensing Centre, Kuala Lumpur, Malaysia.

Hastings, D. A. and Di, Liping (1994) *Characterizing the Global Environment: An Example Using AVHRR to Assess Deserts, and Areas at Risk of Desertification*. National Oceanic and Atmospheric Administration, National Geophysical Data Center, Boulder, CO, USA. http://www.ngdc.noaa.gov/seg/globsys/gisdes2.shtml

Hastings, D. A. and Emery, W. J. (1992) The Advanced Very High Resolution Radiometer (AVHRR): A brief reference guide. *Photogramm. Eng. Remote Sensing.* **58**, 1183–1888.

Hastings, D. A. and Tateishi, R. (1998) Land cover classification: Some new techniques, new source data. *Int. Arch. Photogramm. Remote Sensing* **32**, Part 4, 226–229, 1998.

Kineman, J. J. and Ohrenschall, M. A. (1992) *Global Ecosystems Database Version 1.0 Disc A Documentation Manual*. Key to Geophysical Records Documentation No. 27, National Oceanic and Atmospheric Administration, National Geophysical Data Center, Boulder, CO, USA; 240pp.

Kogan, F. N. (1997) Global drought watch from space. *Bull. Am. Meteorol. Soc.,* **78**, 621–636.

— (2001) El Niño implications for land ecosystems, National Oceanic and Atmospheric Administration, Office of Research and Applications, Camp Springs, MD, USA. http://orbit-net.nesdis.noaa.gov/crad/sat/ surf/vci/elnino.html

Legates, D. R. and Willmott, C. J. (1992) Monthly average surface air temperature and precipitation. Digital raster data on a 30 minute cartesian orthonormal geodetic (lat/long) 360–720 grid. In J. Kineman and M. A. Ohrenschall, *Global Ecosystems Database Version 1.0 Disc-A*. NOAA-EPA Global Ecosystems Database Project.

NOAA-EPA Global Ecosystems Database Project. (1992) US Department of Commerce, National Oceanic and Atmospheric Administration, National Geophysical Data Center, Boulder, CO, USA. GED:1A. 640MB on 1 CDROM.

Tucker, C. J. (2002) *Drought Africa*, National Aeronautics and Space Administration, Goddard Space Flight Center Scientific Visualization Studio, Greenbelt, MD, USA. http://svs.gsfc.nasa.gov/stories/drought/africa.html

Wilhite, D. A. (2000) Preparing for drought: a methodology. In D. A. Wilhite (Ed.), *Drought: A Global Assessment*, Volume 2, Chapter 35. London, Routledge.

6   Atmospheric chemistry in the tropics

GUY BRASSEUR[*], ALEX GUENTHER, AND LARRY HOROWITZ[†]

National Center for Atmospheric Research, Boulder, CO, USA

Keywords
Atmospheric chemistry; ozone; tropics

Abstract
This chapter presents an overview of the chemical processes that determine the budget of tropospheric ozone (O_3) and the formation of the hydroxyl radical (OH), with emphasis on their role in the tropics. These chemical species contribute to the oxidizing power of the atmosphere, which determines the lifetime of many chemical pollutants and several greenhouse gases. Tropospheric O_3 and OH are directly affected by chemical species, including water vapour, carbon monoxide, methane, non-methane hydrocarbons and nitrogen oxides. The emissions of these gases by natural and anthropogenic processes are not sufficiently well quantified to provide accurate estimates of the O_3 and OH budgets. Field campaigns such as the EXPRESSO campaign in central Africa have attempted to improve our quantitative understanding of tropical emissions on the continents. Models will be used to assess the role of isoprene emissions by vegetation, carbon monoxide and nitrogen oxides by large-scale tropical fires and production of nitrogen oxides by lightning activity, which is most intense in tropical thunderstorms.

6.1 INTRODUCTION

Many of the human-induced perturbations in the Earth system involve changes in the chemical composition of the atmosphere. For example, the atmospheric concentrations of carbon monoxide (CO), nitrogen oxides (NO_x), and tropospheric ozone (O_3) have increased in response to fossil fuel consumption, mostly at mid-latitudes in the northern hemisphere, and to biomass burning,

mostly in the tropics. Intensive agricultural practices as well as rapid industrial development have increased the atmospheric burden of radiatively active gases, including carbon dioxide (CO_2), methane (CH_4), nitrous oxide (N_2O), and chlorofluorocarbons (CFCs). These species, by trapping infrared radiation in the atmosphere, contribute to the so-called 'greenhouse effect', and hence are believed to generate substantial global warming of the Earth system. In addition, the chlorofluorocarbons, used for various domestic and industrial applications, provide the largest source of inorganic chlorine compounds in the stratosphere and are known to be responsible for ozone depletion (including the formation of the Antarctic ozone hole) in the stratosphere. The enhancement of the aerosol load of the atmosphere associated with coal burning (e.g. sulphate aerosols) and biomass burning (e.g. soot) provide an additional forcing on the climate system. By reflecting a fraction of the incoming solar radiation, sulphate aerosols tend to cool the Earth surface in regions where large amounts of particles are present. Absorbing aerosols like soot tend to warm the planet. An additional climate effect arises from the fact that some aerosol particles (which play the role of cloud condensation nuclei) may modify the optical properties of the clouds, leading to additional climate forcing. This indirect effect, however, is poorly quantified.

The interactions between climate and atmospheric chemistry are complex because they involve interactions with physical processes in the atmosphere and biological processes at the Earth's surface. Some of these feedbacks may be positive (amplification) or negative (damping). Their importance for long-term climate change remains poorly understood and needs to be assessed through comprehensive Earth system models. For example, climate warming could lead to additional convection and hence to enhanced production of nitrogen oxides (NO_x) by lightning flashes. As will be discussed below, NO_x is a key ingredient for the production of tropospheric ozone, and this latter chemical species, which is an efficient greenhouse gas, enhances global warming (positive feedback). Another example of positive climate feedback

[*]Max Planck Institute for Meteorology, Hamburg, Germany
[†]NOAA Geophysical Fluid Dynamics Laboratory, Princeton, NJ, USA

Climate Change and Africa, ed. Pak Sum Low. Published by Cambridge University Press. © Cambridge University Press 2005.

results from the temperature dependence of biogenic emissions. In a warmer climate, the emissions of natural hydrocarbons, such as isoprene by the vegetation, are enhanced. These compounds also contribute to ozone formation and therefore to enhanced greenhouse warming.

The purpose of this chapter is to briefly review the key processes that determine the formation or the destruction of climatically important gases in the atmosphere, and to show the specific and important role played by photochemical processes in the tropics. The focus will be on the mechanisms that determine the oxidizing power of the atmosphere, and specifically the processes that affect the budget of ozone and the hydroxyl radical (OH) in the troposphere. OH can be regarded as a 'detergent' of the atmosphere because it destroys a large number of pollutants and several greenhouse gases. Ozone and water vapour, as well as carbon monoxide, methane, non-methane hydrocarbons and nitrogen oxides, directly affect the level of OH present in the atmosphere. Because the abundance of such compounds has been changing in response to human activities (and will probably continue to do so in the future), the oxidizing power of the atmosphere and hence the atmospheric lifetime of many chemical compounds has probably changed (and may further change in the future) as a result of intense agricultural and industrial activities.

6.2 THE IMPORTANCE OF THE TROPICS

From a 'global change' perspective, the tropics play a very important role. First, the tropical regions cover approximately half of the Earth's surface and their weight is considerable when calculating globally averaged conditions. Second, the tropics intercept a large fraction of the solar energy reaching the Earth, and hence tropical photochemistry is particularly active. Third, the biomass on tropical continents is abundant and the level of biogenic emissions in these regions (i.e. natural non-methane hydrocarbons) is very high. Fourth, the temperature of the surface is high, so that temperature-sensitive emissions (i.e. NO_x by soils, other biogenic emissions) are particularly high. Tropical oceans provide large atmospheric sources of organic compounds (e.g. methyl bromide, methyl chloride, acetone, etc.). Fifth, convective activity is often intense in tropical regions, leading to vigorous exchanges of chemical compounds from the surface to the upper troposphere, and perhaps even to the lower stratosphere. The high lightning activity associated with convective storms is an important source of nitrogen oxides in the tropical free troposphere that influence the regional and even the global budget of atmospheric oxidants. Sixth, the high abundance of water vapour encountered in the tropics contributes to high concentrations of OH. Seventh, population growth and economic development in the tropics will undoubtedly produce large perturbations of the chemical composition in these regions. Intense biomass burning activities in Africa, South America and Asia have already perturbed the atmosphere considerably. The rapid economic growth in regions like Asia and South America has caused additional stress on the atmosphere, and will probably continue to do so in the future.

6.3 CHEMISTRY OF TROPOSPHERIC PHOTOOXIDANTS

Solar radiation, which penetrates into the atmosphere (and which is particularly intense in the tropics), initiates complex photochemical processes that determine the oxidizing power of the atmosphere, and specifically determines the concentrations of ozone and the hydroxyl radical. For example, the photolysis of ozone by solar ultraviolet radiation (at wavelengths less than 320 nm):

$$O_3 + h\nu\,(\lambda < 320\,\text{nm}) \rightarrow O(^1D) + O_2 \tag{6.1}$$

leads to the formation of the electronically excited oxygen atom $O(^1D)$, which can react with water vapor:

$$H_2O + O(^1D) \rightarrow 2OH \tag{6.2}$$

to produce hydroxyl radicals. OH is further converted to the hydrogen peroxy radical (HO_2) by reaction with ozone:

$$OH + O_3 \rightarrow HO_2 + O_2 \tag{6.3}$$

and with carbon monoxide:

$$OH + CO \xrightarrow{O_2} HO_2 + CO_2 \tag{6.4}$$

HO_2 can be converted back to OH by reaction with ozone:

$$HO_2 + O_3 \rightarrow OH + 2O_2 \tag{6.5}$$

and with nitric oxide:

$$HO_2 + NO \rightarrow OH + NO_2 \tag{6.6}$$

Note that the cycle involving reactions (6.3) and (6.5) leads to the destruction of two ozone molecules. Reaction (6.1) followed by reaction (6.2) also leads to an ozone loss. Conversely, the cycle involving reactions (6.4) and (6.6) leads to the formation of an ozone molecule since NO_2 produced by (6.6) is rapidly photolysed back to NO during daytime:

$$NO_2 + h\nu \rightarrow NO + O \tag{6.7}$$

and the oxygen atom rapidly recombines with O_2:

$$O + O_2 + M \rightarrow O_3 + M \tag{6.8}$$

to form an ozone molecule.

Until 1970, it was believed that the presence of ozone in the global troposphere was provided primarily by intrusion of ozone-rich air masses from the stratosphere, and that the only significant destruction of this molecule resulted from its deposition on the surface (e.g. Junge, 1962). The current picture is that, in addition to

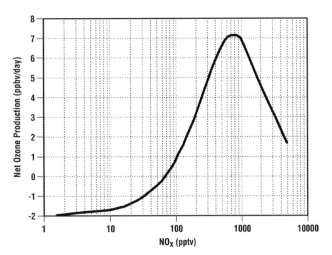

Figure 6.1. Net ozone photochemical production (ppbv/day) represented as a function of NO_x mixing ratio (pptv) for conditions representative of the earth's surface.

stratospheric influxes and surface deposition, the global budget of ozone is considerably affected by photochemical production and destruction mechanisms (Crutzen, 1971, 1973, 1974). The rate-limiting step for ozone production is provided by reaction (6.6) and hence depends on the abundance of NO and of HO_2. Since HO_2 is provided by the oxidation of carbon monoxide (reaction (6.4)), CO can be regarded as a 'fuel' for the ozone production. Other sources of HO_2 or analogous organic peroxy radicals, such as CH_3O_2 or more complex RO_2 radicals, are provided by the oxidation of methane:

$$CH_4 + OH \xrightarrow{O_2} CH_3O_2 + H_2O \tag{6.9}$$

$$CH_3O_2 + NO \xrightarrow{O_2} CH_2O + HO_2 + NO_2 \tag{6.10}$$

$$CH_2O + h\nu \xrightarrow{2O_2} CO + 2HO_2 \tag{6.11}$$

or of other hydrocarbons, such as isoprene (C_5H_8) and other terpenes, which are produced in large quantities by forests and savannahs. Note that reaction (6.10) plays a role similar to reaction (6.6) with the methyl peroxy radical (CH_3O_3) replacing the hydrogen peroxy radical (HO_2).

The role of nitrogen oxides ($NO_x = NO + NO_2$) is to 'catalyse' the formation of ozone. Figure 6.1 shows the net ozone production rate (production minus destruction) calculated as a function of the NO_x concentration. For NO_x mixing ratios less than approximately 60 pptv (regions far away from pollution sources), ozone is subject to weak photochemical destruction, while for NO_x mixing ratios larger than this threshold value (weakly polluted regions), substantial photochemical production takes place. For very high NO_x concentrations (highly polluted regions such as urban areas or close to biomass burning activities), large amounts of HO_x are converted to nitric acid (HNO_3), and the production rate of ozone is significantly reduced.

As described above, the production of O_3 depends on the presence of NO_x and HO_x (OH and HO_2) radicals. These radicals are not consumed by the O_3 production reactions and, once formed, can catalyse the production of many molecules of O_3. Termination of this catalytic cycle occurs when these radicals are deactivated and removed from the atmosphere. Removal of NO_x occurs by conversion to HNO_3, which is relatively inactive chemically in the troposphere and, because of its high solubility, can be efficiently removed from the atmosphere by wet deposition.

Conversion of NO_x to HNO_3 can proceed either by gas-phase oxidation of NO_2 by OH:

$$NO_2 + OH + M \rightarrow HNO_3 + M \tag{6.12}$$

or by the hydrolysis of N_2O_5 on sulphate aerosols (occurring primarily at night):

$$NO_2 + O_3 \rightarrow NO_3 + O_2 \tag{6.13}$$

$$NO_2 + NO_3 + M \leftrightarrow N_2O_5 + M \tag{6.14}$$

$$N_2O_5(g) + H_2O(l) \rightarrow 2HNO_3 \tag{6.15}$$

Reaction (6.12) also acts to remove one molecule of HO_x from the atmosphere, if HNO_3 is scavenged by precipitation. Other removal pathways for HO_x include the self-reaction of HO_2:

$$HO_2 + HO_2 \rightarrow H_2O_2 + O_2 \tag{6.16}$$

and reactions of HO_2 with OH and organic peroxy radicals:

$$HO_2 + OH \rightarrow H_2O + O_2 \tag{6.17}$$

$$HO_2 + RO_2 \rightarrow ROOH + O_2 \tag{6.18}$$

The hydrogen peroxide (H_2O_2) produced by reaction (6.16) is quite soluble, and can be removed from the atmosphere by wet deposition, or by reaction with OH:

$$H_2O_2 + OH \rightarrow H_2O + HO_2 \tag{6.19}$$

both resulting in a permanent loss of 2 HO_x radicals.

In summary, the budget of ozone and the photochemical mechanisms affecting OH and HO_2 radicals in the troposphere are directly controlled by the presence of water vapour, nitrogen oxides and carbon-containing species (such as CO, CH_4 and non-methane hydrocarbons) in the atmosphere. The spatial distribution and seasonal evolution of these compounds is determined by their source and sink processes. Quantitative estimates of these processes are given in Tables 6.1, 6.2 and 6.3, in the case of nitrogen oxides, carbon monoxide, and methane, respectively. The data in these tables suggest that the role of the tropics in the global budget of these compounds is particularly important. For example, emissions of methane by rice paddies, of NO by soils, and of NO and CO by biomass burning are primarily tropical sources. The destruction of CH_4 and CO by the hydroxyl radical takes place also to a large extent at low latitudes. Note that, within the uncertainties on the estimated sources and sinks, the global budgets can be balanced;

Table 6.1. Global budget of NO_x in the troposphere.

	Tg N yr^{-1}
Sources	
Fossil fuel combustion	20 (14–28)
Biomass burning	12 (4–24)
Release from soils	20 (4–40)
Lightning discharges	5 (2–20)
NH_3 oxidation	3 (0–10)
Ocean surface	<1
Aircraft	0.5
Injection from the stratosphere	0.1 (0.6 total NO_y)
Total sources	64 (25–112)
Sinks	
Wet deposition of NO_3^- (land)	19 (8–30)
Wet deposition of NO_3^- (ocean)	8 (4–12)
Dry deposition of NO_y	16 (12–22)
Total sinks	43 (24–64)

Based on Logan, 1983; Davidson, 1991; IPCC, 1994.

Table 6.2. Global budget for carbon monoxide (Tg CO yr^{-1}).

	Magnitude
Sources	
Biomass burning	300–900
Fossil fuel burning	300–600
Vegetation	50–200
Oceans	6–30
Methane oxidation	400–1,000
NMHC oxidation	300–1,000
Total	1,400–3,700
Sinks	
Chemical loss (OH)	1,400–2,600
Uptake by soils	150–500
Total	1,550–3,100

Based on Khalil and Rasmussen, 1990; Bates *et al.*, 1995.

Table 6.3. Estimated sources and sinks of methane in the atmosphere (Tg CH_4 yr^{-1}).

Sources or sinks	Range	Likely
Natural		
Wetlands		
Tropics	30–80	65
Northern latitude	20–60	40
Others	5–15	10
Termites	10–50	20
Ocean	5–50	10
Freshwater	1–25	5
Geological	5–15	10
Total		160
Anthropogenic		
Fossil fuel related		
Coal mines	15–45	30
Natural gas	25–50	40
Petroleum industry	5–30	15
Coal combustion	5–30	15
Waste management system		
Landfills	20–70	40
Animal waste	20–30	25
Domestic waste treatment	15–80	25
Enteric fermentation	65–100	85
Biomass burning	20–80	40
Rice paddies	20–100	60
Total		375
Total sources		535
Sinks		
Reaction with OH	405–575	490
Removal in stratosphere	32–48	40
Removal by soils	15–45	30
Total sinks		560
Atmospheric increase	35–40	37

From IPCC, 1994, 1996.

in the case of methane, however, the observed change in the atmospheric abundance of this gas must result from an imbalance between sources and sinks.

6.4 AN ASSESSMENT OF THE EMISSIONS OF OZONE PRECURSORS IN TROPICAL REGIONS

The *Dynamique et Chimie Atmosphérique es Forêt Equatoriale* (DECAFE) studies in the 1980s established that Central Africa exhibits some of the most dynamic, yet poorly understood biosphere–atmosphere interactions on earth (Lacaux *et al.*, 1995 and references therein) and contributes substantially to the formation of tropospheric oxidants. The Experiment for Regional Sources and Sinks of Oxidants (EXPRESSO) was an international and multi-disciplinary investigation of the processes controlling the chemical composition of the tropical troposphere above this region. The EXPRESSO study focused on the region encompassing the savannahs of the Central African Republic (CAR) in the north (8° N) to the tropical forests of the Republic of Congo (2° N) in the south.

The overall goal of this research programme was to better understand the impact of this region on the global atmosphere. Field measurements associated with EXPRESSO, including aircraft and above-canopy tower sampling, were conducted in 1996 and are described by Delmas *et al.* (1999) and references therein.

6.4.1 Characterizing emission sources

The two major surface emission sources in the central African region are biomass burning, which emits a wide variety of compounds, and vegetation foliage, which is a large source of volatile organic compounds (VOCs). Both of these sources occur under natural conditions but are strongly influenced by human activities. Relatively little of the biomass burning in central Africa is associated with wildfires and so it is not considered a natural source. Although biogenic VOC emissions are typically considered to be a natural source, emissions are strongly dependent on the type and amount of vegetation within a landscape, which can be dramatically changed by human activities (e.g. deforestation).

Estimates of biomass burning emissions can be based on measurements of active fires, area burned, vegetation type, and fuel moisture content. All of these factors can be monitored using remote sensing technology. Extensive efforts were made during EXPRESSO to develop and evaluate methods for satellite monitoring of vegetation fires in central Africa (Grégoire *et al.*, 1999 and references therein). Daily fire maps were produced for the entire EXPRESSO domain throughout 1996. The results demonstrate that this region is one of the largest fire belts in the world. A large increase in fires was observed for 1996, relative to 1993. The large day-to-day variations observed by this study demonstrate the importance of using satellite monitoring to develop emission inputs for chemistry and transport models, rather than climatological data.

The EXPRESSO study included a number of experiments designed to improve our ability to predict biogenic VOC emissions from central Africa. A variety of flux measurement techniques were used to characterize fluxes on small (enclosure systems that isolate individual leaves) to large (relaxed eddy accumulation systems that integrate above-canopy fluxes across hundreds of metres to tens of kilometres) scales. The results of these experiments were incorporated into an isoprene emission model that predicts hourly emissions on a spatial scale of 1 km^2 (Guenther *et al.*, 1999 and references therein). The model uses procedures that are suitable for estimating global emissions but uses regional measurements to parameterize the model. The annual central African isoprene emission predicted by this model (35 Tg C) is only 14% less than that predicted by an earlier model, but emissions for a specific location and time can differ by a factor of five or more.

6.4.2 Interactions between sources

A key question addressed by the EXPRESSO study was whether or not there is any interaction between the trace gases emitted by the two emission sources (biomass burning and vegetation foliage). In particular, efforts were made to determine if the nitrogen oxides emitted by biomass burning can interact with the VOC emitted

by vegetation. Earlier models had suggested that biomass burning occurs primarily in the savannah (in the north), while vegetation emission occurs primarily in the forest (in the south). Significant meteorological features of this region include three permanent anticyclones with air masses that converge along the Intertropical Convergence Zone (ITCZ) and the Intertropical Oceanic Confluence. The result is a potential barrier to mixing of the trace gases emitted from the savannah with those emitted from the forest (in the south).

The EXPRESSO observations indicate that there is likely to be significant interaction between the two emission sources for two reasons. First, the vegetation VOC emission rates for at least some savannah locations were found to be as high as that observed in the forest (Guenther *et al.*, 1999). Second, there appears to be significant exchanges between the air masses above the savannah and forest surfaces (Delmas *et al.*, 1999). This includes horizontal exchanges, induced by diurnal temperature gradients at the savannah-forest interface, and vertical exchanges associated with a variety of processes.

6.4.3 Impact of emissions

Biomass burning and biogenic VOC emission from central Africa provide elevated trace gas levels on regional scales (Delmas *et al.*, 1999). For example, surface CO concentrations are in the range of 250 to 500 ppb, while the surface concentration of odd nitrogen species ($NO_y = NO + NO_2 + HNO_3$) ranges from 4 to 10 ppb. These values are characteristic of heavily polluted regions. The influence of biomass burning can be seen throughout the entire region. The primary impact of biogenic VOC emissions may be to reduce oxidants in the boundary layer. On a global scale, however, the products of VOC oxidation in the boundary layer of this region may serve to increase oxidants. Chemical impacts of tropical sources can be estimated from model studies.

6.5 SENSITIVITY OF TROPICAL EMISSIONS OF NO$_x$, CO AND ISOPRENE ON TROPOSPHERIC OZONE: A MODEL STUDY

6.5.1 Model description

In order to quantify the importance of tropical sources (including isoprene by vegetation, nitrogen oxides by lightning, and carbon monoxide and other chemical compounds by biomass burning) on the formation of tropospheric oxidants, we use a three-dimensional chemical-transport model, Intermediate Model for the Annual and Global Evolution of Chemical Species (IMAGES), developed by Müller and Brasseur (1995). This model provides the global distribution of approximately 60 chemical compounds, including ozone, HO$_x$, NO$_x$, sulphur oxides, methane and several non-methane hydrocarbons. The spatial resolution is 5° in longitude and latitude, with 25 terrain-following σ levels in the vertical between the surface and the lower stratosphere (50 mb). The model accounts for approximately 150 chemical and photochemical reactions. Large-scale advective transport is driven by monthly mean winds derived from the meteorological analyses by European Centre for Medium-Range Weather Forecast (ECMWF). Wind variability at timescales shorter than one month is parameterized as a mixing process, with diffusion coefficients derived from ECMWF wind variances. The transport equation is solved using the semi-Lagrangian algorithm of Smolarciewicz and Rasch (1991). Mass exchanges at subgrid scales are parameterized. Vertical mixing in the boundary layer is represented by diffusion, while convective transport is expressed following the scheme of Costen *et al.* (1988). The distribution of cloud updrafts is represented following the distribution of cumulonimbus provided by the International Satellite Cloud Climatology Project (ISCCP) (D2 climatology). Wet scavenging of soluble species and dry deposition on the surface are also parameterized. Trace gas emissions are based on the inventories established by Müller (1992). The lightning source of NO is distributed according to Müller and Brasseur (1995). The results provided by the IMAGES model have been evaluated by Müller and Brasseur (1995, 1999) and by Friedl (1997) through an extensive comparison with observational data.

6.5.2 Impact of isoprene and terpene emissions

The release of isoprene and other biogenic compounds provides the 'fuel' needed to photochemically produce ozone when sufficiently high levels of NO are present in the atmosphere. The chemical degradation of these biogenic hydrocarbons also leads to the formation of carbon monoxide. In the model, we use the emission of isoprene by various ecosystems, as described by Guenther *et al.* (1995), with a total release of 500 Tg/yr. Large concentrations of isoprene are predicted by the IMAGES model in the boundary layer over tropical forests (in Africa and South America) and, during summertime, over the forested regions of the northern hemisphere (e.g. south-eastern United States). The contribution of isoprene and terpene emissions to the zonally averaged concentration of carbon monoxide (CO) is represented in Figure 6.2.

When the isoprene and terpene sources are taken into account, zonal mean CO is enhanced by as much as 20% in the tropics and 15–18% in the southern hemisphere. Changes in the northern hemisphere are smaller (6–9%) because, in these regions, the abundance of CO depends primarily on anthropogenic (fossil fuel) sources. It is interesting to note that, in spite of the short lifetime of biogenic hydrocarbons and specifically of isoprene (less than 1 day), the impact of this compound on CO extends to high altitudes (10–15 km), especially in the tropics where convective transport of CO and of intermediate organic compounds formed during the

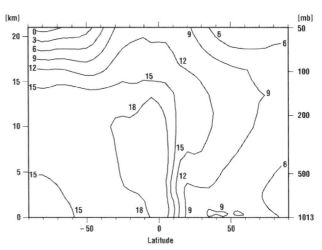

Figure 6.2. Contribution (per cent) of isoprene emissions (total of 500 Tg/yr) on the zonally averaged concentration of carbon monoxide in July.

decomposition of isoprene is vigorous. The impact of isoprene and terpene emissions on ozone (during July in the boundary layer) is shown in Figure 6.3

As expected, the ozone increase associated with these emissions is most pronounced over the continents, and particularly over the tropical forest of South America, Africa and northern Australia (25–30% increase) and, during summertime, in the south-eastern United States and in southern China (20–28% increase). Changes in zonally averaged ozone concentrations near the surface are typically 10–12% in the tropics and northern mid-latitudes, and 3–8% in the southern hemisphere. They decrease with altitude to reach 4% at 7 km in the northern mid-latitudes and at 13 km in the tropics (see Figure 6.3b).

6.5.3 Impact of biomass burning

Biomass burning is conducted extensively in the tropical savannah during the dry season and occurs also in relation with tropical deforestation. At mid and high latitudes, large areas are also burned during summertime; these fires are either accidental or ignited by lightning. The release of NO_x, CO, and other chemical compounds by these fires is large and impacts the global distribution of ozone precursors. We assume in the present model study that biomass burning provides a global NO_x emission of 11 Tg N/yr.

The change in the zonally averaged CO concentration produced by biomass burning during October is shown in Figure 6.4. CO concentrations increase by about 70% in the vicinity of the tropical fires near the surface. CO is transported upward by convective cells, so that enhancements of 40–50% are predicted up to 14 km altitude in the tropics. In the southern hemisphere and at northern mid and high latitudes, biomass burning causes CO to increase by 30% up to 10 km altitude. Perturbations in the concentration of NO_x in the planetary boundary layer are largest in the vicinity

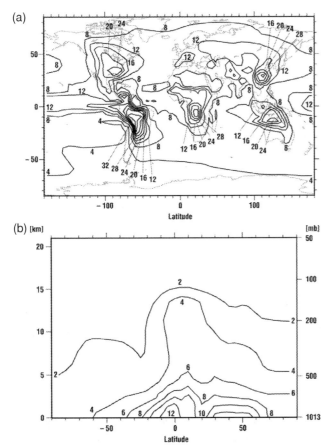

(a)

(b)

Figure 6.3. Relative impact (per cent) of isoprene emissions (total of 500 Tg/yr) on the concentration of ozone in July: impact on (a) boundary layer ozone ($\sigma = 0.9$), (b) zonally averaged ozone.

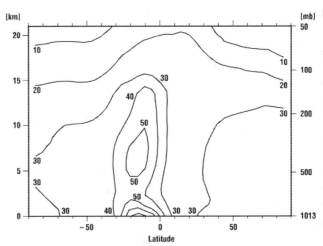

Figure 6.4. Relative impact (per cent) of biomass burning emissions on the zonally averaged concentration of carbon monoxide in October.

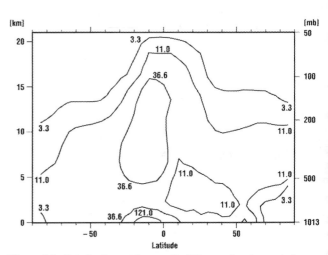

Figure 6.5. Relative impact (per cent) of biomass burning emissions on the concentrations of $NO_x(NO + NO_2)$ in October: impact on the zonally averaged concentrations.

of the sources (more than 200% near the tropical fires, and more than 80% near the boreal fires of Canada), but their magnitude decreases rapidly away from the burning areas since the lifetime of NO_x is only of the order of a day near the Earth's surface. Nitrogen oxides can, however, be transported as peroxyacetyl nitrate (PAN). PAN is a thermally unstable reservoir of species, capable of transporting NO_x over long distances in the cold upper troposphere, and then subsiding and decomposing to release NO_x. Zonally averaged concentrations of NO_x (Figure 6.5), are enhanced by 40–100% in the tropical upper troposphere, and by 10–35% between 5 and 10 km at mid-latitudes in the northern hemisphere. Similar values are derived in the southern hemisphere.

The changes in boundary layer ozone during October due to biomass burning are depicted in Figure 6.6a.

As expected, the largest ozone increases are found over tropical continents (150% over South America and Africa, and 100% over Northern Australia). Increases of about 10% are predicted at northern high latitudes during this period of the year, but are even somewhat higher during August and September when fires are most frequent in the boreal regions. The change in the zonally averaged ozone concentration (Figure 6.6b) is typically 30% in

the tropical free troposphere and 6–10% at mid-latitudes in both hemispheres.

6.5.4 Impact of lightning

Finally, we investigate the changes in the chemical composition of the troposphere in response to lightning. The global amount of NO produced by these electrical discharges has been estimated to be in the range 2–20 Tg N/yr. In the present model, we adopt a working value of 5 Tg N/yr, but as we calculate the impact of this source on ozone, one should be aware of the large uncertainty associated with this value.

The zonally averaged change in the NO_x concentration calculated (see Figure 6.7) when lightning sources are taken into account is larger than 30% in the tropics and reaches even more

(a)

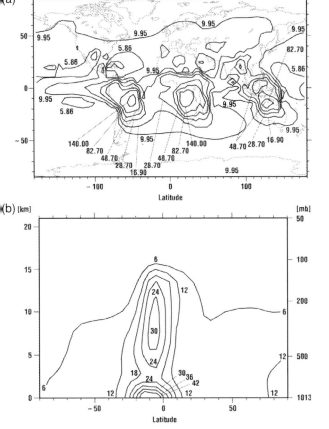

(b)

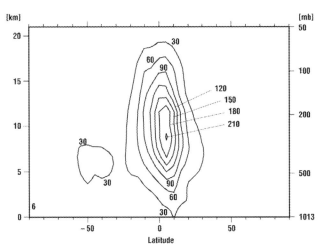

Figure 6.7. Relative impact (per cent) of NO$_x$ lightning sources (total production of 5 Tg N/yr) on the zonally averaged concentration of NO$_x$ (NO + NO$_2$) in July.

Figure 6.6. Relative impact (per cent) of biomass burning emissions on the concentration of ozone in October: impact on (a) boundary layer ozone, (b) zonally averaged ozone.

than 150% near the ITCZ where convection and the associated electrical activity are intense.

The corresponding variation in zonal mean ozone is shown in Figure 6.8, with changes reaching 35% near the equator between 4 and 8 km altitude. The magnitude of the ozone enhancement is reduced away from the equator; at mid-latitudes it reaches 10–15% in the southern hemisphere and 5–10% in the northern hemisphere during July. OH concentrations increase by typically a factor of 2 in the tropics (not shown) and the concentration of CO is reduced by 15–20% in the southern hemisphere and 5–15% in the northern hemisphere in July.

6.6 SUMMARY

Emissions of trace constituents, such as nitrogen oxides, carbon monoxide, methane and non-methane hydrocarbons, have a large impact on the formation of ozone and other oxidants in the tropical atmosphere. The magnitude of these emissions remains poorly quantified. Field campaigns such as the EXPRESSO project in the Central African Republic and in the Republic of Congo have helped quantify some of these sources and their relations with physical parameters such as temperature, light intensity, soil mois-

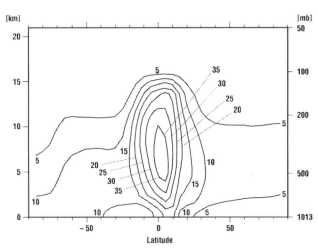

Figure 6.8. Relative impact (per cent) of NO$_x$ lightning sources (total production of 5 Tg N/yr) on the zonally averaged concentration of ozone in July.

ture, etc. Models such as IMAGES provide estimates of the regional and global impacts of these emissions in the boundary layer and in the free troposphere.

The model highlights the importance of biomass burning and lightning activity for ozone formation near the surface and in the free troposphere, particularly in the tropics. These sources also affect the concentration of OH and therefore the oxidizing power of the Earth's atmosphere (including the lifetime of some greenhouse gases). To illustrate the effect of ozone precursor sources on the oxidizing power (global OH concentration), Table 6.4 provides estimates of the global lifetime of methane and methyl chloroform for different conditions. These two compounds are destroyed primarily by the OH radical in the atmosphere.

Future estimates of climate change will have to account not only for the radiative impact of increasing concentrations of

Table 6.4. Global lifetime (years) of methane and methyl chloroform in the atmosphere.

Model case	Methane CH_4	Methyl chloroform CH_3CCl_3
Standard case	8.29	4.15
No biogenic hydrocarbons	8.21	4.12
No biomass burning	8.68	4.30
No lightning	9.91	4.85

CO_2 (and other long-lived greenhouse gases), but also for the impact of ozone and particles produced by large tropical and extra-tropical fires. As industrialization progresses in tropical regions, emissions of ozone precursors from fossil fuel combustion will become increasingly important as well. Comprehensive models will have to account for feedbacks between climate change and the release of ozone precursors (intensity of thunderstorms, frequency of fires, changes in soil and vegetation emissions, etc.). They will also have to consider inter-annual variations in surface emissions (associated with climate variability and episodic events such as El Niño).

ACKNOWLEDGEMENTS

The authors wish to thank J.-F. Müller and C. Granier for their contributions in the development of the IMAGES model. They are also grateful to H. Rhode and M. Lawrence for their review of the manuscript. The National Centre for Atmospheric Research is operated by the University Corporation for Atmospheric Research under sponsorship of the National Science Foundation.

REFERENCES

Bates, T. S., Kelly, K. C., Johnson, J. E. and Gammon, R. H. (1995) Regional and seasonal variations in the flux of oceanic carbon monoxide to the atmosphere. *J. Geophys. Res.*, **100**, 23,093–23,101.

Costen, R. C., Tennille, G. M. and Levine, J. S. (1988) Cloud pumping in a one-dimensional photochemical model. *J. Geophys. Res.*, **93**, 15,941–15,954.

Crutzen, P. J. (1971) Ozone production rates in an oxygen-hydrogen-nitrogen oxide atmosphere. *J. Geophys. Res.*, **76**, 7,311–7,327.

    (1973) A discussion of the chemistry of some minor constituents in the stratosphere and troposphere. *Pure Appl. Geophys.*, **106–108**, 1,385–1,399.

    (1974) Photochemical reactions initiated by and influencing ozone in unpolluted tropospheric air. *Tellus*, **XXVI**, 47–57.

Davidson, E. A. (1991) Fluxes of nitrous oxide and nitric oxide from terrestrial ecosystems. In J. E. Rogers and W. B. Whitman (Eds.), *Microbial Production and Consumption of Greenhouse Gases: Methane, Nitrogen Oxides, and Halomethanes,* pp. 219–235. Washington, DC, American Society for Microbiology.

Delmas, R., Druilhet, A., Cros, B. *et al.* (1999) Experiment for Regional Sources and Sinks of Oxidants (EXPRESSO): An overview. *J. Geophys. Res.*, **104**, 30,609–30,624.

Friedl, R. E. (Ed.) (1997) *Atmospheric Effects of Subsonic Aircraft: Interim Assessment Report of the Advanced Subsonic Technology Programme.* NASA Ref. Publ. 1400, 168 pp.

Grégoire, J.-M., Pinnock, S., and Dwyer, E. (1999) Satellite monitoring of vegetation fires for EXPRESSO: outline of activity and relative importance of the study area in the global picture of biomass burning. *J. Geophys. Res.*, **104**, 30,691–39,700.

Guenther, A., Hewitt, C. N., Erickson, D. *et al.* (1995) A global model of natural volatile organic compound emissions, *J. Geophys. Res.*, **100**, 8,873–8,892.

Guenther, A., Baugh, B., Brasseur, G. *et al.* (1999) Isoprene emission estimates and uncertainties for the Central African EXPRESSO study domain, *J. Geophys. Res.*, **104**, 30,625–30,639.

IPCC (1994) *Climate Change 1994: Radiative Forcing of Climate Change and an Evaluation of the IPCC IS 92 Emission Scenarios.* J. T. Houghton L. G. Meira Filho, J. Bruce. *et al.* (Eds.), Intergovernmental Panel on Climate Change, Cambridge, Cambridge University Press.

    (1996) *Climate Change 1995: The Science of Climate Change.* J. T. Houghton L. G. Meira Filho B. A. Callander, N. Harris, A. Kattenberg and K. Maskell (Eds.), Intergovernmental Panel on Climate Change, Cambridge, Cambridge University Press.

Junge, C. E. (1962) Global ozone budget and exchange between stratosphere and troposphere. *Tellus*, **XIV**, 363–377.

Khalil, M. A. K. and Rasmussen, R. A. (1990) The global cycle of carbon monoxide: Trends and mass balance. *Chemistry*, **20**, 227–242.

Lacaux, J., Cachier, H. and Delmas, R. (1995) Biomass burning in Africa: an overview of its impact on atmospheric chemistry. In P. Crutzen and J. Goldammer (Eds.), *Fire in the Environment: The Ecological and Climatic Influence of Vegetation Fires.* New York, J. Wiley and Sons.

Logan, J. A. (1983) Nitrogen oxides in the troposphere: global and regional budgets, *J. Geophys. Res.*, **88**, 10,785–10,807.

Müller, J.-F. (1992) Geographical distribution and seasonal variation of surface emissions and deposition velocities of atmospheric trace gases, *J. Geophys. Res.*, **97**, 3,787–3,804.

Müller, J.-F. and Brasseur, G. (1995) IMAGES: a three-dimensional chemical transport model of the global troposphere, *J. Geophys. Res.*, **100**, 16,445–16,490.

    (1999) Sources of upper tropospheric HO_x: a three-dimensional study, *J. Geophys. Res.*, **104**, 1,705–1,715.

Smolarkiewicz, P. K. and Rasch, P. J. (1991) Monotone advection on the sphere: an Eulerian versus semi-Lagrangian approach, *J. Atmos. Sci.*, **48**, 793–810.

7 Natural and human-induced biomass burning in Africa: an important source for volatile organic compounds in the troposphere

RALF KOPPMANN[1], KRISTIN VON CZAPIEWSKI[2,*], AND MICHAEL KOMENDA[1,**]

[1]Forschungszentrum Jülich, Germany
[2]York University, Toronto, Canada

Keywords

Biomass burning; volatile organic compounds; emissions; emission factor; combustion efficiency; emission ratio; ozone

Abstract

Biomass burning is the burning of living and dead vegetation. 90% of all biomass-burning events are thought to be initiated by human activities. Human-induced fires are used for land management, such as shifting cultivation, agricultural expansion, deforestation, bush control, weed and residue burning, harvesting practices, and forest and grassland management. A considerable amount of biomass is burnt worldwide in household fires for cooking and heating purposes. 75% of the world's population is using wood as the main energy source. Furthermore, accidental fires, which are related to population density, political situation and poor management practices, contribute to global biomass burning. Natural fires are grassland and forest fires mainly induced by lightning.

It is estimated that 8,700 teragram (Tg) of dry matter are burnt globally per year. 66% of the biomass is burnt in savannah and agricultural fires, while biofuel and tropical forests contribute about 15% each to the total biomass burnt globally. 42% of all fires occur in Africa, 26% in South America, and 32% are distributed over the rest of the world.

Biomass burning is an important source of greenhouse gases, such as carbon dioxide (CO_2), methane (CH_4), and nitrous oxide (N_2O). Beyond that, considerable amounts of chemically active gases such as nitrogen oxide (NO), carbon monoxide (CO) and volatile organic compounds (VOC) are released into the atmosphere, especially in tropical and subtropical regions. From correlations of the emissions of VOC with those of CO for different types of fires, it can be estimated that the total carbon released in form of VOC (CH_4 not included) ranges between 25 Tg C/yr and 48 Tg C/yr. The oxidation of VOC in the presence of nitrogen oxide leads to the formation of ozone, which also acts as a greenhouse gas and has an impact on the radiative forcing of the troposphere.

7.1 INTRODUCTION

Figure 7.1 shows the types of biomass burnt based on an estimate of 8,700 Tg dry matter per year (Andreae, 1991). 66% of the biomass is burnt in savannah and agricultural fires, while biofuel and tropical forests contribute about 15% each to the total biomass burnt globally. The contribution of boreal forests is low compared to other areas, however, the biomass burnt in boreal forests increased by a factor of 10 during the last 20 years (Levine, 1996). The geographical distribution of all fire events is given in Figure 7.2. 42% of all fires occur in Africa, 26% in South America, and 32% are distributed over the rest of the world.

Savannah fires are the single largest source of biomass burning emissions worldwide (Crutzen and Andreae, 1990; Andreae, 1991; Andreae 1993a; Hao and Liu, 1994). The occurrence of savannah fires is controlled by the seasonal cycle of the wet season, during which biomass is produced, and the dry season, during which the biomass is turned into highly flammable material. Although in most cases fires are human-induced, natural savannah fires are also common, when lightning strikes the dry vegetation.

Africa contains about two thirds of the world's savannah regions, African savannah fires alone account for 30% of the biomass burning emissions in the tropics worldwide (Hao and Liu, 1994). 90% of African savannah fires are believed to be human-induced. Agricultural practices include an annual burning of savannah vegetation. The reasons are clearing of dead or unwanted vegetation, fertilization, hunting practices, improvement of access for collection of food and other vegetation products (Andreae *et al.*, 1996). As a result of these practices, Andreae (1993a) estimated that 820 million ha of savannah area is burnt annually, resulting in the

* Current address: Applied Biosystems, LC-MS Support, Darmstadt, Germany
** Current address: Schwarz BioSciences, Pharmaceutical Development Group, Monheim, Germany

Climate Change and Africa, ed. Pak Sum Low. Published by Cambridge University Press. © Cambridge University Press 2005.

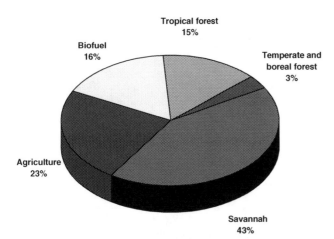

Figure 7.1. Types of biomass burnt (after Andreae, 1991).

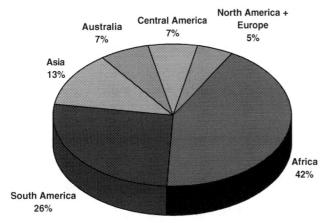

Figure 7.2. Geographical distribution of biomass burning (after Andreae, 1993b, Hao and Liu, 1994).

combustion of about 3,400–3,700 Tg dry matter of biomass per year. Hao *et al.* (1996) estimated that the biomass burnt annually in Africa is about 2000 Tg dry matter, and the area exposed to fires in African savannahs covers about 440 million ha.

In spite of the obvious importance of emissions from savannah fires to the troposphere, there are still only a relatively small number of studies that have investigated the emissions from savannah fires.

The first measurements on biomass burning emissions in Africa were made in West Africa by Delmas (1982). He studied the regional atmospheric chemistry over the Congo forest. A number of following studies indicated that biomass burning in sub-Sahelian savannahs have far-reaching impact on the atmospheric environment (Cros *et al.*, 1991; Helas *et al.*, 1992; Lacaux *et al.*, 1993; Swap *et al.*, 1996).

In the last two decades a growing number of studies were published that deal with field measurements of emissions from biomass burning in Africa (Delmas, 1982; Bonsang *et al.*, 1991; Rudolph *et al.*, 1995; Lacaux *et al.* 1995; Koppmann *et al.*, 1997; Andreae, 1997). The results of various field experiments in Africa

showed that during the biomass burning season, the concentrations of CO and organic trace gases in the planetary boundary layer were considerably enhanced compared with those in the wet season (cf. Rudolph *et al.*, 1992).

Meanwhile, laboratory experiments have provided more information about the parameters that determine the emission factors, such as different burning stages and fuel types (Lobert *et al.*, 1991; Czapiewski, 1999). It has been shown that the burning of organic material in oxygen-deficient fires leads to the emission of methane, non-methane hydrocarbons and a variety of partially oxidized organic compounds. In some cases the emissions of medium and higher molecular weight organic compounds exceed those of light non-methane hydrocarbons. Thus, these emissions contribute significantly to the local budgets of organic trace gases in the regions of biomass burning and may have a considerable impact on the formation of ozone in these areas. Very recent studies showed the significant abundance of higher molecular weight VOC in biomass burning plumes (Czapiewski, 1999).

While savannah and forest fires have been studied intensively, little is known about the contribution of small household fires to global biomass burning emissions. Based on data from the 1980s, 15% of the global energy requirements are met by the burning of biofuel (Hall, 1991). 75% of the world's population use wood as the main source of energy (Marufu *et al.*, 1997). In Africa, the use of fuelwood and charcoal supplies 90% of the energy in private households and 60% of the total energy consumption (Brocard and Lacaux, 1998). As reported by Andreae (1991), the per capita consumption of fuelwood in Africa is about 475 kg per year. Recent estimates showed that this value may be significantly underestimated since official statistics only account for sales data of fuelwood and do not consider collected biofuel. Marufu *et al.* (1997) reported a typical annual consumption of fuelwood of the order of 950 kg per capita and a consumption of agricultural waste of the order of 130 kg per capita for rural areas in Zimbabwe. With increasing population in Africa the consumption of biofuel will also increase considerably. This may have, on the other hand, an impact on the availability of biofuel in wildfires. As a consequence, a change in the combustion characteristics can be expected in the future.

7.2 ORGANIC COMPOSITION OF BIOMASS BURNING EMISSIONS

Vegetation material is made up of cellulose, hemicelluloses, lignin, proteins, nucleic acids, amino acids and volatile substances (Franke, 1989). During the first phase of a fire the volatile compounds such as aldehydes, alcohols and terpenes are released. In the following phase thermal cracking of the fuel molecules occurs. Higher molecular weight compounds are decomposed into a variety of volatile organic compounds (Lobert and Warnatz, 1993). In previous field studies only the light hydrocarbons (up to four carbon atoms) were investigated. However, a variety of higher molecular weight organic trace gases are also emitted from biomass burning. They include alcohols, aldehydes, ketones, carboxylic acids, esters, ethers, furanes, etc. These partly oxidized compounds are predominantly emitted during the smouldering phase of fires (Czapiewski, 1999).

In order to properly characterize the emissions of trace gases from biomass burning, some parameters must be defined. The *emission ratios*, ER, of a trace gas are typically normalized to a reference compound such as CO or more often CO_2. The emission ratio of a compound X normalized to CO_2, for example, is then given by

$$ER = \frac{\Delta[X]}{\Delta[CO_2]} = \frac{[X]_{Plume} - [X]_{Background}}{[CO_2]_{Plume} - [CO_2]_{Background}}$$

where $[]_{Plume}$ is the measured concentration in the plume and $[]_{Background}$ the measured background concentration of the corresponding compound.

Combustion efficiency, CE, is also commonly used. Ward and Hardy (1991) defined the combustion efficiency as the ratio of carbon emitted as CO_2 to the total amount of carbon emitted:

$$CE \equiv \frac{\Delta[CO_2]}{\Delta[CO_1] + \Delta[CO] + \Delta[CH_4] + \Delta[VOC]}$$

where Δ denotes the difference between the concentration inside the plume and background concentration. By definition, the combustion efficiency is the fraction of fuel carbon emitted by the fire that is completely oxidized to CO_2.

A third parameter, which is widely used, is the *emission factor*, EF. The emission factor is expressed in units of mass of a compound emitted per unit mass of carbon burnt (typically expressed in grams per kilogram carbon burnt, g/kgC).

$$EF = \frac{\text{Mass of emitted compound}\,[g]}{\text{Mass of burnt carbon}\,[kg]}$$

To convert this emission factor to units of grams of a compound per kilogram fuel burnt, this value has to be multiplied by the mass fraction of carbon in the fuel (typically, the carbon content of woody fuels is of the order of 50%).

CO_2 is the dominant carbon species emitted from all fires followed by CO. The emission ratio of CO relative to CO_2, $\Delta CO/\Delta CO_2$, is used to characterize the different burning stages. During the flaming phase this ratio is relatively low, in general 5–10%. During the smouldering phase of a fire more CO is emitted and the ratio of $\Delta CO/\Delta CO_2$ increases. Typical values observed during this phase vary between 10 and 15%. Also, the emissions of organic compounds vary as a function of the burning stage. The correlation of an organic trace gas with CO_2 and CO allows the estimation of the emission ratios of that trace gas as a function of the type of fire and the different burning processes, as well as the burning stages.

A list of identified volatile organic compounds is given by Bonsang *et al.* (1991), and an overview of the most important higher molecular weight VOC and their emission ratios relative to CO_2 for the different types of fires are given by Koppmann *et al.* (1997) and Czapiewski (1999). In general, the mixing ratios decrease with increasing number of carbon atoms within any group of homologues. This was also observed in laboratory studies carried out with tropical wood, which showed very similar emission patterns (Czapiewski, 1999). From these data, emission ratios are calculated on a parts per billion (ppb) carbon basis, and the contribution of these compounds to the total emission of organic trace gases and relative to CO and CO_2 can be derived.

As an example, results of VOC measurements in biomass burning plumes are shown here for large savannah fires in Africa investigated during the FOS/DECAFE (Fire of Savannahs/*Dynamique et Chimie Atmosphérique es Forêt Equatoriale*) in the Ivory Coast and SAFARI (Southern African Fire-Atmosphere Research Initiative) campaigns. The latter campaign offered the opportunity to investigate different types of fires (an uncontrolled wildfire, a controlled fire in the Kruger National Park, and two sugar cane fires). Table 7.1 summarizes the emission ratios observed in these field experiments for CO, CH_4, and the sum of organic trace gases relative to CO_2, so as to enable a comparison with typical emission ratios reported in the literature. As mentioned above, the emission ratio of CO relative to CO_2 allows the characterization of the burning conditions of the different fires.

These data show that agricultural fires and large savannah fires like that in the Kruger National Park were predominantly flaming fires with relatively low CO emission ratios. This is in agreement with the results from those samples of the FOS/DECAFE campaign, which were taken during the flaming phases of the fires. In contrast, the samples taken during the smouldering phases show high CO emission ratios coinciding with high CH_4 and VOC emission ratios. For the wildfire encountered during SAFARI, smouldering combustion was likely to be the dominant process. For this fire the highest emission ratios of CH_4 and VOC were observed, indicating that the biofuel was extremely moist at the time of the fire. All measurements show that the emission ratios of VOC exceed those of methane, sometimes by a factor of 2. Our results obtained during SAFARI show further that the emission ratios of organic compounds with a carbon number of 5 or more added up to at least half of the contribution of light organic compounds to the total amount of carbon emitted as VOC.

The emission ratios of the total measured organic compounds (except CH_4) relative to CO_2 ranged between $(1.37 \pm 0.09)\%$ (ppbC/ppbCO$_2$) and $(0.17 \pm 0.07)\%$ (ppbC/ppbCO$_2$) (see Table 1). These values are in agreement with measurements of fires under laboratory conditions as reported by Manø (private communication) and Czapiewski (1999), who derive average emission ratios of the sum of organic compounds from their experiments of 0.84% and 0.3%, respectively. Lobert *et al.* (1991) report an average of 1.19% based on 10 different experiments. The ratio found for the savannah fires is in good agreement with the

Table 7.1. Emission ratios of CO, CH_4, and the sum of VOC relative to CO_2 obtained for different types of fires investigated during FOS/DECAFE and SAFARI.

	Comments	$\Delta CO/\Delta CO_2$ %	$\Delta CH_4/\Delta CO_2$ %	$\Delta\Sigma VOC/\Delta CO_2$%
Bonsang *et al.*, 1991, 1995	Savannah, Ivory Coast, January 1989, average	11.04 ± 4.39	0.54 ± 0.24	0.81 ± 0.38
FOS/DECAFE	Flaming phase	6.95 ± 1.62	0.37 ± 0.11	0.62 ± 0.23
	Smouldering phase	15.42 ± 1.91	0.78 ± 0.20	0.77 ± 0.46
Bonsang *et al.*, 1995	Savannah, Ivory Coast, January 1991, average	11.44 ± 8.00	0.60 ± 0.43	1.08 ± 0.70
FOS/DECAFE	Flaming phase	5.44 ± 1.63	0.42 ± 0.16	0.69 ± 0.37
	Smouldering phase	14.03 ± 2.73	0.78 ± 0.30	1.08 ± 0.30
Koppmann *et al.*, 1997 SAFARI	Wildfire, mnaily wood	9.36 ± 1.30	1.03 ± 0.17	1.37 ± 0.09
Koppmann *et al.*, 1997 SAFARI	Kruger National Park,	5.29 ± 0.71	0.28 ± 0.07	0.74 ± 0.16
Koppmann *et al.*, 1997 SAFARI	Sugar cane fire #1	4.38 ± 2.09	0.26 ± 0.22	0.39 ± 0.29
	Sugar cane fire #2	2.19 ± 0.28	0.11 ± 0.09	0.17 ± 0.07

laboratory measurements, while the ratio found for the wildfire exceeds the laboratory values. The emission ratios of the predominantly flaming sugar cane fires are at the lower end of the results of the laboratory studies.

7.3 LABORATORY STUDIES

Biomass burning experiments were conducted under defined conditions in the laboratory at the burning apparatus of the Max-Planck-Institute for Chemistry in Mainz, Germany (Czapiewski, 1999). For details of the apparatus, see Lobert *et al.* (1991). In these experiments different types of fuelwood typically used in tropical and subtropical regions were investigated. The mass loss of the fuel, exhaust temperature and exhaust flow rate were continuously monitored. VOC were analyzed from grab samples in stainless steel canisters taken over the course of the fire. The samples were analyzed for CO, CO_2, CH_4 and VOC in the laboratory by GC-FID and GC-MS.

Figure 7.3 shows the exhaust temperature and the mass loss rate, and Figure 7.4 shows the course of CO_2 and CO for a typical fire. The maximum of CO_2 emissions represents the flaming phase, and the maximum of CO emissions represents the smouldering phase of the fire. Both phases can be clearly distinguished. Figure 7.5 compares the course of the CO emissions with that of the sum of all measured VOC, indicating that VOC are emitted in parallel to CO and thus predominantly during the smouldering phase of a fire. The same behaviour was also observed for fires investigated in field studies.

The emissions of VOC are influenced by a variety of parameters, the most important of which are the mass, the water content,

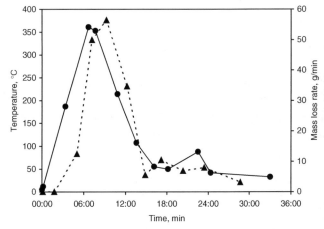

Figure 7.3. Exhaust temperature (solid line) and mass loss rate (dashed line) during a typical laboratory fire.

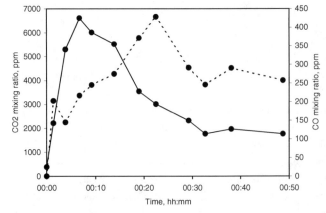

Figure 7.4. Course of CO_2 (solid line) and CO (dashed line) mixing ratios in the exhaust gas of a typical laboratory fire.

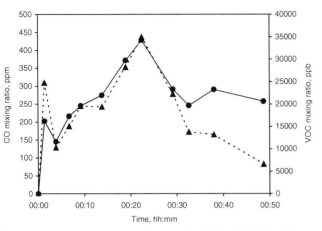

Figure 7.5. Course of CO (solid line) and VOC (dashed line) mixing ratios in the exhaust gas of a typical laboratory fire.

the surface/volume ratio and the species of the fuelwood used for a fire. Additional parameters, which may also influence the emissions, were not investigated in detail. These are the amount of leaves or bark burnt with the fuelwood, the piling of the fuelwood, which affects the oxygen supply, and operation of the fire (stoking, adding new fuel, and extinguishing the smouldering fuel).

Different species of fuelwood were burnt under similar conditions (amount of fuel, water content and burning efficiency). Table 7.2 summarizes the results. While the emission factors of CO_2 vary by about 20%, the emission factors of CO, CH_4, and VOC show variations of a factor of two.

Table 7.3 shows the results of different burning experiments with the same fuelwood with varying water content. In one experiment the water content was reduced to 3.4% by drying the wood, and in another experiment the wood was moistened to a water content of 16%. The results show that the emission factors increase dramatically with increasing water content of the fuel. For CO the emission factor increased by about a factor of two, when the water content was increased by a factor of 4, while the emission factor of VOC increased by a factor of 8.

In one experiment a typical household fire was simulated. In this case the fire was stoked several times and fuelwood was added to initiate more flaming. With the increase in the ratio of flaming to smouldering phases, the emission of VOC decreased by a factor of 2.5.

Figure 7.6 shows the contribution of the different classes of VOC to the total VOC emissions averaged over 14 different burning experiments in the laboratory and 15 different burning experiments under field conditions. A surprising result was the characteristic emission pattern of VOC emitted from biomass burning which seems to be more or less independent from the type of fire or the burning conditions. Alkenes contribute 30% to the total VOC emission. Alkanes, aromatic compounds and alkynes make

Table 7.2. Emission factors of CO_2, CO, CH_4 and VOC for different types of fuelwood.

	Musasa *Brachystegia spiciformis*	Chingunguru *Lopolaena cornifolia*	Muhacha *Uapaca kirkiana*
Combustion efficiency	0.92	0.90	0.89
Water content (%)	13.7	11.0	13.2
Emission factor (g/kg)			
CO_2	1580.0	1868.0	1788.0
CO	74.0	119.0	125.0
CH_4	4.1	9.1	6.9
VOC	3.2	4.2	6.1

Table 7.3. Emission factors of CO_2, CO, CH_4 and VOC for one type of fuelwood with different water contents.

	Musasa, dried	Musasa	Musasa, moist
Combustion efficiency	0.89	0.86	0.83
Water content (%)	3.4	8.5	16.0
Fuel load (kg)	2.0	1.9	2.1
Emission factor (g/kg)			
CO_2	1405.0	1389.0	1209.0
CO	79.0	110.0	126.0
CH_4	6.0	11.0	15.0
VOC	9.8	19.5	25.5

up 18%, 12% and 10%, respectively. 35% of the VOC emitted from biomass burning are unsaturated, and 30% are oxygenated compounds. The contribution of higher molecular weight VOC ($>C_5$) is, on average, 30% to the sum of all VOC.

A comparison of emission ratios measured under laboratory and field conditions (Table 4) shows that the comparability for individual fires is low. Averaging over a large number of fires, however, allows a comparison of emission factors of individual groups of VOC.

Thus, knowing the amount and the typical water content of the fuelwood, the amount of VOC emitted can be estimated. If we further include demographic studies of the fuelwood consumed per capita per year, then it is possible to make a rough estimate of the global VOC emissions from these types of fire.

7.4 CONTRIBUTION TO GLOBAL VOC EMISSIONS

Air samples collected in the vicinity of or even in biomass burning plumes are unlikely to be representative for biomass burning on a

Table 7.4. Comparison of emission ratios (given in ppb C/ppb CO_2) obtained in laboratory and field measurements.

	Field study		Laboratory study	
	Zimbabwe[a]	Nigeria[b]	Chingunguru, dried	Chingunguru
$\Delta CO/\Delta CO_2$, %	8.05	10.47	8.71	12.44
$\Delta VOC/\Delta CO_2$, %	1.53	2.19	0.25	1.03
Water content, %	dry[c]	moist[c]	4.00	11.00
Combustion efficiency	0.94	0.89	0.95	0.90

[a]Fuel was collected during the dry season.
[b]Fuel was collected during the rainy season.
[c]Only estimated moisture data available.

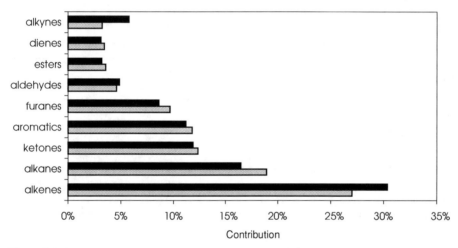

Figure 7.6. Contribution of different classes of VOC to the total VOC emissions. Averages of laboratory fires (black bars) and household fires investigated in Nigeria and Zimbabwe (grey bars) are compared.

global scale, since they represent only a very limited fraction of the areas burnt globally each year. Therefore, any extrapolations of emission ratios derived from those measurements to global scales are highly uncertain. In order to get a rough estimate on the impact of biomass burning emissions on the atmospheric environment, it is nevertheless worthwhile to compare the emissions of organic trace gases with global budgets.

Table 7.5 shows the emissions from biomass burning with all other anthropogenic emissions. Table 7.6 summarizes the emission of total carbon species and CO from the various types of biomass burning based on compilation of literature data (Andreae, 1991, 1993a; Hao and Liu, 1994; Andreae *et al.*, 1997; and references therein). Both laboratory and field studies showed that VOC are predominantly emitted during the smouldering phase of a fire and are better correlated with CO than with CO_2. Therefore, we base our extrapolation here on $\Delta(VOC)/\Delta(CO)$ emission ratios. If

we use the emission ratios from agricultural fires as a lower limit (mean $\Delta(VOC)/\Delta(CO) = 0.08\%$) and those from the uncontrolled savannah fire as an upper limit (mean $\Delta(VOC)/\Delta(CO) = 0.15\%$), then the estimated total carbon released in form of VOC (CH_4 not included) ranges between 25 and 48 Tg C/yr. The higher molecular VOC contribute up to 50% to that value.

This rough estimation of the source strength of organic compounds based on our measurements is a factor of 2 higher than the results of previous estimates by Bonsang *et al.* (1991), who give 12 Tg/year, which was derived from measurements over large savannah fires during the FOS/DECAFE project. This value is below the lower limit of the emission range based on our estimates, because Bonsang *et al.* (1991) based their estimates on measurements of light non-methane hydrocarbons only. From laboratory studies Lobert *et al.* (1991) calculate a carbon release in form of VOC of 42 Tg/year, a value at the upper end of our estimation

Table 7.5. Best guess estimates of gaseous and particulate emissions from global biomass burning and all anthropogenic sources.

Species	Biomass burning contribution[a]	All anthropogenic sources[b]	% due to biomass burning
CO_2	13500	33700	40.1
CO	680	1600	42.5
CH_4	43	275	15.6
VOC	42	100	42.0
H_2	16	40	40.0
NO	21	70	30.0
N_2O	1.3	5.5	23.6
NH_3	6.7	57	11.8
SO_2	4.8	160	3.0
COS	0.21	0.38	55.3
CH_2Cl	1.1	1.1	100.0
CH_3Br	0.019	0.11	17.3
Total PM[b]	90	390	23.1
Carbon	60	90	66.7

[a]Data given in Tg of species per year (Levine, 1996).
[b]Particulate matter.

range. These results indicate that the emissions of biomass burning contribute significantly to the local and regional budget of organic compounds in the troposphere.

7.5 IMPACT OF VOC ON THE FORMATION OF OZONE

At about the time when the first studies on biomass burning in African savannahs were carried out, evidence grew that the fires may be the main cause for a tropical ozone maximum spanning from Africa across the Atlantic to South America (Fishman et al., 1990; Cros et al., 1991; Andreae et al., 1993a). Additionally, African fires were thought to be the source of considerable air pollution, even over the remote Atlantic Ocean. Koppmann et al. (1992) reported elevated mixing ratios of CO and a number of light hydrocarbons in the marine boundary layer south of the equator.

In order to estimate the impact of VOC on the formation of ozone in biomass burning plumes, a number of model studies have been carried out (Thompson et al., 1996; Chatfield et al., 1996; Mauzerall et al., 1998; Poppe et al., 1998). Based on the SAFARI data given above, Poppe et al. (1998) conducted a simple model calculation using the Regional Acid Deposition Model (RADM2) published by Stockwell et al. (1990). The calculations were initialized with measured mixing ratios from the wildfire where the

highest VOC emission ratios were observed during the SAFARI campaign and done for fixed solar zenith angle of 24 degrees and 30 hours sunlight.

Without considering dilution, ozone-mixing ratios (initialized with a background value of 50 ppb) in the plume are shown in Figure 7.7. If only CH_4 and CO as oxidizable carbon compounds are taken into account, ozone reaches a maximum mixing ratio of 180 ppb after 30 sunlit hours. Taking into account the measured VOC, the ozone-mixing ratio increases steeper and earlier, because the ozone formation is now driven by the degradation of the hydrocarbons that react faster with hydroxyl (OH) radicals than CO and CH_4. Ozone mixing ratios increase rapidly to values around 90 ppb within the first two hours after the emissions, followed by a further slower increase to about 280 ppb. Thus, the mixing ratio of ozone in the plume is about 30% larger if the impact of VOC is included with the peak-mixing ratio reaching about 30 hours after the emission.

Under realistic conditions, the ozone formation is substantially influenced by transport processes, causing a dilution of the trace gases within the plume (Chatfield and Delany 1990; Chatfield et al., 1996). In order to account for this, dilution with surrounding air was included in this simulation mixing in background air with fixed mixing ratios of $[CO]^0 = 100$ ppb, $[CH_4]^0 = 1.7$ ppm, $[O_3]^0 = 50$ ppb, and $[VOC]^0 = 5$ ppbC. Measurements of air samples collected outside the plumes gave CO_2 mixing ratios, which were on average 10 ppm above the global background. The dilution was treated in a way that the CO_2 mixing ratios measured in the plumes was finally reduced to the CO_2 mixing ratio observed in the air outside of the plumes.

In the case of dilution, ozone mixing ratios increased to 135 ppb (excluding VOC) and 195 ppb (including VOC) 30 hours after the emission. The initial NO_2 mixing ratio of 30 ppb was taken from the average of the measured data that varied between 6–60 ppb for different plume passages. The amount of ozone produced depends strongly on the NO_2 mixing ratio. Clearly, better temporally and spatially resolved NO_x data are necessary to further assess the impact of biomass burning to the global ozone budget.

7.6 CONCLUSIONS

A large variety of organic trace gases were found in all samples collected in the plumes of biomass fires. Alkenes and alkanes were the most abundant VOC emitted by biomass burning besides methane. VOC from biomass burning also included a variety of oxygenated compounds such as alcohols, aldehydes, ketones, carboxylic acids, esters, ethers, furanes and others. Biomass burning is a major source for VOC in the troposphere, especially in subtropical and tropical regions. The observed emissions showed a considerable variation depending on the amount and type of fuel,

Table 7.6. Emissions of total carbon and CO from various biomass-burning sources.

Source	Biomass burnt (Tg (dry matter)/year)	Carbon released (Tg/year)	CO emission ratio (%)	CO emission (Tg C/year)
Tropical forest	1260	567	10.9	62
Extra-tropical forest	1150	518	11.2	58
Savannah	3690	1661	6.2	103
Biomass fuel	1940	873	8.3	73
Charcoal	20	42	20.0	8
Agricultural waste in fields	850	383	7.2	28
Total	8910	4044	8.2	332

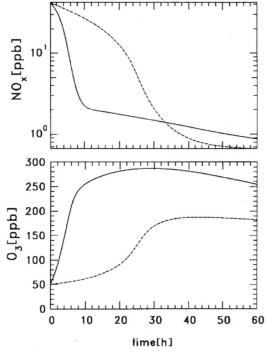

Figure 7.7. Time dependence of ozone and NO_x mixing ratios. Initial conditions were taken from the wildfire record. Solid (dashed) curves include (exclude) the VOC. The gas phase chemistry RADM2 was used. No dilution was applied.

water content of the fuel, the burning conditions and stages of the fires. Carbon monoxide, methane and most of the VOC were emitted during the smouldering phase of a fire, and thus the emission of most organic compounds was correlated with the emission of CO. The emission ratio of the sum of light VOC was typically of the same order of magnitude or even higher than that of methane. About 50% of the emitted VOC were higher molecular weight hydrocarbons with more than five carbon atoms. Oxygenated hydrocarbons were the predominantly emitted VOC (about 40% of the sum of VOC). Unsaturated hydrocarbons contributed about 20% to the total VOC emission. Laboratory studies carried out to investigate the parameters influencing the emissions of VOC and to study the behaviour of household fires showed that the emissions of VOC depended strongly on the humidity of the fuel, the amount of fuel, and the handling of the fuel during the fire. Carbon dioxide was predominantly emitted during the flaming phase of a fire. Except alkynes, all VOC were emitted predominantly during the smouldering phase of a fire. The emission of VOC increased along with those of CO and methane with decreasing burning efficiency. The burning efficiency varied with mass, humidity, surface-to-volume ratio and the type of the burnt biofuel. Surprisingly, the pattern of VOC did not change significantly, even if different types of fuelwood were burnt or burning conditions were changed.

Simple model calculations showed that the formation of ozone in aging biomass burning plumes is substantial, and this has consequences for the global ozone budget. The simulation for a wildfire investigated during the SAFARI campaign showed that the presence of VOC increases the maximum concentration of ozone formed in the biomass-burning plume by up to 40%. Of special importance is the observation that due to VOC, ozone formation in biomass burning plumes is very rapid and starts much earlier if VOC chemistry is considered. The efficiency of ozone formation depends critically, and in a complex way, on the organic composition of biomass burning emissions and the dilution of the plume.

In view of the substantial amount of nitrogen oxides (NO_x) in the emissions from biomass burning, it can be expected that in regions with substantial use of biofuel for cooking and heating these emissions will result in significantly enhanced ozone concentrations.

7.7 OUTLOOK

Field experiments can only investigate casual biomass burning events and thus they are only snapshots of the situation at a certain place and time on earth. Future research should focus on the improvement of remote sensing data to gain information about fire frequencies and distributions, fuel loads, CO and other trace gas emissions. A compilation of more comprehensive data sets will lead to a better knowledge of the organic composition of biomass burning emissions, burning efficiencies and scaling of VOC emissions relative to CO. Based on these data, a more precise estimate of the contribution of biomass burning emissions to the VOC budget should be available. The results should lead to a better understanding of the impact of these emissions on atmospheric chemistry and biogeochemical cycles.

ACKNOWLEDGEMENTS

We would like to thank the colleagues of the Biogeochemistry Department of the Max Planck Institute for Chemistry, Mainz, Germany, for logistical and technical support and for collecting air samples during the field studies in Africa. We would also like to thank an anonymous reviewer for valuable comments and suggestions.

This research is a contribution to the International Geosphere–Biosphere Programme (IGBP) core project International Global Atmospheric Chemistry (IGAC), and it is part of the Biomass Burning Experiment (BIBEX). The Max-Planck Society and the German Ministry for Education and Research supported this work within the programme *Spurenstoffkreisläufe* under grant No. 105252 A 2b-6.

REFERENCES

Andreae, M. O. (1991) Biomass burning: its history, use, and distribution and its impact on environmental quality and global climate. In *Global Biomass Burning: Atmospheric, Climatic, and Biospheric Implications*, pp. 3–21. Cambridge, MA, MIT Press.

(1993a) The influence of tropical biomass burning on climate and the atmospheric environment. In R. S. Oremland (Ed.), *Biogeochemistry of Global Change: Radiatively Active Trace Gases* pp. 113–150. New York, Chapman and Hall.

(1993b) Global distribution of fires seen from space. *EOS Trans. AGU*, **74**, 129.

(1997) Emissions of trace gases and aerosols from savannah fires. In B. W. van Wilgen, M. O. Andreae, J. G. Goldammer and J. A. Lindesay, (Eds.) *Fire in the Southern African Savannah: Ecological and Environmental Perspectives*, pp. 161–183. Johannesburg, South Africa, Witwatersrand University Press.

Andreae, M. O., Atlas, E., Cachier, H., *et al.* (1996) Trace gas and aerosol emissions from savannah fires. In J.S. Levine (Ed.), *Biomass Burning and Global Change*, Cambridge, MA, MIT Press.

Bonsang, B., Lambert, G. and Boissard, C. (1991) Light hydrocarbon emissions from African Savannah burnings. In J. S. Levine (Ed.), *Global Biomass Burning: Atmospheric, Climatic, and Biospheric Implications*, pp.155–166. MIT Press. Cambridge, MA.

Bonsang, B., Boissard, C., Le Cloarec, M. F., Rudolph, J. and Lacaux, J. P. (1995) Methane, carbon monoxide and light non methane hydrocarbon emissions from African savannah burning during the FOS/DECAFE experiment. *J. Atmos. Chem.*, **22**, 149–162.

Brocard, D. and Lacaux, J. P. (1998) Domestic biomass combustion and associated atmospheric emissions in West Africa. *Glob. Biogeochem. Cycl.*, **12**, 127–139.

Chatfield, R. B. and Delany, A. C. (1990) Cloud-vented tropical smog can produce high ozone, but ill-simulated dilution leads models to O_3 overprediction. *J. Geophys. Res.*, **95**, 18,473–18,488.

Chatfield, R. B., Vastano, J. A., Singh, H. B. and Sachse, G. (1996) A general model of how fire emissions and chemistry produce African/Oceanic plumes (O3, CO, PAN, smoke) in TRACE-A. *J. Geophys. Res.*, **101** (D19), 24,279–24,306.

Cros, B., Ahoua, B., Orange, D., Dombele, M. and Lacaux, J. P. (1991) Tropospheric ozone on both sides of the equator in Africa. In J. S. Levine (Ed.), *Biomass Burning and Global Change*, Cambridge, MA, MIT Press.

Crutzen, P. J. and Andreae, M. O. (1990) Biomass burning in the tropics: impact on atmospheric chemistry and biogeochemical cycles. *Science*, **250**, 1,669–1,678.

Czapiewski, K.v. (1999) Untersuchungen zu Emissionen von flüchtigen organischen Verbindungen aus Biomasseverbrennung. Ph. D. thesis, Universität Köln.

Delmas, R. (1982) On the emission of carbon, nitrogen, and sulphur in the atmosphere during bush fires in intertropical savannah zones. *Geophys. Res. Lett.*, **9**, 761–764.

Fishman, J., Watson, C. E., Larson, J. L. and Logan, J. A. (1990) Distribution of tropospheric ozone determined from satellite data. *J. Geophys. Res.* **95**, 3,599–3,617.

Franke, W. (1989) *Nutzpflanzenkunde.* Stuttgart, Germany, Thieme-Verlag.

Hall, D. O. (1991) Biomass energy. *Energy Policy*, 711–737.

Hao, W. M. and Liu, M-H. (1994) Spatial and temporal distribution of tropical biomass burning. *Glob. Biogeochem. Cycl.*, **8**, 495–503.

Hao, W. M., Ward, D. E, Olbu, G. and Baker, S. P. (1996) Emissions of CO_2, CO and hydrocarbons from fires in diverse African savannah ecosystems. *J. Geophys. Res.*, **101**, 23,577–23,584.

Helas, G., Lacaux, J. P., Delmas, R. *et al.* (1992) Ozone as biomass burning product over Africa. *Fesesnius. Environ. Bull.*, 155–160.

Koppmann, R., Bauer, R., Johnen, F. J., Plass, C. and Rudolph, J. (1992) The distribution of light nonmethane hydrocarbons over the Mid-Atlantic: results of the Polarstern cruise ANT VII/1. *J. Atmos. Chem.* **15**, 215–234.

Koppmann, R., Khedim, A., Rudolph, J. *et al.* (1997) Emissions of organic trace gases from savannah fires in southern Africa during SAFARI 92 and their impact on the formation of tropospheric ozone. *J. Geophys. Res.* **102**, 18,879–18,888.

Lacaux, J. P., Cachier, H. and Delmas, R. (1993) Biomass burning in Africa: An overview of its impact on atmospheric chemistry. In P. J. Crutzen and J. G. Goldammer, (Eds.), *Fire in the Environment: The Ecological Atmospheric, and Climatic*

Importance of Vegetation Fires, pp. 159–191. Chichester, UK, John Wiley and Sons.

Lacaux, J. P., Brustet, J. M., Delmas, R. *et al.* (1995) DECAFE 91: Biomass burning in the tropical savannahs of Ivory Coast. An overview of the field experiment Fire of Savannas (FOS/DECAFE 91). *J. Atmos. Chem.*, **22**, 195–216.

Levine, J. S. (Ed.) (1996) *Biomass Burning and Global Change, Vol. 1, Remote Sensing, Modelling and Inventory Development, and Biomass Burning in Africa*, pp. xxxv–xliii. Cambridge, MA, MIT Press.

Lobert, J. M. and Warnatz, J. (1993) Emissions from the combustion process in vegetation. In P. J. Crutzen and J. G. Goldammer (Eds.), *Fire in the Environment: The Ecological, Atmospheric, and Climatic Importance of Vegetation Fires*. Report of the Dahlem Workshop, Berlin, 15–20 March, 1992, New York, John Wiley and Sons.

Lobert, J., Scharffe, M. D. H., Hao, W. M., *et al.* (1991) Experimental evaluation of biomass burning emissions: nitrogen and carbon containing compounds. In J. S. Levine (Ed.), *Global Biomass Burning: Atmospheric, Climatic, and Biospheric Implications*, pp. 289–304. Cambridge, MA, MIT Press.

Marufu, L., Ludwig, J., Andreae, M. O., Meixner, F. X. and Helas, G. (1997) Domestic biomass burning in rural and urban Zimbabwe – Part A. *Biomass and Bioenergy*, **12**, 1, 53–68.

Mauzerall, D. L., Logan, J. A., Jacob, D. J. *et al.* (1998) Photochemistry in biomass burning plumes and implications for tropospheric ozone over the tropical south Atlantic. *J. Geophys. Res.*, **103**, 8,401–8,423.

Poppe, D., Koppmann, R. and Rudolph, J. (1998) Ozone formation in biomass burning plumes: influence of atmospheric dilution. *Geophys. Res. Lett.*, **25**, 3,823–3,826.

Rudolph, J., Khedim, A. and Bonsang, B. (1992) Light hydrocarbons in the tropical boundary layer over tropical Africa. *J. Geophys. Res.*, **97**, 6,181–6,186.

Rudolph, J., Khedim, A., Koppmann, R. and Bonsang, B. (1995) Field study of the emission of methyl chloride and other halocarbons from biomass burning in western Africa. *J. Atmos. Chem.*, **22**, 67–80.

Stockwell, W. R., Middleton, P., Chang, J. S. and Tang, X. (1990) The second generation regional acid deposition model chemical mechanism for regional air quality modelling. *J. Geophys. Res.*, **95**, 16,343–16,367.

Swap, R., Garstang, M., Macko, S. A., *et al.* (1996) The long-range transport of southern African aerosols to the tropical south Atlantic. *J. Geophys. Res.*, **101**, 23,777–23,791.

Thompson, A. M., Pickering, K. E., McNamara, D. P. *et al.* (1996) Where did tropospheric ozone over Southern Africa and the tropical Atlantic come from in October 1992? Insights from TOMS, GTE TRACE-A and SAFARI 92. *J. Geophys. Res.*, **101**, 24,251–24,278.

Ward, D. E. and Hardy, C. C. (1991) Smoke emissions from wildland fires. *Environ. Intern.*, **17**, 117–134.

8 Biomass burning in Africa: role in atmospheric change and opportunities for emission mitigation

EVANS KITUYI[*,1], SHEM O. WANDIGA[1], MEINRAT O. ANDREAE[2], AND GÜNTER HELAS[2]

[1]University of Nairobi, Kenya
[2]Max Planck Institute for Chemistry, Mainz, Germany

Keywords

Biomass burning; greenhouse gas emissions; emission mitigation

Abstract

A review of available literature published on biomass burning and trace gas emissions in Africa reveals household biofuel use, land use and land-use change to be the most important trace gas emission sources in Africa, contributing about 4% to the overall global CO_2 budget. This may not be significant in so far as altering global climate through temperature rise is concerned. However, through the contribution of about 35% of the global photochemical ozone formation, biomass burning in Africa significantly influences important atmospheric processes. Although the total greenhouse gas emissions from Africa are very low compared to those of other continents, countries on the continent could still contribute to global greenhouse gas mitigation efforts through ways that could simultaneously deliver urgent development needs.

8.1 INTRODUCTION

Many African governments have for long underplayed their countries' contribution to regional and global greenhouse gas (GHG) emission levels. The general assumption that their countries contribute insignificant amounts of these gases – responsible for the earth's warming – has influenced their hard stance, witnessed at international negotiations, against any meaningful roles in concerted efforts to mitigate GHG emissions. But intensive atmospheric research work on Africa over the past two decades (e.g., SAFARI 2000, www.safari2000.org) have confirmed the significant contribution that biomass burning has made

to the regional and global tropospheric trace gas budgets. The land-use and energy sectors dominate African GHG emissions (UNEP, 1998).

Country-level data to verify this position and inform appropriate responses is, however, still scanty and unreliable. Only a few countries have, with difficulty, reported some data for gaseous emissions from key sectors in their National Communications to the United Nations Framework Convention on Climate Change (UNFCCC). There is great need to collect, appropriately package and communicate available national GHG emission data to country delegations to the regular international climate change negotiations, as well as policymakers. This way, the African governments will understand and appreciate their national contribution to the global budgets and take some responsibility (as expected of all Parties to the UNFCCC) in the concerted global effort to mitigate climate change via common but differentiated means.

Advances in science and technology have, over time, availed appropriate options that could help African countries mitigate climate change through emission reduction, avoidance of emissions and carbon sequestration. This is in addition to existing indigenous emission mitigation approaches that remain undocumented but that could be further developed and replicated.

Through a survey of literature and analysis of data from various sources, this chapter evaluates the significance of trace gas emission quantities typical of various biomass-burning sources in Africa for regional and global atmospheric chemistry and climate. Furthermore, it seeks to identify sectors within African countries' economies in which sustainable emission mitigation options are viable. The chapter finally proposes the role African countries could play towards the concerted global efforts to abate greenhouse gas emissions, and highlights the development opportunities available for developing countries through relevant international instruments of environmental governance, and the challenges these countries need to address.

*Currently at African Centre for Technology Studies, Nairobi.

Climate Change and Africa, ed. Pak Sum Low. Published by Cambridge University Press. © Cambridge University Press 2005.

8.2 BIOMASS EMISSION BUDGETS FOR AFRICA

8.2.1 Biomass burning a source of atmospheric emissions

It is widely acknowledged that GHGs are accumulating in the atmosphere due to human activities (IPCC, 2001). Although contributions to these emissions from Africa are still considered insignificant at the moment, the rapidly increasing energy consumption and land-use change in developing countries is expected to result in higher GHG emission levels from these countries. Any measures to mitigate GHG emissions demand good knowledge of the relevant sources, sinks and reservoirs, as stated in the UNFCCC.

There is relatively low industrial activity in Africa. Apart from industry, key sources of trace gas emissions in the region include fuel combustion and land-use activity. A significant proportion of the land-use activity involves the burning of vegetation – mainly savannah-burning, forest clearing for settlement and cultivation, agricultural waste combustion and biofuel use. Although most fires are human-induced, natural wild fires also do occur.

The significant role played by biomass burning emissions in influencing the composition of the atmosphere was revealed more than two decades ago (Crutzen *et al.*, 1979). Prior to that, large-scale air pollution had always been associated with anthropogenic activities in industrialized regions of the world, mainly in the northern hemisphere. Later, after this important revelation, satellite measurements of tropospheric ozone showed that a large amount of ozone pollution comes from tropical Africa (Fishman, 1988; Fishman *et al.*, 1990). Furthermore, elevated carbon monoxide (CO) and methane (CH_4) levels were observed on a seasonal basis in the southern hemisphere (Marenco *et al.*, 1989).

Results from subsequent work (Cros *et al.*, 1988; Helas *et al.*, 1995; Helas and Pienaar, 1996) have pointed to vegetation burning on the African continent during the dry season (Cahoon *et al.*, 1992) as being responsible for these observations. Helas (1995) estimates that about 40% of global biomass burning activity occurs in Africa, of which about 40% involves savannah burning (Andreae, 1991). The contribution by various biomass burning types to the overall global source strength are presented in Figure 8.1, which is based on data from Andreae (1991). About 42% of all fires are reported to occur in Africa, and 26% in South America, while the rest of the world accounts for the remaining portion (Koppmann *et al.*, 2005).

The emitted gases that influence the radiation balance of the earth and, thereby, climate may be divided into three groups: (i) gases that have a direct effect on climate due to their radiative properties, including carbon dioxide (CO_2) and nitrous oxide (N_2O); (ii) gases that are not important GHGs in their own right, but which indirectly influence radiative forcing by having an impact

Table 8.1. Best guess estimates of gaseous and particulate emissions from global biomass burning and all anthropogenic sources.

Species	Biomass burning contribution (Tg yr^{-1})	All anthropogenic sources (Tg yr^{-1})	% due to biomass burning
CO_2	13,500	33,700	40.1
CO	680	1,600	42.5
CH_4	43	275	15.6
VOC	42	100	42.0
H_2	16	40	40.0
NO	21	70	30.0
N_2O	1.3	5.5	23.6
NH_3	6.7	57	11.8
SO_2	4.8	160	3.0
COS	0.21	0.38	55.3
CH_3Cl	1.1	1.1	100.0
CH_3Br	0.019	0.11	17.3
Total PM^a	90	390	23.1
Carbon	60	90	66.7

[a] Particulate matter

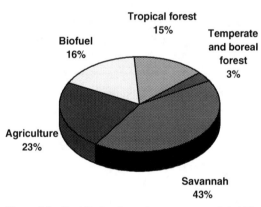

Figure 8.1. Contribution by various sources to global biomass burning (Andreae, 1991).

on the chemical and physical processes in the atmosphere; such gases include nitrogen oxides (NO_x), carbon monoxide (CO) and non-methane hydrocarbons (NMHC); and (iii) gases that possess the ability to affect climate both directly and indirectly, such as methane (CH_4) (IPCC, 1994b).

Emissions from biomass burning fall into all these categories and the magnitude of the quantities emitted is dependent upon, inter alia, the type of fuel and associated combustion efficiencies (Andreae, 1993; Lacaux *et al.*, 1994; Brocard *et al.*, 1998; Marufu, 1999; Kituyi *et al.*, 2001b). Table 8.1 reports on the best guess estimates of gaseous and particulate emissions from global

Table 8.2. Regional biomass burning trace gas emissions in Tg C or N per annum.

Region	CO_2	CO	CH_4	NMHC	NO_x
Africa	337	96	6.2	13.1	3.1
Latin America	345	53.4	4.5	8.5	1.5
India	410	45	5.5	10.7	1.0
China	339	31	1.3	7.8	0.59
East Asia	187	19.8	2.3	4.3	0.39
Oceania	11	1.1	0.11	0.23	0.04
Middle East	36	8.5	0.9	1.5	0.22
Former USSR	15	8.9	0.87	1.43	0.22
Canada	24	2.1	0.2	0.3	0.05
USA	134	10.7	1.0	1.8	0.34
OECD Europe	60	8.2	0.78	1.43	0.21
Eastern Europe	7.6	2.0	0.19	0.38	0.05
Japan	0.25	0.5	0.05	0.23	0.02
Global total	1906	287	24	52	8

Source: Marufu, 1999.

biomass burning and all anthropogenic sources based on Levine (1996).

The breakdown of regional biomass burning-related trace gas emissions of significance to climate and atmospheric chemistry is presented in Table 8.2. The major regions contributing to the total include Africa, Latin America (mainly Brazil), India and China, which together account for 75% of the total CO_2 from biomass fires.

8.2.2 Biomass burning emissions and Africa

The African and global emission budgets for trace gases from biomass burning and all other sources combined are presented in Table 8.3. Biomass burning in this table encompasses sources such as biofuel, savannah, agricultural waste and forest fires. Other known emission sources include biogenic, ocean, lightning and industry. Biomass burning (combining biofuel and other forms of biomass burning) on the continent emerges as the largest emitter, contributing about 60% of the total. The limited industrial activity in the region accounts for 23%.

For Africa, major uncertainties still remain in the availability and quality of much of the biomass burning emission concentration data. For some gaseous species, there are no data. Table 8.3 therefore presents the key emission species for which reliable data for the region are available. According to these data, CO_2 from biomass burning in Africa accounts for 4% of all global sources.

Since woodfuel is extracted unsustainably in most parts of Africa, its combustion is assumed to lead to net CO_2 emissions. Despite the assumption that savannah and agricultural waste fires do not contribute to net CO_2 emissions, biomass burning is the largest source of CO_2 in Africa. The biomass burning contribution of 337 Tg C also represents 18% of the global biomass burning CO_2 total of 1,906 Tg C/yr. Industrial activity in Africa accounts for a paltry 2% of all global industrial-based CO_2 (6,450 Tg C/yr) and 1.5% of the total from all sources. These results highlight the significance of biomass burning emissions in Africa to the overall emissions from the continent. Biofuel use accounts for the largest amount of CO_2 emissions (32%), while other forms of biomass fires account for 28%. Other sources, such as biogenic and oceans, account for the remaining 17% of the total 566 Tg C emitted annually from Africa. The findings above reveal the order of priority by source, which should be fully considered in any proposals on emission abatement on the continent.

Biomass burning in Africa is also the largest source for CO, where about 90% is associated with biomass burning emissions. The 96 Tg C/yr of CO emitted annually on the continent constitutes 20% of the global all-source total. About 6.2 Tg C/yr of CH_4 are also emitted each year from biomass burning processes in Africa – representing 2% of all global methane sources. Most CH_4 originates from biogenic and industrial emissions, which together account for 93% of the global sum. Other sources include rice paddies, municipal wastes, termites and enteric fermentation, whose average contributions to the global CH_4 budget are documented (IPCC, 1994a).

Likewise, 13.1 Tg C/yr of non-methane hydrocarbons (NMHCs) from biomass burning is reported (Table 8.3), which is equivalent to 3% of the global total for all sources. CO, CH_4 and NMHCs are all typical products of incomplete combustion processes, the most common being pyrolytic charcoal production. Global charcoal production from vegetation, estimated at 0.3–0.7 Pg C/yr (Andreae, 1993) releases an average 1.8 Tg C/yr of CH_4 in a range of 1.4– 2.8 Tg C/yr (Delmas, 1994). This occurs mainly in the tropical zones and includes charcoal formed from forest fires, which is left in the soils. Koppmann *et al.* (this volume) reports the emission of 8 Tg C/yr of CO from the annual conversion of 20 Tg dry wood to charcoal for fuel use.

Specific details of the quantities of NMHCs emitted by various biomass burning processes are not readily available. However, globally, biomass burning represents 45%, 37%, 32% and 5% of the total sources of ethylene (C_2H_4), ethane (C_2H_6), propylene (C_3H_6) and other NMHCs respectively (Granier *et al.*, 1996). The contributions from fuelwood for C_2H_4, C_2H_6, C_3H_6, and other NMHCs to the global total are 2.4, 0.8, 0.7 and 1.1 Tg C/yr, respectively. Each of these levels represents about one third of the total biomass burning source strength (Granier *et al.*, 1996). It is reasonable, therefore, to expect high concentrations of NMHCs (mainly those in the C_2–C_6 range) from charcoal production processes in Africa.

Nitrogen oxides are also emitted in significant levels by biomass burning processes. For Africa, 3.1 Tg N/yr of NO_x (NO + NO_2)

Table 8.3. Comparison of regional and global emission budgets of important atmospheric trace gases. All values in Tg C or N per year.

Gaseous species	Africa		Global	
	Biomass burning[a]	All sources[b]	Biomass burning[a]	All sources[b]
C (CO_2)	337	565.9	1,906	8,351
C (CO)	96	112	287	482
C (CH_4)	6.2	18.1	24	380
C (NMHC)$_c$	13.1	123	52	534
N (NO_x)	3.1	7.1	8	40
N (N_2O)	—	—	0.1–1[c]	13–16[c]
N (NH_3)	—	—	0.48[d]	4.4[d]

Unless otherwise indicated, all other values in the columns under Africa and Global are derived from Marufu (1999).

[a]Includes biofuel, savannah, agricultural waste and deforestation fires.
[b]Includes biomass burning, industrial, biogenic, ocean and lightning.
[c]*Source*: Bouwman *et al.* (1995).
[d]*Source*: Andreae (1993).

are emitted. This is equivalent to 8% of the global NO_x emitted from all other sources. NO_x from biomass burning becomes more significant on the continent since it forms about 40% of the total from all other African sources.

Nitrous oxide (N_2O) is an important GHG, but its concentration levels and patterns in various regions are yet to be well understood. Bouwman *et al.* (1995) attribute the witnessed growth in atmospheric N_2O to a large number of poorly known minor sources, which include agricultural soils, soils under natural vegetation, aquatic sources, biomass burning, fossil fuel combustion, industrial sources and traffic. They report the global sum from all sources to lie in the range 13–16 Tg N/yr, of which 0.1–1 Tg N/yr comes from biomass burning. Andreae (1993) estimates the biomass burning contribution to be about 15% of the total source strength, and responsible for 20–30% of the atmospheric increase. Another major cause of N_2O increase in the atmosphere is its secondary formation within smoke plumes, through the oxidation of NH_3 by OH (Bouwman *et al.*, 1995).

8.2.3 National and subregional emission source strengths

One of the key problems hampering accurate estimation of trace gas emissions in the African region is lack of biomass activity data. Despite this scenario, data have been made available for some of the countries and subregions on the continent. Data available for a number of trace carbonaceous and nitrogenous emissions from biomass burning for countries including Uganda, South Africa and Zimbabwe, as well as subregions such

as West Africa are presented in Table 8.4. These are compared with African and global estimates presented in the same table.

With a higher population of 38 million in 1997 (World Bank, 1999) and the consequent higher total national consumption in South Africa, the higher emission quantities reported for CO_2, CO and NO_x (Helas and Pienaar, 1996) are expected. However, the CH_4 level is more than three times less than that for Kenya. This is attributed to the rarity of charcoal availability and use in South Africa, where coal is utilized instead. The much lower population of Zimbabwe, 11 million (World Bank, 1999), and the understanding that no charcoal is either produced or used in the country (Marufu *et al.*, 1997) partially explain the large difference from the results for each gas considered.

Uganda's population of 20 million (World Bank, 1999) was about two thirds that of Kenya in 1997. This could largely explain the lower emission levels for Uganda. Studies on household energy use in Uganda reveal biofuel use patterns with regard to fuel and stove types (Bachou, 1991) that are similar to those reported for Kenya (Kituyi *et al.*, 2001a). However, further studies into the biofuel consumption rates and patterns of Uganda are required to account for the lower levels with respect to Zimbabwe's. The West African subregion consumes a total annual fuelwood quantity of 102 Tg (Brocard *et al.*, 1998), leading to the reported combined high emissions for all the countries in that region.

CO_2 emissions from major sources are closely associated with population growth. In order to predict the future emission situation, it is necessary to define the atmospheric increase of CO_2 in terms of the per capita country or regional contribution. In 1989, Kenya's overall per capita annual CO_2 emission was 0.29 tonne,

Table 8.4. Comparison of annual biofuel emissions for some countries and sub-region to regional and global levels. All values are in Tg C or N per year.

	Kenya[a]	Zimbabwe[b]	Uganda[c]	South Africa[d]	West Africa[e]	Africa[f]	Global[f]
C (CO_2)	9.70	4.56	3.8	17.9	41.7	181	1401
C (CO)	0.92	0.44	0.35	1.7	3.4	16.3	77
C (CH_4)	0.10	0.015	0.06	0.03	0.24	0.8	6
C (NMHC)	0.06	0.024	0.005	—	0.27	1.3	18.7
N (NO)	0.019	0.005	—	—	—	—	—
N (NO_x)	0.021	—	0.011	0.05	—	0.4	1.4
N (N_2O)	0.0009	—	0.003	—	—	—	—

[a]Kituyi (2000)
[b]Marufu *et al.* (1999)
[c]GEF/UNEP (1994)
[d]Helas and Pienaar (1996)
[e]Brocard *et al.* (1998)
[f]Marufu (1999)

about 25% of the then global average of 1.16 tonnes (Othieno, 1991). From the current total national CO_2 emissions of 12.5 Tg C/yr from all documented sources in Kenya (Kituyi, 2000), and the estimated 1997 population of 29.1 million, an annual per capita CO_2 emission rate of 0.43 tonne per person is determined for the country. This represents a rough increase of about 48% over the eight-year period – an average of 0.02 tonne per capita per year.

This national per capita rate of 0.43 is also comparable to the 0.4 tonne per capita reported for Kenya in 1995 by Claussen and McNeilly (1998), who associate the growth in emissions from developing countries with the various factors affecting the energy intensity of an economy. According to their country-by-country analysis, the oil-rich Middle Eastern nations have the highest per capita CO_2 emissions in the world. Topping the list is Qatar with 53 tonnes per capita, followed by the United Arab Emirates (UAE) at 31 tonnes per capita. Other important contributions, according to the analysis, are 19.4, 12.1, and 10.3 tonnes per capita for the USA, Russia and Germany, respectively.

Using the same country-by-country data of Claussen and Mc-Neilly (1998), the average per capita CO_2 emissions (in tonnes) for five regions in Africa in 1995 were established and these are presented in Figure 8.2. According to this figure, northern African countries lead with annual per capita CO_2 emissions of 2.9 tonnes, and those of western African countries imply the least, at 0.5 tonne. South Africa is the sole country on the continent with an annual per capita emission level higher than 3 tonnes, reported to be 7.3 tonnes per capita. This amount is similar to those of European countries, such as Italy and Greece (each about 7.3 tonnes per capita) and even higher than others, including Switzerland (5.5 tonnes per capita). The exclusion of South Africa from the region's per capita average calculation leads to a realis-

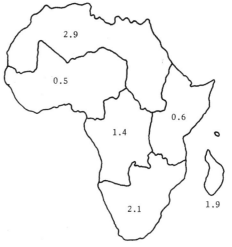

Figure 8.2. CO_2 emissions (tonnes) per capita for various regions of Africa for 1995.

tically lower annual CO_2 emission of about 0.5 tonne per capita for the continent. This is indeed a significant change, down from 1.3 tonnes per capita per year.

An increase in energy intensity in African countries may lead to a corresponding increase in the per capita CO_2 emission rate. The use of per capita CO_2 emission rights has been proposed at various forums as one innovative way towards achieving global equity in climate change mitigation.

8.2.4 An assessment of trace gas sinks

Knowledge of the quantities and diversity of trace gas sinks available on the continent is necessary in determining overall net atmospheric emissions. For Africa, a major challenge is the absence of such information in most countries. Even in cases where some data exist on, say, forests, the carbon sequestration potentials are still uncertain. However, it is clear that non-forest trees on farmlands, with the smallest carbon intake per hectare (ha), have the largest share of carbon intake. The indigenous forest cover has a smaller contribution to carbon intake per ha when compared to the plantation forests (MRT, 1998). The large difference between the CO_2 uptake potentials derived by the IPCC methodology and national methodologies used in the preparation of National Communications to the UNFCCC may be attributed to, inter alia, the highly varying estimates of vegetation areas, the mean annual increments (MAI) and standing stocks.

The other important CO_2 sinks include abandoned, previously managed lands, which in Kenya absorb 32 Gg C/yr (MRT, 1998). This mainly results from biomass regrowth and soil absorption. This sink may not be important, owing to the pressure for land in Kenya as with other countries in Africa – a situation that would not allow for prolonged periods of land abandonment.

Other previously unknown CO_2 sinks have been reported. Seiler and Crutzen (1980) reported charcoal production to be an indirect global atmospheric CO_2 sink. Its capacity to act as a sink is derived from its resistance to microbial decay and characteristic long life in soils. The charcoal includes black carbon production for fuel use and that resulting from forest and savannah fires, some of which is buried in the soil for long periods. Kuhlbusch and Crutzen (1995) reported a pioneering estimate of the global annual black carbon production from vegetation fires in the range of 50–270 Tg C yr^{-1}. They further report that the formation of black carbon by vegetation fires might significantly reduce the net carbon flux of 1,100–2,300 Tg C yr^{-1} from changes in land use by 2–25%.

According to Andreae (1993), global fire-induced carbon sequestration by charcoal formation and burial could reduce, by as much as 10%, the over 7,600 Tg C released annually to the atmosphere by human activity. Given these data and other meagre and ill-documented statistical information on source/sink balances, it is extremely difficult to estimate the net carbon release rates for any year. This difficulty is even more apparent at the country level.

Although the major sink for CH_4 is the reaction in the troposphere with OH (accounting for about 86% of the total atmospheric sink), removal by soils is also a significant sink, removing about 6% (15–45 Tg CH_4/yr) of the atmospheric burden (IPCC, 1994b; Schlesinger, 1997). Sinks for CO are similar to those described for methane, i.e., OH reaction (between 1,400 and 2,600 Tg yr^{-1}), stratospheric destruction and soil uptake (between 150 and 500 Tg yr^{-1}) (Brasseur *et al.*, 2005). Although there is much uncertainty, the known sinks of N_2O include stratospheric destruction, soil microbial activity and the rate of increase of the N_2O itself in the atmosphere (Schlesinger, 1997). However, more research is needed to refine the budget for atmospheric N_2O.

The observations above serve to further reiterate the need for more precise evaluations of sources and sinks by countries. Apart from the uncertainties associated with the use of the IPCC default guidelines for country GHG inventory development and update, the most urgent need lies in the improvement of the quality of activity data, where comprehensive national methodologies would be the best option.

8.3 IMPLICATIONS FOR ATMOSPHERIC ENVIRONMENT

8.3.1 Tropospheric ozone pollution

Other than its production from chemical feedbacks involving CH_4, the production of tropospheric O_3 in the vegetation smoke plumes from tropical areas is based predominantly on photoreactions involving unsaturated hydrocarbons, mainly C_2–C_4 alkenes (Bonsang *et al.*, 1995) and other short-lived precursor gases CO and NO_x (IPCC, 1994b). The levels often reach values typical of industrialized regions (Andreae, 1993) and are highly spatially variable, both regionally and vertically, making assessment of global long-term trends extremely difficult (IPCC, 1994b). The highest concentrations, in the range of 50–100 ppb, are usually found in discrete layers at altitudes of 1–5 km (Andreae *et al.*, 1994; Andreae *et al.*, 1988; Andreae *et al.*, 1992). Elevated mixing ratios of O_3 in photochemical smog layers, due to high NO_x-to-hydrocarbon ratios, at altitudes between 1.5 and 4 km over western Africa were measured during a local burning season in the region (Helas *et al.*, 1995). They reported average mixing ratios of 29 ppb in relatively clean air masses, while averages up to 88 ppb were measured in polluted air masses.

At ground level, concentrations are normally lower and show a pronounced daily cycle, with a minimum at night and a maximum near midday. This cycle is controlled by the balance of O_3 sources and sinks (Andreae, 1993). At night, O_3 consumption by deposition to vegetation and by reaction with NO emitted from soils reduces the concentration of ozone near the ground. During the day, photochemical O_3 formation and downward mixing of O_3-rich air from higher altitudes replenish it. Differences do exist between the O_3 vertical profiles for the dry and wet seasons (Helas *et al.*, 1995). In view of the sharp increase of O_3 with altitude, frequently observed in the tropics, the risk of vegetation damage in high mountainous regions (where the ground surface intersects the levels where the highest O_3 concentrations are encountered) is high.

Available data show that ambient concentrations of ozone are causing both visible injury and economic damage to crops (Tonneijck, 1989), forest and natural ecosystems (Heck *et al.*, 1998). It is also clear that injurious effects of ozone on vegetation frequently occur as a result of cumulative exposures over many days, weeks or months rather than during a few hours on peak ozone days (Heck *et al.*, 1998). Helas *et al.* (1995) compare the production rate of O_3 from biomass burning (a source strength of

8–13 Tmol/yr) to the downward stratospheric input, which is – besides O_3 production from fossil fuel emissions – the other important source of tropospheric ozone. From this point of view, the tropospheric input of O_3 from biomass burning on a global scale cannot be neglected.

A recent model estimate of tropospheric ozone over Africa (in the region below 100 hPa) gave 26.3 Tg O_3, of which 15.5% was attributed to biomass burning (Marufu, 1999). The author further attributes 0.75 Tg of this sum to biofuel burning, a contribution not strictly bound to biofuel use in Africa alone due to transport. However, he indicates that about 35% of the estimated global biomass burning-related O_3 abundance hails from Africa.

8.3.2 Perturbation of tropospheric oxidant cycles

The large-scale increase in tropospheric ozone in the tropics, expected to persist in the future, indicates a fundamental change in the chemical behaviour of the troposphere. Many gases, mainly hydrocarbons, are continuously emitted into the atmosphere from natural and anthropogenic sources. A build-up of these gases in the atmosphere is prevented by a self-cleansing mechanism, whereby these substances are slowly destroyed photochemically to CO_2 (Andreae, 1993). The key molecule responsible for this oxidation process is the hydroxyl radical (OH). The reaction chains involved (Crutzen, 1995) are such that ozone and OH are consumed when the concentration of NO_x is low, which is normally the case for unpolluted troposphere. According to Crutzen (1995), increases in CH_4 and CO in the atmosphere could lead to global decreases in OH and O_3 and, later, to a further increase in CH_4 and CO as a result of a feedback mechanism.

Injection of large amounts of NO_x from biomass burning may counteract this feedback, since hydrocarbon oxidation in the presence of elevated amounts of NO_x produces additional amounts of O_3 and OH in the presence of light of about 400 nm. Whereas considerable uncertainty persists over the net effect of these changes, it appears likely that biomass burning leads to pronounced regional enhancements of O_3 and OH levels, while on a global scale, it contributes to decreasing average OH concentrations. This would lead to a build-up of many gases that are otherwise removed from the atmosphere by OH. However, until the various contributions to NO_x and O_3 levels in the free atmosphere can be better quantified, the relative importance of the various NO_x sources (e.g., biofuel burning) in perturbing tropospheric O_3 quantities cannot be reliably estimated.

8.3.3 Radiative forcing of climate change

The relative importance of various atmospheric gases in driving the enhanced greenhouse gas effect depends on their pattern of spectral absorption, concentration in the atmosphere, rate of change in concentration above pre-industrial levels and longevity in the atmosphere. Increasing concentrations of a number of trace gases in the atmosphere (mainly CO_2, CH_4, N_2O, O_3 and CFCs) are thought to cause an increase in the global temperatures through the *greenhouse effect* (Ramanathan et al., 1985). An increase in the concentrations of these gases is consequently associated with an increase in radiative forcing. The direct radiative forcings associated with increases in the concentrations of these gases since pre-industrial times, and the average annual rate of increase have all been estimated from climate models (IPCC, 1994b). CO_2 is currently responsible for over 60% of the *enhanced* greenhouse effect (UNEP, 1997).

Methane from past emissions currently contributes 15–20%, while N_2O, CFCs and O_3 contribute the remaining 20% of the enhanced greenhouse effect. Photochemical O_3 production from pyrogenic emissions accounts for about 50% of the increase in O_3 in the atmosphere (Andreae, 1993). Ozone soundings (Delany et al., 1985) show that the effect of O_3 production from biomass burning is quite pronounced even in the middle and upper troposphere, where it contributes most significantly to greenhouse warming. Estimates by Andreae (1993) point to contributions of 20% and 25% of CH_4 and N_2O, respectively, from biomass burning to the tropospheric increase in these gases between 1940–1980. Given the increase in human activity in energy and land-use activities in the tropics after 1980, these ratios may now be higher.

In general, GHG emissions have already disturbed the global energy budget by about 2.5 Wm^{-2} (IPCC, 1994b; UNEP, 1997). If no remedial action is taken to reduce the emissions, existing climate models predict a global warming of between 1.4 and 5.8 °C between 1990 and 2100 (IPCC, 2001). These models are only able to predict patterns of change for the continental scale. Predicting how climate change will affect the weather in a particular region is much more difficult. Thus, the practical consequences of global warming for individual regions or countries remain very uncertain. However, the enormity of the risks associated with the non-abatement of CO_2 emissions on a global scale have been set out in the IPCC assessment reports.

8.4 EMISSION MITIGATION

8.4.1 Role Africa could play in GHG reduction

Industrialized countries have historically been held responsible, since they as a group have some of the highest per capita energy use and also have benefited from emitting vast quantities of GHGs over the last century. However, the emissions in developing countries, led by a few large emitters, will increase in the future as they grow and develop and thus these countries can be viewed as having increased future responsibility. In general, countries differ in the options available to them to directly reduce their emissions.

Some developing countries already use resources very efficiently and have fewer cost-effective alternatives available to them. Others are less efficient. Many, however, have cost-effective options to reduce their future emissions below a business-as-usual baseline. Such opportunities should be capitalized on to achieve the environmental and economic benefits that can accompany the carbon reducing activity, e.g. energy efficiency upgrades, etc.

It is clear from the foregoing that, although climate change issues require concerted action at the global level, successful resolution of the problems they raise will ultimately depend on the domestic policies, institutions and processes of individual countries. Most sub-Saharan African countries have extremely low national incomes, low emissions (both historically and for the predictable future) and, consequently, relatively few opportunities to reduce emissions. Such countries should not be required to take action until their situations change. However, where feasible, they should be encouraged to decrease the carbon intensity of their energy sources and increase the energy efficiency of end use. This would entail investment decisions more than additional expenditure, on the assumption that existing and future funding mechanisms (including investment by other countries through the Clean Development Mechanism (CDM) defined under Article 12 of the Kyoto Protocol will fill the gap.

More specific and reliable country data are needed for countries in order to verify the actual GHG contribution by Africa with a higher degree of accuracy. However, any net emissions are bound to be minute with respect to those of industrialized nations in the northern hemisphere. This raises the important, often-asked question of whether or not the continent should really take part in any emission mitigation efforts. From this point of view, Parties in the continent are exempt from commitments provided in the Kyoto Protocol. From a moral perspective, however, it is imperative that African countries participate in GHG emission reduction efforts within a common, but differentiated framework. This should be oriented as to take advantage of opportunities for financing and technology transfer provided within the framework of the UNFCCC and its Kyoto Protocol.

8.4.2 Choice of sectors for GHG mitigation activities: the criteria

It is now clear that emissions from traditional cooking in Africa are significant enough to influence the tropical and subtropical atmosphere. Compared to CO_2, CO and NO_x emissions from wild fires of Africa, the source strengths from cooking are approximately 30–40% of the extensive savannah and tropical forest fires (Marufu, 1999). However, the contribution from savannah and forest fires to the national total in a country such as Kenya appears insignificant within the official country inventory. This implies that the best GHG mitigation strategies need to be tailor-made

on a country-by-country basis, since the contribution from savannah and forest fires gains enormous importance in western and southern African countries.

To achieve such inventories, complete, more precise and updated energy and land-use/forestry emission inventories for African countries are mandatory. For instance, the assessment in this chapter reveals the domestic biofuel energy sector to be the most significant target sector for such efforts in Kenya. For most of the population, this may involve altering the daily fuel use characteristics and other traditional habits, which may significantly influence the composition and quantities of the trace emissions. Although this also applies to most other African countries, in some of them the emphasis would be on changes in land-use and forestry systems.

8.4.3 The challenges

The total requirement for biomass energy is unlikely to decrease in the near or medium term in most African countries unless there is a dramatic switch to commercial fuels by a large share of the population (Scurlock and Hall, 1990). There exists no current evidence of this, since urbanization is accompanied by continued reliance on biomass fuels, particularly charcoal, and the population increase remains high. Due to poor chances of access to cleaner, more convenient commercial fuels in the medium term, most people in the developing countries will have to rely on biomass as the fuel of development. Bioenergy has also been proposed as one source that will play a central role in future, more sustainable energy scenarios (Hall and Scrace, 1998). The challenge, therefore, is for developing countries to formulate appropriate and effective mitigation strategies that will fulfil their development needs while satisfying the objectives of the UNFCCC. Such strategies should be developed on the basis of accurate data, should demonstrate quantifiable results, be cost-efficient and have the capacity to transfer the associated technologies, not only within one country but to the entire continent (UNEP, 1998).

One recently proposed option relevant for Kenya (Kituyi and Helas, 1999) was based on the reported observation that a switch from a traditional to an improved biofuel stove by a household cuts back on emissions of carbon and nitrogen, the fuel amounts and monetary costs (Kituyi, 2000). It was proposed that, under the CDM, Kenya could undertake an improved stove dissemination programme in collaboration with an industrialized country partner. The partner would, in turn, meet the emission limitation and reduction commitments. The improved wood and charcoal stoves would be distributed free of charge or at subsidized rates in a project involving 25% of the population of firewood stove users (591,000) and charcoal stove users (435,000) in the first year. This scenario would imply total annual fuel savings of 0.35 million tonnes of round wood equivalent (1.2% of total annual consumption) and emission reductions of 0.15 Tg of carbon (1.1%) and 1,300 tonnes nitrogen (1.1%) at the end of the first project year. Besides, an average family of six switching from traditional to improved stove would realize annual savings of US$15 for charcoal users and US$11 for firewood users.

The majority of the rural households in Kenya would wish to switch to improved stoves mainly as a status symbol (Kituyi *et al.*, 2001a), but would not be able to afford it owing to the high costs involved. Commissioning of such a cost-efficient strategy would therefore be met with less risk of failure. The potential for successful transfer of this option to neighbouring countries has been demonstrated. Other countries where vigorous projects for improved stove dissemination have recently been implemented include Botswana, Burkina Faso, Burundi, Ghana, Malawi, Mali, Rwanda, Sudan, Tanzania, Uganda, Zambia and Zimbabwe (Walubengo and Kimani, 1993). Except for the large cook stove programmes in India and China, the varied efforts to introduce improved cook stoves to East Africa have been the most extensive (Kammen, 1995). The problems encountered and the lessons learnt in East Africa are now being considered in the development and distribution of cook stoves worldwide (Kammen, 1995). Potential exists, therefore, for extensive emission mitigation projects in all developing countries, provided the relevant data are made available. Also, the above-mentioned regional transfer of improved stove technologies has proved more successful compared to the failed attempt to disseminate biogas, wind, solar and photovoltaic technologies in these countries, for documented reasons (Walubengo and Kimani, 1993).

8.5 CONCLUSION

It is clear that fires from the household energy and land-use sectors (including savannah, forest clearing and agro-wastes) are the most important GHG emission sources in Africa, contributing about 4% to the global overall CO_2 budget. The significance of these sources is, however, exemplified in their role in global photochemical ozone formation, to which they contribute as much as 35%. This formation is mainly dependent upon biomass fire emissions of the main ozone precursors – NMHCs, CO and NO_x. Given the current status of available data on sources and sinks, it is not clear, however, whether the continent is indeed a net CO_2 absorber or emitter. However, the importance of participating in emission reduction efforts by all countries is appreciated in this chapter. Such efforts prescribed for Africa need to be country-based, besides being within the framework of activities included in the household energy and land-use/forestry sectors. The choice of the best mitigation options must be informed by reliable activity data that are mostly not documented for most countries.

From the point of view of collective responsibility, African countries need to participate in emission reduction efforts, using options that will allow them access to basic but sustainable economic development. Such opportunities are provided for in the UNFCCC and its Kyoto Protocol. However, the starting point for the continent should be the development of comprehensive national GHG inventories and the improvement of the quality of activity data (where some data already exist) required for GHG budget estimation.

ACKNOWLEDGEMENTS

The authors are indebted to the Max Planck Institute for Chemistry, Mainz, Germany, and the German Academic Exchange Service (DAAD) for supporting this work. They also appreciate the valuable comments of Professor Peter Brimblecombe and Dr Pedro Sanchez, which significantly added value to this manuscript.

REFERENCES

Andreae, M. O. (1991) Biomass burning: its history, use and distribution and its impact on environmental quality and global climate. In J. S. Levine (Ed.), *Global Biomass Burning*, pp. 3–28. Cambridge, MA, MIT Press.

(1993) The influence of tropical biomass burning on climate and the atmospheric environment. In R. S. Oremland (Ed.), *Biogeochemistry of Global Change: Radiatively Active Trace Gases*, pp. 113–150. New York, Chapman and Hall.

Andreae, M. O., Browell, E.V., Garstang, M. *et al.* (1988) Biomass burning emissions and associated haze layers over Amazonia. *J. Geophys. Res.*, **93** (D2), 1,509–1,527.

Andreae, M. O., Chapuis, A., Cros, B., *et al.* (1992) Ozone and Aitken nuclei over equatorial Africa: airborne observations during DECAFE 88. *J. Geophys. Res.*, **97** (D6), 6,137–6,148.

Andreae, M. O., Anderson, B. E., Blake, D. R., *et al.* (1994) Influence of plumes from biomass burning on atmospheric chemistry over the equatorial and tropical South Atlantic during CITE 3. *J. Geophys. Res.*, **99** (D6), 12,793–12,808.

Bachou, S. A. (1991) Improved cook stoves in Uganda. In *Resources: Journal for Sustainable Development in Africa*, Vol. 2, pp. 36–38, Nairobi, Kenya, KENGO.

Bonsang, B., Boissard, C., LeCloarec, M. F., Rudolf, J. and Lacaux, J. P. (1995) Methane, carbon monoxide and light nonmethane hydrocarbon emissions from African savannah burning during the FOS/DECAFE Experiment. *J. Atmos. Chem.* **22**, 149–162.

Bouwman, A. F., Hoek, K. W. V. d. and Olivier, J. G. J. (1995) Uncertainties in the global source distribution of nitrous oxide. *J. Geophys. Res.*, **100** (D2), 2,785–2,800.

Brasseur, G., Guenter, A. and Horwitz, L. (2005) Atmospheric chemistry in the tropics. This volume.

Brocard, D., Lacaux, J. and Eva, H. (1998) Domestic biomass combustion and associated atmospheric emissions in West Africa. *Global Biogeochemical Cycles*, **12**, 127–139.

Cahoon, D. R., Stocks, B. J., Levine, J. S., Cofer, W. R. and O'Neill, K.P. (1992) Seasonal distribution of African savanna fires. *Nature*, **359**, 812–815.

Claussen, E. and McNeilly, L. (1998) *The Complex Elements of Global Fairness.* Washington, DC, Pew Center on Global Climate Change, 36pp.

Cros, B., Delmas, R., Nganga, D., Clairac, B. and Fontan, J. (1988) Seasonal trends of ozone in equatorial Africa: experimental evidence of photochemical formation. *J. Geophys. Res.*, **93**, 8,355–8,366.

Crutzen, P. J. (1995) On the role of ozone in atmospheric chemistry. In *7th BOC Priestley Conference on The Chemistry of the Atmosphere: Oxidants and Oxidation in the Earth's Atmosphere*, A.R. Bandy (Ed.). The Royal Society of Chemistry, pp. 3–22.

Crutzen, P.J. and Andreae, M.O. (1990) Biomass burning in the tropics: Impact on atmospheric chemistry and biogeochemical cycles. *Science*, **250**, 1,669–1,678.

Crutzen, P. J., Heidt, L. E., Krasnec, W. H. and Seiler, W. (1979) Biomass burning as a source of atmospheric trace gases CO, H_2, N_2O, NO, CH_3Cl and COS. *Nature*, **282**, 253–256.

Delany, A. C., Haagensen, P., Walters, S., Wartburg, A. F. and Crutzen, P.J. (1985) Photochemically produced ozone in the emission from large-scale tropical vegetation fires. *J. Geophys. Res.*, **90** (D1), 2,425–2,429.

Delmas, R. (1994) An overview of present knowledge on methane from biomass burning. *Fert. Res.*, **37**, 181–190.

Fishman, J. (1988) Tropospheric ozone from satellite total ozone measurements. In I.S.A. Isaksen (Eds.), *Tropospheric Ozone: Regional and Global Scale Interactions*, pp. 111–123. Dordrecht, Reidel.

Fishman, J., Watson, C. E., Larsen, J. C. and Logan, J. A. (1990) The distribution of tropospheric ozone determined from satellite data. *J. Geophys. Res.*, **95**, 3,599–3,617.

Granier, C., Hao, W. M., Brasseur, G. and Muller, J. F. (1996) Land-use practices and biomass burning: impact on the chemical composition of the atmosphere. In J.S. Levine (Ed.), *Biomass Burning and Global Change: Remote Sensing, Modelling and Inventory Development, and Biomass Burning in Africa*, pp. 140–148. Cambridge, MA, MIT Press.

Hall, D. O. and Scrace, J. I. (1998) Will biomass be the environmentally friendly fuel of the future? *Biomass and Bioenergy*, **15**, 357–367.

Hao, W. M., Ward, D. E., Olbu, G., Baker, S. P. and Plummer, J. R. (1996) Emissions of trace gases from fallow forests and woodland savannas in Zambia. In J. S. Levine (Ed.), *Biomass Burning and Global Change: Remote Sensing, Modelling and Inventory Development, and Biomass Burning in Africa*, pp. 361–369. Cambridge, MA, MIT Press.

Heck, W. W., Furiness, C. S., Cowling, E. B. and Sims, C. K. (1998) Effects of ozone on crop, forest and natural ecosystems: assessment of research needs. In *EM: Magazine for Environmental Managers*, pp. 11–22. Pittsburg, PA, A&WMA.

Helas, G. (1995) Emissions of atmospheric trace gases from vegetation fires. *Phil. Trans. Roy. Soc. Lond. A* **351**, 297–312.

Helas, G. and Pienaar, J. J. (1996) Biomass burning emissions. In G. Held, B.J. Gore, A.D. Surridge, G.R. Tosen, C.R. Turner and R.D. Walmsley (Eds.), *Air Pollution and its Impacts on the South African Highveld*, pp.12–15. Environmental Scientific Association and National Association for Clean Air.

Helas, G., Lobert, J., Scharffe, D., *et al.* (1995) Ozone production due to emissions from vegetation burning. *J. Atmos. Chemi.*, **22**, 163–174.

IPCC (1994a) *Climate Change 1994: Radiative Forcing of Climate Change and an Evaluation of the IPCC IS92 Emission Scenarios.* J.T. Houghton, L. G. M. Filho, J. Bruce, H. Lee, B.A. Callander, E. Haites, N. Harris and K. Maskell (Eds.). Cambridge, Cambridge University Press, 339pp.

(1994b) *Radiative Forcing of Climate Change. The 1994 Report of the Scientific Assessment Working Group of IPCC: Summary for Policymakers.* Intergovernmental Panel on Climate Change, 28pp.

(2001) *Third Assessment Report: Summaries for Policymakers from the Working Groups.* Intergovernmental Panel on Climate Change (IPCC). [http://www.ipcc.ch/]

Kammen, D. M. (1995) From energy efficiency to social utility: lessons from cook stove design, dissemination and use. In J. Goldemberg and T.B. Johanson (Eds.), *Energy as an Instrument for Socio–economic Development*, pp. 50–62. New York, UNDP.

Koppmann, R., von Czapiewski, K. and Komenda, M. (2005) Natural and human induced biomass burning in Africa: an important source for volatile organic compounds in the troposphere. This volume.

Kituyi, E. N. (2000) Trace gas emission budgets from domestic biomass burning in Kenya. Ph.D. Thesis, University of Nairobi, Kenya.

Kituyi, E. and Helas, G. (1999) Bioenergy in developing countries: opportunities for energy saving and emission reduction. *Energy and Agriculture Towards the Third Millennium*, pp.171–178. Athens, Greece, Agricultural University of Athens.

Kituyi, E., Marufu, L., Huber, B. *et al.* (2001a) Biofuel consumption rates and patterns in Kenya. *Biomass and Bioenergy*, **20**, 83–99.

Kituyi, E., Marufu, L., Wandiga S. O. *et al.* (2001b) Emission characteristics of CO and NO from biofuel fires in Kenya. *Energy Conversion and Management*, **42**, 1,517–1,542.

Kuhlbusch, T. A. J. and Crutzen, P. J. (1995) Toward a global estimate of black carbon in residues of vegetation fires representing a sink of atmospheric CO_2 and a source of O_2. *Global Biogeochemical Cycles*, **9**, 491–501.

Lacaux, J. P., Brocard, D., Lacaux, C., *et al.* (1994) Traditional charcoal making: an important source of atmospheric pollution in the African tropics. *Atmos. Res.*, **35**, 71–76.

Levine, J. S. (1996) *Biomass Burning and Global Change*, Cambridge, MA, MIT Press.

Lopez, R. (1999) *Incorporating Developing Countries Into Global Efforts for Greenhouse Gas Reduction.* RFF Climate Brief No. 6, January. Washington, DC, Resources for the Future, 8pp.

Marenco, A., Macaigne, M. and Prieur, S. (1989) Meridional and vertical CO and CH_4 distributions in the background troposphere (70 N–60 S; 0–12 km altitude) from scientific aircraft measurements during the STRATOZ III experiment (June 1984). *Atmos. Environ.*, **23**, 185–200.

Marufu, L. (1999) Photochemistry of the African troposphere: The influence of biomass burning. Ph.D. Thesis, University of Utrecht, the Netherlands.

Marufu, L., Ludwig, J., Andreae, M. O., Meixner, F. X. and Helas, G. (1997) Domestic biomass burning in rural and urban Zimbabwe – Part A. *Biomass and Bioenergy* **12**, 53–68.

Ministry of Research and Technology (MRT) (1998) *Compendium of Inventory of Greenhouse Gas Emissions and Sinks in Kenya*, Kenya Country Study on Climate Change Project, Ministry of Research and Technology. 64pp.

Okoth-Ogendo, H. W. O. (1995) Environmental governance and development policy. In H. W. O. Okoth-Ogendo and J. B. Ojwang (Eds.) *A Climate for Development: Climate Change Policy Options for Africa*, pp. 239–251. Nairobi, Kenya, African Centre for Technology Studies/Stockholm Environment Institute.

Othieno, H. (1991) Energy contribution to climate change: the case of Kenya. In S. H. Ominde and C. Juma (Eds.), *A Change in the Weather: African Perspectives on Climate Change*, Vol. 5. Nairobi, Kenya, African Centre for Technology Studies, 210pp.

Ramanathan, V., Cicerone, R. J., Singh, H. B. and Kiehl, J. T. (1985) Trace gas trends and their potential role in climate change. *J. Geophys. Res.*, **90** (D3), 5,547–5,566.

Schlesinger, W. H. (1997) *Biogeochemistry: An Analysis of Global Change.* San Diego, USA, Academic Press. 558pp.

Scurlock, J. M. O. and Hall, D. O. (1990) The contribution of biomass to global energy use. *Biomass*, **21**, 75–81.

Seiler, W. and Crutzen, P. J. (1980) Estimates of gross and net fluxes of carbon between the biosphere and the atmosphere from biomass burning. *Climatic Change*, **2**, 207–247.

Tonneijck, A. E. G. (1989) Evaluation of ozone effects on vegetation in the Netherlands. In T. Schneider, S. D. Lee, G. J. R. Wolters and L. D. Grant (Eds.), *Atmospheric Ozone Research and its Policy Implications*, pp. 251–260. Elsevier Science.

UNEP (1997) *Climate Change Information Kit.* UNEP Information Unit for Conventions (IUC).

(1998) *The Clean Development Mechanism and Africa. New Partnerships for Sustainable Development: The Clean Development Mechanism Under the Kyoto Protocol.* UNEP Collaborating Centre on Energy and Environment, 16pp.

UNFCCC (1997) *Kyoto Protocol to the United Nations Framework Convention on Climate Change.* Bonn, Germany, UNFCCC, 23pp.

Walubengo, D. and Kimani, M. J. (1993) Dissemination of RETS in Africa. In D. Walubengo and M. J. Kimani (Eds.), *Whose Technologies? The Development and Dissemination of Renewable Energy Technologies (RETS) in Sub-Saharan Africa.* Nairobi, Kenya, KENGORWEPA.

World Bank (1999) *World Development Report 1998/1999.* Washington, DC, The World Bank, 251pp.

9 Soil micro-organisms as controllers of trace gas emissions over southern Africa

LUANNE B. OTTER[1] AND MARY C. SCHOLES[2]

[1]*Climate Research Group, University of the Witwatersrand, South Africa*
[2]*School of Animal, Plant and Environmental Sciences, University of the Witwatersrand, South Africa*

Keywords

Southern Africa; methane; nitric oxide; nitrous oxide; savannas; soil micro-organisms

Abstract

Soils contribute to the budgets of many atmospheric trace gases by acting as sources or sinks. The most important trace gases include methane (CH_4), nitrous oxide (N_2O) and nitric oxide (NO), which are both consumed and produced by soils. In principle, one has to distinguish between the processes known to produce or consume these gases that are probably irrelevant at the low concentrations typical of atmospheric trace gases, and the processes that really play some role in the gas exchange between soil and atmosphere. The absolute values of total budgets and percentage contributions by soils should not be taken for granted, because the individual source and sink strengths are highly uncertain. There are a number of reasons for this uncertainty. The fluxes of trace gases between soil and atmosphere can be measured with some reliability by using various approaches, but it is not trivial to estimate atmospheric budgets from field fluxes. The problem is that fluxes generally show a high variability with respect to site and time. Integration of fluxes over larger areas and extended periods does not necessarily solve the problem, since each individual flux event is caused by deterministic processes that change in a non-linear way when conditions change even slightly. In addition, soils are presently looked upon as a macroscopic system, although the function is controlled predominantly on a microscopic level, i.e. the level of the micro-organisms. This chapter aims to present results from site-specific chamber measurements taken in the field and incubations in the laboratory. These studies are compared with modelling approaches, integrating the flux over time and space. Section 9.7 deals with impacts of biogenic trace gases on the atmospheric chemistry of the southern African region.

9.1 INTRODUCTION

The atmosphere, which is a mixture of gases, surrounds the earth and supports all forms of life on earth. It consists mainly of nitrogen, oxygen and variable amounts of water vapour. It has been recognized for many years that the composition of the atmosphere is a product of, and controlled by, the activity of the biosphere. Without the inputs from the biota, the atmosphere would have proceeded towards thermodynamic equilibrium represented by a carbon dioxide (CO_2) atmosphere (Lovelock, 1979). The presence of oxygen and nitrogen are linked to processes such as photosynthesis and microbial activity. CO_2 and nitrogen are removed from the atmosphere during the processes of photosynthesis and nitrogen fixation. These are in turn balanced by the inputs of CO_2 and nitrogen gases, such as nitrous oxide (N_2O) and nitric oxide (NO), via the processes of respiration, decay, nitrification and denitrification.

There are a variety of other biological processes, many of them being microbiological, which lead to the release or uptake of trace gases such as hydrocarbons, methane (CH_4), carbon monoxide (CO), ammonia and reduced sulphur compounds. The trace gas component of the atmosphere is less than 0.1%; however, it is a very important component of the atmosphere. The increased levels of CO_2 and trace gases have led to global warming, the destruction of the stratospheric ozone layer, the increase in tropospheric ozone, changes in the density of clouds in the troposphere and aerosols in the stratosphere (Crutzen, 1979; Graedel and Crutzen, 1993). These trace gases are not only radiatively active greenhouse gases with warming potentials, but are also chemically reactive and can change the atmospheric composition. All these dramatic changes in our atmosphere can have serious consequences for the habitability of the earth.

Most of the atmospheric trace gases undergo cycles that have significant biospheric components, such as oceans, vegetation, soils or animals, or are dominated by biospheric processes. Soils are particularly important as they are the habitat for micro-organisms and contribute to the budgets of many atmospheric trace

Climate Change and Africa, ed. Pak Sum Low. Published by Cambridge University Press. © Cambridge University Press 2005.

Table 9.1. Contribution of soil to some of the global cycles of atmospheric trace gases.

Trace gas	Mixing ratio (ppbv)	Lifetime (days)	Total budget (Tg yr^{-1})	Contribution (%) of soils Source	Contribution (%) of soils Sink	Importance in the atmosphere
H_2	550	1,000	90	5	95	Insignificant
CO	100	100	2,600	1	15	Tropospheric chemistry
CH_4	1,700	4,000	540	60	5	Greenhouse effect; tropospheric and stratospheric chemistry
N_2O	310	60,000	15	70	?	Stratospheric chemistry; greenhouse effect
NO	<0.1	1	60	20	?	Tropospheric chemistry

Adapted from Conrad, 1996

gases, including H_2, CO, CH_4, N_2O and NO (Table 9.1) (Conrad, 1996a). Microbial processes are not only important sources of trace gases but are also important sinks.

Globally, the anthropogenic and biogenic sources of NO_x ($NO + NO_2$) have been suggested to be similar (15–29 Tg N yr^{-1} and 6–18 Tg N yr^{-1}, respectively) (Delmas et al., 1997). Soil emissions are the major source of biogenic NO_x emissions, contributing as much as 40% to the global NO_x budget (24–54 Tg N yr^{-1}) (Potter et al., 1996a; Delmas et al., 1997). Globally, soils are estimated to consume 17–23 Tg CH_4 yr^{-1} (Potter et al., 1996b). Consumption of CH_4 (due to the oxidation of atmospheric CH_4 by bacteria in the soil) has been shown to occur in well-drained soils of grasslands (Mosier et al., 1991), forests (Keller et al., 1986; Steudler et al., 1989; Adamsen and King, 1993; Castro et al., 1995; Whalen and Reeburgh, 1996; MacDonald et al., 1998), tundra (Whalen and Reeburgh, 1990), savannas (Seiler et al., 1984; Zepp et al., 1996), deserts (Striegl et al., 1992) and agricultural land (Keller et al., 1990; Tate and Striegl, 1993; Flessa et al., 1995). Potter et al. (1996b) indicated that 40% of the global CH_4 consumption occurs in relatively dry, warm ecosystems.

It is not only the savanna areas of Africa that are important for CH_4 budgets, but also the natural wetland regions that are CH_4 sources. Natural wetlands contribute 20–25% to the global CH_4 source of 500–550 Tg yr^{-1} (Conrad, 1997). In Africa, wetland areas are estimated to cover 355×10^3 km^2, and of this 50% is flood plain and 40% is swamps and marshes (Aselmann and Crutzen, 1989); however, CH_4 flux data for flood plains in particular are lacking. CH_4 fluxes are higher over inundated areas and shallow waters (flood plains, fens and marshes) than over deeper open waters (Morrissey and Livingston, 1992; Wassmann and Thein, 1996). The wetland areas in Africa are mostly inundated rather than open water, and have the potential to produce large amounts of CH_4 because of the warm temperatures. The surrounding upland systems to the wetlands and flood plains are the most important areas in which CH_4 consumption occurs. It is estimated that 60%

of the total annual global CH_4 production is oxidized microbially, often in areas adjacent to the zone of production (Reeburgh et al., 1993).

Tropical savannas have been identified as an important source area for NO emissions and CH_4 fluxes, and since savannas cover approximately 11.5% of the global land surface (Scholes and Hall, 1996), they have the potential to contribute up to 40% of the global soil NO_x emissions (Davidson, 1991). Savannas cover 65% of Africa, and their tropical position makes them extremely important in terms of global trace gas fluxes and budgets. However, the numbers of studies in Africa are limited and many more studies are required to improve regional as well as global budgets.

This chapter investigates soil micro-organisms and how they influence the production and consumption particularly of NO, N_2O and CH_4, and how environmental factors influence emission. Data collected in a South African savanna will be used to illustrate these points and the relationships. The various methods used for measuring trace gas fluxes from soils are also discussed, and comparisons between measured and modelled fluxes from a South African savanna will be made.

9.2 TYPES OF SOIL MICRO-ORGANISM

The definition of micro-organisms as described in Conrad (1996a) is all organisms that are too small to be separated from the soil matrix by mechanical tools, such as bacteria, algae, protozoa and also microfauna smaller than about 1 mm. Bacteria are classified based on nutritional patterns, oxygen needs and symbiotic relationships (Miller and Donahue, 1990). Autotrophic bacteria produce their own food through the synthesis of inorganic materials. Heterotrophic bacteria derive carbon and energy directly from organic substances. Fungi and most bacteria are heterotrophs. The most important group of autotrophic soil bacteria are those that oxidize ammonium to nitrites and then to nitrates. This process, known as nitrification, involves two microbial physiotypes that both use CO_2 as the principal carbon source for biomass formation, and obtain the energy from the oxidation of inorganic nitrogen. Denitrification, on the other hand, is an anaerobic process in which heterotrophic denitrifying bacteria utilize nitrate and nitrite for their growth and reduce them to molecular nitrogen. Both NO and N_2O can be produced via nitrification and denitrification.

Bacteria are then further divided into chemo- and phototrophs (photoautotrophs, photoheterotrophs, chemoautotrophs, chemoheterotrophs), depending on where they obtain their energy. Phototrophs obtain their energy from the sun, while chemotrophs obtain their energy from the oxidation of inorganic substances or organic matter. Micro-organisms are also classified either as aerobic organisms if they require a free gaseous oxygen source, such as methanotrophs, or as anaerobic if they do not require free oxygen and can use electron acceptors other than oxygen, such as the methanogens, which produce CH_4.

Algae are chlorophyll-bearing organisms that develop best in moist, fertile soils. They are not important decomposers of organic matter but are producers of new photosynthetic growth. Algal growth on the soil can produce considerable amounts of organic material. Protozoa are unicellular organisms, which ingest bacteria, fungi and other bacteria, thereby controlling the microbial population and hastening the recycling of plant nutrients.

9.3 MECHANISMS FOR THE PRODUCTION AND CONSUMPTION OF TRACE GASES

Conrad (1996a) proposes that three different categories of soil processes play a role in trace gas exchange: chemical processes, enzymatic processes and microbial processes. Chemical processes involve the production of H_2 when moist samples are brought into contact with iron- or steel-containing compounds, CO through thermal decomposition of humic acids and other organic materials (Conrad and Seiler, 1985; Schade et al., 1999), N_2O by decomposition of hydroxylamine, and NO through the decomposition of nitrite (Chalk and Smith, 1983; Blackmer and Cerrato, 1986). Soil enzymatic processes refer to the abiontic enzymes which are defined as extracellular enzymes, enzymes bound to inert soil particles, and enzymes within dead cells (Conrad, 1996a). Abiontic soil hydrogenases, which are responsible for the oxidation of atmospheric H_2, are the only enzymes at this stage that are important for soil–atmosphere trace gas exchange. Soil catalases are proposed to be involved in the decomposition of N_2O.

The majority of trace gas exchanges between soils and the atmosphere are, however, due to micro-organisms. Two microbial processes control emissions from a particular ecosystem: production and consumption (or oxidation). Soils both produce and consume NO_x, N_2O and CH_4 so the observed flux is a result of the simultaneous production and consumption of these two processes (Conrad, 1994). The compensation-mixing ratio is a point at which the production rate is equivalent to the consumption rate so that the net flux is zero. Consumption occurs if ambient mixing ratios are higher than the compensation-mixing ratio.

9.4 METHODS FOR MEASURING SOIL TRACE GAS FLUXES

Fluxes of trace gases between the soil and the atmosphere can be measured by using a variety of approaches, from laboratory to field measurements. Scaling up is a difficult task due to the variability with respect to space and time. Furthermore, there is the great variability in soil micro-organisms that complicates things further, as they behave differently under the various environmental conditions. A comparison between the data from each of the techniques discussed below will be shown in Section 9.5 when the effects of environmental factors on the fluxes are discussed.

9.4.1 Laboratory techniques

Laboratory incubation methods for measuring trace gas fluxes are often used, as they are useful in investigating the effects of the environment on the fluxes. This is because one can manipulate the environment, such as the temperature or soil moisture and determine the relationship between these factors. The emission rates of CH_4, NO and N_2O under various environmental conditions have been measured in the laboratory (Remde et al., 1989; Drury et al., 1992; Rudolph and Conrad, 1996; Koops et al., 1997; Otter et al., 1999).

There are different kinds of incubation techniques which all have their advantages and disadvantages. The first division is between closed incubation systems and dynamic incubation systems. In closed systems, gas samples are periodically collected from the headspace of the soil chamber. The flux is calculated by measuring the change in concentration of the gas with time. Closed incubation systems allow for the accumulation of the gaseous product, which makes the measurement of small fluxes easier, and equipment is fairly simple and relatively easy and inexpensive to prepare. The disadvantage is that intermediate gaseous products are kept in contact with the soil for longer, which could lead to further chemical reactions and produce products that are not normally found in the field. Furthermore, by allowing the accumulation of trace gases within the chamber, the concentration gradient between the soil and the atmosphere is increased, which may lead to a decreased trace gas flux.

In dynamic incubation systems, purified or background air is continuously drawn through the chamber and forced to flow over the soil surface. The gas flux from the soil surface can be calculated from the concentration difference between the air from the inlet and the outlet, flow rate, and surface area. The use of dynamic chambers eliminates the gas accumulation problems by continuously removing accumulated gases. Dynamic chambers simulate field conditions better than closed systems and also allow for the continuous measurement of trace gases. The disadvantage of laboratory dynamic chambers, however, is that by continuously flushing the chambers with (dry) air, the soil water content of the samples in the chambers is reduced, which will in turn influence the trace gas fluxes. It is also difficult to measure N_2O fluxes from dynamic chambers because of the low concentrations, so the higher the flow rate, the more difficult it is to detect N_2O (Hutchinson and Andre, 1989; Remde *et al.,* 1989; Flessa and Beese, 1995). Higher flow rates are, however, better for the removal of accumulated gases. Closed incubation systems are most often used in laboratory studies but a few dynamic systems have been developed (Hutchinson and Andre, 1989; Remde *et al.,* 1989; Yang and Meixner, 1997). Bollmann *et al.* (1999) recently compared the NO production, consumption and turnover rates and showed that NO consumption rate constants and production rates were lower in a closed system than in a dynamic system. This indicates that consumption became saturated due to the accumulated NO in the closed systems. It was, however, suggested that for oxic conditions the closed system was cheap, easy and reliable in the measurement of NO turnover parameters.

In addition to the above, there are also soil chambers and soil microcosms. In soil chambers, a soil sample is placed in a chamber and, if it is a dynamic system, the air flows over the soil surface (Remde *et al.,* 1989; Yang and Meixner, 1997). The disadvantage of this method is that the soil structure is disturbed. A microcosm, on the other hand, is a core of soil that is removed from the field with as little disturbance of the soil structure as possible (Hantschel *et al.,* 1994). In these systems the air flows through the soil column instead of over the surface. This type of incubation technique is useful for understanding processes involved in the production of trace gases, but chambers are more suitable for determining flux rates.

9.4.2 Flux measurements in the field

Fluxes from the soil surface to the atmosphere are measured in the field by means of chambers. Chamber methods disturb the soil–plant–atmosphere continuum, yet they are useful, cheap and efficient. Chamber techniques are generally used to determine field trace gas fluxes. Data from these analyses are often used for model validation. Field measurements can, to a limited extent, be manipulated by the addition of fertilizer or by irrigation. Chambers can also be used to measure CH_4 consumption in the field, as well as fluxes from the surface of water. The design of chambers and their limitations have been discussed in detail by a number of authors (Hutchinson and Mosier, 1981; Jury *et al.,* 1982; Sebacher and Harriss, 1982; Mosier, 1989; Schutz and Seiler, 1989; Hutchinson and Livingston, 1993) but are briefly discussed here.

In the field, chambers are placed over the soil and are pushed into the ground. Care needs to be taken when pushing chambers into the soil as this causes pressure changes in the soil. This problem can be overcome by installing collars in the soil that are open to the atmosphere (Mosier, 1989). The chamber is then placed over the collar during measurements. As with the incubation techniques, there are closed or open (or dynamic) chambers that follow a principle similar to that described in Section 9.1. In addition to the disadvantage of the closed chambers discussed above, the closed field chambers also alter the atmospheric pressure fluctuations that are normally found at the soil surface; boundary layer resistance at the soil–atmosphere interface may be higher inside the chamber than outside the chamber; and temperature and humidity changes in the soil and under the chamber can occur. It is therefore important to keep the measuring period as short as possible and expose the enclosed area to natural conditions at regular intervals. To further reduce temperature and humidity changes, chambers can be covered with a reflective material. The main advantage of open field chambers is that they maintain environmental conditions close to those outside the chambers.

9.5 FACTORS CONTROLLING THE PRODUCTION AND CONSUMPTION OF TRACE GASES BY MICRO-ORGANISMS

Although production and consumption may occur simultaneously in the soil, the relative magnitudes of these processes differ under varying environmental conditions (Galbally, 1989; Saad and Conrad, 1993; Conrad, 1996a, 1997). This makes extrapolation from soil to global scale difficult; therefore, an understanding of the environmental controls of trace gas fluxes is critical.

9.5.1 NO and N_2O fluxes

The factors controlling or regulating the fluxes of NO and N_2O have been reviewed in a number of papers (Firestone and Davidson, 1989; Galbally, 1989; Davidson, 1991, 1993; Williams *et al.,* 1992; Meixner, 1994; Conrad, 1996b). Firestone and Davidson (1989) take a conceptual approach and illustrate the regulation of NO in two different ways. First, there is the 'hole-in-the-pipe' model that they propose. This model illustrates three levels at which regulation of NO and N_2O fluxes in soils can occur (Figure 9.1).

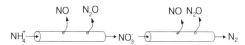

Figure 9.1. Diagram taken from Firestone and Davidson (1989) illustrating the conceptual 'hole-in-the-pipe' model of the levels of regulation of nitrification and denitrification and thus the nitrogenous trace gas fluxes.

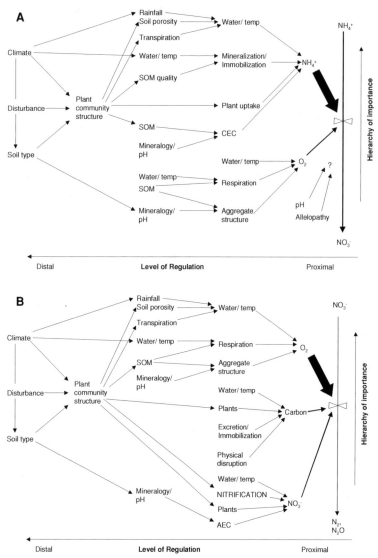

Figure 9.2. A diagram showing the distal and proximal controls of nitrification (A) and denitrification (B) (from Firestone and Davidson, 1989).

At the first level of regulation are the factors affecting rates of nitrification and denitrification, in other words the amount of N cycling through the ecosystem. Ecosystems with high nitrogen availability are expected to have higher N trace gas emissions, and soils such as agricultural soils that are fertilized with nitrogen generally show increased NO emissions. Second are the factors affecting the relative proportions of end products produced, represented by the size of the holes in the pipes. Factors affecting the ratio of NO and N_2O include acidity, temperature, soil moisture content, organic carbon availability, NO_3^- availability and O_2 availability. At the third level are the factors affecting gaseous diffusion though the soil and to the atmosphere.

The model of Robertson (1989) relates the immediate physiological (proximal) controls on processes (in this case nitrification

and denitrification) to the progressively larger scale (distal) controls (Figure 9.2).

Most enzymatic processes are temperature regulated, with the rate of the process increasing exponentially with temperature. The production and consumption of NO occurs mainly via enzymatic processes; therefore, soil temperature is an important controlling factor. NO fluxes have been shown to have this characteristic increase with temperature (Slemr and Seiler, 1984; Anderson and Levine, 1987; Williams *et al.*, 1987, 1988; Williams and Fehsenfeld, 1991; Scholes *et al.*, 1997; Martin *et al.*, 1998), with an approximate doubling of the NO emission rate for each 10 °C rise in temperature (Williams *et al.*, 1992). NO emissions are shown to decline when soil temperatures increase above 35 °C and are near zero as soil temperatures approach freezing point (Williams

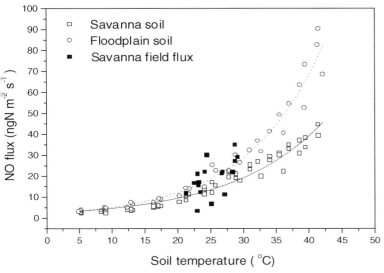

Figure 9.3. Soil NO fluxes measured in the laboratory using a dynamic incubation system show that emissions increase with temperature in sandy savanna soils and clayey flood plain soils (modified from Otter *et al.*, 1999). As a comparison, the emissions measured in the field are shown in solid squares. Field measurements cover only a limited temperature range, as one cannot manipulate the environment to the same extent as in the laboratory.

et al., 1992). There are, however, a few investigations that show that soil temperature has little or no effect on NO flux (Johansson *et al.*, 1988; Cárdenas *et al.*, 1993; Neff *et al.*, 1995), particularly in tropical regions where there are low temperature variations. Laboratory incubation studies show that NO emissions from savanna soils from South Africa increase with temperature (Figure 9.3) (Otter *et al.*, 1999). Field NO fluxes from the same site also show a positive relationship with soil temperature (except under very dry conditions), which supports the laboratory data (Scholes *et al.*, 1997).

Although soil temperature is an important controlling factor, other factors also influence the flux and can often be limiting, thus obscuring the temperature effect. This can cause a different response of NO to soil temperature in the wet and dry seasons, with no temperature response during the dry season and a positive response in the wet season (Meixner *et al.*, 1997).

Soil moisture controls NO production and consumption by regulating the rate of nitrification and denitrification through its effect on substrate supply to microbes and gaseous diffusion. NO emissions have been shown to increase with increasing soil moisture (Slemr and Seiler, 1984; Anderson and Levine, 1987; Johansson and Sanhueza, 1988; Johansson *et al.*, 1988; Williams and Fehsenfeld, 1991; Scholes *et al.*, 1997). Some studies show that there is an increase in flux with soil moisture, but only to optimal soil moisture, after which the flux rate declines (Davidson, 1991, 1992; Cárdenas *et al.*, 1993). One of the reasons for this is that diffusion of gases through water is about 10,000 times slower

than through air (Galbally and Johansson, 1989; Davidson and Schimel, 1995), thus at high soil moistures the diffusion of O_2 into the soil and NO out of the soil is limited. Furthermore, the influence of soil moisture on O_2 availability affects nitrification and denitrification rates, as nitrification requires O_2 for oxidation processes, and denitrification generally occurs in the absence of O_2 or at very low O_2 partial pressures.

Water-filled-pore-space (WFPS) is suggested to be the most appropriate parameter to express soil water content, as it accounts for the variation in total porosity and therefore is comparable between soils of different textures (Davidson, 1993; Davidson and Schimel, 1995). The calculation of WFPS requires soil bulk density values (Linn and Doran, 1984), which are not often given in manuscripts, making comparisons between different soil types difficult. In addition, bulk density values from sieved soils in laboratory studies are also questionable. In the laboratory data presented in Figure 9.4, WFPS was calculated using the sieved bulk density values for comparative purposes between the different soil samples measured in the laboratory.

Skopp *et al.* (1990) found that generally the optimal soil water content for aerobic processes, such as nitrification, was 60% WFPS, whereas for anaerobic processes, such as denitrification, the optimum was >80% WFPS. The optimal value of 60% WFPS represented the intersection of increasing availability of organic carbon and inorganic nitrogen and decreasing availability of O_2 with increasing soil moisture. Nitrification is thought to be a more important source of NO than is denitrification when the WFPS

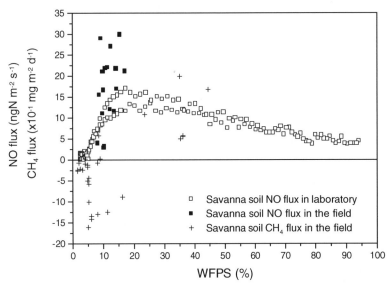

Figure 9.4. Savanna soil NO emissions, as measured in the laboratory, increase with soil moisture to field capacity after which fluxes decline as the soil moisture continues to saturation. The field NO fluxes are also shown as a comparison, indicating that field fluxes also increased with moisture but to a greater degree than found in the laboratory study. Soil CH_4 consumption occurs in dry soils and as moisture increases CH_4 consumption declines and emissions start to occur (field data).

<60% due to the availability of O_2, and that the opposite is true when the WFPS >60% as the availability of O_2 declined (Davidson, 1993). Maximum NO emissions are therefore thought to occur at field capacity, but this field capacity may not be around 60% WFPS.

Cárdenas *et al.* (1993) and Martin *et al.* (1998) found that the maximum NO emissions occurred at field capacity and the field capacities of the soils varied between 32 and 66% WFPS, as field capacities vary greatly with soil type and bulk density, which differs from the generalized value of 60% WFPS that is referred to so often. Bollmann and Conrad (1998) also found the maximum flux to correspond with the field capacity of the soil. In the South African savanna, NO emissions were lowest at a soil moisture of <3% and >30%. The highest flux rates occurred between 28.3 and 56.5% (Figure 9.4), which is a similar range to that found in the field study (32.3–40.4% WFPS) (Scholes *et al.*, 1997). The flood plain soils differed in that the maximum NO flux rates were between 33.9 and 56.4%. As in the other studies, these maximum fluxes correspond fairly well to the field capacities of the various soil types.

NO emissions have also been shown to increase sharply after the wetting of very dry soil (Davidson *et al.*, 1991, 1993; Meixner *et al.*, 1997; Scholes *et al.*, 1997; Martin *et al.*, 1998). Subsequent additions of water may produce further increases in emissions, but these levels are much lower than those measured after the first wetting of dry soils (Slemr and Seiler, 1984; Scholes *et al.*, 1997). The reasons for this pulse of NO are still unclear.

Evidence shows that the availability of organic and inorganic nitrogen in soils influences the emission rate of NO. Emissions have been shown to increase with fertilization with NO_3^-, NH_4^+ or urea (Johansson, 1984; Slemr and Seiler, 1984; Williams *et al.*, 1992; Cárdenas *et al.*, 1993). Most of these studies show a short-term increase in NO after fertilization; however, Martin *et al.* (1998) observed increases in NO fluxes 5–12 years following nitrogen application. It is still unclear whether the NO_3^- or NH_4^+ concentration is more important as a predictor of emission rates, as the literature shows correlations between NO flux rate and soil NO_3^- concentrations (Williams *et al.*, 1988; Williams and Fehsenfeld, 1991; Cárdenas *et al.*, 1993), as well as NH_4^+ concentrations (Slemr and Seiler, 1984; Anderson *et al.*, 1988; Levine *et al.*, 1988; Hutchinson *et al.*, 1993). NO emissions in southern African savannas appear to be related to nitrification and NO_3^- (Parsons *et al.*, 1996; Serça *et al.*, 1998). Further studies using gross mineralization rates together with ^{15}N isotopes will help elucidate the question of substrate availability.

Scholes *et al.* (1997) and Otter *et al.* (1999) showed that soil physical characteristics have a stronger influence over NO emissions than the soil chemical characteristics. Coarse-textured soils generally show higher NO emissions than fine-textured soils (Martin *et al.*, 1998) due to better aeration, which allows for the easy escape of NO from the soil to the atmosphere. In fine-textured soils, gaseous diffusion is the dominant factor controlling emissions, and the more compact the soil (i.e. the higher the clay content), the slower the diffusion of gases from the soil to the

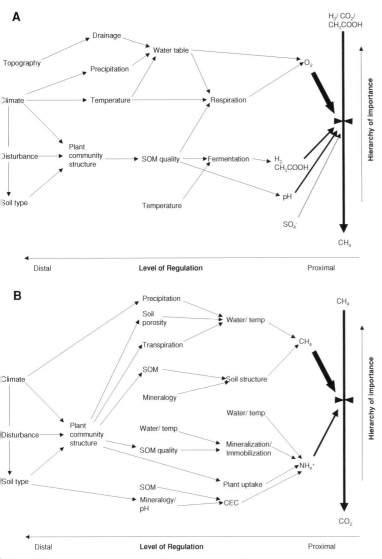

Figure 9.5. A schematic diagram of the proximal and distal controls on methanogenesis (A) and methane consumption in uplands (B) (from Conrad, 1997).

atmosphere. The effect of soil texture on NO emissions is not often reported, and it is therefore an area of research that needs to be expanded.

9.5.2 CH₄ fluxes

The conceptual model of Robertson (1989) for describing the controlling factors of NO and N_2O fluxes was adopted by Schimel *et al.* (1993) and Conrad (1997) for controls on methanogenesis and methane consumption (Figure 9.5).

Many studies have been conducted on CH_4 emissions because of its importance as a greenhouse gas. CH_4 production is also a microbial process, and thus CH_4 fluxes were thought to increase with temperature. In upland regions, some studies show an increased flux with temperature; however, other studies show that the de-

pendence of CH_4 flux on temperature is often very low (Born *et al.*, 1990; Crill, 1991; Koschorreck and Conrad, 1993; Crill *et al.*, 1994; Castro *et al.*, 1995), and sometimes no dependence is found (Keller *et al.*, 1983; Steudler *et al.*, 1989). It is suggested that this could be an indication that consumption and production are occurring at greater soil depths where temperature is more constant. Otter and Scholes (2000) showed that there is no clear relationship between soil temperature and the CH_4 flux from South African savannas. On the other hand, they show that CH_4 emission rates from shallow water on a flood plain are exponentially related to sediment temperature.

In very dry soils, less than 5% WFPS, CH_4 fluxes were near zero (Figure 9.4). This has previously been observed by Nesbit and Breitenbeck (1992) and Kruse *et al.* (1996). It is suggested to be due to the CH_4-oxidizing bacteria being water-stressed.

CH$_4$ consumption declines as moisture increases between 5 and 20% WFPS, which is supported by the literature (Whalen *et al.*, 1991; Adamsen and King, 1993; Castro *et al.*, 1995; Whalen and Reeburgh, 1996). Furthermore, CH$_4$ emissions increase as soil moisture increases, and as it increases above 20%, CH$_4$ emissions from the soil surface occur. This is supported by the data of Keller and Reiner (1994) and van den Pol-van Dasselaar *et al.* (1997), which showed that CH$_4$ emissions occurred when soils had high moisture content and a low effective diffusivity. The decrease in consumption with an increase in soil moisture is due to a diffusion limitation of substrate supply (Dörr *et al.*, 1993). Alternatively, the number of anoxic microsites could increase as soil moisture increases, leading to greater CH$_4$ production to the point where production exceeds consumption.

Three substrates influence the production of CH$_4$: organic matter, electron acceptors such as nitrate, iron, manganese and sulphate, and nutrients such as N, P and K (Conrad, 1989). Organic matter stimulates the production of CH$_4$ due to the production of fermentative CH$_4$ precursors. Generally, electron acceptors are preferred over bicarbonate and therefore inhibit CH$_4$ production by competition. Plants affect the availability of electron acceptors by providing organic root exudates that deplete electron acceptors, and by providing oxygen, which aids in the regeneration of electron acceptors. Nutrient-stimulated growth affects CH$_4$ production via the provision of organic substrates, by vascular transport of CH$_4$, by reoxidation of CH$_4$ by vascular O$_2$ transport, and by depletion and/or regeneration of nitrate, ferric iron, oxidized manganese and sulphate.

A number of studies have shown that nitrogen addition reduces CH$_4$ uptake by the soil, which is suggested to be due to either direct inhibition of CH$_4$ oxidation by ammonium or to gradual competition displacement of the CH$_4$-oxidizing methanotrophs by CH$_4$-oxidizing nitrifiers. Ammonium oxidation is located in the soil surface layers, whereas CH$_4$ oxidation occurs in deeper soil layers. This separation could be due to the inhibition of CH$_4$ oxidation by ammonium (Conrad, 1996a).

Wetland soils are anoxic but they have a thin oxic layer at the surface due to the penetration of O$_2$ from the surface. The consumption of trace gases occurs in this thin oxic layer. An oxic layer is also found around the roots of plants due to the release of O$_2$ from the roots. The rhizosphere is therefore an important site for the oxidation of CH$_4$, ferrous iron, ammonium and sulphide. As one moves further away from the root or soil surface, the O$_2$ concentration declines rapidly, which leads to redox stratification. This stratification occurs in a vertical direction, from soil to water surface, and in a radial dimension around roots. In deep anaerobic soils there is no O$_2$ or other electron acceptors other than CO$_2$ and H$^+$. This zone is dominated by fermentation and methanogenesis, and thus CH$_4$ and CO are produced. There is a flux of H$_2$ from the methanogenic zone into the SO$_4$-reducing zone as sulphate

reducers consume H$_2$, then into the Fe-reducing zones and into the NO$_3$-reducing zone (Conrad, 1996a). The NO$_3$-reducing zone is where H$_2$ concentrations are kept at low partial pressures by denitrifiers or DNRA bacteria, and it is where NO and N$_2$O are produced and consumed.

9.6 MODELLING FLUXES AND MODEL VALIDATION

One of the major problems with developing regional and global trace gas budgets is the substantial spatial and temporal variation in the fluxes from soils. The CH$_4$ fluxes from South African savannas shown in Figure 9.6 demonstrate the seasonal variation in the fluxes, varying from consumption to production. Furthermore, CH$_4$ fluxes can also be used to demonstrate spatial variation, first by looking at the standard error shown at each point, which is an indication of variation over a couple of metres. Second, if one considers a slightly larger spatial scale, the low-lying areas may become flooded during a wet season, leading to a large CH$_4$ production from the water surface (Figure 9.7), while consumption may be occurring in the surrounding upland savanna. These CH$_4$ source and sink changes may not be important for global CH$_4$ budgets, but they are likely to be highly significant at the regional scale. Most regional and global budgets are extrapolated from flux measurements collected from sites that are assumed to be representative of the region.

Accounting for all the variability can be overwhelming; however, it is important for a number of reasons to understand the variability. Variability can tell us something about timing and distribution of fluxes. This is useful as in some cases the episodic high levels of trace gases can be more important in controlling

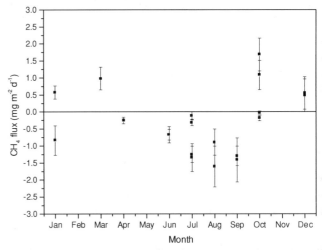

Figure 9.6. The seasonal variation in CH$_4$ fluxes from a South African savanna during 1997 is shown here, indicating that upland savanna soils are not only sinks of CH$_4$, but also can be a source during warm summer months (from Otter and Scholes, 2000).

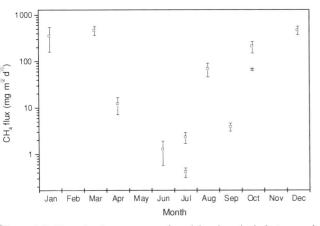

Figure 9.7. There is a large amount of spatial and particularly temporal variability in CH_4 fluxes from flooded soils, with fluxes differing by orders of magnitude each month.

processes than are globally averaged fluxes or concentrations. A good example of this is the effects that the first rains have on NO fluxes. After the first rain there is a sharp peak in NO emissions (Scholes *et al.*, 1997). This peak coincides with the development of new leaves during spring, and these leaves emit hydrocarbons (Guenther *et al.*, 1996). Simultaneous production of increased NO and hydrocarbons are thought to contribute significantly to the high ozone levels detected off the west coast of Africa during spring (Fishman *et al.*, 1991). Furthermore, timing and magnitude of biogenic gas fluxes reveal the nature and dynamics of source ecosystems. Finally, it is only through the understanding of the control of fluxes that we can predict how anthropogenic and natural disturbances of global climate change will affect global fluxes in the future.

There are several different strategies for extrapolating fluxes that have been discussed by a number of authors (Hicks, 1989; Matson *et al.*, 1989; Conrad, 1997). First, fluxes can be estimated using empirical relationships between ecosystem classes and fluxes. This approach is used when there are easily identifiable differences in the ecosystem structure, which coincide with differences in fluxes. A flux estimate for the region is obtained by multiplying the aerial extent by the measured flux for that region. Matson *et al.* (1989) gave the example of Matthews and Fung (1987), who used global databases of vegetation, soil and inundation to stratify wetlands into various groups. Methane fluxes from these various wetlands were used to calculate global methane emissions. Methane emissions from a southern African savanna and flood plain system were calculated in a similar manner by Otter and Scholes (2000). Savannas are estimated to consume an average of 0.04 g m^{-2} yr^{-1}, with southern African savannas consuming a total of 0.23 $\pm$ 0.082 Tg CH_4 yr^{-1}. Saturated and flooded sites were estimated to produce 25.3 and 57.2 g m^{-2} yr^{-1}, respectively. Southern African flood plains are estimated to

produce between 0.2 and 10 Tg CH_4 yr^{-1} (excluding the effects of vegetation-mediated emissions), and therefore produce more CH_4 than the savannas consume.

A second approach is to calculate emission estimates by coupling the classification system with an understanding of controls. This approach is similar to that described above but also incorporates spatial and temporal information. In other words, the factors controlling the processes, such as the climate and topography, which indirectly control methane fluxes as they control whether a site is inundated, saturated or dry. Thus methane emissions can be extrapolated using these variables.

The last approach is to use mechanistic simulation models. This method can reflect the interactions of the controlling variables both spatially and temporally. These types of models can be used to model seasonal changes and can also incorporate daily variations. The data input for simulation models are usually variables such as climate, soils, vegetation and topography, and therefore, to obtain flux estimates from these, one needs to know the relationship between the trace gas flux and the environmental variable.

NO emissions have been modelled and extrapolated for a South Africa savanna (Otter *et al.*, 1999). Firstly, a simulation model approach was used to simulate soil temperature and moisture for the savanna by using soil, vegetation and climate data. The relationships between NO emissions and soil moisture and temperature were measured in the laboratory using a soil incubation system. These were then coupled with the simulated soil moisture and temperature data. The model was run for one year and it showed the daily variation in NO emissions from the South African savanna (Figure 9.8). The annual flux was then multiplied by the area of savannas in southern Africa to give an annual NO emission estimate for southern African savannas. This is a very useful technique, as it shows day-to-day variation and includes information about controlling factors. The limitation is that the output is only as good as the input, and so the accuracy of the emission data is dependent on the simulated soil temperature and moisture. The annual average NO flux from southern African savannas is estimated to be 0.15 $\times$ 10^{-3} kg N m^{-2} yr^{-1}. These annual savanna NO emission estimates agree well with estimates from other savanna studies. Southern African savannas are, according to this study, estimated to produce 1.0 $\pm$ 0.12 Tg N yr^{-1}.

Kirkman *et al.* (2001) added a spatial component to this type of approach to estimate NO emissions for the whole of Zimbabwe. This model is more advanced in that it incorporates spatial climate, vegetation, soil and terrain data in the soil temperature and moisture calculations. NO emissions have been shown to undergo exchange within the canopy (Jacob and Bakwin, 1991), therefore a canopy reduction factor was also incorporated into this model. On a countrywide basis, the NO fluxes were estimated to range from 0.1 to 0.4 ng N m^{-2} s^{-1} in the dry season and from 3.7 to 10.9 ng N m^{-2} s^{-1} in the wet season. Zimbabwe was estimated

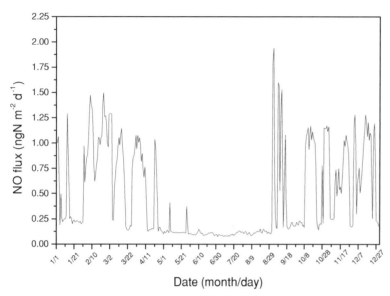

Figure 9.8. The modelled daily variation in NO flux from a nutrient-poor savanna in South Africa for the year 1997 (adapted from Otter *et al.*, 1999). This flux includes the effects of soil temperature and moisture.

to produce 32.9 Gg N yr^{-1}. Simulation models, such as these, are very useful for running scenarios and predicting how emissions may change in the future. Kirkman *et al.* (2001) predicted that a doubling of the current agricultural land would result in an addition of almost 2 Gg N yr^{-1}.

Another very important point about modelling and extrapolating emission data is model validation. It is very important to obtain measurements in the field that can validate the modelled data. In the case of the savanna study discussed above, the capacity of the model to predict field savanna NO fluxes was tested using the field data of Scholes *et al.* (1997). The comparison is quite good (average $r^2 = 0.87$); however, it should be noted, as with many models, that this model has been tested only on sandy soils. Soil type can have an effect on NO emissions, thus the model developed may not be valid with a different type of soil, and hence further validation would be required. One type of model may not be suitable for all cases, so an appropriate model needs to be chosen and the output must be validated. It should also be noted that the NO emissions given by the model described here shows the emission rate at the soil surface. It does not necessarily correspond to the amount of NO going into the troposphere above the tree canopies. This would need to be confirmed by using aircraft measurements and chemical transformation models.

Conrad and Dentener (1999) discuss the application of the compensation point in terms of scaling up the fluxes. This concept is useful in that compensation points exist for a number of trace gases, and also it includes the effects of environmental controlling factors, as the compensation point varies with these factors. This concept, with its mathematical formulations described in detail in Conrad and Dentener (1999), needs further evaluation in the future.

9.7 IMPACTS OF BIOGENIC TRACE GASES ON THE ATMOSPHERIC CHEMISTRY OF THE SOUTHERN AFRICAN REGION

Savannas are a highly significant source of trace gases because of both biomass burning and biogenic emissions. Due to the seasonal and regional concentration of these emissions, their impact is even more conspicuous. Pyrogenic emissions have, in the past, been suggested as the main source of ozone precursors responsible for the southern African tropospheric ozone maximum that extends over most of south central Africa, as well as over the adjacent Atlantic Ocean in August and October. However, evidence is now emerging that biogenic sources may be playing a much bigger role than previously thought. Pyrogenic, biogenic and industrial sources all contribute significant amounts of ozone precursors to the atmosphere over southern Africa, but each gas is dominated by one source. In the case of methane, the major uncertainty relates to the magnitude of the source due to seasonal swamps in south central Africa. Methane data presented in this chapter have, to some extent, helped to reduce this uncertainty. Nitric oxide was formerly thought to originate mostly from combustion, either of fossil fuels, or vegetation. The high NO emission rates from soils measured in South Africa during the SAFARI-92 campaign (Levine *et al.*, 1996; Parsons *et al.*, 1996) are supported by data from savannas in Brazil, as well as from SAFARI-92 aircraft observations of NO soon after the first rains.

Modelling of NO emissions at a regional scale, using both field incubation data and temperature response curves (Figures 9.3, 9.4 and 9.8), have confirmed the view that soils represent a major NO source. The timing of the ozone cloud is not entirely consistent with the peak of the biogenic NO emissions, but the period of production (October–June) is broad enough that soils could be significant contributors to the phenomenon, especially in the early spring.

Emissions of volatile organic compounds (VOC), which are high from savanna tree species (Guenther *et al.*, 1996), combine with CO (from a variety of sources) and NO to form the starting point of photochemical smog that is often blamed on motor vehicle emissions. It is clear that biogenic sources of NO and VOCs, and pyrogenic sources of CO, can contribute to high background levels of these pollution precursors, and will need to be considered when environmental receiving capacity limits are set for the industrial emissions of these gases.

It is difficult to conclusively apportion the sources of these gases as to their relative contribution to the springtime ozone maximum. It seems probable that the phenomenon results from a combination of the tail end of the pyrogenic emissions and the start of the biogenic emissions. If this is the case, it has been a feature of the atmospheric chemistry of southern Africa for millions of years; however, concern exists about the impacts that current and future land-use changes may have on these emissions and their atmospheric implications (Scholes and Scholes, 1998).

9.8 CONCLUSIONS

The fluxes of trace gases between soil and atmosphere can be measured with some reliability by using various approaches. Variability is high between different sites and seasons. Process level studies on the controllers of these emissions together with modelling approaches have helped reduce the uncertainties of the magnitude of the fluxes. Understanding soil biology and the non-linear way in which biological systems respond to change is essential for being able to make reasonable predictions of the impact of biogenic trace gases on atmospheric chemistry.

ACKNOWLEDGEMENTS

We would like to thank the National Research Foundation and the University of the Witwatersrand, South Africa, for providing the funding for much of this research. Dr Frans Meixner and Dr Wenxing Yang are thanked for their assistance in the laboratory and their constructive discussions. We would also like to thank the reviewers, Dr Günter Helas and Professor Jochen Rudolph, for their valuable comments.

REFERENCES

Adamsen, A. P. S. and King, G. M. (1993) Methane consumption in temperate and subarctic forest soils: Rates, vertical zonation and responses to water and nitrogen. *Appl. Environ. Microbiol.*, **59**, 485–490.

Anderson, I. and Levine, J. (1987) Simultaneous field measurements of biogenic emissions of nitric oxide and nitrous oxide, *J. Geophys. Res.*, **92**, 965–976.

Anderson, I. C., Levine, J. S., Poth, M. A. and Riggan, P. J. (1988) Enhanced biogenic emissions of nitric oxide and nitrous oxide following surface biomass burning. *J. Geophys. Res.*, **93**, 3,893–3,898.

Aselmann, I. and Crutzen, P. J. (1989) Freshwater wetland: Global distribution of natural wetlands, rice paddies, their net primary productivity, seasonality and possible methane emissions. *J. Atmos. Chem.*, **8**, 307–358.

Blackmer, A. M. and Cerrato, M. E. (1986) Soil properties affecting formation of nitric oxide by chemical reactions of nitrite. *Soil Sci. Soc. Am. J.*, **50**, 1,215–1,218.

Bollmann, A. and Conrad, R. (1998) Influence of O_2 availability on NO and N_2O release by nitrification and denitrification in soils. *Global Change Biology*, **4**, 387–396.

Bollmann, A., Koschorreck, M., Meuser, K. and Conrad, R. (1999) Comparison of two different methods to measure nitric oxide turnover in soils. *Biol. Fertil. Soils*, **29**, 104–110.

Born, M., Dörr, H. and Levin, I. (1990) Methane consumption in aerated soils of the temperate zone. *Tellus*, **42B**, 2–8.

Cárdenas, L., Rondon, A., Johansson, C. and Sanhueza, E. (1993) Effects of soil moisture, temperature and inorganic nitrogen on nitric oxide emissions from acidic tropical savanna soils. *J. Geophys. Res.*, **98**, 14,783–14,790.

Castro, M. S., Steudler, P. A., Melillo, J. M., Aber, J. D. and Bowden, R.D. (1995) Factors controlling atmospheric methane consumption by temperate forest soils. *Global Biogeochem. Cycles*, **9**, 1–10.

Chalk, P. M. and Smith, C. J. (1983) Chemodenitrification. *Dev. Plant Soil Sci.*, **9**, 65–89.

Conrad, R. (1989) Control of methane production in terrestrial systems. In M. O. Andreae and D. S. Schimel (Eds.), *Exchange of Trace Gases between Terrestrial Ecosystems and the Atmosphere*, pp. 39–58., New York, John Wiley and Sons.

(1994) Compensation concentration as critical variable for regulating the flux of trace gases between soil and atmosphere. *Biogeochemistry*, **27**, 155–170.

(1996a) Soil micro-organisms as controllers of atmospheric trace gases (H_2, CO, CH_4, OCS, N_2O and NO). *Microbiol. Rev.*, **60**, 609–640.

(1996b) Metabolism of nitric oxide in soil and soil micro-organisms and regulation of flux into the atmosphere. In J. C. Murrell and D. P. Kelly (Eds.), *Microbiology of Atmospheric Trace Gases: Sources, Sinks and Global Change Processes*, pp. 167–203. Berlin, Springer-Verlag.

(1997) Production and consumption of methane in the terrestrial biosphere. In G. Helas, J. Slanina, and R. Steinbrecher (Eds.), *Biogenic Volatile Organic Compounds in the Atmosphere*, pp. 27–44. Amsterdam, SPB Academic Publishing, copyright owner Backhuys Publishers, Leiden,

Conrad, R. and Dentener, F. J. (1999) The application of compensation point concepts in scaling of fluxes. In A. F. Bouwman (Ed.), *Approaches to Scaling Trace Gas Fluxes in Ecosystems*, London, Elsevier Science.

Conrad, R. and Seiler, W. (1985) Characteristics of abiological CO formation from soil organic matter, humic acids and phenolic compounds. *Environ. Sci. Technol.*, **19**, 1,165–1,169.

Crill, P. M. (1991) Seasonal patterns of methane uptake and carbon dioxide release by a temperate woodland soil. *Global Biogeochem. Cycles*, **5**, 319–334.

Crill, P. M., Martikainen, P. J., Nykänen, H. and Silvola, J. (1994) Temperature and N fertilization effects on methane oxidation in a drained peatland soil. *Soil Biol. Biochem.*, **26**, 1,331–1,339.

Crutzen, P. J. (1979) The role of NO and NO_2 in the chemistry of the troposphere and stratosphere. *Ann. Rev. Earth Planet. Sci.*, **7**, 443–472.

Davidson, E. A. (1991) Fluxes of nitrous oxide and nitric oxide from terrestrial ecosystems. In J. E. Rogers and W. B. Whitman (Eds.), *Microbial Production and Consumption of Greenhouse Gases: Methane, Nitrogen Oxides and Halomethanes*, pp. 219–235. Washington, American Society for Microbiology.

Davidson, E. A. (1992) Sources of nitric oxide and nitrous oxide following wetting of dry soil. *Soil Sci. Soc. Am. J.*, **56**, 95–102.

(1993) Soil water content and the ratio of nitrous oxide to nitric oxide emitted from soil. In R. S. Oremland (Ed.). *The Biogeochemistry of Global Change: Radiative Active Trace Gases*, pp. 368–386. New York, Chapman and Hall.

Davidson, E. A. and Schimel, J. P. (1995) Microbial processes of production and consumption of nitric oxide, nitrous oxide and methane. In P. A. Matson and R. L. Harriss (Eds.), *Biogenic Trace Gases: Measuring Emissions from Soil and Water*, pp. 327–357. Oxford, Blackwell Science.

Davidson, E. A., Vitousek, P. M., Matson, P. A., Riley, R., García-Méndez, G. and Maass, J. M. (1991) Soil emissions of nitric oxide in a seasonally dry tropical forest of Mexico. *J. Geophys. Res.*, **96**, 15,439–15,445.

Davidson, E. A., Matson, P. A., Vitousek, P. M. *et al.* (1993) Processes regulating soil emissions of NO and N_2O in a seasonally dry tropical forest. *Ecology*, **74**, 130–139.

Delmas, R., Serça, D. and Jambert, C. (1997) Global inventory of NO_x sources. *Nutrient Cycling in Agroecosystems*, **48**, 51–60.

Dörr, H., Katruff, L. and Levin, I. (1993) Soil texture parameterisation of the methane uptake in aerated soils. *Chemosphere*, **26**, 697–713.

Drury, C. F., McKenney, D. J. and Findlay, W. I. (1992) Nitric oxide and nitrous oxide production from soil: water and oxygen effects. *Soil Sci. Soc. Am. J.*, **56**, 766–770.

Firestone, M. K. and Davidson, E. A. (1989) Microbiological basis of NO and N_2O production and consumption in soil. In M. O.

Andreae and D. S. Schimel (Eds.), *Exchange of Trace Gases between Terrestrial Ecosystems and the Atmosphere*, pp. 7–21. New York, John Wiley and Sons. Figures reproduced with permission.

Fishman, J., Fakhruzzman, K., Cros, B. and Nganga, D. (1991) Identification of widespread pollution in the southern Hemisphere deduced from satellite analysis. *Science*, **242**, 1,693–1,696.

Flessa, H. and Beese, F. (1995) Effects of sugarbeet residues on soil redox potential and nitrous oxide emission. *Soil Sci. Soc. Am. J.*, **59**, 1,044–1,051.

Flessa, H., Dorsch, P. and Beese, F. (1995) Seasonal variation of N_2O and CH_4 fluxes in differently managed arable soils in southern Germany. *J. Geophys. Res.*, **100**, 23,115–23,124.

Galbally, I. E. (1989) Factors controlling NO_x emissions from soils. In M. O. Andreae and D. S. Schimel (Eds.), *Exchange of Trace Gases between Terrestrial Ecosystems and the Atmosphere*, pp. 23–38. New York, John Wiley and Sons.

Galbally, I. E. and Johansson, C. (1989) A model relating laboratory measurement of rates of nitric oxide production and field measurement of nitric oxide emission from soils. *J. Geophys. Res.*, **94**, 6,473–6,480.

Graedel, T. E. and Crutzen, P. J. (1993) *Atmospheric Change: An Earth System Perspective*, New York, W. H. Freeman and Company.

Guenther, A., Otter, L., Zimmerman, P., Greenberg, J., Scholes, R. and Scholes, M. (1996) Biogenic hydrocarbon emissions from southern African savannas. *J. Geophys. Res.*, **101**, 25,859–25,865.

Hantschel, R. E., Flessa, H. and Beese, F. (1994) An automated microcosm system for studying soil ecological processes. *Soil Sci. Soc. Am. J.*, **58**, 401–404.

Hicks, B. B. (1989) Regional extrapolation: vegetation-atmosphere approach. In M. O. Andreae and D. S. Schimel (Eds.), *Exchange of Trace Gases between Terrestrial Ecosystems and the Atmosphere*, pp. 109–118. New York, John Wiley and Sons.

Hutchinson, G. L. and Andre, C. E. (1989) Flow-through system for monitoring aerobic soil nitric and nitrous oxide emissions. *Soil Sci. Soc. Am. J.*, **53**, 1,068–1,074.

Hutchinson, G. L. and Livingston, G. P. (1993) Use of chamber systems to measure trace gas fluxes. In L. A. Harper, A. R. Mosier, J. M. Duxbury and D. E. Rolston (Eds.), *Agricultural Ecosystem Effects on Trace Gases and Global Climate Change*, pp. 63–78. Wisconsin, American Society of Agronomy.

Hutchinson, G. L. and Mosier, A. R. (1981) Improved soil cover method for field measurement of nitrous oxide fluxes. *Soil Sci. Soc. Am. J.*, **45**, 311–316.

Hutchinson, G. L., Livingston, G. P. and Brams, E. A. (1993) Nitric and nitrous oxide evolution from managed subtropical grassland. In R. S. Oremland, *The Biogeochemistry of Global Change: Radiatively Active Trace Gases*, New York, Chapman and Hall.

Jacob, D. J. and Bakwin, P. S. (1991) Cycling of NO_x in tropical forest canopies. In J. E. Rogers and W. B. Whitman (Eds.), *Microbial Production and Consumption of Greenhouse Gases: Methane, Nitrogen Oxides and Halomethanes*, pp. 237–253. Washington, DC, American Society of Microbiology.

Johansson, C. (1984) Field measurements of emission of nitric oxide from fertilized and unfertilized forest soils in Sweden. *J. Atmos. Chem.*, **1**, 429–442.

Johansson, C. and Sanhueza, E. (1988) Emission of NO from savanna soils during rainy season. *J. Geophys. Res.*, **93**, 14,193–14,198.

Johansson, C., Rodhe, H. and Sanhueza, E. (1988) Emission of NO in a tropical savanna and a cloud forest during the dry season. *J. Geophys. Res.*, **93**, 7,180–7,192.

Jury, W. A., Letey, J. and Collins, T. (1982) Analysis of chamber methods used for measuring nitrous oxide production in the field. *Soil Sci. Soc. Am. J.*, **46**, 250–256.

Keller, M. and Reiner, W. A. (1994) Soil-atmosphere exchange of nitrous oxide, nitric oxide and methane under secondary succession of pasture to forest in the Atlantic lowlands of Costa Rica. *Global Biogeochem.Cycles*, **8**, 399–409.

Keller, M., Goreau, T. G., Wofsy, S. C., Kaplan, W. A. and McElroy, M. B. (1983) Production of nitrous oxide and consumption of methane by forest soils. *Geophys. Res. Lett.*, **10**, 1,156–1,159.

Keller, M., Kaplan, W. A. and Wofsy, S. C. (1986) Emissions of N_2O, CH_4 and CO_2 from tropical forest soils. *J. Geophys. Res.*, **91**, 11,791–11,802.

Keller, M., Mitre, E. M. and Stallard, R. F. (1990) Consumption of atmospheric methane in soils of Central Panama: effects of agricultural development. *Global Biogeochem. Cycles*, **4**, 21–27.

Kirkman, G. A., Yang, W. X. and Meixner, F. X. (2001) Biogenic nitric oxide emissions up scaling: an approach for Zimbabwe. *Global Biogeochem. Cycles*, **15**, 1005–1020.

Koops, J. G., van Beusichem and Oenema, O. (1997) Nitrogen loss from grassland on peat soils through nitrous oxide production. *Plant and Soil*, **188**, 119–130.

Koschorreck, M. and Conrad, R. (1993) Oxidation of atmospheric methane in soil: measurements in the field, in soil cores and in soil samples. *Global Biogeochem. Cycles*, **7**, 109–121.

Kruse, C. W., Moldrup, P. and Irversen, N. (1996) Modelling diffusion and reaction in soils: II. Atmospheric methane diffusion and consumption in a forest soil. *Soil Sci*, **161**, 355–365.

Levine, J. S., Cofer, W. S., Sebacher, D. I. *et al.* (1988) The effects of fire on biogenic emissions of nitric oxide and nitrous oxide. *Global Biogeochem. Cycles*, **2**, 445–449.

Levine, J. S., Winstead, E. L., Parsons, D. A. P. *et al.* (1996) Biogenic soil emissions of nitric oxide (NO) and nitrous oxide (N_2O) from savannas in South Africa: The impact of wetting and burning. *J. Geophys. Res.*, **101**, 23,689–23,697.

Linn, D. M. and Doran, J. W. (1984) Effect of water-filled pore space on carbon dioxide and nitrous oxide production in tilled and nontilled soils. *Soil Sci. Soc. Am. J.*, **48**, 1,267–1,272.

Lovelock, J. E. (1979) *Gaia. A New Look at Life on Earth*, London, Oxford University Press.

MacDonald, J. A., Eggleton, P., Bignell D. E., Forzi, F. and Fowler, D. (1998) Methane emission by termites and oxidation by soils, across a forest disturbance gradient in the Mbalmayo Forest Reserve, Cameroon. *Global Change Biol.* **4**, 409–418.

Martin, R. E., Scholes, M. C., Mosier, A. R. *et al.* (1998) Controls on annual emissions of nitric oxide from soils of the Colorado shortgrass steppe. *Global Biogeochem. Cycles*, **12**, 81–91.

Matson, P. A., Vitousek, P. M. and Schimel, D. S. (1989) Regional extrapolation of trace gas flux based on soils and ecosystems. In M. O. Andreae and D. S. Schimel (Eds.), *Exchange of Trace Gases between Terrestrial Ecosystems and the Atmosphere*, pp. 97–108. New York, John Wiley and Sons.

Matthews, E. and Fung, I. (1987) Methane emission from natural wetlands: global distribution, area and environmental characteristics of sources. *Global Biogeochem. Cycles*, **1**, 61–86.

Meixner, F. X. (1994) Surface exchange of odd nitrogen oxides. *Nova Acta Leopoldina*, NF 70(288), 299–348.

Meixner, F. X., Fickinger, Th., Marufu, L. *et al.* (1997) Preliminary results on nitric oxide emission from a southern African savanna ecosystem. *Nutrient Cycling in Agroecosystems*, **48**, 123–138.

Miller, R. W. and Donahue, R. L. (1990) *Soils: An Introduction to Soils and Plant Growth*, New Jersey, Prentice Hall.

Morrissey, L. A. and Livingston, G. P. (1992) Methane emissions from Alaska Arctic tundra: an assessment of local spacial variability. *J. Geophys. Res.*, **97**, 16,661–16,670.

Mosier, A. R. (1989) Chamber and isotope techniques. In M. O. Andreae and D. S. Schimel (Eds.), *Exchange of Trace Gases between Terrestrial Ecosystems and the Atmosphere*, pp. 175–188. New York, John Wiley and Sons.

Mosier, A. R., Schimel, D.,Valentine, D., Bronson, K. and Parton, W. (1991) Methane and nitrous oxide fluxes in native, fertilized and cultivated grasslands. *Nature*, **350**, 330–332.

Neff, J. C., Keller, M., Holland, E. A., Weitz, A. W. and Veldkamp, E. (1995) Fluxes of nitric oxide from soils following the clearing and burning of a secondary tropical rain forest. *J. Geophys. Res.*, **100**, 25,913–25,922.

Nesbit, S. P. and Breitenbeck, G. A. (1992) A laboratory study of factors influencing methane uptake by soils. *Agric. Ecosys. Environ.*, **41**, 39–54.

Otter, L. B. and Scholes, M. C. (2000) Methane emission and consumption in a periodically flooded South African savanna. *Global Biogeochem. Cycles*, **14**, 97–111. Figures reproduced/modified by permission of American Geophysical Union.

Otter, L. B., Yang, W. X., Scholes, M. C. and Meixner, F. X. (1999) Nitric oxide emissions from a southern African savanna. *J. Geophys. Res.*, **104**, 18,471–18,485. Figures reproduced/modified by permission of American Geophysical Union.

Parsons, D. A. B., Scholes, M. C., Scholes, R. J. and Levine, J. S. (1996) Biogenic NO emissions from savanna soils as a function of fire regime, soil type, soil nitrogen and water status. *J. Geophys. Res.*, **101**, 23,683–23,688.

Potter, C. S., Matson, P. A., Vitousek, P. M. and Davidson, E. A. (1996a) Process modelling of controls on nitrogen trace gas emissions from soils worldwide. *J. Geophys. Res.*, **101**, 1,361–1,377.

Potter, C.S., Davidson, E. A. and Verchot, L. V. (1996b) Estimation of global biogeochemical controls and seasonality in soil methane consumption. *Chemosphere*, **32**, 2,219–2,246.

Reeburgh, W. S., Whalen, S. C. and Alperin, M. J. (1993) The role of methylotrophy in the global methane budget. In J. C. Murrell and D. P. Kelly (Eds.), *Microbial Growth on C1 Compounds*, pp. 1–14. Andover, UK, Intercept Inc.

Remde, A., Slemr, F. and Conrad, R. (1989) Microbial production and uptake of nitric oxide in soil. *FEMS Microbiol. Ecol.*, **62**, 221–230.

Robertson, G. P. (1989) Nitrification and denitrification in humid tropical ecosystems: potential controls on nitrogen retention. In J. Proctor (Ed.), *Mineral Nutrients in Tropical Forest and Savanna Ecosystems*, pp. 55–69. British Ecological Society Symposium Series, Oxford, Blackwell Scientific.

Rudolph, J. and Conrad, R. (1996) Flux between soil and atmosphere, vertical concentration profiles in soil and turnover of nitric oxide: 2. Experiments with naturally layered soil cores. *J. Atmos. Chem.*, **23**, 274–299.

Saad, O. A. L. O. and Conrad, R. (1993) Temperature dependence of nitrification, denitrification and turnover of nitric oxide in different soils. *Biol. Fertil. Soils* **15**, 21–27.

Schade, G. W., Hofmann, R-M. and Crutzen, P. J. (1999) CO emissions from degrading plant matter. Part 1: Measurements. *Tellus* **51B,** 889–908.

Schimel, J. P., Holland, E. A. and Valentine, D. (1993) Controls on methane flux from terrestrial ecosystems. In Harper L. A. *et al.* (Eds.), *Agricultural Ecosystem Effects on Trace Gases and Global Climate Change*, pp. 167–182. Wisconsin, American Society of Agronomy.

Scholes, M. C., Martin, R., Scholes, R. J., Parsons, D. and Winstead, E. (1997) NO and N_2O emissions from savanna soils following the first simulated rains of the season. *Nutrient Cycling in Agroecosystems*, **48**, 115–122.

Scholes, R. J. and Hall, D. O. (1996) The carbon budget of tropical savannas, woodlands and grasslands. In A. I. Breymeyer, D. O. Hall, J. M. Melillo and G. I. Agren (Eds.), *Global Change: Effects on Coniferous Forests and Grasslands* (Scope 56), Chichester, New York, John Wiley and Sons.

Scholes, R. J. and Scholes, M. C. (1998) Natural and human-related sources of ozone-forming trace gases in southern Africa. *South African J. Sci.*, **94**, 422–425.

Schutz, H. and Seiler, W. (1989) Methane flux measurements: methods and results. In M. O. Andreae and D. S. Schimel (Eds.), *Exchange of Trace Gases between Terrestrial Ecosystems and the Atmosphere*, pp. 209–228. New York, John Wiley and Sons.

Sebacher, D. I. and Harriss, R. C. (1982) A system for measuring methane fluxes from inland and coastal wetland environments. *J. Environ. Qual.*, **11**, 34–37.

Seiler, W., Conrad, R. and Scharffe, D. (1984) Field studies of methane emission from termite nests into the atmosphere and measurements of methane uptake by tropical soils. *J. Atmos. Chem.*, **1**, 171–186.

Serça, D., Delmas, R., Le Roux, X. *et al.* (1998) Variability of nitrogen monoxide emissions from African tropical ecosystems. *Global Biogeochem. Cycles*, **12**, 637–651.

Skopp, J., Jawson, M. D. and Doran, J. W. (1990) Steady-state aerobic microbial activity as a function of soil water content. *Soil Sci. Soc. Am. J.*, **54**, 1,619–1,625.

Slemr, F. and Seiler, W. (1984) Field measurements of NO and NO_2 emissions from fertilized and unfertilized soils. *J. Atmos. Chem.*, **2**, 1–24.

Steudler, P. A., Bowden, R. D., Melillo, J. M. and Aber, J. D. (1989) Influence of nitrogen fertilization on methane uptake in temperate forest soils. *Nature*, **341**, 314–316.

Striegl, R. G., McConnaughey, T. A., Thorstenson, D. C., Weeks, E. P. and Woodward, J. C. (1992) Consumption of atmospheric methane by desert soils. *Nature*, **357**, 145–147.

Tate, C. M. and Striegl, R. G. (1993) Methane consumption and carbon dioxide emission in tallgrass prairie: effects of biomass burning and conversion to agriculture. *Global Biogeochem. Cycles*, **7**, 735–748.

Van den Pol-van Dasselaar, A., van Beusichem, M. L. and Oenema, O. (1997) Effects of grassland management on the emission of methane from intensively managed grasslands on peat soil. *Plant and Soil*, **189**, 1–9.

Wassmann, R. and Thein, U.G. (1996) Spatial and seasonal variation of methane emission from an Amazon floodplain lake. *Mitt. Internat. Verein. Limnol.*, **25**, 179–185.

Whalen, S. C. and Reeburgh, W. S. (1990) Consumption of atmospheric methane by tundra soils. *Nature*, **346**, 160–162.

(1996) Moisture and temperature sensitivity of CH_4 oxidation in boreal soils. *Soil Biol. Biochem.*, **28**, 1,271–1,281.

Whalen, S. C., Reeburgh, W. S. and Kizer, K. S. (1991) Methane consumption and emission by taiga. *Global Biogeochem. Cycles*, **5**, 261–273.

Williams, E. J. and Fehsenfeld, F. C. (1991) Measurement of soil nitrogen oxide emissions at three north American ecosystems. *J. Geophys. Res.*, **96**, 1,033–1,042.

Williams, E. J., Parrish, D. D. and Fehsenfeld, F. C. (1987) Determination of nitrogen oxide emissions from soils: results from a grassland site in Colorado, United States. *J. Geophys. Res.*, **92**, 2,173–2,179.

Williams, E. J., Parrish, D. D., Buhr, M. P., Fehsenfeld, F. C. and Fall, R. (1988) Measurement of soil NO_x emissions in central Pennsylvania. *J. Geophys. Res.*, **93**, 9,539–9,546.

Williams, E. J., Hutchinson, G. L. and Fehsenfeld, F. C. (1992) NO_x and N_2O emissions from soil. *Global Biogeochem. Cycles*, **6**, 351–388.

Yang, W. X. and Meixner, F. X. (1997) Laboratory studies on the release of nitric oxide from subtropical grassland soils: the effect of soil temperature and moisture. In S. C. Jarvis and B. F. Pain (Eds.), *Gaseous Nitrogen Emissions from Grasslands*, New York, CAB International.

Zepp, R. G., Miller, W. L., Burke, R. A., Parsons, D. A. B. and Scholes, M. C. (1996) Effects of moisture and burning on soil-atmosphere exchange of trace carbon gases in a southern African savanna. *J. Geophys. Res.*, **101**, 23,699–23,706.

PART II

Sustainable energy development, mitigation and policy

10 Biomass energy in sub-Saharan Africa

DAVID O. HALL[1] (deceased) AND J. IVAN SCRASE[2,*]

[1]*King's College London, UK*
[2]*Imperial College London, UK*

Keywords

Biomass energy; sub-Saharan Africa; flow chart; deforestation; health; fuel switching

Abstract

In sub-Saharan Africa available evidence suggests that biomass use for energy has increased roughly in proportion to population growth. With urbanization the biomass energy sector is becoming more commercialized, and consumption of charcoal is increasing (which leads to higher biomass consumption, given the low conversion efficiencies in most charcoal production). Localized fuel scarcity and resource degradation has occurred, but biomass scarcity has been exaggerated in the past in macro-level studies, which overlooked many non-forest sources of biomass fuel. National studies of biomass availability and use in Kenya, Zimbabwe, Sierra Leone, Botswana, Rwanda, South Africa and Uganda indicate that large unused biomass resources exist, and that scarcity often leads to more efficient use and fuel-switching to other forms of biomass rather than to fossil fuels. Consumption of biomass energy in sub-Saharan Africa is not expected to fall in the near future. Therefore there is an urgent need for recognition of the opportunities and problems which dependence on biomass energy represents. The resource is large and potentially sustainable, but biomass use creates health problems and can cause environmental degradation. National and regional energy planning and policies must take into account the importance of biomass energy supply and use, and seek to modernize the sector so that problems can be minimized.

*Formerly Research Associate with Professor David Hall at King's College London

10.1 BIOMASS PRODUCTION AND USE IN SUB-SAHARAN AFRICA: THE PRESENT SITUATION

Biomass currently provides about 85% of energy consumed in sub-Saharan Africa (SSA). This level of dependence on biomass is higher than anywhere else in the world. Industrialized countries and other parts of the developing world have reduced their dependence on biomass, and in the past there has been a general assumption that a similar shift to fossil fuels would take place in Africa. However, there is now an acceptance that biomass consumption in sub-Saharan Africa is not declining and may even be increasing in absolute terms. According to the International Energy Agency, this reflects economic stagnation, growing population and increased use of inefficiently produced charcoal (D'Apote, 1998).

The World Bank (1996) predicts that dependence on biomass in SSA will fall from 85% to 80% by 2010. In comparison, South Asia's dependence is expected to fall from 60% to 43%. These figures are presented with those for other developing regions in Table 10.1.

10.2 PROBLEMS ASSOCIATED WITH BIOMASS USE FOR ENERGY

Biomass provides people with essential energy services on which life and health depend. Cooking and boiling make food and water safe, and fires provide warmth, light and a social focus in many homes. Without adequate biomass energy supplies, the health and well-being of people without access to alternatives is jeopardized. However, the use of biomass for energy causes a number of specific problems, most notably health hazards due to inhalation of smoke from indoor open fires. Also, in the 1970s biomass use for energy was thought to be a major cause of global deforestation. This assumption has now been discredited, and land clearance for agriculture or other uses has been shown to be the driving force in deforestation. However, on a local scale fuelwood scarcity has become an issue, resulting in inefficient use of energy in biomass

Table 10.1. Current and projected use of biomass by region (% of total energy used).

Region	1990	2000	2010
Sub-Saharan Africa	85	83	80
South Asia	60	52	43
East Asia and Pacific	33	26	20
North Africa and Middle East	27	23	19
Latin America and Caribbean	26	22	19

World Bank, 1996.

transportation and charcoal production. The issues of health and fuelwood scarcity are discussed in more detail below.

10.2.1 Pollution and health

Pneumonia is the commonest form of Acute Lower Respiratory Infection (ALRI), and is now the single most important cause of death worldwide of children under five years of age. A number of studies have linked exposure to biomass smoke with ALRI and other respiratory complaints such as bronchitis. Low birth weight has been associated with exposure to carbon monoxide in biomass smoke during pregnancy. Biomass smoke also appears to increase incidence of conjunctivitis. It is certain that exposure to particulates in houses that burn biomass indoors in unventilated conditions can be thousands of times higher than safe maximums as specified by bodies such as the United States Environmental Protection Agency (USEPA), but many of the studies linking health effects with biomass smoke exposure are not entirely conclusive (Bruce, 1998). This is because few actually measure exposure to smoke directly, and because lung disorders, low birth weight and eye infections are more common among low-income groups for a range of reasons such as quality of housing and diet. It is very probable that biomass smoke plays a part in poor health in sub-Saharan Africa, but it is difficult to isolate its effect from other factors. It should also be noted that improved stoves (designed to burn biomass efficiently) do not generally reduce exposure to particulate pollution to levels considered safe by the USEPA, and should not be expected to have any great effect on health unless introduced as part of a wider improvement in living conditions.

10.2.2 Deforestation and fuelwood scarcity

There is no doubt that biomass is used very inefficiently and in large quantities for energy, but the view prevalent in the 1960s to 1980s that biomass-for-energy is a major cause of deforestation has been discredited. The idea of a 'fuelwood gap' was based on data on wood consumption rates compared to productivity of stem wood in forests, but was also based on a false assumption that forest clearance is the only source of fuelwood. A detailed study of

fuelwood procurement in Botswana confirmed the growing consensus that fuelwood is mainly collected in its dead form, largely from non-forest trees, but also from the forest floor and predominantly as a by-product of land clearance for agriculture or other purposes (Kgathi and Mlotshwa, 1997). Openshaw (1998) has found non-forest trees supply 60% of rural households' fuelwood in Kenya, and 40% of urban households' fuelwood in Malawi. This appears to be the case throughout sub-Saharan Africa and around the world. A Food and Agricultural Organization (FAO) study concluded that 'Wood energy use is not and will not be a general or main cause of deforestation' (FAO, 1997). This study found that in Asian countries two thirds of all fuelwood is collected or derived from non-forest trees.

Locally, however, fuelwood scarcity can be a serious problem. This can result where urbanization occurs with little switching to fossil energy or electricity, and can be accelerated if households switch from wood to charcoal. This process of fuel switching can be generalized by a 'fuel preference ladder', with an order of preference for household use beginning with dung as the least favoured, then crop residues, wood, charcoal, kerosene and LPG, which is at the top of the ladder. There is a general assumption that biomass energy use will diminish as consumers become wealthier and move up the preference ladder. In fact, wood scarcity and continuing poverty for many in sub-Saharan Africa means fuel switching has often been down rather than up the ladder, to less-favoured fuels such as dung and crop residues. In situations of wood scarcity, fuel tends to become more commercialized, and consequently consumers use it more carefully and perhaps more efficiently. Transport costs can be a very significant part of the cost of wood fuel, hence charcoal with its lower transport costs per unit of energy may become preferable. This ultimately leads to further scarcity, as charcoal usually is produced very inefficiently. Some of these issues of fuel procurement, switching and scarcity are discussed further in the case studies below.

10.3 CASE STUDIES

10.3.1 Biomass flow charts

Because biomass energy has been seen as a low-status fuel, associated with the past and with poverty, it has been given very little attention in national policies and statistics. In sub-Saharan Africa very few countries have policies on biomass energy, despite its predominance in energy supply. For this situation to improve it is necessary to have a good statistical basis for policy formulation, i.e. there is a need for detailed assessments of all biomass resources and consumption patterns at the national level. In response to this, 'biomass energy flow charts' have been prepared for Kenya (Senelwa and Hall, 1993), Zimbabwe (Hemstock and Hall, 1995) and Sierra Leone (Amoo-Gottfried and Hall, 1999). A brief summary of their findings follows.

Kenya derives 85% of its energy from biomass, with a per capita consumption of 18.6 gigajoules per year (GJ/yr). The remaining 15% is made up largely of imported petroleum fuels, which represent a major drain on foreign exchange reserves. Eighty per cent of Kenya's population lives in rural areas and are nearly 100% dependent on biomass for their energy needs. Urban populations are approximately 75% dependent on biomass.

Total annual above-ground terrestrial biomass production in Kenya was estimated at 2.57 exajoules (EJ), 87% of which is produced on agricultural land; 1.14 EJ of the total is produced in agriculture or forestry (i.e. it is not grasses) and is therefore potentially available for fuel. In fact, only 65% of this potential harvest is collected. Four fifths of the collected biomass is for fuel, and just one fifth for food and other uses. Including losses during and after harvest, and in charcoal conversion, only 47% (0.53 EJ) of the potential harvest reaches an end use.

There is a considerable flow of unused biomass, 0.48 EJ of which is agricultural residues and dung. These could be used as fuel, but obstacles include the difficulty and expense of collecting scattered residues, and these are not popular fuels since they are associated with extreme poverty, and dung burning gives off noxious smoke (bad smells, eye irritation). There is therefore a case for modernization of the use of residues and dung for energy, for use in family biogas plants. We estimate that if only 20% of the available dung were used it could displace 50% of the commercial (fossil) energy consumption in Kenya.

Concerning wood fuel consumption, minor technical improvements could raise the efficiency of charcoal kilns from 16 to 20%, stoves from 20 to 30%, and open fires from 10 to 15%. These improvements would allow the quantity of biomass energy consumed by 23 million people (in 1988) to support a population of 40 million.

In Zimbabwe, Hemstock and Hall (1995) found similar dependence on biomass for energy, and large unused or inefficiently used resource flows. Total annual above-ground biomass production totals 409 petajoules (PJ) (0.4 EJ), 49% of which is utilized; 176 PJ/yr is unused agricultural residues or dung and 33 PJ/yr is unused forestry residues, and 108 PJ/yr of biomass is used directly for fuel, i.e. 13 GJ per capita per year. The Zimbabwe study highlighted the difficulty making reliable estimates of the sustainability of current or projected biomass consumption rates, given the poor quality of data on standing stock and mean annual increments.

In Sierra Leone, Amoo-Gottfried and Hall (1999) found that 41% (54 PJ/yr) of total annual biomass production reaches an end use. Sierra Leone is 80% dependent on biomass for energy. The country has a low population density but rapid deforestation due to unregulated timber extraction, fuelwood collection for domestic use and for tobacco and fish processing, and shifting bush fallow cultivation. Annual biomass use per capita is estimated at 15.8 GJ, 11.8 GJ of which is for energy, almost all of which is provided by

wood. Crop residues and dung represent a large resource: if just 25% of this resource were used for energy, it could provide an additional 13 PJ/yr of energy. We note that technology for using dung (biogas) and residues (e.g. briquetting) is improving rapidly, and deforestation is taking place at an alarming rate, and suggest that Sierra Leone should consider this potential energy resource with some urgency. However, it should be noted that most of the causes will only lead to temporary deforestation, as the trees will regenerate once they have been felled, unless there is a permanent change of land use.

The biomass energy flow chart studies demonstrate that it is possible to identify resource flows and areas where biomass resources are under-exploited. However, preparing these studies is very data intensive, and accurate data are difficult to find. Ideally these studies should be made on a regular basis so that changes through time could be charted, but this would require much greater investment in data collection and analysis than is currently directed to biomass energy issues in Africa.

10.3.2 Biomass consumption and responses to scarcity/fuel switching

Local-level studies have challenged many of the assumptions made about biomass procurement and consumption for energy, and explain why the 'energy transition' from biomass to commercial fuels undergone by industrialized countries is not being repeated in sub-Saharan Africa. The assumption behind the energy transition model is that as incomes rise consumers will switch to commercial fuels. However, it has been shown that in fact fuelwood consumption in many parts of Africa is at the bare minimum level needed for survival. The consequence of this is that there is actually positive income elasticity for biomass, i.e. as incomes rise, people become a little less frugal in their energy use. This was found to be the case in Kenya (Hosier, 1985) and in Botswana (Kgathi and Mlotshwa, 1997).

One must also consider that for large sections of the population of sub-Saharan Africa incomes have not been rising. Inequality and population growth have caused an expansion of the population in extreme poverty and localized fuel scarcity. Therefore, for many, the energy transition may in fact be down the list of fuel preferences, i.e. from wood to dung or agricultural residues. Though often plentiful, without technological improvements these fuels can be unpleasant to use, time-consuming tending the fire, and often carry a social stigma.

These issues are addressed by Marufu et al. (1997) in a study of domestic biomass burning in rural and urban Zimbabwe. This study looked in detail at biomass procurement and use in four rural study areas and one urban area, using a questionnaire survey. Three of the four rural areas had some degree of fuel scarcity. The study found that even within one country, patterns of biomass sourcing and use varied widely: 'Land use patterns, land tenure systems, vegetation types, productivity, social structure, population density and climate can be of major influence on biofuel availability, sourcing patterns and consumption levels'. Consequently, 'the fuelwood situation in sub-Saharan Africa is highly location specific' (Marufu *et al.*, 1997).

In the Zimbabwe study, family sizes ranged from 5.5 to 7.9 persons per household, and a strong inverse correlation was found between family size and fuelwood consumption per capita. This is an important consideration when extrapolating per capita consumption figures to regional or national levels of aggregation.

In rural areas, 100% of households used wood fuel, and maize cobs were used during the harvest season, and some use of dung for fuel was reported. In the urban study all houses had electricity, but 77% used biofuels, mainly to cut costs for cooking and heating. In both rural and urban areas, families had separate, well-ventilated kitchen buildings, and used a hollow in the floor for the fire, and a simple pot stand.

Interesting differences were found between rural areas with fuel sufficiency and those where wood fuel was scarcer. Where fuelwood was plentiful, householders made very little use of any other fuel, and would keep fires burning long after they were needed for cooking. Where plentiful, the fuel was all collected by the users, all as dead wood. In Chilhota, where fuelwood was scarcest, 30% of households used dung as fuel, and almost all used agricultural residues. Wood still provided 85% of energy need, residues provided 12% and dung 3%. Because of wood scarcity, 45% of the wood fuel used was bought in the local market. Interestingly, kerosene was not seen as a fuel of preference because its availability is unreliable; it is too costly and does not provide multiple energy services of cooking, light and heat in the same way as an open fire. However, it was resorted to where fuelwood was scarce. In areas of scarcity it was also noted that fires are quenched immediately after use, to conserve fuel.

The preference in all study areas was for indigenous tree species for fuel, because of better splitting and burning characteristics. When indigenous species were not available, wood from exotic pine and eucalyptus trees was used. In the urban study residues from exotic trees used in the timber industry were a significant source of fuel. We concluded that the rate of consumption of biofuels was a function primarily of family size and availability, with income and other factors playing little part.

10.4 SCENARIOS OF FUTURE BIOMASS ENERGY PRODUCTION AND USE IN SUB-SAHARAN AFRICA

The Intergovernmental Panel on Climate Change (IPCC) Second Assessment Report includes a biomass intensive energy scenario in which 180 EJ/yr of sustainably produced biomass is used globally by 2050. For Africa in 2025, the IPCC scenario foresees 3.9 EJ/yr of biomass energy derived from forest residues, 3.6 EJ from other residues, and 5 EJ from plantations. In 2050, plantations would provide 14 EJ/yr. In comparison, commercial energy use in Africa today amounts to about 10 EJ/yr.

To assess whether these estimates for plantation energy are realistic, Marrison and Larson (1996) considered population change predictions, plantation yield potentials given rainfall distribution, changes in agricultural crop yields and current land use. The United Nations (1993) baseline prediction of population change in Africa quoted by Marrison and Larson (1996) is that it will increase to 2.5 times the 1990 level by 2025. Given current levels of under-nourishment, food production will need to increase by 2.7 times by 2025 to ensure minimum calorific needs are met. Agricultural yields are assumed to increase at the rate observed in Africa for cereals between 1972–1990, giving a 43% increase in yields between 1990 and 2025 (in 2025 yields are still assumed to be considerably less than 1990 yields in the USA). This means that the area of cropland in Africa will need to increase 2.4 times, from 186 million ha in 1990 to 452 million ha in 2025. Excluding land classified as forest or wilderness by the World Resources Institute, this would still leave 1.1 billion hectares of land available for other uses such as energy plantations.

Marrison and Larson (1996) assume that energy crop yields are a function of available rainfall, with a maximum of 30 dry tonnes per hectare per year (dt/(ha yr)), and an average of 8.3 dt/(ha yr). Assuming only 10% of the available land (111 million ha) were used for plantations, 18.4 EJ/yr of biomass energy could be produced (16.7 EJ of which would be from sub-Saharan Africa). This indicates that the IPCC biomass intensive scenario, with 5 EJ/yr from plantations in Africa in 2025, and 14 EJ/yr in 2050, are well within reason given future demand for land for food. Kenya, Mozambique, South Africa and Zaire could each produce over 1 EJ/yr of plantation biomass on the basis of these calculations. The results were found to be robust in sensitivity analysis, but Marrison and Larson (1996) admit that the data used are very coarse, and that country level analysis is needed.

Hall *et al.* (1993) estimate that a similar area of land suitable for plantations already exists in sub-Saharan Africa in the form of deforested land or forest fallow. They envisage a scenario for 2025 in which 101 million hectares of currently degraded land produce a high average yield of 16 dt/(ha yr), or a total output of 30 EJ/yr of biomass energy. They propose that the region could become a major exporter of biomass-derived fuels.

Plantations therefore appear to offer a viable way to modernize and expand the provision of biomass for energy. Given the typically low efficiencies in conversion, there is also a need for modernization of biomass use. Improved stove programmes have met with varied success. The stoves can raise efficiencies in cooking, but are not popular where other aspects of quality of life such is heat and light are sacrificed. Biogas schemes have a limited potential given the size of the accessible resource. Success has been rather limited due to the problems of collecting dung, the cost of the digester, water availability and maintenance of the biogas systems. For example, a programme of introducing biogas digesters in Rwanda in the 1980s found that half of the digesters had fallen into disrepair by 1989 (Karenzi, 1994). The problems are not necessarily with the technology but often with the way they are introduced, with too little consideration for the needs and perceptions of the users.

In rural sub-Saharan Africa, very few people have access to electricity, and given the remoteness of many villages, decentralized energy generation could be a good option. Senelwa and Sims (1999) consider the potential for small-scale biomass for electricity systems in Kenya. They estimate that 75% of the population of Kenya is rural, and only 1% of rural people have access to electricity supplies. Fuelwood provides 70% of Kenya's energy, with imported oil making up most of the remainder. The lack of access to electricity and commercial energy supplies in rural areas is holding up progress in industry and agriculture and in the improvement of standards of living. Furthermore, importing oil to generate electricity is a drain on foreign currency reserves and leaves the economy vulnerable to oil price shocks. Therefore, Senelwa and Sims (1999) consider sites where small-scale gasifiers could generate electricity from cheap, reliable sources of biomass. In Kenya, many families have reacted to fuel scarcity by planting woodlots and using efficient charcoal stoves. However, the non-forest wood resource was considered too scattered to support small gasifiers, and state forests are too remote and laws prohibit wood harvesting for fuel. Given these considerations, 73 large sawmills in Kenya are proposed as the best sites for electricity generation from timber residues. It is estimated that 12–16% of the national electricity load could be met by installing a 200 kW electricity generator at each of the 73 mills. The sale of the electricity to the grid would be legally possible now that Kenya has initiated an Independent Power Producers scheme.

10.5 CONCLUSION

Biomass provides for about 85% of the energy needs of people in sub-Saharan Africa, and consumption rates are not expected to fall and may increase in decades to come. There are health problems and local deforestation concerns associated with traditional use of biomass for energy, but one must not overlook the essential role biomass plays in providing warmth, safe food and water for people who lack access to other energy forms. Biomass energy has been neglected in national statistics and policies, and insufficient effort has gone into modernizing the use of biomass and efficiently exploiting the large energy resource it represents. Biomass energy flow charts can be used to improve statistics and

allow formulation of policies for efficient resource exploitation, and simple technological improvements such as biogas generation and improved charcoal kilns and wood stoves are a step in the right direction. As technologies such as gasification for electricity improve, there is great potential for modernization of electricity and fuel supply based on biomass energy plantations in sub-Saharan Africa.

ACKNOWLEDGEMENTS

I would like to thank Keith Openshaw and Professor Cherla Sastry for refereeing the paper and for their helpful comments. Ivan Scrase accepts responsibility for any errors, as Professor Hall was not able to comment on later drafts.

REFERENCES

Amoo-Gottfried, K. and Hall, D. O. (1999) A biomass energy flow chart for Sierra Leone. *Biomass and Bioenergy,* **16**: 361–376.

Bruce, N. (1998) Smoke from biomass and its effects on infants. *Boiling Point,* **40**: 2–5.

D'Apote, S. L. (1998) IEA Biomass energy analysis and projections. *Proceedings of IEA/OECD Conference 'Biomass Energy: Data, Analysis and Trends',* Paris, March 23–24, pp. 151–177.

FAO (1997) *Regional Study on Wood Energy Today and Tomorrow in Asia.* FAO Field Document 50, Bangkok, Thailand.

Hall, D. O., Rosillo-Calle, F., Williams, R. H. and Woods, J. (1993) Biomass energy: Supply prospects. In T. B. Johansson, H. Kelly, A. K. N. Reddy and R. H. Williams (Eds.), *Renewable Energy: Sources for Fuels and Electricity,* London, Earthscan.

Hemstock, S. L. and Hall, D. O. (1995) Biomass energy flows in Zimbabwe. *Biomass and Bioenergy,* **8**: 151–173.

Hosier, R. (1985) *Energy Use in Kenya: Household Demand and Rural Transformation.* Stockholm, Beijer Institute.

Karenzi, P. C. (1994) Biomass in Rwanda. In D. O. Hall and Y. S. Mao (Eds.), *Biomass Energy and Coal in Africa,* Chapter 4. London, African Energy Policy Research Network/Zed Books.

Kgathi, D. L. and Mlotshwa, C. V. (1997) Fuelwood procurement, consumption and substitution in selected areas of Botswana. In D. L. Kgathi, D. O. Hall, A. Hategeka and M. B. M. Sekhwela (Eds.), *Biomass Energy Policy in Africa: Selected Case Studies,* Chapter 2. London, African Energy Policy Research Network/Zed Books.

Marrison, C. I. and Larson, E. D. (1996) A preliminary analysis of the biomass energy production potential in Africa in 2025 considering projected land needs for food production. *Biomass and Bioenergy,* **10**: 337–351.

Marufu, L., Ludwig, J., Andreae, M. O., Meixner, F. X. and Helas, G.

(1997) Domestic biomass burning in rural and urban Zimbabwe – part A. *Biomass and Bioenergy,* **12** 53–68.

Openshaw, K. (1998) Estimating biomass supply: focus on Africa. *Proceedings of IEA/OECD Conference 'Biomass Energy: Data, Analysis and Trends',* Paris, March 23–24.

Senelwa, K. A. and Hall, D. O. (1993) A biomass energy flow chart for Kenya. *Biomass and Bioenergy,* **4**: 35–48.

Senelwa, K. and Sims, R. E. (1999) Opportunities for small scale biomass-electricity systems in Kenya. *Biomass and Bioenergy,* **17**: 239–255.

United Nations (1993) *World Population Prospects, 1992 Revision.* New York, UN.

World Bank (1996) *Rural Energy and Development: Improving Energy Supplies for 2 Billion People.* A World Bank Best Practice Paper. Industry and Energy Dept. Report 15912 GLB, July 1996. Washington, DC.

11 Natural resources: population growth and sustainable development in Africa

KEITH OPENSHAW*

International Resources Group, Silver Spring, MD, USA

Keywords

Africa; population pressure and land resources; environmental degradation; productivity; natural resources; sustainable development

Abstract

The nexus between population growth and the demand for agricultural land in the different regions of Africa is examined. Given present trends, the demand for agricultural land may increase by about 75% in 2025 and by more than threefold when Africa's population stabilizes after 2100. A comparison is made between the past demand for agricultural land and the decrease in forest area. In most regions of Africa, there is a close correlation between these two factors. Therefore, without active intervention, the future enlarged agricultural estate will be at the expense of high forests and woodlands. By the time the population stabilizes, Africa could lose half its forests. At present, there is a surplus of wood compared to demand, and this surplus could be used to expand rural opportunities. However, in nearly every country there are pockets of shortages. These pockets will expand if little is done to improve agricultural and silvicultural productivity. Ways of improving productivity are discussed, especially with farm trees; this will release the pressure on forest areas. Also, strategies for meeting demand in wood deficit areas are specified. Land-use change targets are given based on various productivity assumptions and increased food intakes. If the forecasted gains are made, coupled with the expanded use of renewable resources, then the demand for extra farmland may increase by only one-third by 2025 and by one-half when the population stabilizes. These initiatives should be seen as opportunities to alleviate poverty for millions of people by making the rural population active participants in sustainable economic and social development based on indigenous and renewable resources.

*Retired in 2004

11.1 INTRODUCTION

Over the past 50 years, much of Africa has emerged from its colonial past to become a continent of independent nations. However, many conflicts still remain, several as legacies of colonial boundaries, but some as a result of ideologies and others as disputes over valuable natural resources. Many of these resources have been squandered or used to benefit the few. Most governments are either in the hands of the military or are one-party states with a minimum of democracy. Also, since the end of the Cold War, the West and the former Soviet Union have lost interest in Africa and assistance has been sharply curtailed. In consequence, while the elite has been increasing its wealth, the majority of the population has benefited little from the sale of natural resources or development assistance, and has to pay for it.

In addition, during the last 50 years, the population has increased from about 200 to 820 million. As a result, the percapita area of land, excluding deserts, has decreased from 10.5 hectares (ha) in 1950 to 2.5 ha today. By the time the population stabilizes, in about a hundred years' time, there will be less than 0.75 ha per person! This increase in population has and will continue to put pressure on the land and its natural resources to supply food, fuel and building materials, etc. In consequence, there has been a steadily increasing loss of forest area,[1] with an average decline of about 1% per year over the last 20 years. Much of this

---

[1] The general term 'forest' usually incorporates all woody areas over 0.5 ha with a crown cover greater than 10%. Forests can be broadly divided into two categories, namely high or closed forests and woodlands. Natural high forests (and plantations) are regularly found in areas with annual rainfall above 1,000 mm, with mature trees between 20–30 metres high and a crown cover greater than 80%. Woodlands commonly occupy areas with annual rainfalls between 200 and 1,000 mm. Generally, they have more open formations, ranging from 10 to 80% crown cover and the height of mature trees is 10–15 metres. Individual trees are branchy with large crowns, whereas high forest trees tend to have relatively small branches and crowns. In this chapter, the terms 'forest', 'forest land' or 'forest area' are used for any tree formation including high or closed forests and woodlands (see Table 11.4).

Climate Change and Africa, ed. Pak Sum Low. Published by Cambridge University Press. © Cambridge University Press 2005.

deforestation has been blamed on commercial logging for industrial wood products, such as sawnwood and plywood, for both domestic and international markets, and on traded wood fuel (fuelwood and charcoal) (World Bank, 2000).

Therefore, for each country in Africa, a detailed examination was made of the likely causes of deforestation, namely clearing land for increased food production, both from crops and domestic animals, or to supply wood fuel and building materials. It was found that in most countries, there is a surplus of wood, but on the other hand, there is a close correlation between the increased demand for food and the rate of deforestation. There have been some gains in agricultural productivity over the years, but this has not kept pace with population increase. Therefore, much of the additional food production to feed the increased population has come from opening up new forest land. In addition, in order to earn foreign exchange, many governments have encouraged cash crops such as cotton, tobacco, tea and coffee. This has further exacerbated land clearing. This has led to a steady but pervasive decline in the forest area, both in terms of quantity and quality. In turn, this has adversely affected the biodiversity throughout the continent and the water flow of many rivers.

Despite the pressure on the land and natural resources, Africa is still a resource-rich continent, and given more enlightened management, especially through democratic control at the local level, the population could use these resources sustainably for the benefit of the masses. There are two principal driving forces that will determine whether Africa can successfully pursue a policy of sustainable development or continue on the present path of environmental degradation, namely the rate of population increase and the proper use of resources, especially renewable ones. Government policies can and will affect this direction; this, in turn, will be affected by external pressures, both from within the continent and from the rest of the world. However, in order to slow down the rate of deforestation, there has to be a concerted effort to increase agricultural productivity. More controversially, it is argued that the forest resources are being under-exploited throughout the continent. Expanding their use could increase the economic opportunities; this could have a tempering effect on the rate of population increase.

This chapter examines past land-use changes, the likely increase in population until stability and its effect on land use under different options. It explores ways of using renewable resources to keep Africa on a sustainable development path assuming greater political stability and better governance.

11.2 DEFORESTATION: THE DEMAND FOR LAND AND WOOD

The Food and Agricultural Organization (FAO) of the United Nations reported that in 2000 about 9.4 million ha of forests were lost worldwide, down from 12 million ha in 1998. The main cause of forest destruction was attributed to fires (80%), but logging, road building, insects and diseases were also cited as reasons for deforestation. Fire is a principal tool when clearing land for arable and pastoral agriculture and to prepare land for perennial crops such as palm oil, tea and coffee. It is an integral part of the shifting cultivation cycle – wood ash being an important fertilizer for the grain crops. Fire is used when managing woodlands to promote new grass, kill ticks, etc. and to capture animals for bush meat. Most woodland and high forest areas will recover from fire, both wild and intentional, if no land-use changes occur. Thus, some deforestation is temporary, as during the cropping phase in the shifting cultivation cycle or after selective felling.[2]

There is considerable faulty analysis, emotional debate and inaccurate perceptions when discussing deforestation. Some people conclude that when a group of trees is cut down or even when selective felling occurs, this is deforestation. Yet the same reasoning does not apply when a field of maize or wheat is cut; this is not called 'defarming' but harvesting. In most cases, both are harvesting. Deforestation occurs when there is a change of land use, an area fails to recover when it is cleared or it dies through a natural calamity such as insect attack, fire or drought. Some logging operations provide the opportunity for farmers to move in and clear the remaining trees. But this may only ease the clearing process without stopping it.

Degradation is another word that is loosely used. Reducing the forest capital through, say, selective felling usually leads to gaps appearing in the high forest or woodland. If these gaps do not close, this could be considered as degradation. Similarly, if certain plant species in a mixed forest stand are eliminated or severely reduced, this is degradation. However, old trees or densely packed trees may be felled or thinned in order to encourage succession, reduce competition and promote growth. Felling old or moribund trees decreases the growing stock, but generally encourages vigorous growth from young and suppressed trees. Therefore, while the wood capital declines, annual yield may increase. These are management options and should enhance the forest rather than degrade it.

In the 10-year period 1981–1990, FAO estimated that tropical Africa lost about 41 million ha of forests, while only 1.3 million ha of plantations were established (FAO, 1993). Including South Africa and the un-assessed islands, the total net loss may approach 42 million ha. Burundi, Cape Verde, Lesotho and Niger increased

---

[2] Selective felling entails cutting 'commercial' species greater than a minimum diameter (usually greater than 60 cm). Surrounding trees may be damaged during logging, and only about one third to one half of the utilizable wood may be removed, because of minor faults in the discarded logs. In theory, the forest area is left to recover and is only visited again after 20–40 years when the operation is repeated. This is a standard forest practice in tropical areas.

Table 11.1. Sub-Saharan Africa. Forest loss between 1980 and 1990 compared to cropland requirements for the additional population.

| Regions in sub-Saharan Africa | Population increase Million (m) | Forest area lost | | Additional area required for: | | Cropland as a percentage of forest loss |
		Total m ha	Per capita ha	Cereals m ha	All crops m ha	
Middle	16.52	7.44	(0.45)	5.48	7.37	99
Western	53.71	18.76	(0.35)	12.19	16.39	87
Eastern	50.36	12.32	(0.24)	9.47	12.73	103
Southern	8.90	3.20	(0.36)	1.96	2.63	82
S-S Africa	129.49	41.72	(0.32)	29.10	39.12	94

Note: The northern region comprises the countries of Morocco, Algeria, Tunisia, Libya and Egypt. It has been excluded from the above table because much of its extra food requirements are imported and additional land is reclaimed from the desert, not forest areas. Middle Africa consists of Burkina Faso, Cape Verde, Chad, the Gambia, Guinea-Bissau, Mali, Mauritania, Niger, Senegal and Sudan. Western Africa stretches from Guinea to Angola, including the west-coast islands, except Cape Verde. Eastern Africa runs from Ethiopia to Mozambique, including the east-coast islands. Southern Africa consists of Botswana, Lesotho, Namibia, South Africa and Swaziland. The population increase of every country was listed together with the cereal intake and the average cereal yield. An estimate of cereal land required to feed the increased population was calculated country-by-country and summed for the region. This figure was multiplied by 1.344 to obtain an estimate of total area required for all annual crops. This multiplier was derived from Table 11.4 on land use. The net forest area lost between 1980 and 1990 was listed for each country and totalled. 1,000 kcal = 4.18 MJ = 100.75 kg grain.

Sources: World Bank (1990), Tables 1, 2 and 28; kcal supply. World Resources Institute (1994), Table 18.1; FAO (1993), Tables 1a, 4a and 8a.

their forest area; a few retained their area, such as Mauritania, Mauritius and Seychelles; but most countries lost forests, ranging from 10,000 ha in the Gambia to nearly 7.3 million ha in Congo DR.

What is the principal cause of this deforestation – the demand for wood or the demand for agricultural land? If the area of forest lost between 1980 and 1990 is compared to the new agricultural land required to meet the food and other demands of the additional population in sub-Saharan Africa, this accounts for over 90% of the deforestation. This is shown in Table 11.1. This method of estimating additional land for arable agriculture was used because reliable land-use statistics are lacking.

When examining the figures for individual countries, some show a greater demand for arable land than the forest area lost, while others show a much larger loss of forests than can be accounted for by the demand for cereals and total crop land. This indicates that, in part, the statistics must be at fault, although some forests are converted directly to grasslands to accommodate farm animals, which have expanded roughly at the same rate as the human population increase. Forest land is also cleared for dams, roads and buildings, and it is indirectly converted when agricultural land surrounding urban areas is taken over for urban development. Usually, forests are then cleared for crops, etc. in compensation for this loss of agricultural land.

One group of countries where deforestation greatly exceeds the land required for arable agriculture is in West Africa, from

Cameroon to Congo DR. The estimated requirement for cropland is 4.65 million ha, only 40% of the area deforested over the 10-year period 1980–1990, calculated by FAO to be 11.26 million ha. This is an area rich in tropical high forests and has an active commercial logging industry and export trade. Annual production of sawlogs in the 1980s and 1990s was about 5 million cubic metres (m^3) roundwood, of which half was exported, but this is less than 1% of annual growth. Nevertheless, logging could account for some of this deforestation, but if farmers do not claim logged-over areas, they should revert to forests. However, annual fires in some of the drier zones may convert some logged-over areas into grasslands. Again, because the soils in some of these areas are relatively rich, there has been an expansion of cash crops for exports such as oil palm, coffee and cotton. Another country where deforestation exceeds demand for arable agricultural land is Botswana. In this case, the difference may be attributed to land cleared for ranching purposes.

Nature abhors a vacuum. If partially cleared or felled forest land is left alone with little interference from humans, it will revert to its climatic climax (forest) state. This fact has been used by shifting cultivators since time immemorial and copied by professional foresters. Shifting cultivators farm a cleared forest area for two to three years and then 'rest' it for 10–20 years. Usually, it reverts to secondary forest and is managed for its forest products. Gradually the soil regains its former fertility. It is then cleared again and the cycle is repeated with the arable crop being fertilized by the

Table 11.2. Africa 2000: Estimated above-ground woody growing stock and yield on all land types, compared to annual demand for roundwood.

Region	Population 10^6	Growing stock		Annual yield		Annual demand	
		Total 10^6 t	Per capita t $[m^3]$	Total 10^6 t	Per capita t $[m^3]$	Total 10^6 t	Per capita t $[m^3]$
Northern	139	820	6 [8]	40	0.3 [0.4]	14	0.1 [0.2]
Middle	90	4,760	53 [74]	180	2.0 [2.8]	58	0.6 [1.0]
Western	271	38,950	144 [202]	2,640	9.7 [13.6]	295	1.1 [1.6]
Eastern	266	20,700	78 [109]	600	2.3 [3.2]	260	1.0 [1.6]
Southern	54	3,040	56 [78]	100	1.9 [2.7]	32	0.6 [0.9]
Africa	820	68,270	86 [116]	3,560	4.3 [6.0]	659	0.8 [1.2]

Note: Growing stock and yield figures for northern Africa have been estimated based on forest and woodland areas. All the other figures were taken from an estimate of woody biomass in sub-Saharan Africa (World Bank, 1994b). These figures are for above-ground wood and are given in dry tonnes. A multiplying factor of 1.5 has been used to estimate cubic metres. Annual demand figures were taken from the FAO *Forest Products Yearbook 1996*. Projections were made for the period 1996–2000, based on population increase. The figures are given in cubic metres roundwood. These numbers were multiplied by 1.5 to account for under-recording of fuelwood, poles and sawlogs and then divided by 1.5 to obtain dry weight estimates. Net exports are included in the above demand figures. For western and southern Africa they are respectively, 6.0 and 2.5 million m^3 roundwood. Wood for fuel is the dominant end-use. It accounts for 91% of roundwood demand. The regional breakdowns are: Northern 81%, Middle 90%, Western 91%, Eastern 94% and Southern 66% of demand.
Sources: World Bank (1994b), Table 1; FAO (1998b); Author's projections.

nutrient build-up in the soil and from wood ash produced when the forest trash is burnt in situ. Because of population pressure, reducing the fallow period may shorten the shifting cultivation cycle. Thus, soil fertility is decreased and unit crop yields are reduced. Because of this, crops may only be grown for one or two years before the land is returned to tree fallow again; this leads to unsustainable land use. Occasionally, the cropping cycle may continue until the soil is severely exhausted. This could lead to an invasion of weeds such as *Imperata cylindrica* that are difficult to remove. In such cases the land will not revert back to forest. Even on newly converted arable land where fertilizers are used to maintain fertility, the farmer has an additional battle to destroy tree suckers from felled stumps and roots or from self-sown seedlings germinating from seeds dispersed by animals or the wind.

Therefore, deforestation is principally tied to arable and pastoral agricultural clearing (plus some resulting from urbanization, etc.), and not to the use of wood products such as fuelwood, charcoal, poles and sawnwood. In fact in Africa today, the annual growth of wood is over four times demand. The estimated above-ground growing stock and yield plus the annual demand for all kinds of roundwood is given in Table 11.2.

Except for northern Africa, including the Sahel, Africa is a wood-rich continent. It contains nearly 70 billion tonnes of above-ground wood from twigs to trunk (over 100 m^3 per capita), with an estimated yield of some 3.56 billion tonnes of woody biomass (6 m^3 per capita.). The overall yearly demand for all kinds of roundwood is less than 20% of sustainable supply, even after increasing FAO's consumption figures by 50% to allow for under-recording! The western region, which includes the Congo basin, has a huge surplus of nine times demand, whereas in the eastern region, the surplus is twice the demand.

This does not imply that there are no areas of shortage. In northern Africa and some other countries, such as Djibouti, Cape Verde and Lesotho, there are overall deficits; here wood is imported. Deficits also occur in some parts of most countries, usually around population concentrations. These areas are likely to expand if land clearing continues at its present pace and local initiatives to increase supply and mitigate demand are not stepped up. But there is considerable scope to expand production in surplus areas and to pursue a policy of sustainability in deficit zones.

11.3 POPULATION TRENDS

Over the last 50 years, the countries of Africa, like many other developing countries, have witnessed an unprecedented population increase, mainly due to advances in medicine and public hygiene. Since 1950, the population has quadrupled. Because the increase in wealth has not kept pace with this population increase, per capita income has only risen slowly, and in fact over the past 10 years it has stagnated or declined. As income rises, the population growth rate declines, but the anticipated per capita increase in wealth will be very modest over the next decade or so. As a result, the population in Africa is expected to increase by 75% over the next 25 years despite the AIDS epidemic (Table 11.3).

Table 11.3. Population trends in Africa by region.

Region of Africa	1975 (million)	2000 (million)	2025 (million)	Stability (million)
Northern (arid N. Africa)	94	139	200	285
Middle (Sudano-Sahelian)	45	90	155	325
Western	123	271	505	1,050
Eastern	122	266	490	1,020
Southern	29	54	80	170
Africa [% of world pop.]	413 [10]	820 [13]	1,430 [18]	2,850 [25]
World	4,080	6,110	8,120	11,480

Sources: FAO (1986); United Nations (1988) (adjusted); World Bank (1994a).

It is estimated that the population of Africa will exceed 1 billion by 2010 and eventually stabilize after 2100 at about 2.85 billion – three and a half times today's population! In 1975, the population of Africa was about 10% of the world's population. This percentage will increase to 18% by 2025 and reach 25% at the point of hypothetical stability in about 100 years' time. For many decades, population increase will remain the principal determinant behind the use of natural resources.

Sixty-five per cent of the population live in rural areas, and although this has dropped from about 75% in 1975, most people will continue to obtain their livelihood from the land in the foreseeable future. About 50% of the population is under 18 years, 76% will receive primary education, thus by inference, considerable (scarce) resources are devoted to education. Thus, education policies should be tailored to rural development through placing as much emphasis on farming practices and environmental management, etc., as on reading, writing and arithmetic.

The percentage of people below the poverty line averages 32, ranging from 2% in Algeria to 85% in Zambia (World Bank, 1990, 1997). Reducing poverty is a prime concern of national governments and international agencies. In Africa, the average life expectancy is about 51 years compared to a world average of 67. Life expectancy is decreasing slightly, mainly as a result of AIDS and other epidemics; many health facilities are being overstretched because of this and other basic health demands. Thus, financial resources are being extended. Wars and the political situation in several African countries exacerbate this situation, with many people forced into subsistence living or herded into refugee camps.

Apart from extending family planning initiatives, including much more open discussion on AIDS with aggressive policies to prevent its spread, the best way to try to reduce the rate of population increase is to enhance rural opportunities for economic development. This should come through:

• Concerted efforts to improve agricultural productivity;
• Better and sustainable use of natural resources;
• Improved rural access and infrastructural development;
• Appropriate prices for agricultural/natural resources, goods and services.

These will now be discussed together with the consequences of pursuing a policy of 'business as usual'.

11.4 LAND AREA AND LAND USE

Have the countries and individual regions of Africa sufficient land, water, wood and other natural resources not only to meet the basic needs of an expanding population, but also to lift the millions of people out of poverty and provide a better standard of living? Africa's population is going to double in about 30 years' time. Thus, other things being equal, the demand for food, fodder, water and wood will double also. When the population of Africa eventually stabilizes, the demand for these natural resources may be about three and a half times what they are today! This will put tremendous pressure on all resources and could lead to increased land degradation and more conflicts if appropriate actions are not taken. Before discussing this, existing land use is examined. Table 11.4 gives a breakdown of land use in Africa today. It was derived from several sources; some were modified for consistency and compatibility.

Woodlands cover the largest area of Africa as a whole and of every region, except northern and Sudano-Sahelian Africa where deserts dominate. Woodlands are more open than high forests, and the average tree height is 10–15 m. These woodlands have numerous uses. They are home to abundant wildlife, both plant and animal. They provide browse, fodder and shade for domestic animals. They are a source of fuel, poles, timber, fibre and thatch. They furnish fruit, nuts, honey, vegetable oils, gums, waxes, resins, incense, medicines and bush meat. They are a source of agricultural land, mainly on a shifting agricultural cycle, but sometimes cleared permanently for subsistence and cash crops. From an environmental viewpoint, they affect climate positively, are an

Table 11.4. Africa 2000: Land-use by region (units: million hectares; population millions).

Land use	Population	Water area	Cropland	Grassland	High forest	Woodland	Desert	Total area
Northern	139	0.3	26.8	68.2	1.3	12.0	491.5	600.1
Middle	90	5.3	54.6	152.1	1.1	228.5	343.3	784.9
Western	271	10.8	87.5	15.1	283.6	340.5	2.0	739.5
Eastern	266	18.2	57.0	66.3	72.0	392.8	29.0	635.3
Southern	54	0.2	16.1	88.9	2.8	128.3	32.9	269.2
Africa	820	34.8	242.0	390.6	360.8	1,102.1	898.7	3,029.0
%		1	8	13	12	36	30	100

Note: Water bodies are large natural or man-made lakes/dams; excluded are small rivers and small water bodies. Generally, croplands are devoted to annual crops such as grain, pulses, cotton, sugar, etc. Many fields have woody biomass scattered throughout them or along borders. Some areas are under perennial crops such as plant (palm) oils, fruit/nuts, rubber, sisal, coffee, tea, etc. Also, a little cropland and wooded areas are under shifting cultivation. Grasslands include open grasslands, heaths, wooded grasslands and permanent pasture. High forest areas include natural high or closed forests and plantations or woodlots; annual rainfall above 1,000 mm. Some natural forest areas are under shifting agriculture or have an under-storey of cash crops such as cocoa. Woodlands are usually found in areas where the annual precipitation ranges from about 200 to 1,000 mm. They include shrubland, bushland, thicket, low wooded biomass mosaic and all types of woodland where trees dominate. Some of these areas are under shifting cultivation. Desert areas are found in regions with annual precipitation usually below 200 mm. They may not be devoid of plant growth, especially when it rains. Also, some areas have been reclaimed with irrigation. Urban, roads and other built-up areas have been excluded. They represent about 1% of the total area.

Sources: World Bank (1990, 1994a, 1994b, 1997); FAO (1986, 1993, 1998a). Information from these various publications was modified for compatibility.

important gene pool, a vast store of organic carbon and are vital for water resource flow and availability.

However, it is the high/closed forest areas that are the first target for conversion to agriculture. These forests are generally closed formations, with the dominant trees 20–30 m tall. They are in areas with rainfall of 1,000 mm or more, or in low rainfall areas where groundwater accumulates. Also, the soils tend to be deeper and more fertile than woodland soils. Only in western Africa from Cameroon down to Congo DR are there large reserves of tropical high forests. In other areas, there are patches or islands of natural forests, usually in upland regions. These are vital watersheds, but are under threat from an expanding population requiring agricultural land. Out of an estimated 361 million ha of high forests, 3 million ha (1%) are plantations, over half of which are in eastern Africa.

An estimated 242 million ha of land (8%) are under permanent arable agriculture. In addition, there are perennial (agricultural) cash crops – fruit, nuts, palm oils, rubber, beverages, etc. – that are sometimes classified as forests, and there are forest areas worked on a shifting cultivation cycle. Thus, the total area under 'agricultural' crops may be about 300 million ha, or 10% of Africa's land. All regions except northern Africa have sufficient agricultural land to at least meet the requirements for cereals. As a group, the northern countries are net importers of cereals, paid for mainly from oil and gas revenues. However, there are some countries such as Chad, Mozambique and Ethiopia, which do not grow sufficient

grain to meet even the World Food Programme's grain ration standard of 176 kg per person per year. The average yearly intake of grain ranges from 161 kg per person in Mozambique to 363 kg per person in Libya. Clearly an effort has to be made to increase food production, not only to improve the food intake of the population, but also to meet the demands of the projected increase in population. This is a major challenge for decision makers. Can sufficient resources be channelled into increasing agricultural productivity, or will the demand for more food be met by converting additional forest land (high forest and woodland) to agriculture? This is now discussed.

11.5 FARMLAND: POSSIBLE LAND-USE TRENDS

As detailed above, most expansion in grain production has come not from increasing yields, but from expanding the area under arable agriculture, principally at the expense of forest land. There are exceptions, notably in Egypt, where irrigation, double cropping, improved seeds and adequate and correct fertilizer applications give an annual grain production of about 5.7 tonnes (t) per ha. This compares to a continental average of about 1.06 t/ha and a low of 340 kg/ha in Mozambique (World Resources Institute, 1994). If this trend continues, then the area required for grain production will be about 315 million ha by 2025 and 630 million ha when the population stabilizes. Not only that, land will be required for other food and cash crops such as root crops, pulse, plant oils,

Table 11.5. Agricultural land requirements by 2025 and the population stability date with no increases in productivity or grain intake.

| | | | Land required for grain crops | | | | | |
| | | | 2000 | | 2025 | | Stability (2100+) | |
Region	Average grain intake kg/yr	Average production kg/ha	Pop. m	Area m ha	Pop. m	Area m ha	Pop. m	Area m ha
Northern	314	1,608	139	27.1	200	39.1	285	55.7
Middle	221	666	90	29.9	155	51.4	325	107.8
Western	212	934	271	61.5	505	114.6	1,050	238.3
Eastern	210	1,117	266	50.0	490	92.1	1,020	191.8
Southern	286	1,299	54	11.9	80	17.6	170	37.4
Africa	233	1,059	820	180.4	1,430	314.8	2,850	631.0
Land required for all crops				242.0		422.0		846.0
Estimated cropland including perennial cash crops and shifting cultivation				300.0		523.0		1,049
Percentage of total land area				10%		17%		35%

Note: The estimate of land required to satisfy the demand for grain consumption is calculated by determining a weighted regional average for yield and grain intake from country data. Then the regional population totals are multiplied by the average grain intake and divided by the average per hectare production. Example: Northern region, year 2000. 139 million (m) × 314 kg/1,608 kg/ha = 27.1 million (m) ha. For 2025 and 'stability', the estimated land for total crop production and total agricultural land, including shifting cultivation and perennial cash crops, was made using the same proportional increases as for 2000. This is 1.34 times the grain area for all permanent cropland (see Table 11.4) and 1.24 times all permanent cropland for all arable land, including shifting cultivation and perennial cash crops. 1,000 kcal = 4.18 MJ = 0.75 kg grain.
Sources: World Bank (1990), Tables 1, 2 and 28; World Resources Institute (1994), Table 18.1.

sugar, cotton, etc. Thus, the cropland requirements may be about 422 million ha in 2025 and 846 million ha when the population stabilizes at an estimated 2.85 billion. In addition, shifting cultivation and perennial cash crops may bring the total to 523 and 1,049 million ha respectively by those two dates. This is shown in Table 11.5.

There is some inconsistency between Tables 11.4 and 11.5. While the total land required for all farm crops in 2000 is the same at 242 million ha, the estimated area for cereals in the northern region in Table 11.5 is more than the total area in Table 11.4, indicating that this region imports some of its requirements. But this does not invalidate the macro picture for Africa.

By the year 2025, if there are no regional increases in agricultural productivity, an additional 180 million ha of forest land may be cleared to grow all annual crops, and up to 223 million ha may be cleared if shifting cultivation and perennial cash crops are included. In addition, some high forests and woodlands will be cleared for pastoral agriculture. About 750 million ha of high forests and woodlands could be cleared for crop production by the time the population stabilizes. Individual countries, the African continent and the world as a whole cannot afford to let this happen. There has to be a concerted effort not only to increase agricultural productivity, but also to improve the average food intake of the

population. If the demand for additional agricultural land is to be curtailed, an increase in productivity has to more than offset direct and indirect increased grain consumption. Table 11.6 forecasts the areas of land required for arable and pastoral agriculture by 2025 and 2100, assuming different increases in agricultural productivity.

If, by 2025, overall agricultural productivity could be increased to 1.7 t/ha of grain equivalent – a level already achieved in Gabon today – then, even allowing for a 30% increase in average grain consumption, the requirements for cereal land would only be 250 million ha. This is some 65 million ha less than the amount of land needed assuming that there were no increases in productivity and average grain consumption remained static. Assuming that by the time of population stability average cereal yields could be the same as in Europe today, namely 3.7 t/ha, then even allowing for a 55% increase in per capita grain intake, the land area required for cereal production is only 277 million ha. This is 354 million ha less than the area needed assuming that productivity and grain intake remains constant!

In order to stabilize agricultural land at today's level, allowing for a gradual increase in grain intake, cereal productivity will have to increase to 2.38 t/ha by 2025 and to 5.69 t/ha by the time Africa's population stabilizes (Table 11.6). This is possible, given

Table 11.6. Africa: agricultural land required for food self-sufficiency and export crops with different levels of productivity and grain consumption.

| | | | | Per capita | | Area required for: | | |
| | | | | | | | | |
Yield assumptions	Year	Population m	Grain yield kg/ha	intake kg/yr	Cereal m ha	Permanent agriculture m ha	Total agriculture m ha	Agriculture land Total %
Average yield in	2000	820	1,059	233	180	242	300	10
Africa today	2025	1,430	1,059	233	315	422	523	17
	2100	2,850	1,059	233	631	846	1,049	35
Gabon today	2025	1,430	1,714	300	250	330	409	13
Europe today	2100	2,850	3,700	360	277	365	453	15
Stabilized area at 2000 level	2025	1,430	2,380	300	180	242	300	10
	2100	2,850	5,690	360	180	242	300	10

Note: The cereal area is for domestic consumption only (both human and animal) and does not include exports. The land area for permanent agriculture includes grain crop exports, land for pulse, root crops, sugar, annual vegetable oils, non-cereal animal feed, horticultural crops, etc. Total agricultural area includes the above plus perennial cash crops and area under shifting cultivation. For all land options in 2025 and when population stabilizes, it is assumed that agricultural productivity on all lands will increase at the same rate as that on cereal land. The average grain productivity of 5.69 t/ha, assumed when the population stabilizes in about 100 years time, is about the average grain yield in Egypt today (5.7 t/ha). 1,000 kcal = 4.18 MJ = 100.75 kg grain.
Sources: World Bank (1990), Tables 1, 2 and 28; World Resources Institute (1994), Table 18.1.

the correct motivation, education, training and technical input. It also means that alternative employment opportunities would have to be found for the people not directly employed in agriculture.

Much of Africa's agriculture is rain-fed with little mechanization and low inputs of mineral fertilizers. The average application of mineral fertilizers is about 20 kg/ha, compared to 123 kg/ha in Asia and 192 kg/ha in Europe. Distribution systems are poor and most farmers have little money to spend on inputs. Therefore, farmers depend on organic fertilizers, including wood ash, crop rotations/intercropping and the fertility build-up during the tree phase of the shifting cultivation cycle. More could be done to increase soil fertility and enhance texture, without heavy reliance on chemical fertilizers, and to reduce poor farming practices, especially on sloping lands. Nitrogen-fixing trees planted in fields (agroforestry) can double yields,[3] while at the same time reduce erosion and provide small-diameter wood. The correct and timely applications of manure and/or mulch both improve crop yields and soil texture, especially if improved varieties are used. With irrigation, double and triple cropping is possible, but water management may be critical. Many soils in Africa are too acidic and

require lime applications to release mineral elements locked up in the soil. All these initiatives require relatively little monetary input, except for irrigation.

In addition, advantage should be taken of the wood surplus in many parts of most African countries. In areas where there are actual or pending shortages, steps could be taken to meet demand through increased supply and/or demand mitigation.

11.6 WOOD AND OTHER BIOMASS ENERGY AND SUSTAINABLE DEVELOPMENT

Wood fuel is the largest end-use of roundwood, and about 90% of households in Africa use it as the principal cooking fuel. Fuelwood is used in many rural industries, such as brick and lime burning, beer brewing, bread making, fish smoking, jaggery production, tannin manufacturing, tea and timber drying, tobacco curing, etc., and it is the feedstock for charcoal production – an important urban fuel. Fuelwood and charcoal are the dominant fuels in practically all sub-Saharan African countries. They are also the principal traded fuels and provide numerous opportunities for rural employment and off-farm income. Yet many people regard the reliance on biomass fuels as an example of underdevelopment. The World Bank has frequently stated that 'two billion people are cooking with wood and two billion people are without electricity' (World Bank, 1999). As if with access to electricity, rapid development will be assured. Supplying these people with electricity will

---

[3] The author visited the ICRAF field station at Zomba, Malawi. On maize fields intercropped with *Sesbania* spp. and *Tephrosia* spp., maize yields increased from 800 kg/ha to 1.6 t/ha from year 3 onwards. The annual yield of stick wood was about 1 t (air-dry). Thus total yield more than doubled.

cost billions of dollars especially tapping the hydro potential of many countries, [4] and of itself, will not guarantee development or cooking fuel. It is the lack of opportunities, especially in rural areas, that prevent people from improving their living standards and being able to afford electricity, even if it is available. National and international policies contribute to this state of affairs, through subsidies on food and imported fuels, poor infrastructure, trade barriers, government regulations, inadequate governance, lack of transparency, insecure land rights, conflicts, monopoly producer boards and other inappropriate development policies.

Most of the common alternative forms of energy to wood are derived directly or indirectly from fossil fuels (coal, kerosene, LPG, etc.). While they may be more convenient to use, they cannot sustain development indefinitely because their reserves are finite. They are also the major contributors to atmospheric greenhouse gas (GHG) accumulation, accounting for about 80% of net GHG accumulation. Another contributor to GHGs is the destruction of forests for agricultural development (about 15%); this should be a compelling reason to increase agricultural productivity so as to curtail forest loss. Evidence is accruing that global warming is occurring as a direct consequence of GHG accumulation, and the consequences of such warming may have adverse effects on millions of people and could result in regional if not global disasters. Continued use of fossil fuels, until they are exhausted, may be calamitous for humankind.

Apart from direct use, wood and other forms of plant and animal materials can be turned into more convenient solid, liquid and gaseous fuels – charcoal and densified wood, methanol (wood alcohol) and ethanol, biogas and producer gas. Most importantly, they are conditionally renewable, provided they are managed properly without overexploitation, and significantly, they do not contribute to greenhouse gas accumulation when sustainably managed. Globally, about 100 billion tonnes of carbon circulate between plant biomass and the atmosphere each year (the carbon cycle). (Hall and Rao, 1994) Of this total, about 50 billion tonnes are potentially available as fuel from biomass (Hall and Rao, 1994). This is equivalent to about eight times the amount of carbon that is burnt in fossil fuels annually. Much of the carbon in plant material decays or is destroyed in wild fires without being used, and only about 2.5 billion tonnes are used as energy or fibre. Given the correct incentives, especially the control and management of local forest resources, the people of Africa

could use wood to maintain, if not expand, its dominant position as a sustainable energy resource and a greenhouse gas neutral fuel.

11.7 NATURAL RESOURCES: OPTIONS AND OPPORTUNITIES

Africa is at the crossroads. Should it continue to copy the 'Western Economic Development Model' that is based on the use of non-renewable energy resources? Should it pursue grandiose projects, many of which have turned out to be economic burdens rather than saviours? Should the elite try to get rich at the expense of the general public, or should they use their talents for the benefit of the nation and the continent?

The Western model is non-sustainable and indeed the developed countries have pledged to reduce their dependency on fossil fuels, principally because they realize that their continued use may lead to environmental disasters. And besides, these fuels are finite, and alternatives have to be expanded or developed. One alternative that is being promoted is 'clean' hydrogen; this has about 3.5 times the energy of carbon per unit weight and when burnt it reverts back to water vapour, a more or less stable greenhouse gas. Work is being done on microbes that will break down water into hydrogen and oxygen, but at present the only practical method of large-scale hydrogen production is through the electrolysis of water. Therefore, considerable investments will have to be made in electrical generation; there has to be an adequate supply of water and there has to be a carrier for hydrogen, which is a very explosive fuel as a gas. Thus, as a household fuel and a low cost fuel for rural industries, it is a non-starter. Likewise, electricity, even from renewable resources, is an expensive energy form and cannot be stored in large quantities. The cooking habits of people necessitate that standby capacity is available to satisfy peak demand. This is a luxury that most African governments cannot afford. They are struggling even to connect and provide a reliable supply to urban consumers, let alone the vast majority of rural people who have no electricity at present, except through batteries. Again, most people with electricity do not use it for cooking purposes, as it is expensive when compared to alternatives and the supply is generally unreliable. It is also an expensive boiler fuel for industry.

Western energy planners are talking about 'low-carbon' technology by sequestering the carbon that is produced as a result of burning coal and other fossil fuels. It could be sequestered in oceans, deep saline formations, depleted oil and gas reservoirs and in coal seams (*The Economist*: 6–12 July 2002), but not trees or soils for some unexplained reason. This same issue of *The Economist* recommends that 'governments everywhere must send a signal that carbon is going out of fashion – with the best way through a carbon tax.' This completely ignores the fact that there

---

[4] There is considerable hydro potential in several central and eastern African countries, but many existing hydro-electrical generating schemes are not working at their full potential. This is because watersheds are not being properly protected, principally through farming on steep slopes in these areas, causing erosion and flash flooding in the rainy season and poor stream flow in the dry season. This has resulted in interrupted power supply to many cities in Ethiopia, Kenya, Malawi, Tanzania, Zambia and Zimbabwe. This points to the importance of protecting the natural resource base.

is a vast untapped source of renewable carbon that is produced each year by plants through photosynthesis. What is more, if this carbon is not used, it decays and reverts back to carbon dioxide. As mentioned previously, the potentially available carbon produced by photosynthesis is about eight times that burnt annually in fossil fuels. Using more of this renewable carbon is probably a much cheaper option than the alternatives of sequestering carbon from fossil fuels in the oceans, etc., or pursuing hydrogen and other energy alternatives. In addition, the technology to use this carbon directly or in more convenient solid, liquid and gaseous forms is well understood and available. Already over 90% of Africa's energy is from renewable resources, principally biomass, and if managed properly these resources could be expanded mainly through the efforts of the rural people. This is a golden opportunity to promote rural development through appropriate technology and self-reliance.

Africa should capitalize on its natural resources, while protecting its environment, especially its wildlife habitat. But in order to do this, it has to place much more emphasis on increasing agricultural and silvicultural productivity. First and foremost, agricultural yields must increase if deforestation is to be curtailed and eventually stopped or even reversed. Enhancing soil fertility (and texture) is key to this initiative, but not necessarily only with chemical fertilizers. Nitrogen-fixing plants, including trees, should assist in this endeavour as well as manure, mulch, wood ash and lime. If farmers do not have tenure security, this should be guaranteed as a right. Other necessary or desirable initiatives include:

- Involving farmers in planning and decision-making at the local level;
- Enhancing extension efforts;
- Improving access to goods, services and markets;
- Providing market intelligence;
- Removing subsidies from food to increase the farm-gate price;
- Stopping marketing boards fixing low producer prices so that farmers rather than boards receive the greatest rewards.

In other words, there have to be enabling activities to assist the farmers rather than trying to control them.

If things continue as they are, Africa could lose about 20% of its high forests and woodlands by 2025 (280 million ha), and by the time its population stabilizes, over half of its forest lands (800 million ha) may have been converted to arable and pastoral agriculture. This is alarming and could be cataclysmic. Such a trend will have profound and adverse impacts on the local habitat, the natural vegetation, the indigenous animal population and the water resources. In turn, this will have a negative effect on the continental and global environment. Thus, it is imperative to place far more emphasis on rural development. This should be viewed not as a problem, but as an excellent opportunity to lift millions of people out of poverty by contributing to Africa's economic and social development, while at the same time ensuring environmental integrity.

Because of lack of control over high forests and woodlands, local people often regard the trees on these areas as expendable resources. This is exacerbated by wood energy having to compete with electricity and fossil fuels, which are generally subsidized. This reduces the incentive to manage the trees and increases the tendency to mine them instead. Through ownership and participation in planning and managing the local forests, villagers will not only protect these areas, but also husband them sustainably for a multitude of forest products. The attitude of many forest services to local communities has to change from one of being adversarial – regarding them as the destroyers of forests – to one where they assist them in managing their resources. Putting guards and a fence around forests will not protect them if the local people have few other opportunities. Cooperation rather than confrontation is a prescription for resource protection, not destruction.

Even though Africa is a forest-rich continent, shortages of wood are occurring in most countries, especially in areas of high population density. Surveys should be undertaken in such areas to examine the demand/supply situation and propose least-cost strategies to satisfy demand sustainably. Options include managing better what is there already, planting trees on farms to assist with agricultural productivity, plus various other regeneration initiatives. Energy efficiency interventions, such as improved stoves, could not only reduce demand, but also reduce indoor and outdoor pollution. In areas where there are biomass surpluses, opportunities to use these surpluses should be actively pursued. Subsidies on energy should be removed so that biomass can compete with fossil fuels and electricity on equal terms; this should promote the planting of trees and improve the management of wood resources. A carbon tax on fossil fuels and 'carbon trading' could also boost the growing and use of biomass. All such interventions would generate local employment and stimulate economic development using renewable resources. Rural people may then be in a position to afford electricity!

In summary, Africa has to capitalize on its land and natural resources, especially renewable ones, using them as the engines for development. Resources must be managed sustainably and, therefore, increased food production has to come from increasing yields on existing agricultural lands, rather than clearing land to expand output. This latter course may be a recipe for disaster. In order to promote this, more resources have to be channelled into rural areas to improve the infrastructure and communications. It should go without saying that governments must work with people and ensure that they have land tenure and/or have rights o manage forest areas and benefit from them. These initiatives may need some external help, but Africa has to get away from aid dependency and become more self-reliant.

African governments should recognize that most children will have to work on the land when they leave school. Thus, farming practices, resource management and environmental education should be expanded or introduced into school, colleges, adult education and extension courses.

Africa is already relying on indigenous biomass for its food, fuel and shelter requirements. The opportunities to expand these uses are substantial, especially if they are coupled with other renewable resources such as water, wind and solar energy. In the long run, sustainable development can only be achieved using renewable resources. Africa can be and, because of its comparative advantage, should be at the forefront of this revolution.

ACKNOWLEDGEMENTS

This chapter is dedicated to the memory of Jorgen von Ubisch, a Norwegian forester who spent many years in East Africa expanding opportunities for rural people, and to the late Sir Kenneth Alexander, former Chancellor of the University of Aberdeen (Scotland), who mentored my progress for many decades. I would like to thank the two reviewers Professor Cherla Sastry of the University of Toronto, Canada, and Ivan Scrase of Imperial College of Science, Technology and Medicine, London, UK. They made useful comments and suggestions to improve the paper. However, any errors and mistakes are my sole responsibility.

REFERENCES

The Economist 6–12 July 2002. Articles on Carbon Sequestration, Coal and the Global Environment. London, The Economist Business Group.

FAO (1986) *Atlas of African Agriculture – The Next 25 years.* Rome, Food and Agricultural Organization of the United Nations (FAO).

   (1993) *Forest Resource Assessment 1990: Tropical Countries.* Forestry Paper 112. Rome, FAO.

   (1998a) *Agricultural Production Yearbook 1996.* Rome, FAO.

   (1998b) *Forest Products Yearbook 1996.* Rome, FAO.

   (2001) *Global Forest Resource Assessment 2000.* Rome, FAO.

Hall, D. O and Rao, K. K (1994). *Photosynthesis.* Cambridge, Cambridge University Press.

UN (1988) *World Demographic Estimates and Projections, 1950–2025.* New York, United Nations.

World Bank (1990) *World Development Report.* Washington, DC, World Bank.

   (1994a) *World Development Report.* Washington, DC, World Bank.

   (1994b) *Estimating Woody Biomass in Sub-Saharan Africa.* Washington, DC, World Bank.

   (1997) *World Development Report.* Washington, DC, World Bank.

   (1999) *Proceeding of the 1999 Energy Week April 6–8.* Washington, DC, World Bank.

   (2000) (Operations Evaluation Department) *A Review of the World Bank's 1991 Forest Strategy and its Implementation.* Washington, DC, World Bank.

World Resources Institute (1994) *World Resources 1994–95: A Guide to the Global Environment.* Washington, DC, Oxford University Press and World Resources Institute (WRI).

12 Sustainable energy development and the Clean Development Mechanism: African priorities*

RANDALL SPALDING-FECHER[1,†] AND GILLIAN SIMMONDS[2,‡]

[1]ECON Analysis (Oslo), South Africa
[2]University of Bath, UK

Keywords

Africa; energy; climate change; Clean Development Mechanism; sustainable development; technology transfer; capacity-building

Abstract

The Kyoto Protocol to the United Nations Framework Convention on Climate Change (UNFCCC) created a new possibility for North–South cooperation in mitigating climate change through joint projects under the Clean Development Mechanism (CDM). Much work remains to be done to clarify whether and how the CDM can meet its dual goals of contributing to sustainable development in developing countries and assisting industrialized countries meeting their commitments to emissions reduction and limitation. The purpose of this chapter is to highlight some key issues related to the development of the CDM from the perspective of African energy development, and to discuss options for addressing those issues. We begin by reviewing some key indicators of energy in Africa. We then discuss five areas within the negotiations where the design of the CDM and associated institutions can contribute to sustainable energy development in Africa: broadening the scope of CDM projects to include regional projects and institutional development; operationalizing sustainable development in project selection criteria; promoting technology transfer; ensuring a fair distribution of credits and other benefits from CDM projects; and building capacity to both implement projects and formulate enabling policy for the CDM.

*This chapter was completed in early 2001 before the agreements on the CDM were reached in July 2001 in Bonn and in November 2001 in Marrakech.
†Formerly of University of Cape Town, South Africa
‡Currently at CBI, London, UK

12.1 INTRODUCTION

The Kyoto Protocol to the United Nations Framework Convention on Climate Change (UNFCCC) created a new possibility for North–South cooperation in mitigating climate change through joint projects. The Clean Development Mechanism (CDM), described in Article 12 of the Protocol, allows industrialized countries to purchase 'certified emissions reductions' (CERs) from projects in developing countries which mitigate climate change. The CDM builds on the Activities Implemented Jointly (AIJ) pilot phase, and involves investors exchanging capital and technology for emissions reductions from joint projects in developing countries. What is significant is that the objective of the CDM, as stated in the Protocol, is to contribute to sustainable development and the overall objectives of the Convention, as well as assisting industrialized countries in meeting their emissions reduction targets (UNFCCC, 1997).

If well constructed, the CDM will bring great attention and resources to sustainable development in developing countries through an emphasis on avoided future emissions, while contributing to reaching the emission reduction targets for industrialized countries. The energy sector is both a major contributor to greenhouse gas (GHG) emissions and a powerful instrument for local economic and social development (Goldemberg and Johansson, 1995; Reddy et al., 1997). Financing more 'sustainable' energy projects and systems in Africa, therefore, should be a priority for the CDM. Moving towards sustainability not only involves cleaner technologies but also, as described below in more detail, widening access to affordable energy, capacity-building, and fostering regional cooperation in energy development. Whether the CDM can facilitate this process depends on how many of the details of the CDM rules and institutional structures are elaborated within the international negotiations. The purpose of this chapter is to highlight some of the most important issues from the perspective of African energy development, and to discuss options for addressing those issues.

Climate Change and Africa, ed. Pak Sum Low. Published by Cambridge University Press. © Cambridge University Press 2005.

We begin by reviewing some key indicators of energy in Africa, so as to provide a context for the discussion of climate change funding mechanisms. Following this, we discuss five areas where the design of the CDM and associated institutions can contribute to sustainable energy development in Africa: broadening the scope of CDM projects; operationalizing sustainable development in project selection criteria; promoting technology transfer; ensuring a fair distribution of credits and other benefits from CDM projects; and building capacity to both implement projects and formulate enabling policy for the CDM.

12.2 AFRICAN ENERGY SECTOR: TRENDS AND EMISSIONS

12.2.1 African energy sector trends

The African energy sector is a critical driver for development on the continent. Yet, to date, the sector has been plagued by problems which reflect the economic and environmental problems of many African countries: frequent power and fuel cut-offs; low access to commercial fuels and electricity; financially precarious energy sector institutions; and a chronic lack of infrastructural investment.

The lack of access to sufficient, affordable and environmentally sustainable commercial energy is reflected in key energy indicators. Biomass continues to be the largest energy source, providing half of sub-Saharan energy. Per capita commercial energy use, the lowest in the world, has actually been falling in recent years. [1] Even though commercial energy use is low, energy intensity (expressed as energy use per unit of GDP) is more than three times that of the industrialized Organization for Economic Cooperation and Development (OECD) countries (IEA, 2000a). This means that even the existing energy resources could be used more efficiently to provide better services. Beyond existing resources, however, there is an urgent need to increase available energy sources to meet basic development requirements and promote economic development.

Untapped commercial energy resources – hydropower, oil and gas – are significant, but concentrated in a few countries, necessitating better energy transport infrastructure. In contrast, renewable energy sources, particularly solar, are abundant and well distributed, but major financial and other barriers to their use remain unresolved. Electricity generation is still limited outside of a few countries, with South Africa producing 50% of Africa's electricity. Total African generating capacity excluding South Africa is only one eighteenth of that of Europe, for a region with almost twice the population (IEA, 2000b; ESKOM, 1998).

The investment requirements to extend affordable energy access to all of Africa are staggering. As an example, South Africa's ambitious accelerated electrification programme has succeeded in bringing more than 3.3 million homes onto the grid in the 1990s (NER, 1999; NER, 2000a). The programme, funded largely by the national utility, ESKOM, and municipal distributors, has cost more than 8 billion Rands (US$ 1 billion at 2001 exchange rates) (NER, 2000b). Which other African countries would be able to generate the investment domestically to provide electricity to the tens of millions throughout the continent without access – even utilizing an optimal mix of grid and off-grid technologies? Most of the funding for energy investments will of necessity come from abroad, and the CDM provides a potential mechanism to both increase investment flows to Africa and direct those flows toward more 'sustainable' energy projects. Whether this is possible depends on the design of CDM projects, and the overall role of the CDM as a global institution. These are discussed in Sections 12.3 and 12.4 below.

12.2.2 African energy sector emissions

Given the small size of the commercial energy sector, it is not surprising that Africa's contribution to global emissions of greenhouse gases is minimal. As shown in Figure 12.1 total energy sector emissions of CO_2 from Africa (approximately 730 million tonnes) were just over 3% of world emissions in 1998, even though Africa has 13% of the world's population (IEA, 2000a). Africa excluding South Africa accounted for less than 2% of world energy CO_2 emissions (IEA, 2000a). By contrast, the European Union's commitments under the Kyoto Protocol imply a reduction of 600 million tonnes of CO_2 by 2008–2012, relative to business as usual, or as much as Africa's total current emissions (IEA, 2000b). Clearly then, Africa's participation in the flexible mechanisms under the Kyoto Protocol must be based on avoiding emissions in the future rather than reducing current emissions.

In terms of per capita emissions, Africa is a quarter of the world average and less than 10% of the OECD average (Figure 12.2)

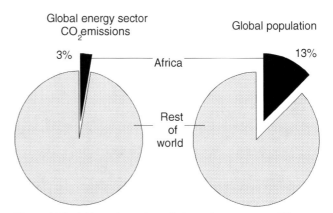

Figure 12.1. African share of population and energy sector CO_2 emissions, 1998 (Source: IEA 2000a).

---

[1] Total primary energy supply per capita in Africa dropped from 0.638 tonne of oil equivalent in 1990 to 0.636 tonne in 1998 (IEA 2000a).

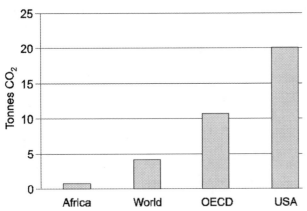

Figure 12.2. Per capita energy sector emissions, 1998 (tonnes CO_2) (*Source*: IEA 2000a).

- Structural energy sector reform to enhance performance, attract investment to the sector, and speed up widening access to commercial fuels, particularly electricity;
- Access to cleaner technologies for conventional fossil fuel systems;
- Regional energy trade and interconnections;
- A greater range of technological choice, financing mechanisms, and technology support for rural and decentralized systems;
- Sustained capacity-building for policy analysis and implementation;
- Regional cooperation on standards, technology development and policy;
- Improved forest management, cook stoves, and charcoal conversion methods to ensure sustainable and environmentally sound use of traditional fuels.

(IEA, 2000a). The notable exception to the low emissions per capita is South Africa, with per capita energy sector emissions of 8.5 tonnes. This is on a par with many European countries, despite the fact that much of the South African population is living below the poverty line. This is related not only to lower efficiency of energy use in some sectors, but more importantly to the structure of the South African economy and heavy dependence on coal as a carbon-intensive energy source (Spalding-Fecher *et al.*, 2000).

This vast disparity in emissions between the 'haves' and the 'have nots' highlights the rationale for the 'common but differentiated responsibilities' mentioned in the UNFCCC and Kyoto Protocol. Simply put, Africa has very little responsibility for historic, or even current, emissions, so any efforts to mitigate climate change must not place a burden on African countries. On the contrary, developing countries with low per capita emissions are implicitly (and unwillingly) subsidizing the wealthier countries – providing them with free 'ecological space'. International mechanisms to deal with climate change should be structured to compensate for this contribution – for example, by providing local benefits in addition to global benefits.

Energy sector projects to address GHG emissions, by contrast, have tended to focus on a much narrower range of issues. Typical examples of energy sector GHG mitigation projects are presented in Table 12.1. In addition, biomass and traditional fuel accessibility are linked to afforestation projects, forest management projects, or reforestation, depending on how these projects are implemented.

Typical mitigation projects do not address many key areas for the African energy sector. For CDM to address the real issues in the African energy sector, the scope of CDM projects must be broader than that conceived within the AIJ pilot phase, which comprised mostly site-specific emissions reduction projects and a large number of forestry projects. An example of how the CDM could be conceived more broadly, addressed in more detail below, is the need for *regional* energy infrastructure. CDM projects must be more than technology development and project-level implementation efforts. The CDM will be most effective in reducing future energy sector emissions if it addresses the capacity, policy, financing, and technology issues that are critical to the development of a 'sustainable' energy sector in Africa.

12.3 EXPANDING CDM PROJECT ELIGIBILITY

Article 12 of the Kyoto Protocol defines the CDM as a *project-based* activity (Yamin, 1998). Goldemberg and Reid (1998) recall that a proposal at Kyoto to replace the term 'projects' in Article 12 of the Protocol with the text 'programmes and policies' was presented and rejected. If this rejection can be interpreted to mean that CDM project activities are restricted in their scope to project-level implementation efforts, then the CDM will not fully meet the development needs of the African energy sector.

Building an energy sector that can fuel African development requires a range of approaches and strategies, which include the following (Farinelli, 1999):

12.3.1 Regional energy infrastructure

In contrast to industrialized countries, throughout much of Africa the major contributors to final demand of commercial energy are liquid fuels for transport and electricity for all sectors. For electricity, this is particularly problematic because generation capacity and potential power development are distributed unevenly. Regional electricity inter-connections are therefore one of the most important tools for improving access to commercial energy while reducing environmental impact. In southern Africa, for example, greater subregional trade in electricity could substitute hydropower from Zambia, Mozambique and the Democratic Republic of Congo for southern coal-fired power stations,

Table 12.1. Typical greenhouse gas abatement projects

Energy	Transportation and energy
End-use efficiency improvements in household, industry and service sectors	Efficiency improvements for vehicles
Improved transmission systems	Switch to fuel systems with lower emissions
Fuel substitution	Improved transport system efficiency
Renewable energy technologies, particularly decentralized ones	Modal shifts
Improved centralized supply technologies from fossil fuels, nuclear and renewables	Managed transport demand

Figure 12.3. The Southern African power grid, 1997.

reducing both energy cost and related emissions (see Figure 12.3). A 1993 Southern African Development Community (SADC) energy project, which focused on five countries in the region, found that cooperation in power sector development could result in savings of more than US$ 1.5 billion over a period of 15 years (SADC, 1993) – and this excluding South Africa, by far the largest producer and consumer of electricity on the continent. More re-

cently, the benefits of regional integration and cooperation in generation expansion have been estimated at more than US$ 2 billion over 20 years (Graeber and Spalding-Fecher, 2000). Even optimizing *current* generation capacity within SADC could save US$ 80–100 million per year, according to a Purdue University study (Sparrow *et al.*, 1999). Yet these investments are not happening because of a wide range of political and market barriers. Significant investments in transmission infrastructure and maintenance are a prerequisite for taking advantage of regional cooperation. In addition, strong coordinating centres that can control supply switching, and ongoing capacity-building in utilities and governments to manage a regional power system are necessary, not to mention greater political stability. The CDM could help to remove the barriers to these regional infrastructure initiatives, and support not only specific projects (such as interconnector lines), but also programmes and policies to support integration.

The Southern African Development Community, African Development Bank, and other African organizations could use CDM funding as an opportunity to develop stronger regional infrastructures, policy coordination, and regional economic integration. These large-scale initiatives would arguably do more for both local sustainable development and the global environment than a host of micro-level project initiatives. While we acknowledge that calculating the direct emissions reductions from these programmes may be more difficult than for specific projects, this is not an insurmountable problem and a challenge that the international research and policy community should address urgently.

12.4 OPERATIONALIZING SUSTAINABLE DEVELOPMENT CRITERIA AND INDICATORS

Although the inclusion of 'promoting sustainable development' as one of the goals of the CDM is, in theory, a step forward in promoting climate change mitigation initiatives that support development, in practice this is not easy to implement. There is no operable definition of sustainable development provided within the Kyoto Protocol or other international agreements. Rather, the challenge for countries, individually and collectively, is to develop criteria that are measurable, in line with the goals of the CDM and UNFCCC, and reflect national and international development priorities. This must also be linked to the broader discussion of sustainable development criteria as part of global implementation of Agenda 21, and national sustainable development plans. This section provides some suggestions on how to resolve some of the current debate around this complex issue.

12.4.1 The politics of sustainability criteria

If we accept the dual goals of the CDM, then sustainable development must be elevated to the same level as emissions reduction in assessing which projects would be eligible for CDM activities. The question that remains unanswered is: should sustainable development be explicitly included in the project eligibility criteria of the CDM, or should individual host governments be responsible for assessing the contribution of the project to sustainable development?

In the international climate change negotiations, few countries – industrialized or developing – have been willing to support standard measures for sustainable development on CDM projects. The option of international standards did not even appear in the text that negotiators took into the Sixth Conference of the Parties, for example (Spalding-Fecher et al., 1999; UNFCCC, 2000). The arguments against such global standards are that they infringe upon national sovereignty by taking away the right of developing countries to judge what is good for their development and what is sustainable, and that they make operationalizing the requirements for prospective projects difficult.

Few developing countries, however, have articulated national development plans that seriously tackle the complex social, economic and environmental issues involved in sustainable development. While this is clearly not an easy task, there is a valid concern that without international guidance on sustainable development criteria – or at least a basket of possible criteria from which to choose – the immediate economic benefits of proposed CDM projects will overshadow any sustainable development criteria in the evaluation of project eligibility.

As one possible solution, some analysts are suggesting that responsibility for determining project compliance with sustainable development criteria be shared between the host country and an international authority (UNCTAD, 1998). General sustainable development criteria could be applied globally, but would be interpreted within the national context and be driven by national concerns and objectives.

To ensure that sustainable development indicators are effectively applied as criteria throughout the CDM project cycle – development, approval, implementation, monitoring, verification and certification – it is critical that developing countries put appropriate institutions and processes in place for CDM project approval, as well as monitoring and evaluation of compliance. The process should involve the participation of stakeholders, as well as expert review, to determine the sustainable development impacts of proposed projects both before and after implementation.

12.4.2 Defining sustainable development in the CDM

The classic definition of sustainable development is development that meets the needs of the present generation without compromising the ability of future generations to meet their own needs (WCED, 1987). This, however, does not provide an operational definition for CDM projects. To meet the sustainable development requirements of African countries, the following sustainability aspects need to be addressed in the context of national development priorities (Thorne and La Rovere, 1999; James and Spalding-Fecher, 1999).

Economic sustainability: Africa is suffering from the economic ravages of the 1980s and 1990s. Declining overseas development assistance, inadequately compensated by increases in foreign direct investment, together with the impacts of structural adjustment programmes, have seriously affected the balance of payments and growth of these countries (see World Bank, 2000). The economic sustainability of developing countries should be addressed through contributions to both micro- and macroeconomic sustainability. Contributions to sustainable economic development include economic growth, balance of payments and foreign exchange benefits.

Social sustainability: One of the key development needs in developing countries is uplifting the poor through poverty eradication, employment creation and a more equitable distribution of resources. Improving the socio-economic status of the poorer portion of the population will contribute to this development goal.

Environmental sustainability: Energy consumption and use in developing countries has impacts both globally, in terms of climate change resulting from greenhouse gas emissions, and locally, in terms of severe health impacts resulting from other air pollutants. Reducing these air pollutants will have positive impacts on health and the local environment, as will efforts to reduce water pollution and land degradation. Managing biomass resources and agricultural soils more sustainably for multiple uses should also be a priority.

Technological sustainability: 'Cleaner development paths' in Africa are often constrained by lack of access to finance and technological resources. To achieve sustainable technological development, it is imperative to transfer not only the technological hardware but also the necessary human resource skills and institutional capacity to assess, apply, maintain and further develop technologies. Ultimately, the objective should be to reduce the dependence on imported technologies in developing countries.

Institutional sustainability: Sustainable institutional arrangements are an essential part of the process of laying the foundations for economic, social, environmental and technical sustainability (Thomas *et al.*, 1999). The implementation of project activities focused on achieving sustainable development has often failed as a result of the lack of institutional support structures and networks. Furthermore, barriers to the transfer of environmentally sound technologies require intervention from institutions that can disseminate information and serve as intermediaries.

12.4.3 Measuring sustainability for the CDM

Measuring sustainable development is one of the most difficult areas for CDM projects or, for that matter, any development projects. To ensure that sustainable development re-

mains a criterion by which projects are selected and evaluated, there is a need to translate broad criteria for sustainability into indicators of sustainability. These will allow countries to rank available project options in terms of their own sustainable development priorities and will provide a basis for monitoring the implementation of the projects and evaluating project compliance (Thorne and La Rovere 1999). While generic indicators cannot be applied across all projects, and indicators should be adapted on a project-by-project basis, a range of indicators that could be applicable to different CDM projects are presented below. It is important to bear in mind that, just as the definitions of sustainable development can vary between countries and groups within countries, so the basket of indicators appropriate for a given project will vary. The following indicator options are drawn in part from the work of Thorne and La Rovere (1999) and James and Spalding-Fecher (1999).

The economic performance of the project can be assessed using indicators such as cost-benefit ratios, net present value, payback periods, rate of return, and cost per tonne of greenhouse gases abated. The economic cost-effectiveness of the project should account for the full life-cycle costs of the project and include training, information dissemination, monitoring, verification and other transaction costs.

The contribution of the project to macro-economic sustainability can be measured in terms of the effect on GDP, impact on foreign exchange and reduction of direct government investments made possible by the foreign private investment in the CDM project. These estimates, however, require sophisticated macro-economic models and projects of a size that will affect the macro-economy. In practice, the more direct micro-economic effects are likely to be easier to measure.

The social aspects of sustainable development are possibly the most difficult to measure. Social development includes both quantitative and qualitative aspects. Examples of quantitative indicators of social development are impacts on local employment, extent and appropriateness of technology transfer, impacts on health status, impacts on awareness of environmental issues, and impacts on resource distribution (for example income distribution).

While it is important that indicators used to assess or measure the contribution of a CDM project to social sustainability are measurable, to ensure that the project does indeed make a contribution, it is equally important to assess aspects that are not easily quantified. Qualitative aspects of social development include impacts on social structures, the extent of public participation and contribution to community empowerment, and effects on local cultures. In addition, in practice it may only be possible to provide a qualitative assessment of the previous list of more quantitative indicators. Effective measurement of qualitative aspects of sustainable social development requires the participation of local communities and other stakeholders.

Indicators to assess the contribution of the project to environmental sustainability include the contribution of the project to mitigating climate change (that is, calculation of the emissions reduced or avoided) and the contribution of the project to local environmental sustainability, such as avoided emissions of local pollutants, impacts on biodiversity and impacts on the sustainable natural resource development and management.

The contribution of the project to technological sustainability should be assessed in terms of its contribution to the technological self-reliance of the country. A simple indicator suggested by Thorne and La Rovere (1999) assesses technological development in terms of the impact on foreign currency expenditures with the technology. This does not account for skills and capacity transfer, an important component of technological self-reliance (technology transfer is discussed in more detail in Section 12.5). An additional issue would be how the project promotes local development of technology.

Building institutional capacity must be a part of the assessment of a CDM programme. As with technological sustainability, the long-term goal should be that host countries have the local institutional and administrative capacity to implement and monitor projects independently. Moreover, capacity-building should benefit not only the implementation of climate change projects, but also the ability of government and social institutions to respond fully to the challenges of sustainable development – from poverty alleviation to environmental and economic management. Qualitative indicators are needed to assess the impact the CDM projects have on these administrative and institutional capacities, and how effective CDM partnerships are in transferring 'soft' skills in public management.

12.5 THE CDM: AN INSTRUMENT FOR TECHNOLOGY TRANSFER

The need for transfer of environmentally sound technology and resources from developed to developing countries is clearly stated in both the UNFCCC and the Kyoto Protocol. While the current contribution of developing countries to global emissions is small, significant growth in energy consumption is essential to fuel needed economic and social development in these countries. As energy consumption in developing nations grows, so too will emissions of greenhouse gases; in time, these emissions will become significant contributors to climate change. While efforts are already under way in several African countries to take actions to combat climate change – such as pursuing solar and biomass energy systems and shifting to natural gas and more efficient forms of power generation – the pace of clean development in Africa is limited by lack of access to finance and technological resources (Davidson, 1998). The challenge for Parties to the UNFCCC is, therefore, to develop tools to optimally transfer and assimilate

technology to promote clean development in non-Annex I countries. The CDM provides one such tool to promote and facilitate the transfer of technology to developing countries, contributing to developmental and environmental goals. The CDM should not replace existing commitments related to financial assistance and technology transfer, but be seen as an additional and complementary means of fulfilling obligations under the Convention.

Past mechanisms to transfer technology from developed to developing countries have met with limited success. Evidence from the energy sectors of developing countries has shown that many imported technologies do not reach their designed operational efficiencies and that efficiencies deteriorate over the productive life of the technology (Mathur *et al.*, 1998). In fact, rather than resulting in technology adoption, technology transfer projects have often resulted in the increased technological dependence of developing countries on developed nations. This is largely due to the focus on transferring a product, rather than on the skills and capacity required to manage the technology throughout its life cycle. The failure of these mechanisms to secure the transfer and assimilation of technology in developing countries can be attributed to several factors (Davidson *et al.*, 1995; South African Government, 1999; Lennon, 1999):

- Technology transfer priorities have been set by industrialized country donors and have not matched the needs of developing countries.
- Technologies have been imported wholesale from developed countries and have not been adapted to the local environmental, socio-economic, cultural, security and skills conditions of the recipient country. Effective technology transfer requires appropriate administrative and legal frameworks, institutions, and research and development to 'indigenize' the technology and promote its assimilation.
- Technology transfer has involved the employment of experts from outside the developing country, with minimal involvement of local skills or development of indigenous capacity. Where knowledge has been transferred, this has often been limited to the installation and operation of the technology.
- Support infrastructure for the efficient performance of the technology is inadequate. This includes institutional support structures, physical infrastructure, financing to stimulate demand for the technology, and operation, maintenance and service skills.

For the CDM to be an effective tool for the transfer and assimilation of technology in the energy sector, developing countries should use their CDM focal points to evaluate which projects are deemed good for the country and to promote projects which encourage technology transfer. The following criteria could be applied (Mathur *et al.*, 1998; South African Government, 1999; Lennon, 1999):

- CDM projects should contribute to achieving national or local development goals and be assessed in terms of their ability to meet the technological requirements for energy sector development.
- CDM project proposals should elaborate plans to adapt technologies to local characteristics. Technologies should be compatible with needs, skills, training, finances and the environment of the people who will be using them. If technologies are not wholly compatible, plans for adaptation of the technology to the local characteristics and for the transfer of skills and capacity should be elaborated.
- CDM projects should transfer soft technology, such as training and education on the appropriate use and maintenance of the technologies that could be used in the project. Skills transferred should sustain the technology throughout its life cycle, from design through construction, operation, maintenance, upgrading and decommissioning. The training should also promote replication of similar projects in the sector.
- CDM projects should involve local institutions and stakeholders in all stages of project development and implementation.
- Other support infrastructure, including enabling policy necessary for the successful transfer and assimilation of the technology, should be in place.

The identification and prioritization of the technological needs for the development of the energy sector has high transaction costs, and few African countries are likely to have the available resources to do so rapidly. A parallel technology transfer mechanism is required to provide developing countries with the necessary information to assess the appropriateness of potential technology transfer projects – an issue that is currently under discussion in the climate change negotiations, separate from the CDM. As one example, the technology transfer mechanism proposed by the South African Government (1999) in its submission to the UNFCCC provides an approach to ensuring that technology transfer is need-driven. Developing countries could be assisted to develop strategies for technology transfer, adaptation, application and assimilation, which would be funded by sources outside of the CDM, but could also form part of the criteria against which CDM project activities could be assessed. The process would begin with the UNFCCC, or some other competent international technical body, identifying generic mitigation technologies in different sectors. Developing nations, with financial and technical assistance funded through the Convention process, would then identify gaps between their current technology base and the technologies required for optimal performance in terms of mitigation of GHGs. This would include core technologies required on a sectoral basis. They would then test the desirability of these technologies against national priorities, and develop strategies for the highest priority technologies (see also Lennon, 1999).

12.6 SHARING OF CREDITS

The Kyoto Protocol authorizes Annex I countries to use credits generated from CDM projects to meet their emission reduction commitments. No mention is made of credit acquisition and use by non-Annex I countries, but Article 12.3 of the Protocol does specify that non-Annex I countries must benefit from CDM project activities. One argument has been that credits from CDM projects will accrue to the investor party only, because developing countries do not have reduction commitments and will therefore not be interested in sharing in the credits. This argument also assumes that the positive externalities from the project will be sufficient as incentive for host country participation in CDM project activities (Michaelowa et al., 1999).

It may be true that developing countries desperate for foreign investment will find the positive externalities from CDM projects – including some measure of economic growth, technology transfer, contribution to sustainable development, capacity-building and local empowerment – sufficient. Even if these benefits are significant, as suggested by a report from the World Resources Institute (Austin et al., 1999), is such a distribution of project benefits equitable? The marginal cost of avoiding emissions in Africa, current or future, is likely to be much lower than in the OECD, yet investors are in essence claiming that all of the 'economic surplus' (the difference in marginal costs between countries) should go to the investor, with the host country only receiving enough money to cover their direct expenses.

A wide range of African experts have advocated some form of credit sharing should exist within the CDM (e.g. UNDP, 1998; Friends of the Earth Ghana, 2000). The CDM allows for Annex I countries to achieve emissions reductions in developing countries at a lower cost than if those reductions were made domestically. African countries feel that they should receive a share of these gains. With increasing pressure on developing countries to voluntarily take on emissions targets, there is a valid concern amongst developing countries that low-cost mitigation options will be exploited in the initial years of CDM activity, leaving only high cost options available when they confront future commitments (Hamwey and Szekely, 1998). Credits accruing to developing countries through these CDM activities could be banked for use against possible future commitments or could potentially be sold on the credit trading market to provide additional capital for development (TERI, 1999; Michaelowa et al., 1999).

The implementation of the CDM is an attempt to bring together the divergent interests of industrialized and developing country parties. Unequal levels of expertise and awareness between the two parties could result in a divergence in negotiating capacities and lead to a situation of exploitation and an inequitable partition of credits (UNCTAD, 1998). The experience of African governments negotiating with multinational corporations on energy projects is testimony to the difficulties faced by developing country partners in such negotiations (Spalding-Fecher et al., 1999). To ensure a 'fair' distribution of credits, capacity needs to be built in developing countries to negotiate favourable terms, or negotiating support needs to be provided to these countries (UNCTAD, 1998). In addition, broad guidelines for credit sharing laid out by the CDM could prevent exploitation of developing country parties and promote an equitable distribution of gains. Linked to this is the debate on whether there should be a mechanism to ensure that all regions have an equitable share of CDM benefits, rather than just those few countries that already receive the most foreign direct investment (French, 1996; Sokona et al., 1998). Although few negotiators favour explicit quotas, mechanisms to promote broader geographical distribution of CDM projects (such as dedicated multilateral CDM funds) will be essential.

Investors in CDM projects are interested in minimizing their cost per credit unit. Credit-sharing will raise the price of credits, making CDM projects more expensive. This, in turn, could lower the demand for CDM projects. In addition, with the investor looking for hosts where the net costs after credit sharing are lowest, credit sharing could result in competition between potential host countries (Michaelowa and Dutschke, 1998; Michaelowa et al., 1999). The combination of reduced demand for CDM projects and increased competition between host countries to attract investors would serve to reduce the bargaining or negotiating powers of developing countries. To counter the potential for competition between developing countries for CDM projects, the CDM could set guidelines for a uniform credit-sharing ratio. Without such guidelines, it is difficult to see how project activities would meet the sustainable development objectives of the CDM. The credit-sharing ratio adopted needs to be carefully considered in the light of international activity in the credit trading market. The gains of selling or banking credits need to be balanced against the loss in demand for CDM projects.

The distribution of accrued credits within the host countries also needs to be considered. While some Annex I countries are already setting up domestic emissions trading regimes to increase private interest in emissions reduction opportunities, few developing countries are likely to implement such complex systems in the medium term. Some African governments have therefore argued that domestic rules on the sharing of credits between government and implementing entities are necessary, particularly if private parties in developing countries have access to the international emissions trading market (UNDP, 1998). An appropriate analogy here is the 'resource rents' that governments collect on the private use of public resources, such as forests, minerals and petroleum. It is standard practice for governments to place a levy on private use of these goods, so that the public receives some of the benefit from the private use of these publicly owned resources (Varian, 1996). In practice, however, these levies are often much less than the windfall profits that accrue to private developers of public resources (e.g. Blignaut et al., 2000). Developing country governments should carefully consider who should receive any windfall profits from the creation of a global carbon market and credits from CDM projects.

12.7 CAPACITY-BUILDING REQUIREMENTS

One of the most significant barriers to the successful implementation of the CDM in Africa is the absence of the institutional capacity and administrative infrastructure required to develop and implement climate change policy. Without strong institutional structures, it is difficult for African countries to develop a strategic vision for their involvement in the international climate change arena, and more specifically, the CDM. The constraints on framing or shaping their involvement in the CDM create difficulties for Africa in ensuring that developmental needs and energy priorities are given sufficient weight. Furthermore, this lack of institutional capacity raises the transaction costs and risks associated with conducting CDM projects in Africa. The creation and maintenance of strong institutions and efficient structural linkages between these institutions is thus essential for the implementation of the CDM in Africa.

In most developing countries, the climate change office or focal point has limited capacity and is restricted to an informational role. For African countries to be proactive in the CDM market, they will need institutions with much stronger capacity and responsibilities related to climate change and CDM projects. The scope of work for these climate change/CDM offices would include the following areas:

- Develop climate change policy, with input from relevant government and non-governmental stakeholders, and define national CDM goals within the context of national and regional policy processes;
- Coordinate interdepartmental cooperation at the national government level to integrate CDM policy with national and international economic, environmental and social policies;
- Formulate national negotiating positions and contribute to regional and international debate;
- Develop procedures, criteria and guidelines for CDM project approval, registration, monitoring and evaluation – including expert reviews in the project approval and evaluation processes;
- Develop sustainable development criteria that will be used for evaluating potential CDM projects in the context of national and local development priorities;

Develop the technical basis for evaluating projects, including possible multi-project baselines that will facilitate efficient and cost-effective development of small-scale projects;

• Engage in proactive project identification by creating a portfolio of possible projects, marketing them to investors and disseminating information to potential project partners and stakeholders;

• Facilitate stakeholder involvement in the project selection and evaluation processes to anticipate and assess the impact of CDM project activities on local communities and other affected stakeholders.

To carry out these tasks, governments will need to build a wider range of skills than many climate change offices now possess. Some of the high priority capabilities would include the following skills:

• Project and process management, including policy development, project and programme development or approval, and monitoring and oversight;

• Policy analysis, including international and domestic policies in the economic, environmental, energy and trade sectors;

• International negotiations and an understanding of international law;

• Research project management;

• Facilitation of stakeholder input.

12.8 CONCLUSIONS

Much work remains to be done to clarify whether or not and how the CDM can meet its dual goals of contributing to sustainable development in developing countries and assisting industrialized countries in meeting their commitments to emissions reduction and limitation. This chapter has raised a number of issues around the structure and operation of the CDM to meet the energy development needs of the African continent.

Energy plays a key role in African development, and the CDM can contribute to sustainable development through investment in the energy sector if it focuses on African energy priorities, which are broader than one-off technology and demonstration projects, and addresses regional initiatives not typical of greenhouse gas mitigation projects. A prerequisite for CDM projects to contribute to sustainable development is the need for practical and effective sustainable development criteria and indicators for project selection and monitoring. The criteria must go beyond mere host country approval and demonstrate how projects will affect economic, environmental, social, technological and institutional development in the host country.

The parallel discussions of technology transfer mechanisms under the UNFCCC and the Kyoto Protocol should also provide a basis for ensuring that CDM projects are part of meaningful technology transfer. This will include the 'soft' technology of know-how and skills as well as technological hardware, and will aim to create an enabling environment for the host country to assimilate new technologies and skills.

Many developing country governments, including African ones, support credit-sharing between host and investor countries – and potentially between governments and implementing entities – as a way to share the benefits of CDM projects more equitably between the North and South. Without credit-sharing and some type of international rules to protect those with limited capacity to negotiate, developing countries may be left only barely covering the costs of their projects, while investors reap windfall profits from cheaper reductions in the South. In addition, developing countries considering taking on voluntary commitments are wary of giving away all of the least expensive abatement options through early CDM projects. The opportunity to keep some of the credits and put forward unilateral CDM projects could encourage more action in developing countries.

Finally, implementing the CDM will place even greater strain on many developing country governments already struggling to contribute to the climate change negotiations and assess their own domestic climate change and development priorities. Even though governments will most likely not be directly involved in implementing CDM projects, effectively promoting, managing, and approving CDM projects will require significant additional capacity. Some of the tasks and skills have been described here, and include not only developing policy and evaluating projects, but also managing outside research, engaging more effectively in the negotiations, facilitating stakeholder input, and analysing the implications of international developments for domestic climate change and CDM policy.

ACKNOWLEDGEMENTS

The authors would like to thank Ian Rowlands for his early and constructive comments on this chapter. Comments from Harald Winkler and Khorommbi Matibe were also appreciated. Any errors are solely the responsibility of the authors.

REFERENCES

Austin, D., Faeth, P., Da Motta, R. S. *et al.* (1999) *How Much Sustainable Development Can We Expect From the CDM?* Washington, DC, World Resources Institute.

Blignaut, J., Hassan, R. and Lange, G-M. (2000) *Natural resource accounts for minerals: a Southern Africa country comparison.*

Presented at 2nd International Environment and Development Economics Conference. Stockholm, Sweden.

Davidson, O. (1998) Energy initiatives in Africa for cleaner development. In J. Goldemberg and W. Reid (Eds.), *Promoting Development While Limiting Greenhouse Gas Emissions: Trends and Baselines*, pp. 75–84. New York, United Nations Development Programme.

Davidson, O., Maya, R. and Zhou, P. (1995) In H. Okoth-Ogendo and J. Ojwan (Eds.), *A Climate for Development: Climate Change Policy Options for Africa*. Stockholm, Stockholm Environment Institute.

ESKOM (1998) *Annual Report 1998*. Sandton, South Africa, ESKOM.

Farinelli, U. (1999) *Energy as a Tool for Sustainable Development*. New York, United Nations Development Programme and Brussels, European Commission.

French, H. (1996) Private finance flows to Third World. In L. Brown, C. Flavin and H. Kane (Eds.), *Vital Signs 1996: The Trends That Are Shaping Our Future*, pp.116–117. Washington, DC, Worldwatch Institute.

Friends of the Earth Ghana (2000) *Resolutions from the African Preparatory Workshop on Climate Change Towards the COP6*. October 28–30. Accra, Ghana.

Goldemberg, J. and Johansson, T. (1995) *Energy as an Instrument for Socio-economic Development*. New York, United Nations Development Programme.

Goldemberg, J. and Reid, W. (1998) Greenhouse gas emissions and development: a review of lessons learned. In J. Goldemberg and W. Reid (Eds.) *Promoting Development While Limiting Greenhouse Gas Emissions: Trends and Baselines*, pp.1–13. New York, United Nations Development Programme.

Graeber, B. and Spalding-Fecher, R. (2000) Regional integrated resource planning and its role in the regional electricity cooperation and development in Southern Africa. *Energy for Sustainable Development*, **4** (2), 32–37.

Hamwey, R. and Szekely, F. (1998) Practical approaches in the energy sector. In J. Goldemberg (Ed.) (1998) *Issues and Options: The Clean Development Mechanism*, pp.119–135. New York, United Nations Development Programme.

IEA (International Energy Agency) (2000a) *CO_2 Emissions from Fuel Combustion 1971–98: 2000 Edition*. Paris, IEA.

(2000b) *World Energy Outlook 2000*. Paris, IEA.

James, B. and Spalding-Fecher, R. (1999) *Evaluation Criteria for Assessing Climate Change Mitigation Options in South Africa*. Report for the South African Climate Change Country Study. Energy and Development Research Centre, University of Cape Town, South Africa.

Lennon, S. (1999) Technology transfer and its role in optimizing development whilst contributing to achieving the objectives of the United Nations Framework Convention on Climate Change. *J. Energy in Southern Africa*, **10** (2), 48–51.

Mathur, A., Bhandari, P. and Srikanth, S. (1998) Effective technology transfer: issues and options. In T. Forsyth (Ed.), *Positive Measures for Technology Transfer Under the Climate Change Convention*, pp.35–46. London, Royal Institute of International Affairs.

Michaelowa, A. and Dutschke, M. (1998) *Creation and sharing of credits through the Clean Development Mechanism under the Kyoto Protocol*. Paper presented at the Dealing with Carbon Credits after Kyoto workshop, May 28–29, 1998. Callantsoog, The Netherlands.

Michaelowa, A., Begg, K., Dutschke, M., Matsu, N. and Parkinson, S. (1999) Crediting and credit sharing. In R. Dixon (Ed.), *The UN Framework Convention on Climate Change Activities Implemented Jointly (AIJ) Pilot: Experiences and Lessons Learned*, pp.89–104. Dordrecht, The Netherlands, Kluwer.

NER (National Electricity Regulator) (1999) *Lighting up South Africa, 1997/8*. Sandton, South Africa, NER.

(2000a) *Annual Report 1999/2000*. Sandton, South Africa, NER.

(2000b) *Lighting up South Africa: A Century of Electricity Serving Humankind*. Johannesburg, South Africa, Open Hand.

Reddy, A., Williams, R. and Johansson, T. (1997) *Energy After Rio: Prospects and Challenges*. New York, United Nations Development Programme.

SADC (Southern African Development Community) (1993) *Regional Generation and Transmission Capacities including Interregional Pricing Policies*. Phase II, Final Technical Report, SADC Energy Project AAA 3.8.

Sokona, Y., Humphreys, S. and Thomas, J-P. (1998) *The Clean Development Mechanism: What Prospects for Africa?* www.enda.sn/energie/CDM2.html

South African Government (1999) *Paper No. 6: South Africa: Technology transfer*. FCCC/SBSTA/1999/MISC. 5.

Spalding-Fecher, R., Simmonds, G. and Matibe, K. (1999) The Clean Development Mechanism and energy development in Africa: Climate change funding for sustainable development. *J. Energy in Southern Africa*, **10** (2), 34–42.

Spalding-Fecher, R., Williams, A. and Van Horen, C. (2000) Energy and environment in South Africa: charting a course to sustainability. *Energy for Sustainable Development* (Special issue on South Africa) **4** (4), 8–17.

Sparrow, F., Masters, W., Yu, Z., Bowen, B. and Robinson, P. (1999) *Modelling Electricity Trade in Southern Africa: Second Year Final Report to the Southern African Power Pool*. Purdue University, USA.

TERI (Tata Energy Research Institute) (1999) *Clean Development Mechanism: Issues and Modalities*. New Delhi, TERI.

Thomas, J-P., Sokona, Y. and Humphreys, S. (1999) *After Buenos Aires: A Development and Environment NGO Perspective*. www.enda.sn/energie/cc

Thorne, S. and La Rovere, E. (1999) *Criteria and indicators for the appraisal of Clean Development Mechanism (CDM) projects*. Paper presented at Fifth Conference of the Parties to the Framework Convention on Climate Change, 26 November 1999, Bonn, Germany.

UNCTAD (United Nations Conference on Trade and Development) (1998) *Design and Implementation of the Clean Development Mechanism*. A draft concept paper of the *Ad Hoc* International

Working Group on the Clean Development Mechanism. Geneva UNCTAD.

UNDP (United Nations Development Programme) (1998) *New Partnerships for Sustainable Development: The Clean Development Mechanism and the Kyoto Protocol*. African Regional Workshop, Accra International Conference Centre, Accra, September 21–24, 1998, Ghana.

UNFCCC (1997) *Kyoto Protocol to the United Nations Framework Convention on Climate Change*. FCCC/CP/1997/7/Add. 1.

(2000) *Mechanisms Pursuant to Articles 6, 12 and 17*. Text by the chairmen. October 26. FCCC/SB/2000/10/Add. 2.

Varian, H. (1996) *Intermediate Economics: A Modern Approach*. New York, Norton.

WCED (World Commission on Environment and Development) (1987). *Our Common Future*. Oxford, Oxford University Press.

World Bank (2000) *World Development Report 2000/01*. London, Oxford University Press.

Yamin, F. (1998) Operational and institutional challenges. In J. Goldemberg (Ed.), *Issues and Options: The Clean Development Mechanism*. New York, United Nations Development Programme. USA.

13 Opportunities for clean energy in the SADC under the UNFCCC: the case for the electricity and transport sectors

PETER P. ZHOU

EECG Consultants, Gaborone, Botswana

Keywords

Greenhouse gas emissions; electricity sector; transport; development; decarbonization; energy efficiency; clean energy and its financing; regional planning

Abstract

This chapter presents opportunities for application of clean energy technologies in the power and transport sectors in the Southern African Development Community (SADC) region, which are the commercial energy sectors responsible for the largest share of greenhouse gas (GHG) emissions in the region. The region depends mainly on coal-based electricity and imported oil-dependent transport sectors, causing high GHG emissions and socio-economic impacts.

Most of the SADC countries have poor economies and, through power and transport development projects, have incurred debts which they cannot service. Hence, shifting to more benign energy technology solutions would not be easily realizable. Opportunities for funding clean energy projects under the United Nations Framework Convention on Climate Change (UNFCCC) should therefore be explored.

Examples of clean energy solutions in the electricity and transport sectors that could be aligned with the UNFCCC requirements are presented, among them are decarbonization, energy efficiency and waste management technologies, and regional planning practices. Some clean energy options analysed for some countries in the region can be implemented at costs of less than US$10/tonne of carbon avoided and generally at negative costs, implying that it will be beneficial to the region to adopt these options, aside from their GHG reduction potential.

The relevant activities already being promoted under the SADC energy agenda are recognized and additional strategies are recommended. Apart from technical solutions, SADC countries require energy policies that are supported by strong research and development, institutional and legal frameworks, information base, and catalytic elements.

13.1 INTRODUCTION

The energy sector worldwide is the largest emitter of GHGs and it is no exception in the Southern African Development Community (SADC) states. Although biofuels dominate energy balances of most SADC countries (greater than 50%) (Kgathi and Zhou, 1995), the fossil fuels in the form of coal and petroleum are the key commercial energy fuels used in the electricity and transport sectors.

GHG emissions (CO_2 equivalent) from the electricity sector alone amounted to 201 million tonnes in 1994 from the SADC countries on the mainland of Africa (Batidzirai and Zhou, 1998). In comparison, the total energy sector emissions for the Netherlands (165.7 Mt) and Ireland (31.7 Mt) combined, in 1992 were 197.4 Mt (OECD/IEA, 1994). The largest emissions were from South Africa (94%), Zimbabwe (4%) and Botswana (1%), where large coal reserves exist and are used in electricity generation.

Over 40% of the presently mined coal in the SADC region is for electricity generation. By 1999 the regional installed capacity was 46,020 MW with a demand of 35,013 MW (SADC, 2000). Of this capacity, South Africa alone had 42,994 MW, including capacity in reserve and under construction. This regional installed capacity is dominated by coal generation, which exceeds 75% for the whole of SADC and is 87% for South Africa. Botswana, though it has a small installed capacity, has 100% thermal capacity, followed by Zimbabwe and Swaziland with about 65%. An assessment by Southern Centre (1999) showed that regional electricity demand could rise to nearly 60,000 MW by 2020 and will entail an electricity expansion plan with the same high proportion of coal-electricity generation.

The regional transport sector emitted about 70 million tonnes CO_2 equivalent during the same period considered for the

Climate Change and Africa, ed. Pak Sum Low. Published by Cambridge University Press. © Cambridge University Press 2005.

Table 13.1. Energy reserves relevant for electricity generation in the SADC region

Country	Coal recoverable reserves (Mt)	Hydro potential (MWe)[b]	Oil (Mt)[a]	Natural gas (MCM)[a]	Uranium (tonnes)[b]
Angola	—	16,000	734–1,156 {37}	45,310–107,600 (566)	
Botswana	17,000 (1.02)	0	0	0	
Lesotho	0	450	—	0	
Malawi	2	900	0		
Mozambique	9,000 (0.07)	12,500 [2,182]	0	56,634 (57)	
Namibia	0	1,800	0	84,950	101,000
South Africa	55,000 (250)	3,500 [668]	9,000 {1.24}	22,650 (1,400)	426,000
Swaziland	5,000 (0.3)	600	0		
Tanzania	2,000 (0.004)	6,000 [400]	0	28,320	
Zambia	0	8,000 [1,700]	0		
Zimbabwe	2,000 (4.6)	13,300 [666]	0		
Zaire	97[a]	100,000 [2,500]	27.2		
Total	90,099 (256)	163,050	9,761–10,183	237,860–300,160	527,000

The bracketed values in the table refer to the annual consumption of coal in Mt in 1998 (column 2), the hydro installed capacity in MWe in 1998 (column 3), the annual oil production in Mt in 2001 (column 4); and the annual natural gas consumption in MCM in 2001 (column 5), respectively, for the countries indicated (column 1) where data are available. For example, (1.02) refers to 1.02 Mt of annual coal consumption in 1998 in Botswana; [668] refers to 668 MWe installed hydro capacity in 1998 in South Africa; {37} refers to 37 Mt of annual oil production in 2001 in Angola; and (556) refers to 556 MCM of the annual natural gas consumption in 2001 in Angola.
[a]International Energy Database, January 2001.
[b]Batidzirai and Zhou (1998).
Source: *SADC Energy Annual Report* July 1999–June 2000 unless otherwise indicated.

electricity sector (Zhou, 2000), mainly from the road transport energy. Over 80% of transport energy in the SADC countries is provided by petroleum products, which, apart from Angola (and to a small extent South Africa), are all imported. The transport energy demand is growing at a rate faster than the rates of growth of the other economic sectors (Zhou, 2000). The fuel demand is aggravated by the prevalent use of passenger cars or light-duty vehicles (LDV), use of old vehicles and poor road conditions in some countries. The response has been meeting additional energy demand with additional supply, a situation begging for options to reduce demand.

Such dependence on the two most polluting fossil fuels is bad news for both GHG emissions and local pollution. The current climate change initiative, which seeks to reduce the rate of GHG emissions into the atmosphere, could be an opportunity for SADC

countries to shift to cleaner energy solutions. The challenge for the SADC states is, therefore, to identify energy supply and demand projects that can qualify for UNFCCC financing mechanisms and can also assist in meeting the sustainable development objectives of these states.

13.2 SADC ENERGY RESOURCE ENDOWMENTS AND UTILIZATION

For the electricity sector, there are still abundant proven reserves of coal in the SADC region amounting to over 90 billion tonnes distributed in the various member states (Table 13.1). At the current mining rate of about 250 million tonnes per year, these reserves can last for over 350 years.

The SADC region therefore has vast coal resources that can serve the region for a long time, but continued dependence on coal for electricity generation contradicts the current global thinking of decarbonizing the world economies to reduce GHG emissions.

On the other hand, the SADC region has vast hydro potential and natural gas (Table 13.1) resources, most of which are still not tapped.

Hydroelectricity resources are concentrated in the countries of the north, with the Democratic Republic of Congo (DRC) having the bulk of the resources, estimated at 100 GW. The other major hydroelectricity resources are on the Zambezi River basin bordering Zimbabwe, Mozambique and Zambia (33.8 GW) and in Angola (16 GW). To date, only about 5% of the estimated 163 GW has been exploited, implying a large untapped potential.

The only setback with hydroelectricity in the region is the repeated drought that results in low hydrological flows. The large dams are also increasingly becoming unpopular worldwide, due to destabilization of human settlements and animal habitats. The Inga Dam in the DRC, however, has the advantage that the hydro-site does not require a dam to generate hydropower. Another important element of hydroelectricity being vigorously investigated is the potential for meeting smaller and isolated electricity demand from mini- and micro-hydropower systems, but such sites are also limited in distribution.

About 240–300 billion cubic metres of natural gas have been discovered in five of the twelve SADC countries on the mainland, namely Angola, Mozambique, Namibia, Tanzania and South Africa. Gas explorations continue in the region, with Angola committed to developing projects to exploit significant gas reserves linked to deep-water oil discoveries. The Kudu gas field developed by the Shell-led consortium is at the centre of ambitious plans for power generation, but the proposed 750 MW Kudu gas power plant in Namibia is facing delays in its implementation due to a dispute over gas pricing with some utilities involved in the project (SADC, 2000). The consortium is also considering establishing another gas power plant in the Western Cape after the Kudu plant (*Business Times*, undated, http://www.btimes.co.za). Application of gas electricity generation can be increased as more gas and infrastructure come on line.

Coal-bed methane is a new gas in the region, and Zimbabwe and South Africa are at the forefront of investigating its potential. If successful, other countries with coal resources may follow. The bulk of natural gas and coal-bed methane could be used to diversify dependence on coal, and gas is increasingly becoming a popular alternative to coal for electricity generation globally.

Namibia and South Africa have major uranium deposits, but these are mainly exploited for mineral exports, although South Africa operates a 1,840 MW nuclear plant. Uranium use for nuclear electricity generation has little or zero public acceptance in the region. Some advances are, however, being made in nuclear

Table 13.2. Solar energy intensity in selected SADC countries

Country	Malawi	Botswana	Angola	Lesotho
Average solar energy intensity (MJ/m^2/day)	21.1	20.5	18.3	19.5

Source: UNEP/Risoe (1995).

technology that is more environmentally benign, e.g. the Pebble Bed technology being investigated by the Electricity Supply Kommission (ESKOM) of South Africa.

The SADC region is also endowed with huge renewable energy resources in the form of solar and wind energy and, to a lesser extent, mini- and micro-hydro power.

The region has up to 3,000–3,500 hours of sunshine per year, with energy intensity of 20 MJ/m^2 per day (5.6 kWh/m^2 per day) in some countries, as shown in Table 13.2.

The use of solar energy has been only in photovoltaic systems and solar water heaters at the domestic and institutional levels. The region has in excess of 20,000 solar PV systems. In Zimbabwe, the Global Environment Facility (GEF) project alone installed 9,000 PV systems, and the ESKOM-Shell initiative aimed to install another 6,000 PV systems in South Africa. Other countries have similar national, donor- or multilateral-funded projects.

Centralized PV systems are still of mini-grid scale and are not yet linked to national electricity grids. Larger systems at the research and development stage may soon be unveiled in South Africa by ESKOM.

Wind energy is currently widely used for water pumping, but scope for wind electricity generation exists, especially along the coastal areas. Wind electricity capacity of the order of 30 MW is being planned for the coast of Namibia and similarly for South Africa. ESKOM has plans to put up wind power plants of 10 MW, but these still have to be demonstrated.

Cost-effective exploitation and utilization of the hydro potential, natural gas and renewable energy resources can be realized through cooperative efforts by SADC countries and could dovetail into the UNFCCC GHG reduction agenda for funding. The varied geographical locations of these resources mean that adequate interconnections are needed to provide a good network for sharing electricity and gas to meet the regional electricity demand. SADC is already active in putting up the electricity interconnections, but efforts with regard to gas pipelines and infrastructure are still minimal. There are two major gas pipelines planned for the region. One will link Kudu (in Namibia) to Western Cape in South Africa, and another will be along the Maputo Corridor to move gas to South Africa from Namibia and Mozambique, respectively.

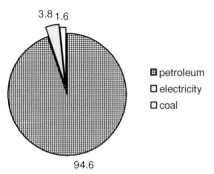

3.8 1.6

- petroleum
- electricity
- coal

94.6

Figure 13.1. Share of transport energy fuels/sources in SADC (adapted from Zhou, 2000).

Apart from benefiting from the economies' regional infrastructure through joint investments, countries of the region can also share experiences and risks. The larger regional projects are also likely to attract more significant funding from UNFCCC financing mechanisms than individual national projects, which can be small. The oil reserves that are relevant for the transport sector are not well known, but current estimates indicate total proven reserves of about 10 billion tonnes. The International Energy Agency (International Energy Database, 2001) puts Angola's reserves at between 761 and 1,183 Mt. Data presented by Mbendi Information for Africa (http://www.mbendi.com) based in South Africa, show estimated reserves of 53–73 million barrels or about 9,000 Mt from current exploration activities in South Africa. Consumption of oil products for the SADC region is 23,716 kt, 98% of which is consumed in South Africa (SADC, 2000). The rest of the countries import their petroleum products. Petroleum products form about 100% of the transport energy, except in South Africa and Zimbabwe, where electricity and some coal are used in rail transport (Figure 13.1). Coal usage in transport is, however, dwindling.

There is potential to use natural gas in compressed (CNG) form in transport, resulting in GHG reduction, and now that the natural gas is available in the region, this would also reduce cost to the region in the form of petroleum imports. In Mozambique, Sasol of South Africa is doing feasibility studies on gas to be used in vehicles and is also already transforming gas into liquid fuels (SADC, 2000).

Another limitation of natural gas use is the high price, which can be up to six times that of coal in South Africa (ESKOM, personal communication), if pegged in line with international gas prices. Creating a widespread infrastructure and an established market can boost further demand for natural gas in the transport sector. The GEF financial resources could also be used to establish the gas market and to bring down gas prices.

Biomass fuels for transport use are presently limited in the region. There is a limited production of ethanol in Zimbabwe and Malawi, the former prompted by the embargo/sanctions on the former Rhodesia. Ethanol is sometimes blended with petrol in Zimbabwe, with up to 13% ethanol content. When the petrol price falls, the ethanol is exported to markets in Brazil and Europe. Further potential to produce ethanol is being investigated in Zambia, where rainfall and land are suitable for sugar-based production. Potential for ethanol production also exists in the other wetter countries like the Democratic Republic of Congo and Angola, but the political instability in these countries currently prevents the realization of this potential.

13.3 ENERGY CONSUMPTION PATTERNS

This section presents the key consumers of electricity and transport energy in the SADC region, with a view to identifying those activities where clean energy technologies can be applied.

Electricity in South Africa, Zimbabwe and Botswana (Figure 13.2) is mainly used in industry and commerce (52%), followed by mining (26%) and in the domestic sector (15%). Transport (4%) and agriculture (3%) consume the least electricity.

In this regard, the situation would tend to suggest that considerable electricity savings could be realized in industry and commerce, mining and the domestic sector. Since most of the regional electricity is produced from thermal electricity plants, the demand-side energy technologies would involve mostly energy efficiency to achieve both economic gains and GHG reduction in those sectors.

With respect to transport modes, the road sub-sector is the largest consumer of transport energy in the SADC region. On average, road transport consumed 72% of the transport oil energy in the 1990s, but it was 92% in South Africa, 94% in Botswana, 83% in Zimbabwe and 81% in Zambia (Figure 13.3).

Poor road conditions also aggravate the fuel demand in some countries of the region. Only 11% of the total road network was paved in 1995, and 54% of the paved roads, or only 6% of the total road network, was in good motorable condition.

The high road energy share is also a result of large fleets of private cars and commercial vehicles. Private cars and commercial vehicles together constitute over 90% of the regional fleets (Zhou, 1997), thereby accounting for most of the road transport energy consumption. The fleets continue to increase at a fast rate, with implications for further energy demand.

Rail transport is not as large a consumer of energy as the road sector. Emphasis for rail transport has been export-oriented with the major traffic activities, linking centres of economic activities with seaports or other countries but with limited intra-country network, thus resulting in low rail density for the region. The average rail density for the SADC countries on the mainland in 1999 was 7.8 km/1,000 km^2, up from 6.5 km/1,000 km^2 in 1990, which is a small increase over a decade (Zhou, 1997) Only the railway network densities in South Africa (18.7 km/1,000 km^2), Swaziland (17 km/1,000 km^2) and Zimbabwe (7.9 km/1,000

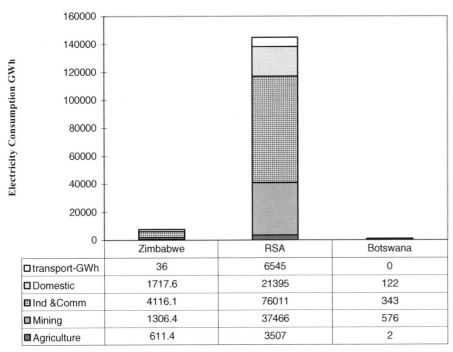

	Zimbabwe	RSA	Botswana
☐ transport-GWh	36	6545	0
☐ Domestic	1717.6	21395	122
⊞ Ind &Comm	4116.1	76011	343
▦ Mining	1306.4	37466	576
▨ Agriculture	611.4	3507	2

Figure 13.2. Electricity consumption by sector for selected SADC countries in 1993 (Southern Centre, 1999).

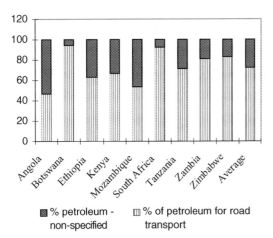

■ % petroleum -   ▥ % of petroleum for road
non-specified   transport

Figure 13.3. Observed transport energy share of total transport oil consumption in 1992 (from Zhou, 2000).

km^2) were larger than the average. In comparison, the average road network in 1995 was 134 km/1,000 km^2 with South Africa (419 km/1,000 km^2), Swaziland (289 km/1,000 km^2), Zimbabwe (219 km/1,000 km^2) and Lesotho (173 km/1,000 km^2) exceeding that average.

There is no emphasis on increasing the rail network, and there has been a considerable shift from rail to road transport in the region. Some donor support in the transport sector has gone into rail rehabilitation where required, otherwise emphasis has been channelled to improve the efficiency of the road sector.

In terms of energy consumption, rail transport has lower energy intensity per tonne-kilometre (t-km) or passenger-kilometre (p-km) than road transport. A fully laden diesel train averages 0.23 MJ/t-km, while a truck trailer consumes 0.85 MJ/t-km (ETSU, 1995). An electric train consumes about 0.27 MJ/t-km, but it has more carrying capacity than a diesel train. This means that the fuel intensity of rail transport is only about 27–31% that of road transport. The implication is a significant energy saving (up to 73%) when, for instance, rail freight is substituted for road freight.

13.4 ENERGY AND DEVELOPMENT ISSUES

Electricity plant investments have contributed to the regional debt, which is in excess of 50% of Gross Domestic Product (GDP). For the transport sector, petroleum import bills are estimated to range between 20 and 40% of the national export earnings (Teferra, 1996). This means that energy supply expansion will continue to expend the scarce foreign incomes and thus limit allocation of financial resources to other critical development and social sectors.

Donor funding has also been slow coming into the region, as reflected by the SADC electricity projects approved as early as 1984 that are still waiting for foreign funding.

Foreign Development Investment (FDI) in sub-Saharan Africa (SSA) has fallen sharply in the last 20 years to about US$2.2

Table 13.3. Access to electricity.

Country	% Urban households connected to electricity	% Rural households connected to electricity
Angola	No data	No data
Botswana[a]	50	6
Lesotho	3	—
Malawi	19	1
Mozambique	28	<1
Namibia	No data	No data
South Africa[b]	80	27
Swaziland	11	—
Tanzania	3	—
Zambia	37	2
Zimbabwe[b]	84	7
DRC	No data	No data

Sources: UNEP/Risoe (1995); [a]Zhou (2001); [b]Zhou *et al.* (1999).

billion per year, more than halving from 6.6% in the latter half of the 1980s, to 3% in the first half of the 1990s.

The countries of the region themselves have shown very limited capacity to provide funds for their own energy expansion development, and this emphasizes the need to explore for additional financing under the UNFCCC.

Investments in the electricity and transport sectors are lagging behind demand for the services. Access to electricity by households is still low in the region, particularly in the rural areas, where over 70% of the SADC population lives. In all SADC countries rural household connections to electricity fall below 10% (Table 13.3), except in South Africa (27%), where a huge subsidy was spent on rural electrification between 1996 and 2000 (Davis, 1998).

Rural access to electricity is low in most SADC countries, because potential consumers are required to pay in advance for connection costs, which the majority cannot afford.

In the transport sector, access to motorized transport is also low; and even car ownership of 43 vehicles/1,000 people falls far below the world average of 103 vehicles/1,000 people (Eriksen, 1995). Mass transport systems for both road and rail are limited in extent and quality as a result of the low infrastructure density and poor transport planning and management systems.

International decision-making related to climate change, as established by the UNFCCC, is already influencing the pattern of investments in the energy sector worldwide. In the electricity sector, the electricity market in Europe is expected to be 40% gas-fired by 2020, up from the current 15% (*The Economist*, 1999). The SADC countries, through their utilities, should be able to read these market and technological changes in order to ensure sustained operations.

Deregulation of the electricity market with the incoming of independent power producers will bring competition, and so the prices of electricity may also go up to meet shareholder profits, a situation that may bring both positive aspects with respect to reliability of supply, but negative aspects in terms of reduced access to electricity by the low-income groups. Under the circumstances, exploitation of renewable energies (solar and wind) becomes an option to supply isolated demand centres of these low-income groups.

Possible reduction in hydrological flows predicted for the region (Hulme, 1996), which can negatively impact particularly on hydroelectricity generation, if climate change occurs, should be a cause for exploring more reliable hydropower resources further north in the DRC, where hydrological flows are not likely to be affected by the predicted droughts.

In the transport sector, the SADC region has always been vulnerable to interrupted petroleum supply, since the fuel is imported and countries often keep only 90-day reserves. During the conflict in Kuwait, there was a potential threat of failing to procure fuels from the contracted sources, thus impinging on the security of supply. Under conditions of high risk, like war and uncertainty on the markets, the oil prices also rise sharply. The availability of natural gas in the region complemented by ethanol production, if exploited for this market, could offer more security of supply and cleaner energy sources to the region.

Both the electricity and transport sectors are the major contributors to air pollution in locations with power plants and in cities, respectively. Besides reducing the GHG emissions, it is imperative to sustain the local air quality, particularly for health reasons, hence the need to shift to cleaner energy sources.

13.5 BASELINES AND CLEAN ENERGY SOLUTIONS

The study conducted for the SADC electricity utilities by Southern Centre (1999) showed a possible regional electricity supply scenario increasing from 197,000 GWh in 1997 to 282,000 GWh in 2020 (Figure 13.4), with GHG emissions expected to increase by 14% to 230 Mt of CO_2 equivalent emissions by 2020.

Sustainable energy solutions in the electricity sector can be realized on both the supply and demand side by using decarbonization, energy efficiency, waste management and regional electricity trading.

In the electricity sector, key decarbonization efforts on the supply side entail substituting future coal generating capacity with either hydropower or natural gas.

GWh

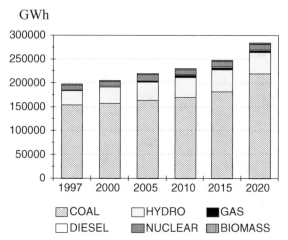

Figure 13.4. Projected electricity supply by fuel in GWh (from Southern Centre, 1999).

Hydropower remains the most developed renewable resource capable of replacing the bulk of thermal electricity. Although hydroelectricity is known to emit some CH_4, its estimated GHG level is about a tenth of coal (Batidzirai and Zhou, 1998); therefore hydropower is a plausible substitute for coal in electricity generation. The displacement of people is counterbalanced to a large extent by the ancillary benefits of having dams in the form of water storage (for other consumptive uses) and recreational facilities. To minimize negative socio-economic impacts, hydro-development for the region should focus on sites like the Inga Dam and micro- and mini-hydro sites.

The abundance of natural gas warrants its use in the region. A switch to gas from coal allows the use of high-efficiency and low-capital investment Combined Cycle Gas Turbine (CCGT) technology that reduces GHG emissions. Gas turbines have efficiencies of up to 60% compared with current high coal plant efficiencies of 48.5%. Natural gas plants have lower upfront/investment costs of US$300–500/kW, compared to US$1,300–1,750/kW for coal plants, and shorter lead times to realize gas plant use of two years, compared with three or more years for coal plants. Natural gas is also finding its way into cutting-edge fuel cell technology as a source of hydrogen for electricity generation.

Various options of generating electricity from waste are available. The option most practised in the region is the use of bagasse (sugar waste) in electricity plants, alternatively with coal or other fuel, depending on the availability of bagasse. Swaziland and Mauritius have significant experience of this technology, with capacities of 43 MW and 21.7 MW, respectively. South Africa has 105 MW of bagasse capacity (but 29 MW was in use in 1998). Further to the north, sugar industries in Tanzania, Kenya and Uganda are investigating the feasibility of utilizing their bagasse for electricity generation (Maya, 1999).

Urban waste has previously attracted funding from the GEF for a pilot project in the city of Dar es Salaam in Tanzania, involving generation of electricity from municipal solid waste. The project, however, failed to proceed even after the project design, for reasons that are not yet public. The potential for replication of such projects is high, considering the high waste output and the need to properly manage it in the region. Zimbabwe and Angola have investigated the use of urban solid waste to generate electricity through pyrolysis, whereby the waste is gasified and then the gas is used as fuel for gas turbines.

Generation through recovery of landfill gas has been analysed (Table 13.4) and appears to be cost-effective.

Analysis of central PV systems for Zimbabwe indicated that the cost of implementation is still higher than those for other cleaner energy options. A small centralized PV system (5 kW) is operating on a mini-grid scale in Botswana, but its GHG reduction potential has not been evaluated.

Table 13.4 shows the results of the options analysed in the region for the electricity sector supply side and the related costs of implementation.

The hydropower option is hindered by high upfront costs and a low rate of return, so this is unlikely to attract private sector investments. Individual governments are also unlikely to afford it under the present economic setbacks, hence the need for joint efforts to source funding. Natural gas plants are hindered by unavailability of gas infrastructure but, due to the low capital investment requirements, independent power producers may venture into this generation technology if the electricity prices allow a profitable rate of return. Landfill gas plants require further refinement of applicability and markets for the electricity generated. Centralized solar PV is hindered by both technology and high upfront costs that are even higher than those for hydropower per kW.

There is a slowly growing renewable energy industry for solar energy. Centralized PV and thermal solar plants will soon be demonstrated in South Africa. Botswana is planning to install a 1 MW demonstration pilot solar thermal plant, based on the parabolic trough technology. An effective regional renewable policy is required to support the growth of this energy industry that can reduce demand for fossil fuels. Table 13.5 shows various renewable energy technologies, cost of investment and potential GHG reduction per kW (displacing coal plants), based on a Financing Energy Services for Small Enterprises (FINESSE) study conducted by Southern Centre (1997) for Southern Africa.

The indications are that solar water heaters have the largest potential to reduce emissions per kW installed, followed by micro-hydro systems and solar PV systems. This is not surprising, considering that water heating takes 45–50% of household electricity consumption. It was estimated that peak electricity demand could be cut by 4 MW by switching to solar water heaters in Namibia alone (Zhou, 1998).

Table 13.4. Electricity sector decarbonization options and costs of GHG reduction.

Option	US$/tonne carbon avoided	Source
Hydropower	−2.25	Nziramasanga *et al.* (1998)
Natural gas plant	16.7	Nziramasanga *et al.* (1998)
Waste/landfill gas	−11.9	UNEP/Southern Centre (1993)
Solar/centralized PV	44.2	UNEP/Southern Centre (1993)

Table 13.5. Renewable energy technologies and their potential GHG reduction.

Renewable energy technology	Value sold (US$)	Installed/ saved kW	CO_2 avoided (tonnes)	Average (t/kW per yr)	Cost per tCO_2 avoided (US$)	Cost per kW (US$)
Solar home systems	764,824	1,623	43,772	27	18	472
Institutional PV electrification	412,878	875	23,597	27	18	472
PV pumping	3,510,813	745	20,092	27	18	471
Domestic solar water heaters	609,678	120,000	10,000,000	83	0.1	5
Micro-hydro system	339,186	396	15,380	39	22	857
Mechanical wind pumps	714,000,000	278,000	541,860	2	1,318	2,568
Small wind pumps	9,207,708	430	232	1	39,688	21,413
Solar cookers	3,362,400					

Adapted from Southern Centre (1997).

The costs of implementation analysed for a central PV system of US$44.2/tonne of carbon reduced (Table 13.4) and that obtained in the FINESSE project of US$64/tonne of carbon (US$18/tonne CO_2 Table 13.5) are of similar magnitude when presented as per tonne of carbon.

Wind systems have a high price but have the lowest GHG reduction potential per kW, although this depends on the operating time of the system. Wind power, however, has potential for larger capacity than the other renewable energy technology systems. The potential also exists for wind electricity to feed into the regional electricity grid or as bulk wind farm, with an independent mini-grid.

Energy efficiency on both the supply and demand side would offer an opportunity for a significant reduction in GHG emissions. Indications are that energy savings can be made through investments in more efficient technologies and reflection of the true costs of energy supply.

On the supply side, energy efficiency could entail opting for electricity plants of higher efficiency, even if the fuel is coal. Plant efficiencies are expected to improve to beyond 48.5–55% by 2020. The majority of power plants in the region have efficiencies below 35% (Batidzirai and Zhou, 1998).

The supply side can be complemented by cogeneration, where heat and electricity generation can be met from the same energy in large industries with steam generation systems. A cogeneration potential capacity of 1,500 MW was estimated for South Africa by Etzinger (1995).

On the demand side, energy-efficient options are also available through demand-side management (DSM), whereby the electricity utility assists its consumers to use electricity efficiently, e.g. through more energy-efficient devices like electric motors and compact fluorescent lamps (CFLs) and the upcoming light emitting diodes (LED). For instance, it was discovered that implementation of efficient lighting using CFLs could reduce the lighting load by another 40% in Namibia (Zhou, 1998). In this way, the demand for electricity per unit of activity is reduced. South Africa has a huge efficient lighting project, which is funded partly by the GEF, and will distribute some 18 million CFLs (SADC, 1998). Similar projects could be replicated in the other countries.

Based on preliminary utility expansion plans, there is potential for deferring 370 MW through DSM in South Africa (Southern Centre, 1999). The 1993–1995 DSM programme of the World Bank's Energy Sector Management Assistance Programme (ESMAP) for Tanzania showed potential energy

Table 13.6. Energy efficiency options in Botswana's household and industrial sectors.

GHG reduction option	US$/tonne CO_2 avoided
Efficient lighting – household (CFLs)	−107
Prepayment meters – household	−24
Geyser timer-switches – household	−18
Electricity factor correction – industry	−10
Efficient boilers – industry	−8
Efficient motors – industry	0.5

Source: Zhou (1999).

savings of 35,642 MWh (9.65 MW) from power factor correction and implementing efficient lighting and industrial motors, amounting to US$4.5 million (SADC, Energy Management News, 1996), compared to about US$15 million for a new electricity plant of similar size.

Table 13.6 shows the results of clean energy electricity-based options analysed for Botswana's household and industrial (including mining) sectors. The costs of GHG reduction are low and predominantly negative, and thus can be implemented.

Awareness and incentive for application are limited in the household sector, and this limitation is compounded by failure to appreciate the life costs of the options. Industrialists also do not have any incentives to use cleaner energy solutions, as their operations are not currently threatened by either legislation or short supply. The region has low electricity tariffs and the bulk of electricity is still sold below 5 US cents/kWh by ESKOM.

Since the energy efficiency options above can be implemented at negative costs, the region will stand to benefit by implementing them. To date, the approved SADC project on Energy Efficiency in Heavy Industry still requires funding (SADC, 2000). The Industrial Energy Management and Demand Side Management projects, which are both under implementation, have not had a significant influence on the uptake of clean energy technologies and practices in both industry and household sectors, so more effort is still required in this regard.

As part of regional planning, there can be significant efficiency of electricity generation in an environment of free electricity trading. This ensures that the cheaper and cleaner plants are run and that generation for peak demands can be reduced or eliminated, thereby saving on energy and reducing GHG. With respect to GHG emissions, it was estimated that electricity trading between three regional boards in India could realize 7.7 million tonnes of CO_2 reduction (INFORSE, 1997).

Electricity trading under the Southern African Power Pooling (SAPP) arrangements will probably offer a similar opportunity to save GHG emissions from electricity generation to meet peak demand in the individual countries. ESKOM of South Africa aspires to extend its trading regime far north and, in the same vein, also provide cheap and reliable electricity for industries and for growth in the neighbouring countries.

With respect to SAPP, grid extension is receiving significant attention with various planned projects linking the member states. By the end of 2010, all member states should be connected to the regional grid, funding permitting. The extent of electricity sharing among the SADC countries is, however, still low. ESKOM is presently selling only 2–4% of its electricity through SAPP; this is by far the largest share from an individual SADC country at the moment.

For the transport sector in the SADC region, the estimated CO_2 emission would be about 150 Mt in 2020, doubling that estimated for 1994 (Zhou, 2000).

The decarbonization options available in the transport sector are the pre-combustion options, involving substitution of petrol and diesel with compressed natural gas, ethanol, bio-diesel and electric vehicles. Table 13.7 shows some of the options analysed for the region.

Adoption of compressed natural gas is hindered by lack of both gas and infrastructure. Ethanol has limited scope for application, especially when petrol prices are low, and sometimes fails to be competitive with petrol because of high production costs. The technology is, however, resident in the region and can have wide-scale application if production is on a large scale, and it can be competitive with petrol if socio-economic and environmental considerations are included in the fuel pricing.

Electric rail is in use in South Africa and Zimbabwe but could be expanded and extended to other countries. This option would, however, qualify for carbon financing if only hydropower were used.

As an option, it would really reduce the cost of diesel imports currently used to drive trains. The option, however, becomes cost-effective where loads are large, for example, along the regional transport corridors leading to seaports.

A number of energy-efficient transport options exist in the region involving both vehicular (e.g. vehicle efficiency) and infrastructural improvements (paved roads or mass transit systems). Regulatory measures (like vehicle inspection) and fiscal measures (e.g. fuel tax) can also result in efficient use of transport energy.

Table 13.8 shows transport options analysed for Botswana for the transport sector and their costs of reduction.

Paving of roads and maintaining them in good condition does not require any new technology, but requires proper financial allocation and good maintenance and management skills in the countries. This is one of the options that can be implemented easily but has not received attention from planners or finance ministries.

The option of moving road freight to rail would result in less energy intensity per t-km, but rail is not competitive at the moment

Table 13.7. Transport decarbonization options and costs of GHG reduction.

GHG reduction option	US$/tonne carbon	Source
Compressed natural gas	1.37	Zhou *et al.* (1998)
Ethanol	−186.50	UNEP/Southern Centre (1993)
Diesel to electric rail	38.28	Zhou *et al.* (1998)

Table 13.8. Botswana's energy efficiency options for the transport sector

GHG reduction option	US$/tonne CO_2
Paved roads	−142
Road freight to rail[a]	−108
Pipeline[a]	−65
Fuel pricing	0.0
Vehicle inspection	0.7

[a] If introduced in 2030 when loads are large.
Source: Zhou (1999).

because of inefficient handling, long turnaround times and limited infrastructure. The option could be implemented by improving infrastructure, but this element requires financing.

There are operating pipelines in some parts of the region, which save on fuels that would be otherwise used by road and rail tankers and also spare the roads. Pipelines, however, are more cost-effective when large quantities of fuel are moved. The critical limitation for putting up more pipelines where they are needed in the region is cost, and hence some financing is required. The region also has to undertake proper planning concerning where the pipelines will be needed. The pipeline analysed in Table 13.8 was for moving fuel from Pretoria in South Africa to Gaborone in Botswana, some 400 km away.

Fuel pricing is an unpopular measure politically, but has found some applications in the region. This option can fail to reduce fuel demand if it is applied in an environment of poor fuel procurement policies. For example, in cases of fuel shortage, people tend to acquire more fuel than they require at the time, hence the demand may not decrease with increase in prices.

Vehicle inspection is a regulatory measure, and one that may require strengthening of inspection centres and putting up testing equipment and infrastructure. Sometimes the inspection fails to achieve desired results, especially for freight vehicles, due to corruption.

In the transport sector, urban planning would reduce travel demand and hence transport energy, resulting in GHG reduction. Urban planning involves integrated transport planning, which can realize significant fuel savings but which poses a challenge in estimating the energy savings and GHG avoided. Capacity-building

in such an urban planning-transport nexus is thus critical for future exploitation of this option.

13.6 FINANCING UNDER THE UNFCCC AND KYOTO PROTOCOL

More such clean energy options exist in the SADC countries, but what is critically lacking is financing to support the clean energy projects or programmes.

The UNFCCC was realized to bring a culture of cleaner development through emission avoidance, since the technological applications and policy measures required to bring about the reduction on GHG emissions also result in efficient use of resources, such as energy.

For developing countries, the promising UNFCCC provisions to finance the transition to clean development are the GEF and Activities Implemented Jointly (AIJ). Of course, the Clean Development Mechanism (CDM) under the Kyoto Protocol offers similar opportunities.

The GEF is an operating entity of the financial mechanism of the UNFCCC. It channels multilateral funds to finance projects in selected focal areas under various Operational Programmes, including both decarbonization and energy efficiency projects (http://gefweb.org/Operational_Policies/operational_programs/operational_programs.html). It now has an Operational Programme for promoting environmentally sustainable transport (http://gefweb.org/Operational_Policies/Operational_Programs/OP_11_English.pdf). A number of pilot AIJ projects have been implemented in Africa that could reduce greenhouse gases and also achieve an economic development objective.

SADC states, however, have not benefited much in the past phases of GEF and AIJ, probably due to lack of awareness and expertise to package acceptable projects for these financing arrangements.

Africa benefited the least from the GEF's Pilot Phase, compared with Asia and Latin America, even though the population of Latin America is less than that of Africa. The only notable project financed by the GEF in its Phase I in the SADC was the Solar PV Project in Zimbabwe, which installed 9,000 systems (45 Wp) between 1993–1998. This was a barrier-removal project, intended to establish the market (suppliers and consumers) for solar PV technology in the country. The other project would have been the waste-fired electricity plant intended for Tanzania that was to get US$3 million from the GEF, but failed before implementation. Similar barrier-removal, renewable and energy efficiency projects are now being developed in the region under GEF financing, indicating that the possibility exists to use these resources for adopting clean energy technologies. GEF financing is continuing, with the largest replenishment of US$2.92 billion over the next four years (2002–2006). Thus, opportunities still exist for further clean energy projects to be developed for GEF financing. Emphasis should be on implementation projects involving the private sector. Past projects have mainly been managed through governments, a situation that has sidelined the private sector and hindered sustainability of projects.

During the first part of the AIJ, there was one project in Africa, the Burkina Faso Forestry Project, and none in the SADC region. The Gugulethu efficient housing scheme in South Africa, involving some 600 eco-efficient houses in Cape Town, is a more recent AIJ project in the region. More AIJ projects are, however, being approved for Africa during the extension of the AIJ period, for example through United States Initiative on Joint Implementation (USIJI). SADC should also tap into that financing mechanism by aggressively packaging qualifying projects.

The operationalization of the Clean Development Mechanism has started with the appointment of the Executive Board at the seventh session of the Conference of the Parties (COP 7), in Marrakech, and the adoption of the Marrakech Accords (2001), which include decisions relating to CDM project implementation. Various project initiatives have been started to 'learn by doing' and to build capacity in developing CDM projects and project design documents. The special attention given to small-scale projects provides opportunity for participation by all sizes of economies and businesses.

Another financing mechanism for clean energy solutions is the initial capital outlay provided by multilateral lending institutions, such as the European Bank of Reconstruction and Development (EBRD), to assist the creation of Energy Service Companies (ESCOs) in developing countries venturing into the energy efficiency business. This financial arrangement can leverage larger national clean energy investments.

These are some of the opportunities from which the SADC states can benefit, but the modalities of realizing these potential benefits will demand dedicated policies, research and planning by the regional countries.

13.7 THE WAY FORWARD

The National Communications to the UNFCCC that are prepared by the Parties to the Convention should be used as marketing documents for clean energy technology needs of the SADC states.

Past National Communications have been weak in this respect. Emphasis has been given to creation of GHG gas inventories and, legitimately, to vulnerability. But mitigation projects with national economic benefits have not been comprehensively articulated in most National Communications.

National decision makers should also be actively involved in climate change activities as part of the general framework of development, and their climate change-related strategies, programmes and projects should clearly reflect contemporary governments' development priorities.

There are, however, some prerequisites in order for the region to benefit from the UNFCCC in terms of well-established institutions, legal frameworks, information flow and catalytic activities such as research and development and capacity-building.

A supportive institutional framework will require governments and parliaments that plan and promote sustainable development through protection of the environment.

The involvement of Ministers of Environment in the African Ministerial Conference on the Environment (AMCEN) and parliamentarians in the Global Legislators Organization for a Balanced Environment (GLOBE) are examples in which these key players have been in active participation with regard to the UNFCCC and other conventions. For example, the SADC Ministers of Environment have been involved in bringing Clean Technology Initiative (CTI) projects to the region. The regional GLOBE parliamentarians have participated and debated environmental issues facing the region, including climate change.

Private sector institutions, such as Chambers of Commerce and Industry and Transport Associations, could be used as conduits for information on clean energy technologies, helping their members in both industry and transport sectors to take opportunities under the UNFCCC and Kyoto Protocol to finance clean energy initiatives.

The idea of ESCOs is gaining momentum worldwide with respect to providing required skills in the adoption of the clean energy technologies, particularly in industry, so SADC states can also take advantage of EBRD financing to promote dissemination of clean energy technology in the region.

In cases in which governments may be seeking answers to policy questions, think tanks or expert groups should be formed to provide technical and policy-related answers. Similar working groups could be tasked to produce portfolios of clean energy projects and programmes that will qualify for funding under the various available financial mechanisms discussed above.

Governments and parliaments can support UNFCCC efforts in a number of ways that include enacting appropriate policies, improving standards of business operation, and passing legislation to regulate performance of the various energy providers and users. Private sector entities are not yet keen to engage in climate change activities because they feel no threat to their businesses in the absence of strong legislation governing consumption patterns, including that for energy.

With the emergence of Independent Power Producers in the electricity market, an effective legal framework will be required to ensure acceptable energy efficiency and environmental standards in the market. The framework should be coupled with a regulatory body to monitor the energy actors.

Governments have to provide clear signals that encourage investments in cleaner energy sources, such as renewable energy technologies (solar, wind and hydroelectricity) or less polluting fuels, such as natural gas, through fiscal measures, including energy pricing and carbon taxes.

Standards have also been used elsewhere 'to level the playing field' for various business actors. An example is a Building Energy Code that can set standards for thermal efficiency in buildings. Standards may be set in line with international standards, such as those of the International Organization for Standardization (ISO). Energy audits in industrial facilities and cogeneration could be made mandatory, where applicable. Legal conditions have also required labelling and energy standards for energy appliances to assist consumers in their choices of technologies.

Legislation carrying heavy fines and possible closure of premises for failure to comply with pollution have been used in the past and can be used to influence adoption of clean energy technologies by the private sector.

One major barrier to the SADC countries' advancing development through the UNFCCC and the Kyoto Protocol stems from poor information flow about the opportunities available under the Convention and the Protocol, and how to access funding.

Awareness of both the opportunities and the available technologies is critical to the adoption of clean energy solutions. Full participation in the UNFCCC activities and other related forums provides opportunities for an enhanced knowledge base in clean technologies.

Exchange fairs can expose what technologies are on the market, and the countries of the region can exchange experiences through such fairs. The fairs should also be held on a North–South and South–South cooperation basis, thus increasing the scope of information on appropriate technologies. Such fairs have become popular in other developing countries, such as China.

Databases on best practices in the region and elsewhere will be another effective means of disseminating clean technologies. Also of importance are knowledge networks for sharing information between researchers and implementers. Under this network arrangement, there can be provision of advice and guidance to governments on project identification and implementation.

Governments and industries could benefit by sponsoring research and development. Researchers could also be employed to design potential GEF, AIJ and CDM projects and package them for appropriate funding. The process will still require prompt government support to expedite the financial application process with the funding institutions. Private sectors should be encouraged to fund research and development that will assist them in their enterprises, particularly with respect to energy management technologies. Results of research conducted by independent research institutions often do not reach the potential users or government implementers.

Incentives may be found to work better than regulation in clean energy development efforts, hence beneficial opportunities in this area ought to be exposed through demonstration or showcase projects. Incentives could be in the form of awards, as has been practised by ESKOM through their ESKOM Energy Effective Design Awards (EEEDA) (SADC, 1997).

Governments could also budget for seed money dedicated to initiating clean energy programmes or promote uptake of clean energy technological options. The seed fund could be used to identify potential projects, thereby creating portfolios of projects for adoption by the private sector.

13.8 CONCLUSIONS

Continued prevalent use of coal and oil in the energy sector in the SADC region will result in a significant increase in GHG emissions, with implications for high social and environmental costs. Between 1994 and 2020, electricity sector and transport emissions will have increased by over 50% unless cleaner energy technologies are adopted.

Continued investments in coal- and oil-dependent technologies will have significant requirements for investment that is not forthcoming. Some investment could accrue from the UNFCCC mechanisms for more benign energy technologies. These technologies, when carefully selected, could also meet the needs for clean energy by the poor population of the region. A shift in technology in tune with the rest of the world will also ensure that SADC countries are not vulnerable to the effects of technological changes.

There are large reserves of cleaner fuels and opportunities for improving efficient use of energy to which SADC countries can shift.

Notable options are recognized in the electricity and transport sectors, as follows:

- Substitution of coal-based electricity plants with the abundant hydroelectricity and natural gas, complemented by renewable energy sources (solar and wind) and waste-fuelled plants, will result in significant reduction in GHG and also cleaner economic development through proper management of resources and waste.

- There are also many energy efficiency options in the key electricity-consuming sectors consisting of industry and commerce, mining and household sectors.
- Planning measures, such as electricity trading and urban/transport planning, could also make a significant contribution to sustainable development in the SADC region.
- The transport sector has to target the road sector for GHG reduction and improvement of services, as this is the largest energy consumer with consequences of air pollution, congestion and accidents.
- On the pre-combustion side, substitution of petroleum products by compressed natural gas and ethanol (where feasible) hold promise to relieve the region of large import bills and also reduce GHG emissions.
- On the demand side, vehicle efficiency, regulatory measures, such as vehicle inspection and fuel taxes, coupled with infrastructural improvements such as road paving and mass transit systems, have a large potential to improve the efficiency of the transport sector and also reduce GHG emissions.

The majority of these electricity and transport options can be implemented at negative costs and at most below US$10/tonne of carbon, suggesting that there are benefits in adopting these energy technologies; and most of them can be implemented under the GEF, AIJ and CDM arrangements.

The region would, however, need to put its house in order by having an environment conducive to benefiting from the UNFCCC and Kyoto Protocol processes.

ACKNOWLEDGEMENTS

I thank Dr Pak Sum Low for encouraging me to submit the chapter. The comments of Dr Hesphina Rukato and the anonymous reviewer on an early draft of this chapter are acknowledged.

REFERENCES

Batidzirai, B. and Zhou, P. P. (1998) Options for greenhouse gas mitigation: Inventory of GHG emissions from the electricity sector. In R. S. Maya (Ed.), *SADC's Electricity Sector*. Harare, Zimbabwe, Southern Centre publication.

Davis, M. (1998) *Rural Electrification in South Africa*. Energy and Development Research Centre. University of Cape Town, South Africa.

The Economist (1999) 29 May–4 June 1999. Vol. **351** No. 8,121. London.

Eriksen, S. H. (1995) Transport and GHG emissions. In H.W.O. Okoth-Ogendo and J. B. Ojwang (Eds.), *A Climate for Develop-

ment*. ACTS Environmental Series No. 5. Nairobi, Kenya, ACTS Press.

ESKOM (1993) *ESKOM 1993 Annual Report*. South Africa.

ETSU (1995) Mechanisms for improved energy efficiency in transport. *Overseas Development Administration Report*. ETSU/RYCA/18400304/Z3, UK.

Etzinger, A. (1995) Breaking down walls between demand-side management, supply-side management and the environment – ESKOM's Experience. In *Demand-Side Management. Energy and the Environment Series Vol. 4*. Berlin, CDG.

Hulme, M. (Ed.) (1996) *Climate Change and Southern Africa: An Exploration of Some Potential Impacts and Implications for the SADC Region*. WWF International Report, University of East Anglia, UK.

INFORSE (International Network for Sustainable Energy) (1997) Energy from municipal solid waste in Africa. *Sustainable Energy News*, No. 17, Denmark.

International Energy Database (2001) http://www.eri.uct.ac.za

Kgathi, D. L and Zhou, P. P. (1995) Biofuel use assessments in Africa: implications for greenhouse gas emissions and mitigation strategies. In J. F. Fitzgerald *et al.* (Eds.), *African Greenhouse Gas Emission Inventories and Mitigation Options*. Dordrecht, The Netherlands, Kluwer Academic Publishers.

Maya, R. S. (1999) Clean technology options in the SADC region. Paper presented at the Clean Technology Initiative Workshop, 13–17 September, 1999, Gaborone, Botswana.

Mbendi Information for Africa. http://www.mbendi.com

Nziramasanga, N., Batidzirai, B. and Rowlands, I. H. (1998) Regional electricity mitigation options. In I. H. Rowlands (Ed.), *Climate Change Cooperation in Southern Africa*. London, EARTHSCAN Publications Ltd.

OECD and IEA (1994) *Climate Change Policy Initiatives. Vol. 1 OECD Countries*. Paris, OECD.

SADC (1996–1998) *SADC Energy Management News, 1995–98*. Energy Research Institute. University of Cape Town, South Africa.

(2000) *SADC Energy Annual Report, July 1999–June 2000*. ISBN 99912-80-43-X. Luanda, Angola.

Southern Centre (1997) *SADC FINESSE South Africa Country Study*. Southern Centre, Harare, Zinbabwe (or SADC Technical Administration Unit-Luanda, Angola).

(1999) National and regional baselines and mitigation options. In *Options for Greenhouse Gas Mitigation Under Electricity Pooling in Southern Africa*. Southern Centre, Harare, Zimbabwe. Draft Report.

Teferra, M. (1996) Transport in Ethiopia. In M. R. Bhagavan (Ed.) *Transport Energy in Africa*. African Energy Policy Research Network Kenya; ZED Books, UK.

UNEP/Risoe (1995) *Climate Change in Southern Africa. Methodological Development, Regional Implementation Aspects, National Mitigation Analysis and Institutional Capacity-Building in Botswana, Tanzania, Zambia and Zimbabwe*. Phase I. UNEP Collaborating Centre for Energy and Environment. ISBN 87-550-2020-9: Risoe National Laboratory, Denmark.

UNEP/Southern Centre (1993) *UNEP Greenhouse Gas Abatement Costing Studies. Zimbabwe Country Study Phase II.* Risoe National Laboratory, Denmark.

Zhou, P. P. (1997) *Development of an Efficient Transport Sector in East and Southern Africa – opportunities and policy implications under the UNFCCC.* Working paper (2nd Report) AFREPREN, Kenya.

(1998) *Energy Efficiency for Climate Change – Opportunities and Prerequisites for Southern Africa.* Cape Town, GLOBE Southern Africa.

(1999) *Climate Change Mitigation in Southern Africa. Botswana Case Study.* UNEP Collaborating Centre for Energy and Environment. ISBN 87-550-2429-9. Risoe National Laboratory, Denmark.

(2000) *Sustainable Mobility. Perspectives from the Transport Sector in East and Southern Africa.* UNEP Industry and Environment. ISSN 03789993. Vol. 23, No. 1. Paris.

(2001) *Rural Energy Needs and Requirements in Botswana.* Final Report. Botswana, Ministry of Minerals, Energy and Water Affairs.

Zhou, P. P., Rowlands, I. H. and Turkson, J. K. (1998) Other regional options. In I. H. Rowlands (Ed.), *Climate Change Cooperation in Southern Africa.* London, EARTHSCAN Publications Ltd.

Zhou, P. P., Ramaphane, K. M and Mhozya, X. (1999) *Evaluation of the Rural Electrification Collective Scheme in Botswana.* Final Report. Botswana, Ministry of Minerals, Energy and Water Affairs.

14 Regional approaches to global climate change policy in sub-Saharan Africa

IAN H. ROWLANDS

University of Waterloo, Canada

Keywords

Global climate change policy; global climate change response strategies; regional cooperation; regional organizations; sub-Saharan Africa

Abstract

This chapter examines the potential for different *regional approaches* in response to global climate change challenges in sub-Saharan Africa. Working together, governments, businesses, non-governmental organizations and/or other groups in two or more neighbouring countries could develop response strategies that would benefit both the global climate and local development in Africa. In this chapter, the following possible regional response strategies are presented and investigated: countries working together as a regional actor in international negotiations; regional strategies to adapt to global climate change; regional mitigation strategies; and the development of knowledge at the regional level. Of course, not every possible regional approach to global climate change policy will be desirable. Nevertheless, by presenting a broad range of possibilities, this chapter aims to stimulate interest in, and further research on, regional approaches to global climate change policy in sub-Saharan Africa.

14.1 INTRODUCTION

The purpose of this chapter is to examine the potential for different regional approaches in response to global climate change challenges in sub-Saharan Africa. The argument advanced is that it is in African decision makers' interests to complement the usual focus upon *national* approaches with a consideration of how *regional* approaches could be part of a broader portfolio of climate change responses. To investigate this potential, the chapter proceeds in five main sections.

Following this brief introduction, the factors encouraging increased consideration of regional cooperation, generally, are examined. The motivation for neighbouring countries to engage in cooperative action may be economic, political and/or social. Following worldwide trends during the 1990s, the interest in regional endeavours in sub-Saharan Africa has recently been high; many continue to call for sustained, and in places, increased efforts towards regionalism.

In Section 14.3 the role of regional approaches – some formal, some informal – in the climate change negotiations to the end of 1999 is examined. The potential offered by the international agreements for more regional activity in response to global climate change, as well as the relationship between regional approaches to climate change policy and broader efforts at regionalization, are also investigated.

Various kinds of regional cooperation – presented as means to respond to the challenge of global climate change – are presented in Section 14.4. Some are presented by means of specific examples, with real-world references to different parts of sub-Saharan Africa; others are more abstract, identifying patterns of activity that are worthy of consideration in different contexts. Finally, Section 14.5 provides a brief conclusion.

The chapter makes no claim to provide an all-encompassing blueprint for regional responses to global climate change in sub-Saharan Africa. Instead, it is meant to provide a sample of the ways in which regional cooperation in climate change-related issues – among different groups of African countries at different times – could prove beneficial to all involved. In this way, it is hoped that the chapter will stimulate interest, so that decision makers may investigate the issue further.

14.2 ADVANTAGES OF A REGIONAL APPROACH

During the 1990s, there was increased interest in how neighbouring countries could work together to realize mutual benefits – some called this phenomenon the 'new regionalism' (for example,

---

Climate Change and Africa, ed. Pak Sum Low. Published by Cambridge University Press. © Cambridge University Press 2005.

le Melo and Panagariya, 1993). Before considering, more specifically, the potential associated with a regional approach to climate change issues, it is useful to review the possible advantages of adopting a regional approach more broadly. By 'more broadly' here, I refer to a vast range of activities – indeed, anything that requires the coordinated efforts of actors in two or more neighbouring countries.[1] Though economists often restrict 'regional cooperation' to solely 'economic integration', it also includes other actions that are aimed at enhancing the common interests of people in different countries of a region – for example, project coordination, defence treaties or cultural exchanges. Regional cooperation may involve formal organizations (that is, legal entities with offices, staff, equipment and the like) or it may not, instead relying upon informal networks.

In support of this new regionalism, its proponents advance many arguments. One of the first often cited involves economic benefits. Economies of scale can create unit-cost reductions in the goods produced and services provided in a region. For example, by 'avoiding duplication of efforts at the national level in several countries, such [regional] organizations can offer services at a much lower cost' (Tewungwa, 1994). New foreign investment may also be attracted by a regional arrangement (for this would create access to a larger market). Additionally, the cost of managing uncertainty could fall: leaders could make decisions more confident of the actions their neighbours will take, by virtue of having some kind of institutional framework in place. In total, the proponents argue, greater regional cooperation will increase national levels of trade, income and employment (e.g. Asante, 1997; Mwase, 1994).

Other reasons are also advanced to support increased cooperative efforts. Many maintain that if developing countries act together, then they will have greater bargaining strength in negotiations with external actors – be they donors, transnational corporations, international financial institutions or others. Weeks (1996), for example, argues 'regional groupings provide perhaps the only viable vehicle by which developing country governments can exert bargaining influence . . . ' Without such cooperation, external agents may be able to play one country off against another, and so reap the gains that otherwise could accrue to the region itself. Regional organizations may also be able to 'render more efficient service and respond effectively to the problem at hand, as they are to some extent insulated from parochial political influence and maneuvering' (Tewungwa, 1994). Additionally, by 'pooling human resources drawn from a wide variety of countries, regional organizations tend to be better staffed, offering an optimal way of utilizing meagre human resources in the region. Appointments to

bureaucratic positions in regional organizations are subjected to less national political interference.' (Tewungwa, 1994).

Regional cooperation in any one particular area may also lead to increased cooperation in other areas. Referring to southern Africa in particular, Vale and Matlosa (1996) argue that: 'Through the process of re-knitting together, a surprising number of everyday problems will be overcome' Thompson (1992), meanwhile, identifies flexibility as a particular benefit of greater regional action: 'Instead of putting up physical or policy barriers, regional decisions find multiple ways of mobilizing resources for solutions to development problems.'

Notwithstanding all of these assertions, it is important to recognize that regional cooperation will not necessarily generate these or other benefits. Creation of a regional free trade area could, for example, generate more 'trade diversion' than 'trade creation' – that is, it could serve to replace efficiently produced, externally sourced goods with inefficiently produced, internally sourced ones. Regional organizations could also be susceptible to 'capture' by an individual country, or could not be sufficiently supported by participating countries so as to be effective. Moreover, regional activity could increase feelings of alienation amongst groups of individuals and therefore inhibit efforts to advance social sustainability. Consequently, the particular arguments above are meant to highlight the *potential* for regional action to generate net benefits – whether they do or not depends, of course, upon the particular kind of regional action undertaken. Nevertheless, the fact that the potential exists encourages its further consideration.

The 'new regionalism' did not bypass sub-Saharan Africa during the 1990s. Indeed, regional cooperation on the continent was given an even greater push by the conclusion of the Abuja Treaty in 1991 (which built upon the 1980 Lagos Plan of Action): this identifies regional groupings as the building blocks for an African Economic Community by the year 2025 (see, for example, Asante, 1997, Chapter 5). In addition to these regional building blocks, there are numerous other regional groupings around the continent (see, for example, Söderbaum, 1996). Indeed, regional cooperation in Africa has numerous supporters, both inside and outside the continent. For example, the Global Coalition for Africa – 'a North-South forum dedicated to forging policy consensus on development priorities among African governments, their northern partners, and non-governmental groups working in and on Africa' (Global Coalition for Africa, undated) – released the following statement in 1995:

Neighbouring countries, if grouped into well-functioning regional integration schemes, could form the natural markets for non-traditional exports of the member countries. Quality improvements and increased cost-competitiveness can be nurtured within such regional markets that can serve as springboards to wider international markets. Regional integration would also enable African countries to

---

[1] 'Regional' does not necessarily mean 'inter-state'. Regional approaches to global climate change policy could involve coordinated action among businesses, non-governmental organizations or cities in neighbouring countries.

present themselves as larger, more attractive trading and investment partners to the outside world. These and other benefits of integration, as well as the formation and rapid consolidation of trading blocs elsewhere in the world, should compel African policy makers to pursue integration with stronger political commitment. An integrated Africa would have better prospects of becoming a genuine partner to other regional blocs such as the European Union.

(quoted in Rwegasira, 1997)

Regional cooperation, broadly, presently has many supporters in sub-Saharan Africa.

14.3 ROLE OF REGIONAL APPROACHES IN GLOBAL CLIMATE CHANGE

Efforts to address environmental challenges involving two or more countries have usually involved some kind of 'international approach'. Given, moreover, the spatially limited nature of many environmental challenges (at least in perception, if not always in reality), this 'international approach' would often consist of a 'regional approach'. Consequently, I fully acknowledge that regional approaches to environmental problems have been present for many years. In many of these instances, however, the regional approach usually consisted of occasional meetings of representatives from neighbouring countries, followed by periods of action (or inaction, as the case may be) by those same representatives, but within their own individual countries. This was usually the extent of any regional activity.

The developing global regime to address the challenges of global climate change offers – at least, in principle – greater potential for regional approaches. For one, the United Nations Framework Convention on Climate Change (UNFCCC) specifically refers to regional cooperation as a means of mitigating and adapting to global climate change. Among Article 4's commitments are that all Parties shall formulate, 'implement, publish and regularly update national and, where appropriate, *regional* programmes . . . ' (UNFCCC, 1992, Article 4.1(b), emphasis added).

The UNFCCC also allows the participation of 'regional economic integration organizations' (REIOs) in the international regime (in addition to 'states', which are the usual participants in international environmental agreements). References to REIOs include Article 18 on 'Right to Vote', Article 20 on 'Signature' and Article 22 on 'Ratification, Acceptance, Approval or Accession' (UNFCCC, 1992).

The possibility of 'joint communications' is also raised in the UNFCCC:

Any group of Parties may, subject to guidelines adopted by the Conference of the Parties, and to prior notification to the Conference of the Parties, make a joint communication in fulfillment of their obligations under this Article, provided that such a communication includes information on the fulfillment by each of these Parties of its individual obligations under the Convention.

(UNFCCC, 1992, Article 12.8)

The Kyoto Protocol has increased the potential for international responses (including, within that, regional responses) to global climate change. For one, developed countries, collectively, agreed to a 5.2% reduction in their net 1990 greenhouse gas emissions by 2008–2012. To help meet this, it appears that some kind of emissions trading scheme will be established in order to allow them to achieve this goal *jointly*, if they do not do so *individually* (Kyoto Protocol, 1997, Article 6).

Additionally, the establishment of the Clean Development Mechanism (CDM) further suggests that international action will be part of the global response to climate change (Kyoto Protocol, 1997, Article 12). Though the modalities of the CDM have yet to be agreed (and, of course, though the Kyoto Protocol has yet to enter into force), it would appear that this would increase attention upon all kinds of international activities – regional ones included.

Finally, the Kyoto Protocol restates the possibility of regional approaches as part of all countries' broader responses to global climate change. Article 10(a) states that all Parties shall formulate, 'where relevant and to the extent possible, cost-effective national and, where appropriate, *regional programmes* to improve the quality of local emission factors, activity data and/or models which reflect the socio-economic conditions of each Party . . . ' Similarly, Article 10(b) commits all Parties to formulate, 'implement, publish and regularly update national and, where appropriate, *regional programmes* containing measures to mitigate climate change and measures to facilitate adequate adaptation to climate change . . . ' (Kyoto Protocol, 1997, emphasis added).

Given the fact that regional approaches are being – it would appear – increasingly accommodated by the international agreements governing global climate change, along with the broader interest in regional cooperation reported upon in the previous section of this chapter, it is somewhat surprising that regional approaches to climate change policy have received relatively little attention to date, particularly in sub-Saharan Africa. Though there has certainly been some interest demonstrated (and the subsequent section of this chapter provides supporting examples), the following quotation, though made without specific reference to global climate change, nevertheless appears to hold true for the issue I am examining here:

Numerous African states and external aid agencies have adopted sustainable development and regional integration as systematic points of reference. It is, therefore, surprising to find how little recognition has been given to the interplay between these two issue areas, and how reluctant the major international bodies are to address them in an integrated fashion. However, the interdependence of environmental and socio-economic issues and the need for harmonization of efforts at the regional level are becoming increasingly clear.

(Debailleul *et al.*, 1997)

Such reflections motivate the investigation in this chapter of potential regional approaches.

14.4 POTENTIAL REGIONAL APPROACHES

14.4.1 Regional actor

With more than 200 countries now part of the global community and 179 having ratified the Framework Convention on Climate Change, as of 14 June 1999), it is sometimes difficult for individual voices to be heard. As a consequence, countries often group together to articulate and to try to advance a particular position. Perhaps the group known as AOSIS (Alliance of Small Island States) has been most successful in this regard on the climate change issue. Consisting of representatives from 42 small island and coastal countries 'that share common objectives on environmental and sustainable development matters' (AOSIS, 1999), the group has been a vocal proponent of strict emission reduction targets and timetables. The influence they have exerted is probably greater than what these various national leaders could have achieved acting on their own. The European Union is, of course, another example of an international organization (and a regional one, more specifically) that has increasingly exerted influence in international environmental negotiations, the climate change ones included (e.g. Sbragia, 1999). [2]

African leaders have also participated collectively in the negotiations. Prior to the Fourth Conference of the Parties held in 1998 in Buenos Aires, for example, consultations among African ministers of the environment (under the authority of a special consultation of the African Ministerial Conference on Environment) produced a common position for the conference, particularly with regard to the Clean Development Mechanism (UNEP, 1998). At times, moreover, the 'African Group' has spoken with one voice at the conferences themselves (for example, during the debate on CDM at the SBI/SBSTA Joint Session, (Earth Negotiations Bulletin, 1998)). Similar interventions were made during the lead up to the Third Conference of the Parties (Kyoto Protocol, 1997). In August 1997 at the eighth session of the AGBM, five months before the COP3 in Kyoto, the Africa Group of Parties made an historic intervention by declaring the basic principles that should govern the reduction of emissions by both developed and developing countries. This proposal ensured that equity and the principle

of the common but differentiated responsibilities was adhered to by all' (Karimanzira, 1998). Subcontinental groupings have also participated: for example, Southern African Development Community (SADC) opinions were articulated in plenary sessions at the Third Conference of the Parties held in 1997 in Kyoto (Earth Negotiations Bulletin, 1997). [3]

Some, however, argue that even more could be done to develop and to present a common African position (e.g. Mumma, undated). Though there are considerable differences amongst African countries, there are sufficient similarities to be able to find a common negotiating stance on some parts of the global climate change negotiations – the need to ensure that equity remains central to international governance mechanisms is but one example. Indeed, given the challenges associated with having regional organizations serve as 'regional actors' in international environmental negotiations, [4] it makes sense to concentrate upon a limited number of issues that unite regional partners: perhaps the CDM for the continent as a whole and particular concerns for more-restricted sets of African countries are appropriate areas for coordinated policy action (see, for example, the work of ENDA Tiers Monde, 1999).

14.4.2 Regional adaptation strategies

Global climate change, many believe, is under way. Moreover, given the time lag associated with the issue – greenhouse gases emitted today will continue to exercise a positive radiative forcing on climate, thereby increasing temperatures for many years to come – the world is committed to even more climatic changes. Consequently, people around the world need to be planning to – and implementing strategies to – adapt to global climate change. Sub-Saharan Africa is no exception. Indeed, a report on the regional impacts of climate change by the Intergovernmental Panel on Climate Change (IPCC) concluded that, with its widespread poverty and heavy reliance upon agriculture, the 'African continent is particularly vulnerable to the impacts of climate change' (Watson *et al.*, 1997).

Significant changes in forest and rangeland cover; shifts in species distribution, composition and migration patterns; alteration of biome distribution; increased water-stresses; vulnerability of agricultural systems and new challenges to food security; and threatened coastal zones are just some of the specific environmental impacts that would have serious additional environmental, economic and social repercussions in Africa (Zinyowera *et al.*, 1998). Given that environmental change catalyzed by global climate change will occur in areas defined by natural characteristics,

---

[2] Indeed, the language concerning Regional Economic Integration Organizations (noted above) was primarily intended for the benefit of the European Union. Now that it exists, however, presumably any qualifying organization can benefit from it.

[3] For more on the positions of African countries during the international negotiations, see Eleri (1997).

[4] Underdal (1994), for one, argues that the regional organisation must 'have a minimum of internal coherence (unity), autonomy, resources, and also in fact engage in some relevant external activities'.

rather than political borders, it will often make sense to respond in a multinational (or regional) manner. Moreover, as Glantz and Price (1994) argue:

The dislocation of certain kinds of resources as a result of global climate change (e.g. water resources, fish populations, forests and agricultural production) is most likely to occur at the regional level. Thus, the threat, if not the reality, of climate change could provide a catalyst for regional cooperation when countries within a region realize that transboundary changes in the availability (e.g. location or abundance) of their resources require regional cooperation for the rational management of those resources. Interstate cooperation would be necessary for developing regional strategies to cope with climate-change-related resource impacts, as well as ways to deal with changes in bargaining power vis-à-vis various regions.

Of course, adaptation (even if not explicitly labelled as such) coordinated at the regional level is already occurring in sub-Saharan Africa. Institutions aimed at dealing with drought, for example, presently exist. Glantz and Price (1994) also report:

The impacts of climate variability on society and ecosystems for the most part do not respect international borders. Thus, transboundary collaboration on coping with such impacts is usually more effective, as, for example, in the cases of southern African and Sahelian droughts, where regional groups (SADCC (Southern African Development Coordination Community) and CILSS (Permanent Inter-State Committee on Drought Control in the Sahel), respectively) have developed regional tactical and strategic responses, both short-term (e.g. SADCC's pledging conference) and long-term (e.g. CILSS's ongoing environmental programmes).

The continent's various regional water resource management structures provide another set of examples (Debailleul et al., 1997), as do the forest management structures.[5]

The potential impact of global climate change lends even greater encouragement to a regional approach. Bruce (1994), for example, argues that: 'Adaptation to the consequences of global warming, including sea-level rise, is an important priority in many countries, and in some areas a regional approach can be helpful in devising appropriate response strategies. This is especially true, for example, in connection with resolution of water resource conflicts, sea-level rise, and reduction of losses due to natural disasters.' A regional approach to information-gathering and analysis – so as to improve early warning systems to anticipate climate change – is another example. So too is the development of strategic food reserves at the regional level, so as to 'buffer potential increases in the variation of local and national production' (Zinyowera et al., 1998). Encouraging free trade between and among countries is offered by some as another adaptation measure: 'In principle,

free trade allows national surpluses and deficits to be accommodated more efficiently. Thus, supply and price fluctuations are buffered at the global level, widening the potential pool of responses to climate change.' (Zinyowera et al., 1998) Therefore, regional strategies to adapt to the impacts of global climate change are worth continued consideration.

14.4.3 Regional mitigation strategies

Given that sub-Saharan Africa is responsible for such a small relative share of the world's greenhouse gas emissions (perhaps 3%), and that its per capita emissions are well below the global average, incredulity may be the reader's initial reaction to the title of this section: 'Economic growth and poverty alleviation are this continent's main priorities. Why should Africa be considering limitations in its greenhouse gas emissions? Northern countries have caused most of the problem, and they should be working to address it.'

This is true, and it is widely accepted that northern countries 'should take the lead in combating climate change and the adverse effects thereof' (UNFCCC, 1992). The Kyoto Protocol makes this commitment tangible and legally binding, by placing scheduled emission reduction obligations upon northern countries. Efforts must be made to ensure that all countries ratify the Protocol and that these initial targets are met.

This, however, does not – and must not – preclude the study of the potential for climate change mitigation in the South. Indeed, in terms of thinking strategically about southern (developing) countries' participation in efforts to build an international regime to meet the challenge of global climate change, such investigations are critical for two reasons (Rowlands, 1999).

The first reason has a 'short-term' time-horizon. It appears probable that northern countries will use the CDM to help them meet their own reduction targets. As outlined in Article 12 of the Kyoto Protocol, CDM projects are not only meant to help northern countries meet their emission reduction targets, but also designed to assist developing countries in achieving sustainable development. Therefore, CDM (and the associated international investment) should also generate a variety of local benefits. These conceivably include technology transfer and cooperation, employment opportunities, local environmental benefits (for example, reduced air and water pollution), improved national infrastructure, and capacity-building in both the public and private sectors.

As previously noted, the details surrounding implementation of the CDM are still being examined and negotiated, and many important decisions have yet to be made. Nevertheless, given that a system may soon be in place, it makes sense for developing countries to be examining possible CDM projects now.

The second reason for mitigation studies has a ' long-term' time-horizon. Some posit that southern countries will have to cap their

---

[5] Regional efforts at forest management (for example, activities catalysed by the Yaounde Declaration) might also be considered 'regional mitigation strategies' (compare with, for example, Onugu, 1999).

emissions at some point in the future. To successfully meet the challenge of global warming, we need 'stabilization of greenhouse gas concentrations in the atmosphere at a level that would prevent dangerous anthropogenic interference with the climate system' (UNFCCC, 1992, Article 2). Scientists believe that, to achieve this *stabilization in concentrations*, we need a 60% *reduction in emissions*.

Of course, that does not mean that each country reduces by 60%: that would be unfair, because it would entrench present inequalities between North and South. But even if each country were given an 'equal per capita emission entitlement' (something which is favoured by many southern representatives), every country would reach its limit at some point.

These two drivers mean that climate change mitigation in the South will inevitably become an important issue, and probably sooner rather than later. Given this, engagement with the issue is crucial: this will not only allow those in the South to set their own priorities and seek opportunities within the emerging arrangements, but also allow them to push the discussions towards outcomes that they deem to be preferable. [6]

Hydropower development

Africa has extensive potential for hydro-generated electricity – at present, only about 5% of available sites are developed. And though there are certainly challenges associated with the sustainable development of this resource (see, for example, IUCN, 1997; World Commission on Dams, 2000), hydro-generated electricity has the potential to make a contribution to global climate change mitigation efforts. [7] A regional approach also has a role: much of the continent's potential is located in countries with relatively low levels of energy demand. Consequently, in order to exploit more fully the resource, international trade in electricity is needed.

Southern Africa offers perhaps the best example for hydropower development as a regional climate change mitigation option. In the northern part of the region, there exists the Inga site in the Democratic Republic of the Congo (DRC), which has an estimated generating potential of approximately 36,000 to 100,000 MW, far greater than the DRC's recent peak demand of 600 MW (SAD-ELEC and MEPC, 1996). Angola, Mozambique and Zambia offer additional prospects for hydropower development. Indeed, estimates place the total medium-term potential in southern Africa at between 70,000 and 135,000 MW (Batidzirai *et al.*, 1998). In the southern part of the region, meanwhile, there exists considerable demand (peak demand of 32,107 MW in South Africa during the first half of 1999 (ESKOM, 1999)), much of which is presently met by that country's extensive coal resources. Regional cooperation could clearly serve to reduce greenhouse gas emissions (Nziramasanga *et al.*, 1998).

Additional benefits of increased regional cooperation on hydropower development could include lower electricity reserve requirements, a catalyst for regional economic development (by virtue of redressing trade imbalances between South Africa and its regional neighbours), and better local air quality. If not properly managed, however, it could also bring problems: dam construction and operation can cause various kinds of damage and the social and economic effects upon, in particular, the coal industry (as a result of less coal use) would need to be addressed. Nevertheless, the potential for the benefits to more than outweigh the problems has encouraged, for one, the region's utilities: they are working together to construct a Southern African Power Pool. Moreover, their efforts have been largely motivated by non-climate factors. Introduction of concepts like the CDM into the discussions could give their initiative additional, tangible (for example, monetary) support.

Though regional cooperation in southern African appears to offer the potential for the largest greenhouse gas emission reductions, other regions in sub-Saharan African should also consider this kind of trade in electricity. In the Horn of Africa, Ethiopia has considerable hydropower potential – about 30,000 MW, much of which is potentially 'low-cost' and only '1.5% or less [of which] . . . has been exploited so far' (Mariam, 1998a). Among its neighbours, Sudan offers one outlet for the potential power (as does, of course, Egypt). Sudan, however, is also considering how it might develop further its hydropower resources. What is certainly clear is that the two countries (and indeed, all riparian countries along the Blue Nile and the White Nile) should cooperate to maximize the rivers' potential, while minimizing the damaging consequences of hydropower development. Again, this would suggest a continued regional approach to this development and climate change priority. The existing Undugu (Swahili for 'brotherhood') framework (which brings together the 10 Nile riparian countries) may provide an existing institutional forum in which to pursue this (Mariam, 1998b).

Uganda also has the potential to be a significant source of hydroelectric power in the future. The country supplies modest amounts to Kenya (30 MW), Tanzania and Rwanda (5 MW each) ('Tanzania Investigates Power Privatisation' 1999). Given that plans for an 'East African Power Pool' were well-advanced at the end of 1999, this may well increase in the future. Nonetheless, an analysis considering three, four or five countries' electricity demands and supplies together could well produce plans that are more secure,

---

[6] Bruce (1994) has argued that: 'It would be useful to encourage greater efforts by other UN regional economic commissions to foster greenhouse gas limitation strategies within their regions as part of an overall effort at sustainable development planning at the regional level.'

[7] Hydropower is not necessarily free of any greenhouse gas emissions: construction of the dam generates emissions and, depending upon the characteristics of the area that was flooded to create the dam's reservoir, operation of the dam can also generate emissions (see Gagnon and van de Vate, 1997).

more economical and more beneficial than individual countries' plans (for the general case, see Grollman, 1997).

In West Africa, meanwhile, Senegal – a country that has a thermal-dominated electricity generation profile – could look to its neighbours for hydroelectricity. Sarfoh (1990), for example, suggests that Guinea, with hydropower 'potential warranting investigation' of 2,200 MW could be a supplier in the region; Mauritania (with 900 MW) could be another. Indeed, the experience of the *Organization Pour La Mise en Valeur de la Valleé du Senegal* (OMVS, which groups together Mali, Mauritania and Senegal), which built the Diama and Manantali dams, reveals that some kind of regional cooperation has already taken place.

Additionally, though Ghana was extremely hard-hit by drought in 1997/98 (which meant that generation at the 912 MW-rated Akosombo facility was only 300 MW), the country is still considering increased used of hydropower (though within a much more diversified portfolio of energy supply resources). The Bui Hydroelectric Project, for one, could not only be used to help meet domestic energy demand, but could also generate electricity for export to Burkina Faso, Mali and Côte d'Ivoire (EIA, 1999b). [8]

In Central Africa, meanwhile, Cameroon may be interconnected with Chad, so that the latter does not have to rely upon its oil reserves for power generation. 'This interconnection would make it possible to reduce the number of outages in Chad, which forces people to use generator sets at a cost of CFA Fr150 per thermally produced kilowatt-hour . . . while the price per kilowatt-hour produced by a hydroelectric power station would be about CFA Fr40' ('Chad Looks to Local Oil for Cheaper Power', 1999). Reduced use of diesel, moreover, means reduced greenhouse gas emissions.

Finally – and coming 'full circle' in this section – the sheer size of the DRC's potential, along with its central location in the continent, means that it could possibly provide electricity to virtually any part of Africa. Indeed, analyses of the impact of new transmission lines have been examined on routes between the DRC and South Africa, as well as between the DRC and Egypt (African Business, 1995). The notion of a 'Cape to Cairo' grid is certainly on some people's minds. [9]

Natural gas utilization

Sub-Saharan Africa has significant natural gas reserves, the vast majority of which are found in Nigeria (BP Amoco, 1999). Natural gas, when burnt, has a lower greenhouse gas emission rate (per unit of energy produced) than oil, coal or wood. Consequently, the substitution of higher-carbon fuels with natural gas offers a potential means to mitigate global climate change. Moreover, given that natural gas burns more cleanly on many other measures as well – for example, it emits fewer nitrogen oxides and particulates (Parfomak, 1997), which contribute to smog – increased use can also generate a range of local-specific benefits. These include improved air quality, which, in turn, would lead to higher standards of health, increased agricultural yields and greater productivity more broadly.

One of the major drawbacks of natural gas use, however, is that it is relatively expensive to transport. Consequently, unlike oil (which has a global market), natural gas markets tend to be regionally based. The fact that the greatest natural gas trading (at the international level) is conducted between – in order of volume – Canada and the United States; Russia and Germany; and Indonesia and Japan (BP Amoco, 1999) is further testament to this. Therefore, the challenge to those wanting to increase the use of natural gas is to uncover demand centres that are situated close to substantial natural gas reserves.

To determine whether such 'coupling' exists in sub-Saharan Africa, it makes sense to begin with the area's largest natural gas producer: Nigeria. Nigeria, with its 3.5 trillion cubic metres of natural gas reserves (BP Amoco, 1999), has great potential as a natural gas producer and exporter. At this time, however, the vast majority of natural gas is flared, given the lack of market for the product. Climate change mitigation, however, could be the force that encourages greater demand for the product. [10] Greenhouse gas emission reductions would not be achieved by displacing the primary source of electricity in the region (which is, outside of Nigeria, hydropower), but instead to electrify new areas that hitherto had been reliant upon biomass or kerosene for their energy needs. It could also displace diesel generators – not only those that are used where grid-connected hydropower is unavailable, but also those that are used in order to act as a back-up to centrally supplied hydropower in case of power outages. In this way, natural gas, by diversifying the sources of supply in the region, would also be improving the performance of the region's electricity supply system. Job creation, and economic stimulation

---

[8] Ghana, of course, already has experience with regional electricity sales: traditionally (since the early 1970s), about 50 MW (4%) of the total electricity generated from the Akosombo plant in Ghana has been exported to Togo and Benin. Nigeria has also exported electricity from the Kainji hydropower projects (Sarfoh, 1990).

[9] In addition to examination of 'generation' as a regional mitigation strategy, 'transmission' should also be investigated. Given the levels of losses in some parts of the transmission systems throughout Africa (sometimes exceeding 30%, when universal standards are below 8% (Zinyowera et al, 1998)), improving the performance of some existing interconnectors and international transmission networks, particularly when in conjunction with any kind of thermal facility, could also be presented as a regional mitigation strategy.

[10] Although the focus in this chapter is upon *regional* approaches, the reader should recognise that *national* mitigation actions are by no means precluded (for virtually any of the activities investigated in this chapter). Indeed, Nigerian natural gas could be used to displace biomass fuels within Nigeria, the unsustainable use of which contributes to global climate change. For a more general discussion, see Ibitoye and Akinbami (1999).

n general, are other touted benefits of its increased use (Miller, 1998).

Plans to export Nigerian natural gas to regional neighbours were, at the end of 1999, well-advanced. Officials at Chevron have taken the lead in pushing for the so-called West African Gas Pipeline (WAGP). The WAGP would move natural gas from Nigeria through a 1,000-kilometre offshore pipeline to Ghana, Benin and Togo. Upon arrival, the natural gas would be used to fuel power stations (one of which – in Ghana – is already operating). Up to 180 million cubic feet of gas could be moved daily (Alexander's Gas and Oil Connections, 1999). As of the end of 1999, a feasibility study of the project was under way, and supporters hoped that the pipeline would be operational by 2002. [11]

Côte d'Ivoire, which also has significant natural gas reserves, is also considering the regional dimensions of natural gas-fired electricity generation: two 150 MW power facilities were, at the end of 1999, being completed. These will 'address the shortage of power generation capacity in Côte d'Ivoire and will also have a regional impact by enabling electricity exports to neighbouring countries such as Ghana, Togo, Benin, and Burkina Faso' (Country Updates: Côte d'Ivoire, 1999). Of course, these projects were deemed to be economically viable without any additional assistance by virtue of their greenhouse gas emission-reduction potential; by definition, therefore, they are not potential CDM projects. Nevertheless, Côte d'Ivoire's development of facilities for possible electricity export suggests that energy cooperation is clearly on the regional agenda. Additional projects – that are just beyond a state of economic viability at this moment – could be supported by the CDM and consequently become attainable.

Though West Africa dominates sub-Saharan Africa's natural gas scene, other regional climate change mitigation projects involving this resource are conceivable. In southern Africa, for example, natural gas from existing fields in Namibia (Kudu) and Mozambique (Pande) could be used to displace coal as an energy source in South Africa. Though discussions about each were under way, increased linkage to climate change debates could, again, mobilize additional resources (e.g. Nziramasanga et al., 1998). Moreover, many report that Angola has vast natural gas reserves (larger than Nigeria's on some measures (e.g. EIA, 1999a, which quotes the Oil and Gas Journal and World Oil)). If exploited (instead of being flared, or more emphatically, in-

stead of being vented), this could reduce regional greenhouse gas emissions. [12]

Transportation coordination

Although the level of intra-African trade is not particularly high, it appears to be growing. Moreover, as reported above, there exists a range of regional organizations in Africa that are intent at increasing further the value of goods that travel between neighbours. Therefore, we should expect to see greater use of transportation within sub-Saharan Africa – much of which, moreover, will be fossil fuel powered.

Therefore, regionally coordinated efforts to improve the efficiency of transportation, or to decarbonize transportation, could be presented as climate change mitigation options (Zhou et al., 1998). This might include something as modest as road repairs, in the form of paving an unpaved road or repairing a deteriorating road (for it takes less fuel to travel the same distance on a higher-quality road). Indeed, as estimated by Heggie (1995), who examined sub-Saharan Africa as a whole, regular maintenance of roads makes economic sense, irrespective of climate change benefits: 'When a road is not maintained – and is allowed to deteriorate from good to poor condition – each dollar saved on road maintenance increases vehicle operating costs by $2 to $3. Far from saving money, cutting back on road maintenance increases the cost of road transport and raises the net costs to the economy as a whole'. [13]

Alternatively, efforts could be made to move goods off of roads and onto rails. Even if the engines were to continue to be powered by diesel, there would still be significant energy (and hence greenhouse gas emission) savings: reductions of the order of 30% may be realized (Michaelis, 1996). If, however, some kind of carbon-free source were used to generate electricity, which, in turn, was used to move the trains along the tracks, then the reductions would be even higher. A plan to electrify international railways, therefore, could be conceived of as a regional climate change mitigation option. Increased use of sea routes for neighbouring coastal

---

[11] Officials from Chevron have already linked the development to climate change mitigation: 'In short, CDM [Clean Development Mechanism] could provide additional incentives both to companies and host countries to build projects that eliminate flaring. Clearly, projects like the West African Gas Pipeline should be eligible for the emission credits proposed in the CDM. Not only will these projects reduce greenhouse emissions, they'll reduce those emissions at less than half of the estimated cost of alternative carbon-mitigation projects in Europe and the United States.' (Miller, 1998).

[12] Cameroon, Equatorial Guinea, Gabon – each with significant natural gas reserves – might also be able to supply the South African market. The distance from each of these countries, however, might mean that the costs of transportation are prohibitive (and therefore, the option, as a mitigation activity, would be unattractive). Nevertheless, it is probably worth investigating: even if not to transport the natural gas itself, then to transport the electricity generated by power stations in the natural gas-producing countries. (Though losses can be greater when energy is transported through electricity lines, rather than through well-maintained natural gas pipelines, the impact upon global warming is more muted: methane is a particularly potent greenhouse gas, on a molecule-for-molecule basis.)

[13] Of course, the presence of 'better roads' in the region may generate a large increase in transportation, which, in turn, might increase total greenhouse gas emissions. As with all options identified here, more research is certainly warranted.

countries and pipelines to move petroleum and petroleum products to inland countries are two additional possibilities.

The potential impact of these kinds of mitigation options will largely be determined by the anticipated levels (and types) of trade between neighbouring countries. Most intra-African trade is conducted among the countries of the continent's major regional integration organizations (e.g. Lyakurwa *et al.,*1997). Hence it makes sense to begin any examination there. Rwegasira (1997), for one, reports that:

In the Great Lakes Corridor, for instance, a 1995 World Bank/EC study has estimated that potential savings from similar changes [improving services and facilitating trade procedures] in the eight land-locked countries of the area may range from 14–33% of typical values of the commodities being carried, because of the high proportion of transportation costs in the c.i.f. value of imports. Thus, the deepening of regional cooperation in this area is indeed important. Improvements and strategic decisions are also called for in African railways which have comparatively high overheads and low productivity.

Though Rwegasira makes no explicit link to global climate change, it could clearly provide a catalyst to regional transportation improvements.[14]

Development of regional markets

The discussion of mitigation options to this point has focused upon tangible 'projects': power stations, pipelines, transportation corridors, etc. In this section, I direct the reader's attention to, alternatively, *policies* – that is, broader efforts to establish structures that will, in turn, encourage activity to reduce greenhouse gas emissions. To date, such mitigation policies (as opposed to projects) have not received as much attention in discussions about global climate change. This is primarily because it is difficult to quantify associated costs and the amount of greenhouse gas abated. It is one thing to cost, for example, a pipeline, but it is altogether another thing to put a price on the establishment of, for example, a free-trade zone. Although economists do attempt to do so, debates about methodology and operationalization still abound. [15] This, in and of itself, however, should not preclude further consideration. For this reason, I consider it here (following Zhou *et al.*, 1998).

More specifically, I introduce the establishment of a regional free trade area as a climate change mitigation option. Justification is provided by the fact that this could serve to reduce the cost of many 'climate-beneficial' goods and services –

for example, renewable energy technologies or energy efficiency technologies (e.g. Zhou, 1998). By creating a free trade area, not only might production inputs be cheaper (because regional suppliers can supply goods at lower costs), but economies of scale in terms of production runs could also be captured because of the increased size of the regional market.[16] In the end, therefore, an existing producer may well be able to supply the technology more cheaply, and/or new producers may have been encouraged to emerge. Consider just a couple of examples.

Karekezi and Ranja (1997), for example, argue that import duties are an important contributor to the high price of renewable energy technologies (RETs) in Africa, which, in turn, reduce their use. 'Cumulative duty (import duties plus various surcharges on components) on RETs in Malawi is estimated to be as high as 75%. In Zambia, a solar lighting system with a value of US$934 attracted sales tax and import duty amounting to about 70% of the price . . .' (Karekezi and Ranja, 1997). Though effective implementation of a regional trade agreement will not necessarily reduce the selling price of technologies that are imported from beyond the region, it may well encourage development of the capacity within the area. Sarfoh (1990), meanwhile, reports that the 'Chinese and Indians have reached a level that permits them to manufacture most of the equipment needed in their small hydropower development programs, thus enabling them to cut down cost considerably.' He suggests that West Africans could do the same to exploit their resources. [17]

The point is simply that coordinated regional action – in this case, the establishment of a free trade area – may cause the net benefit or cost of that particular mitigation option to shift, as compared with the result of action solely at the national level. Moreover, as already noted, the establishment of a regional free trade area is clearly already a part of many regional organizations' mandates in sub-Saharan Africa. Consequently, what is presented as a 'regional climate change mitigation option' here would appear to fit well with the continent's own development priorities. [18]

Strategically, therefore, it would make sense for officials from Africa's regional organizations to consider how efforts to promote regional integration could be presented as a means to address global climate change. Though the net 'costs' of establishing a regional trade area could well, in the long run, be negative (indeed, this is one of the forces motivating its consideration),

---

[14] Diouf (1990) also notes that: ' . . . the same is true of moves to connect rail systems, especially where track gauges are different, as between the French- and English-speaking countries of West Africa.' Again, a link to the climate change debate could prove beneficial.

[15] This is not meant to suggest that there are no debates surrounding the costing of individual projects! There certainly are.

[16] Regional markets may encourage new technologies: technologies need 'to reach sufficient volume to lower costs to become competitive' (Ishitani and Johansson, 1996).

[17] Here, I am assuming that the production and use of climate-friendly goods would increase as a result of development of a free trade area; similarly, the extent to which the production and use of climate-unfriendly goods increased would also have to be examined.

[18] Additionally, Silveira (1994) argues that: 'Regional organizations may play an essential role in choosing and coordinating the dissemination of technologies that can be beneficial at a larger scale in the region.'

there would inevitably be some kind of start-up costs associated with it: perhaps the tangible costs of supporting organizational development of the free trade area (for example, the establishment of a secretariat or the training of customs officers), as well as the 'side-payments' that might be necessary to keep national treasuries happy in the short-term (perhaps financial compensation for lost revenue from tariffs). These kinds of activities could be presented as climate change mitigation activities.

14.4.4 Regional knowledge development

A regional approach to knowledge could also prove beneficial. This could include, for example, basic education, primary research (including data collection) or innovation. No matter what the focus, there are particular advantages for sub-Saharan Africa to have at least some kind of regional approach to the formation, application and dissemination of knowledge. As in many sectors, 'economies of scale' can be secured: 'Regional cooperation in certain branches of education is the only way of achieving quality instruction at least costs. The present proliferation of national universities, regarded as prestige institutions, is both costly and inconsistent with the desired human resource development.' (Diouf, 1990). Additionally, however, greater understanding of particular phenomena may only be achieved through regional cooperation. Natural processes (either those affecting global climate change, or those potentially affected by global climate change) do not always respect national borders. At a minimum, therefore, data collection (in a standardized manner) in two or more neighbouring countries will be required. To provide specific substance to this, Karekezi and Ranja (1997) argue that: 'Limited access to information on the region's resource base is a major barrier to wider use of RETs and a major cause of contradictory and inconsistent information on RETs.' They argue for the establishment of 'a regional data bank for RETs to enable and to facilitate international comparability and projection'.

Innovation could also benefit from increased regional cooperation. It has been argued that 'common actions in this area might further help to share the costs, risks and benefits associated with experimentation, and provide opportunities for additional exchanges of ideas and experience which would facilitate governments and others in identifying and evaluating key opportunities'. (Annex I Expert Group, 1996). While looking at Africa more specifically, Mytelka (1994) agrees:

Given current financial constraints, transfer of technology from abroad, while vital, cannot fulfil all of Africa's needs. Moreover, it is a costly process which generally requires recurrent expenditures for the import of capital and intermediate goods, management skills, maintenance services and technical know-how that Africa can ill afford at present. It has thus become imperative for the enterprise

sector in Africa to strengthen its ability to solve its own problems and to overcome bottlenecks in production . . . These resources can and must come through a process of collaborative research and development involving other domestic actors including supplier firms, university faculties, engineering consultancy firms and research institutions. The critical density required for such problem-solving innovation is, however, rarely present in any single African country. Regional networking is thus critical here.

There have been relevant regional efforts in this regard to date in sub-Saharan Africa. The West African Economic Community (CEAO), for example, created the Regional Centre for Solar Energy (*Centre Regional d'Énergie Solaire*) in Bamako, Mali (Bundu, 1997). A number of regional interconnection projects have been studied by the *Union des Producteurs et Distributeurs d'Energie en Afrique* (UPDEA) (Girod and Percebois, 1998). Moreover, various regional organizations (for example, the Mano River Union) have supported the training of customs procedures and practices, to help to facilitate their integration efforts (Diouf, 1990). The activities of the Eastern and Southern Africa Management Institute are additional examples, as is the System for Analysis, Research and Training (START) programme, which has a Pan-African component. The point here is simply that location of their situation within a global climate change context could provide additional support for their activities.

14.5 CONCLUSIONS

The purpose of this chapter has been to examine the potential for different regional approaches in response to global climate change challenges in sub-Saharan Africa. After locating the investigation within the broader discussions surrounding the new regionalism, a range of different regional approaches to global climate change were presented: these revolved around the major themes of regional actor, regional adaptation strategies, regional mitigation strategies and regional knowledge development. In some instances, particular candidates for regional action were identified; in other instances, less-specific possibilities were flagged. Of course, not all kinds of regional activity on global climate change are appropriate for sub-Saharan Africa; for example, while inter-country systems of tradable permits are currently being investigated by European (and some 'Umbrella Group') countries, they do not, at this time, make sense for African countries to consider. Instead, the focus of African approaches should be upon those activities that have a good potential to generate both climate and development benefits. By presenting a range of possibilities, the discussion in this chapter was intended to stimulate interest in, and further research on, regional approaches to global climate change policy.

ACKNOWLEDGEMENTS

I would like to thank Randall Spalding-Fecher for his helpful comments on an earlier draft of this chapter. John Turkson also provided valuable comments on an earlier draft. John was killed in an airplane accident in January 2000. John and I worked together at the UNEP Collaborating Centre on Energy and Environment in Roskilde, Denmark. He was a wonderful friend and a great colleague. I would like to dedicate this chapter to him. Notwithstanding the assistance received, I remain responsible for the contents of the chapter.

REFERENCES

African Business (1995) The Great Inga Dream, *African Business.* No. 198, 28 April, London, IC Publications.

Alexander's Gas and Oil Connections (1999) West African gas pipeline project beneficial at all sides. Vol. 3, Issue 2, 2 February, http://www.gasandoil.com/goc/company/cna90659.htm accessed 1 July 1999.

Annex I Expert Group (1996) *Sustainable Transport Policies: CO_2 Emissions from Road Vehicles.* Paris, IEA/OECD.

AOSIS (1999) http://www.aosis.org, accessed 1 July 1999.

Asante, S. K .B. (1997) *Regionalism and Africa's Development: Expectations, Reality and Challenges.* New York, St. Martin's Press.

Batidzirai, B., Nziramasanga, N. and Rowlands, I. H. (1998) Regional electricity demand and supply: developing the baseline. In I. H. Rowlands (Ed.), *Climate Change Cooperation in Southern Africa,* pp. 76–101. London, Earthscan.

BP Amoco (1999) *Statistical Review of World Energy.* London, BP Amoco.

Bruce, J. P. (1994) The climate change issue: policy aspects. In M. H. Glantz (Ed.), *The Role of Regional Organizations in the Context of Climate Change,* pp. 68–73. New York, Springer-Verlag.

Bundu, A. (1997) ECOWAS and the future of regional integration in West Africa. In R. Lavergne (Ed.), *Regional Integration and Cooperation in West Africa: A Multidimensional Perspective,* pp. 29–47. Trenton, NJ, Africa World Press, and Ottawa, Canada, International Development Research Centre.

Chad looks to local oil for cheaper power (1999) *African Energy,* **1,** No. 1.

Country updates: Côte d'Ivoire (1999) *African Energy,* **1,** No. 1.

Debailleul, G., Grenon, E., Kalala, M-M. and Vuillet, A. (1997) The regional dimension of environmental management. In R. Lavergne (Ed.), *Regional Integration and Cooperation in West Africa: A Multidimensional Perspective,* pp. 279–301. Trenton, NJ, Africa World Press, and Ottawa, Canada, International Development Research Centre.

De Melo, J. and Panagariya, A. (1993) Introduction. In J. de Melo and A. Panagariya (Eds.), *New Dimensions in Regional Integration,* pp. 3–21. Cambridge, Cambridge University Press.

Diouf, M. (1990) Evaluation of West African experiments in economic integration. In *The Long-Term Perspective Study of Sub-Saharan Africa, Volume 4. Proceedings of a Workshop on Regional Integration and Cooperation,* pp. 21–26. Washington, DC, World Bank.

Earth Negotiations Bulletin (1997) *Report of the Third Conference of the Parties to the United Nations Framework Convention on Climate Change: 1–11 December 1997.* Vol. 12, No. 76, 13 December. Winnipeg, Canada, International Institute for Sustainable Development.

(1998) *Report of the Fourth Conference of the Parties to the UN Framework Convention on Climate Change.* Vol. 12, No. 97, 16 November. Winnipeg, Canada, International Institute for Sustainable Development.

EIA (1999a) *International Energy Annual.* Washington, DC, Energy Information Administration. http://www.eia.doe.gov/emeu/iea/ accessed 1 July 1999.

(1999b) *Ghana.* Washington, DC, Energy Information Administration. http:www.eia.doe.gov/emeu/cabs/ghana.html accessed 1 July 1999.

Eleri, E. O. (1997) Africa and climate change. In G. Fermann (Ed.) *International Politics of Climate Change: Key Issues and Critical Actors,* pp. 265–284. Oslo, Scandinavian University Press.

ENDA Tiers Monde (1999) *From Joint Implementation to the Clean Development Mechanism: Should African Positions Change After the Kyoto Protocol?* ENDA Tiers Monde, April, Dakar, Senegal.

ESKOM (1999) *ESIVIEW.* http://www.esi-view.org.za accessed 10 July 1999.

Gagnon, L. and van de Vate, J. F. (1997) Greenhouse gas emission from hydropower: the state of research in 1996. *Energy Policy,* **25,** No. 1, 7–13.

Girod, J. and Percebois, J. (1998) Reforms in sub-Saharan Africa's power industries. *Energy Policy,* **26,** No. 1, January 1998, 21–32.

Glantz, M. H. and Price, M. F. (1994) Summary of discussion sessions. In M. H. Glantz (Ed.), *The Role of Regional Organizations in the Context of Climate Change,* pp. 33–54. New York, Springer-Verlag.

Global Coalition for Africa (undated) http://www.gca-cma.org/ eabout.htm accessed 10 July 1999.

Grollman, N. (1997) The energy subregion as a basis for greenhouse policy. *Energy Policy,* **25,** No. 4, March 1997, 459–467.

Heggie, I. G. (1995) *Management and Financing of Road: An Agenda for Reform.* World Bank Technical Paper No. 275, Africa Technical Series, Washington, DC.

Ibitoye, F. I. and Akinbami, J-F. K (1999) Strategies for implementation of CO_2-mitigation options in Nigeria's energy sector. *Applied Energy,* **63,** 1–16.

Ishitani, H. and Johansson, T. B. (1996) Energy supply mitigation options. In R. T. Watson, M. C. Zinyowera and R. H. Moss (Eds.), *Climate Change 1995: Impacts, Adaptations and Mitigation of Climate Change: Scientific-Technical Analyses,* pp. 587–647. Cambridge, Cambridge University Press.

IUCN (1997) *Large Dams: Learning from the Past, Looking at the*

Future. Workshop Proceedings. IUCN, The World Conservation Union, Gland and the World Bank Group, Washington, DC.

Karekezi, S. and Ranja, T. (1997) *Renewable Energy Technologies in Africa.* London, Zed Books in association with AFREPREN and SEI.

Karimanzira, R. P. (1998) *The Road to Buenos Aires.* http://www.globesa.org/KARIMANZIRA.html accessed 1 May 1999.

Kyoto Protocol (1997) *Kyoto Protocol to the United Nations Framework Convention on Climate Change.* FCCC/CP/1997/L.7/Add.1, 10 December.

Lyakurwa, W., McKay, A., Ng'eno, N. and Kennes, W. (1997) Regional integration in sub-saharan Africa: a review of experiences and issues. In A. Oyejide, I. Elbadawi and P. Collier (Eds.), *Regional Integration and Trade Liberalization in Sub-Saharan Africa, Volume 1: Framework, Issues and Methodological Perspectives,* pp. 159–209. New York, St. Martin's Press.

Mariam, H. G. (1998a) assisted by Bogale, W. The status and performance of the power sector in Ethiopia. In L. Khalema-Redeby, H. Mariam, A. Mbewe and B. Ramaseli (Eds.), *Planning and Management in the African Power Sector,* pp. 143–187. New York, Zed Books, in association with AFREPREN.

(1998b) assisted by Bogale, W. Future plans and prospects of the power sector. In L. Khalema-Redeby, H. Mariam, A. Mbewe and B. Ramaseli (Eds.), *Planning and Management in the African Power Sector,* pp. 204–233. New York, Zed Books, in association with AFREPREN.

Michaelis, L. (1996) Mitigation options in the transportation sector. In R. T. Watson M. C. Zinyowera and R. H. Moss (Eds.), *Climate Change 1995: Impacts, Adaptations and Mitigation of Climate Change: Scientific-Technical Analyses,* pp. 679–712. Cambridge, Cambridge University Press.

Miller, C. (1998) The West African Gas Pipeline Project: three short steps to long-term energy. Speech by Project Manager, West African Gas Pipeline Project, Chevron, Second Forum on World Bank Group's Role in the Oil and Gas Sector, Washington, DC, December 2. http://www.chevron.com/newsvs/pressrel/1998/98-12-09.html accessed 1 July 1999.

Mumma, A. (undated) The Poverty of Africa's Position at the Climate Change Convention Negotiations. http://www.meteo.go.ke/africap.html accessed 10 July 1999.

Mwase, N. (1994) Economic integration for development in eastern and southern Africa: assessment and prospects. *IDS Bulletin,* **25,** No. 3, 31–39.

Mytelka, L. K. (1994) Beyond trade: networking for innovation through south-south cooperation. In L. K. Mytelka (Ed.), *South-South Cooperation in a Global Perspective,* pp. 263–271. Paris, OECD.

Nziramasanga, N., Batidzirai, B. and Rowlands, I. H. (1998) Regional electricity mitigation options. In I. H. Rowlands (Ed.), *Climate Change Cooperation in Southern Africa,* pp. 102–119. London, Earthscan.

Onugu, A. J. (1999) Linking climate change, wetlands, biodiversity conservation and the private sector: new energy and development options for West Africa under the CDM. Paper presented at a workshop on Wetlands and the Private Sector, San Jose, Costa Rica, 7–9 May.

Parfomak, P. W. (1997) Falling generation costs, environmental externalities and the economics of electricity conservation. *Energy Policy,* **25,** No. 10, 845–860.

Rowlands, I. H. (1999) Climate change mitigation in Southern Africa. *GLOBE Southern Africa.* Global Legislators' Organisation for a Balanced Environment Southern Africa, Cape Town, South Africa.

Rwegasira, D. G. (1997) Economic cooperation and integration in Africa: experiences, challenges and opportunities. In *Toward Autonomous Development in Africa: Conference Proceedings of the Roundtable on the Emerging African Development Agenda, April 15–16, 1996,* pp. 93–112. Ottawa, Canada, North-South Institute.

SAD-ELEC and MEPC (1996) (Southern African Development through Electricity; Minerals and Energy Policy Research Centre) *Electricity in Southern Africa: Investment Opportunities in an Emerging Regional Market.* London, Financial Times Energy Publishing.

Sarfoh, J. A. (1990) *Hydropower Development in West Africa: A Study in Resource Development.* New York, Peter Lang.

Sbragia, A. M. (1999) The changing role of the European Union in international environmental politics: institution building and the politics of climate change. *Environment and Planning C: Government and Policy,* **17,** No. 1, February, 53–68.

Silveira, S. (Ed.) (1994) *African Voices on Climate Change: Policy Concerns and Potentials.* Stockholm, Stockholm Environment Institute.

Söderbaum, F. (1996) *Handbook of Regional Organizations in Africa*: Uppsala, Sweden, Nordiska Afrikainstitutet.

Tanzania investigates power privatisation (1999) *African Energy.,* **1,** No. 2. Roosevelt Park, South Africa, Resource Publications (PTY) Ltd.

Tewungwa, S. (1994) Regional organizations and environmental change. In M. H. Glantz (Ed.), *The Role of Regional Organizations in the Context of Climate Change,* pp. 141–149. New York, Springer-Verlag.

Thompson, C. B. (1992) African initiatives for development: the practice of regional economic cooperation in Southern Africa. *J. International Affairs,* **46,** No. 1, Summer, 125–144.

Underdal, A. (1994) The roles of IGOs in international environmental management: arena or actor? In M. H. Glantz (Ed.), *The Role of Regional Organizations in the Context of Climate Change,* pp. 153–159. New York, Springer-Verlag.

UNEP (1998) *African Environment Ministers Agree Common Position on the Kyoto Protocol's Clean Development Mechanism.* Nairobi, News Release, 1998/111, 23 October.

UNFCCC (1992). United Nations Framework Convention on Climate Change. Reprinted in *International Legal Materials,* **31,** 1992, 849–873.

Vale, P. and Matlosa, K. (1996) *Beyond and Below: The Future of the Nation-State in Southern Africa.* Centre for Southern African Studies, No. 53, South Africa.

Watson, R. T., Zinyowera, M. C. and Moss, R. H. (1997) *The Regional Impacts of Climate Change: An Assessment of Vulnerability; Summary for Policymakers.* Intergovernmental Panel on Climate Change, November, Geneva, Switzerland.

Weeks, J. (1996) Regional Cooperation and Southern African Development. *J. Southern African Studies*, **22**, No. 11, March, 99–117.

World Commission on Dams (2000) *Dams and Development: A New Framework for Decision-Making.* London, Earthscan.

Zhou, P. (1998) *Energy Efficiency and Climate Change: Opportunities and Prerequisites for Southern Africa.* http://www.globesa.org/ZHOU.html accessed 1 May 1999.

Zhou, P., Rowlands, I. H. and Turkson, J. K. (1998) Other regional mitigation options. In Rowlands I. H. (Ed.), *Climate Change Cooperation in Southern Africa*, pp.139–161. London, Earthscan.

Zinyowera, M. C., Jallow, B. P., Maya, R. S. and Okoth-Ogendo, H. W. O. (1998) Africa. In R. T. Watson M . C. Zinyowera and R. H. Moss (Eds.), *The Regional Impacts of Climate Change: An Assessment of Vulnerability*, pp. 29–84. New York, Cambridge University Press.

15 Energy for development: solar home systems in Africa and global carbon emissions

RICHARD D. DUKE[1,*] AND DANIEL M. KAMMEN[2]

[1]*Princeton University, USA*
[2]*University of California, Berkeley, USA*

Keywords

Renewables; market transformation; photovoltaics; solar home systems; buydown

Abstract

A growing number of rural African households are using small solar home systems (SHS) to obtain better access to lighting, television and radio. Various non-governmental organizations, multilateral institutions and international aid agencies have catalysed these markets, partially motivated by a desire to reduce global carbon emissions. This chapter assesses the carbon mitigation potential of African SHS markets, concluding that direct carbon displacement will be limited. Indirect benefits from helping the global photovoltaics (PV) industry scale up production and bring down costs via the manufacturing experience curve will be larger, but still trivial relative to grid-connected markets. Nonetheless, by 2025 SHS could provide cost-effective basic electricity to a substantial share of rural households, and grid-connected PV could make an important contribution to overall electricity needs in Africa.

15.1 INTRODUCTION

The Kyoto Protocol under the United Nations Framework Convention on Climate Change (UNFCCC) allows for the creation of a Clean Development Mechanism (CDM). Under the CDM, so-called 'Annex' countries that take on binding carbon abatement commitments may be able to partially comply by supporting initiatives that reduce greenhouse gas emissions in 'non-Annex' countries. Solar home systems (SHS) represent one possible arena for generating such trades of money and technology for abatement credits, and Africa is an important part of the current and potential market for SHS.

A number of multilateral, national, private and non-governmental organization (NGO) projects have already targeted SHS in Africa. The World Bank Group's Photovoltaic Market Transformation Initiative (PVMTI) has selected Kenya and Morocco for two of its three geographical focus areas. In addition, the Global Environment Facility (GEF) has recently completed a SHS project in Zimbabwe; it is currently implementing a SHS project in Uganda; and it is actively considering similar efforts in Benin, Cape Verde, and Togo (Kaufman *et al.*, 1999; Duke *et al.*, 2001). All of these are motivated in part by their carbon abatement potential.

Substantial NGO and private sector SHS efforts are also under way in South Africa. The government may rely heavily on SHS for the next phase of its successful (but increasingly expensive) rural electrification programme – granting concessions to rent SHS in defined rural areas to various businesses, including the South African utility ESKOM, Shell Solar, and the Dutch utility Nuon (Kammen, 1999; Anderson and Duke, 2001).

Even complete saturation of the global SHS market would have a negligible direct impact on global carbon emissions. Nonetheless, the CDM, or similar mechanisms, might provide an important boost to SHS markets. As detailed below, carbon abatement credits generated by SHS could significantly reduce the price of solar electricity for rural households in developing countries.

In addition to direct abatement (primarily through displacement of kerosene lighting, battery charging, and to a lesser extent, generators), SHS may also yield indirect carbon emissions reductions. First, SHS are a near-term niche market for photovoltaics (PV). As such, SHS sales help to drive the virtuous cycle between (i) cost reductions from greater PV production experience; and (ii) increased global demand for PV due to those cost reductions (Duke and Kammen, 1999a). As a result of these dynamic effects, the African SHS market itself may marginally contribute to efforts to reduce the global price of PV – though major programmes to subsidize grid-connected residential and commercial markets in

*Current address: McKinsey & Company, Florham Park, NJ, USA

Climate Change and Africa, ed. Pak Sum Low. Published by Cambridge University Press. © Cambridge University Press 2005.

Japan, Germany, and other industrialized countries increasingly dominate global PV markets (Duke, 2002).

Another indirect carbon benefit associated with SHS is that they may delay or displace conventional grid extension. There is no sure-fire technique for estimating the magnitude of this effect, but it appears to be operative in the South African context (Anderson and Duke, 2001).

These indirect market transformation and grid displacement benefits are unlikely to be sufficiently quantifiable to generate certified CDM credits, but they may motivate SHS investments and support from public and private funders interested in promoting carbon abatement.

15.2 LEARNING AND EXPERIENCE CURVES

Learning curves describe the relationship between cumulative production of a manufactured good, such as PV, and the labour inputs necessary per unit produced.[1] During the 1970s, Boston Consulting Group (BCG) generalized the labour productivity learning curve to include all costs necessary to research, develop, produce and market a given product (Boston Consulting Group, 1972). That is, BCG argued that learning-by-doing occurs not only in the narrow sense of labour productivity improvements, but also in associated R&D, overhead, advertising and sales expenses.

These efficiency gains, in conjunction with the benefits from scale economies, often yield cost reductions characterized by an experience curve:

$$UC = a \cdot q^{-b}$$

Where UC = unit cost, q = cumulative production, a = the cost of the first unit produced, and b = the experience parameter.[2] The underlying intuition for this exponential relationship is that there are diminishing returns to experience. Cost reductions are fast initially, but taper off as worker productivity becomes optimized, production is fully scaled up, incremental process improvements are made, and so on.

In addition to distinguishing between learning and experience curves, it is also possible to apply this concept to individual firms or to an entire industry. Table 15.1 illustrates the four different possibilities. If a given firm is able to completely retain the knowledge that it generates from its own production experience, then a firm-specific learning or experience curve approach is appropriate. However, to the extent that learning-by-doing spills over among firms, an industry-wide approach is more applicable.

Spillovers are often substantial since firms routinely poach employees from each other, purchase equipment and other inputs

Table 15.1. A taxonomy of learning-by-doing terms

	Labour Costs Only	All Costs
No spillover	Firm-specific learning curve	Firm-specific experience curve
Perfect spillover	Industry learning curve	Industry experience curve

from the same specialized suppliers, reverse-engineer their competitors' new products and even resort to industrial espionage. Lieberman (1987) discusses empirical evidence of spillovers as high as 60–90% in some cases and summarizes other empirical literature suggesting high spillover rates.

The conventional measure of experience is the progress ratio (Dutton and Thomas 1984; Argote and Epple 1990). For each doubling of cumulative production, the cost per unit decreases by (1 – progress ratio) per cent. Thus, counter-intuitively, higher progress ratios imply slower cost reductions.

Figure 15.1 illustrates experience curves for gas turbines, windmills and PV. The graph shows a tight relationship between cumulative industry-wide production and unit price, indicating that the industry experience curve is an appropriate approximation for PV. It is, however, important to highlight three concerns with this approach.

First, the experience curve for gas turbines is clearly 'kinked' after 1963, underscoring that the slope of experience curves can change abruptly (in this case due to a transition from an active research and innovation phase to one dominated by deployment only). To account for this, we employ a range of progress ratio estimates in this analysis.

Second, unit price is an imperfect substitute for unit costs. Profit margins can and do vary, and this can be one reason for anomalies such as that observed for gas turbines (Boston Consulting Group, 1972). It is preferable to define learning and experience curves using manufacturing cost; however, where these data are unavailable, price provides a legitimate proxy if any of the following conditions hold (Lieberman, 1984):

(1) Price/cost margins remain constant over time.
(2) Price/cost margins change, but in a manner controlled for in the analysis.
(3) Changes in margins are small relative to changes in production costs.

The third condition holds for PV since real module prices have fallen by a factor of 16 since 1975 (Johnson, 2002). Thus, short-term changes in the price/cost margin introduce only small deviations relative to the pronounced long-term cost reduction trend. Moreover, PV module production appears to be characterized by a high degree of innovation spillover, and this suggests that profit

---

[1] This section draws from Duke and Kammen (1999a).
[2] See Hirschman (1964), Argote and Epple (1990), and Badiru (1992) for variants of the equation. Also, Arrow (1962) uses cumulative capital goods investment as the learning proxy.

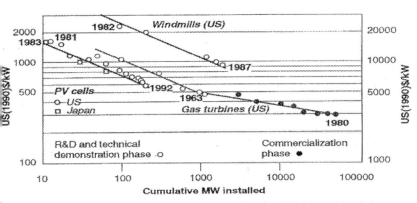

Figure 15.1. Industry-wide experience curve relationships for PV (right scale), wind generators (left scale), and gas turbines (left scale). (*Source*: IIASA/WEC, 1995).

margins in the industry will tend towards a standard competitive rate of return (Duke, 2002).

Finally, there is reason to be concerned about the assumption that cumulative production experience is the sole determinant of unit costs. Hall and Howell (1985) argue that cost reductions are driven by five factors: (i) scale economies; (ii) technological progress; (iii) input price changes; (iv) internal efficiency improvements; and (v) learning-by-doing. Cumulative production unambiguously drives only the latter two factors, but Duke (2002) argues that intensive learning-by-doing is an essential prerequisite for scaling up both manufacturing and delivery mechanisms (e.g. marketing, regulatory interface, installation and maintenance) for energy technologies. There is also evidence that use-inspired process and technological innovations are major drivers of manufacturing cost reductions in a variety of industries (von Hippel, 1988), and Lieberman (1987) suggests that learning effects dominate economies of scale in driving cost reductions. Also, higher levels of cumulative production will tend to drive down key input prices (e.g. for specialized machinery) as suppliers gain production experience and take advantage of scale economies.

In sum, using cumulative production as the sole independent variable is a reasonable and parsimonious approach for the PV case. Moreover, Duke (2002) shows that adding variables for time or current production does not substantially improve the model, while Isoard and Soria (1997) survey multiple empirical analyses, showing that learning effects tend to dominate scale economies across multiple industries, including PV. Similarly, Watanabe (1999) performs an econometric analysis that suggests learning effects drive 70% of long-term price reductions in the Japanese PV industry.

15.3 USE OF EXPERIENCE CURVES FOR ANALYSING PV MARKETS

The Photovoltaic Market Transformation Initiative (PVMTI) is an initiative funded by the International Finance Corporation (IFC)

and the GEF '. . . to significantly accelerate the commercialization, market penetration, and financial viability of PV technology in the developing world.'[3] Project documents do not provide any quantitative estimates of PVMTI's impact on module prices, but a background paper for PVMTI refers to a progress ratio of 0.80 for PV in order to project business-as-usual (BAU) scenario price trends (World Bank Group, 1996) based on an experience curve approach.

Experience curves have been widely applied to analyze PV markets in academic papers, including a number of publications by the authors of this chapter. Duke and Kammen (1999a) model the positive feedback between demand and experience effects in order to examine PVMTI – concluding that the programme is too small to substantially affect global PV module prices, but SHS subsidies are potentially cost-effective if implemented efficiently. Duke and Kammen (1999b) show that restricting PVMTI support to immature/high-potential thin-film PV technologies might increase benefit-cost ratios, but this strategy would be risky and politically difficult. Payne *et al.* (2001) employ experience curves as a 'top-down' cross check on its 'bottom-up' assessment of the cost reductions from scaling up thin-film PV production levels by an order of magnitude. Finally, Duke (2002) considers learning-by-doing spillover as a novel economic rationale for government 'buydowns' of clean energy technologies and quantifies an optimal global 'demand-pull' PV subsidy scheme to compensate for this externality.

Other academic work that has employed experience curves to analyse PV include a benefit-cost assessment of PV commercialization efforts (Williams and Terzian, 1993) and various discussions of PV experience curves (such as Cody and Tiedje 1997; and Neij, 1997). Moreover, policy analysts outside of academia have often employed experience curves to assess PV markets. Examples

---

[3] World Bank Group (1996). Note that PVMTI documentation refers to experience curve analysis to underscore the validity of this 'demand-pull' approach.

Table 15.2. Base case projections of the impact of global SHS sales on PV price.

Year	No-SHS scenario			SHS scenario			
	Annual PV sales in GWp	Cumulative GWp	$/Wp	Annual GWp SHS Sales	Cumulative GWp with SHS	$/Wp	Price effect
2000	0.20	1.3	4.00			4.00	
2001	0.24	1.5	3.78	0.012	1.5	3.77	0.3%
2002	0.29	1.8	3.57	0.014	1.8	3.55	0.5%
2003	0.35	2.1	3.37	0.017	2.2	3.35	0.7%
2004	0.41	2.5	3.18	0.021	2.6	3.16	0.8%
2005	0.50	3.0	3.01	0.025	3.1	2.98	0.9%
2006	0.60	3.6	2.84	0.030	3.8	2.81	1.0%
2007	0.72	4.3	2.68	0.036	4.5	2.65	1.1%
2008	0.86	5.2	2.53	0.043	5.4	2.50	1.2%
2009	1.0	6.2	2.38	0.052	6.5	2.35	1.3%
2010	1.2	7.5	2.25	0.062	7.8	2.22	1.3%
2011	1.5	9.0	2.12	0.074	9.4	2.09	1.3%
2012	1.8	11	2.00	0.089	11	1.97	1.4%
2013	2.1	13	1.89	0.12	13	1.86	1.4%
2014	2.6	15	1.78	0.13	16	1.75	1.4%
2015	3.1	19	1.68	0.15	19	1.65	1.5%
2016	3.7	22	1.58	0.19	23	1.56	1.5%
2017	4.4	27	1.49	0.22	28	1.47	1.5%
2018	5.3	32	1.41	0.27	34	1.39	1.5%
2019	6.4	38	1.33	0.32	40	1.31	1.5%
2020	7.7	46	1.25	0.38	48	1.23	1.5%

include a recent Electric Power Research Institute (EPRI) brief on thin-film PV (Peterson, 1997) and Maycock (1996).

We now turn to a general discussion of the carbon abatement potential of global SHS before specifically considering the potential importance of African SHS markets for climate change policy.

15.4 IMPACT OF SHS SALES ON FUTURE PV PRICE: A STATIC EXPERIENCE CURVE ANALYSIS

It is possible to extrapolate from the historical PV experience curve in order to estimate future PV prices as a function of projected sales growth rates. If the experience relationship holds, faster sales growth will mean more rapid unit cost reductions as the industry 'rides down' the experience curve more quickly. Given a progress ratio of 0.80 and a 2000 wholesale price of about $4.00 per Wp (Nitsch, 1998; Harmon, 2000; Johnson, 2002) and assuming 20% annual sales growth, then module prices will be expected to fall to $1.25 per Wp by 2020, based on cumulative sales of 48 peak gigawatts (GWp).[4]

It is possible to estimate the impact of current and projected SHS sales on future PV prices by subtracting current and projected SHS sales from the overall PV market projections, then using the experience curve to estimate how much higher prices will be in each year if it is assumed that all of these SHS sales cease.

World Bank Group (1998) estimates 1996 SHS sales of 4–13 peak megawatts (MWp). The base case for this analysis assumes 10 MWp for 2000, equivalent to 250,000 SHS sold worldwide with an average size of 40 Wp. Assuming that SHS sales match the projected 20% annual growth rate for the overall PV market, determining the projected impact of SHS sales on global PV module prices involves subtracting 12 MWp of SHS sales from 2001, 14 MWp of projected SHS sales from 2000, and so on. Removing SHS markets from projected overall PV module sales yields an estimated year-2020 PV price of $1.23, or only 1.5% higher than the projection that includes module sales for SHS markets.

In this base case forecast, SHS penetrate 11% of the maximum projected SHS market by 2020.[5] The assumed upper bound of

---

[4] This analysis uses constant 2000 dollars and refers to wholesale module prices and sales volumes for the combined market for both crystalline and amorphous thin-film panels.

[5] It is important to note that this simple approach does not account for experience curve effects for balance of systems equipment, retail distribution, and installation, which collectively account for more than half of typical SHS

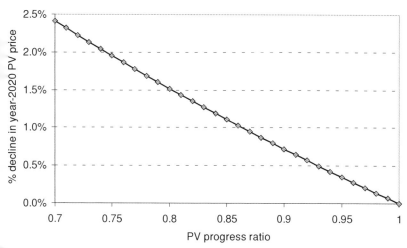

Figure 15.2. Decline in year-2020 PV module prices attributable to SHS markets, assuming base case parameters but varying the PV progress ratio.

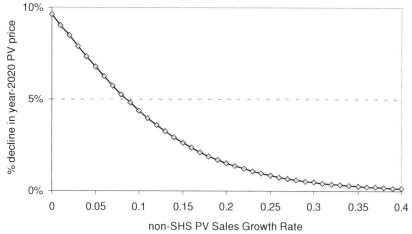

Figure 15.3. Decline in year-2020 PV module prices attributable to SHS markets assuming base case parameters but varying the non-SHS PV sales growth rate.

20,000 MWp of SHS comes from projecting that 400 million households remain unelectrified through 2020 (with population growth roughly keeping pace with grid extension) and each of these homes purchases a 50 Wp system.

The estimated impact of global SHS sales on PV module prices is sensitive to the assumed progress ratio. Figure 15.2 shows the percentage PV price decline attributable to SHS over the period from 2000 to 2020. If the progress ratio were to prove as low as 0.7,

then the model predicts SHS sales would cause a price decline of 2.4% over this 20-year period. However, if the future PV progress ratio worsens, the impact of projected SHS sales on PV prices in this static analysis falls proportionately.

Figure 15.3 shows the negative relationship between the growth rate of non-SHS PV sales and the impact on module prices of the SHS component of the PV market.

Figure 15.4 shows that it is also possible to vary the projected SHS sales growth rate. Holding the base case parameters constant, if annual SHS increase at 30% rather than 20%, by 2020 this yields 40% saturation of the SHS market and a price decline attributable to SHS of about 6%.

In sum, static analysis suggests that SHS markets are unlikely to play a major role in global PV commercialization efforts. The next section discusses the implications of dynamic feedback mechanisms.

---

costs. It is likely that these costs would come down if the number of SHS installed in any given country were to expand rapidly from a small initial base of cumulative experience. On the other hand, batteries are a mature technology and they represent about 30% of life cycle costs for standard 50 Wp SHS (Banks, 1998) and up to 70% for small 10–20 Wp systems such as those typically found in Kenya (based on calculations derived from the lifecycle cost data in Duke *et al.*, 2000).

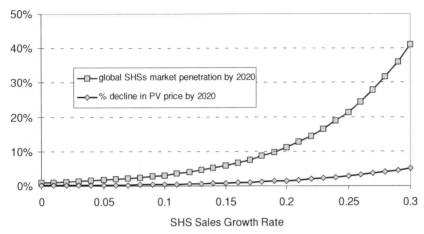

Figure 15.4. Decline in year-2020 PV module prices attributable to SHS markets assuming base case parameters but varying the SHS PV sales growth rate.

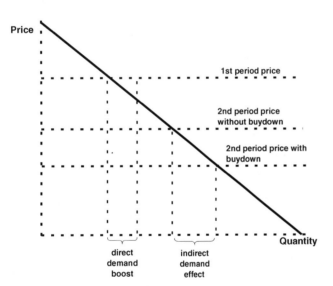

Figure 15.5. Two-period buydown.

15.5 DYNAMIC CONSIDERATIONS

Two important factors driving the diffusion of any new technology are cost reductions through experience effects and the responsiveness of market demand to any such cost reductions. The latter can be characterized as the percentage increase in sales associated with a 1% decline in price, i.e. the demand elasticity.

Anything that boosts PV sales will cause a price reduction via the experience curve. This, in turn, will induce an increase in future sales levels that will further reduce PV prices along the experience curve. This 'virtuous cycle' will likely dampen over time (Colombier and Menanteau, 1997).

Figure 15.5 presents a simplified two-period illustration of the positive feedback effect from a PV buydown. A one-period subsidy artificially inflates demand. As a result of associated ex-

perience benefits, in the second period unit cost is lower and the quantity of PV demanded is higher than it would have been absent the first-period buydown. In the third period this 'indirect demand effect' drives prices down still further via the experience effect, and so on.

It is difficult to quantify the importance of these dynamic effects; however, one analysis suggests that the indirect demand effects of PVMTI may exceed the static benefits from the programme (Duke and Kammen, 1999a). Also, Duke (2002) develops methodologies for determining the optimal long-term subsidy path for demand-pull 'buydown' programmes to help commercialize clean energy technologies like PV.

15.6 CARBON ABATEMENT IMPLICATIONS OF GLOBAL SHS MARKETS

It is important to assess the implications of the scenarios outlined above for CO_2 emissions. In the static base case, direct CO_2 displacement from SHS is unlikely to have an important impact on global emissions. Even if the entire potential market of 400 million households receives SHS, this would displace only approximately 20 million tonnes of carbon equivalent (tC) annually, or about 0.3% of global emissions.[6]

Thousands of rural consumers purchase SHS every year even though they receive no compensation for the value of avoided carbon emissions from kerosene lanterns. Forecasts of expected carbon prices range from about $15 to $350 per tC (Energy

---

[6] This assumes that each 50 Wp SHS displaces about 0.05 tC per year or about 1 tC over a 20-year system lifetime. These figures are based on an analysis of eight countries, taking into account kerosene lighting displacement as well as upstream emissions from fossil fuels, lead-acid battery production and PV module production (Ybema *et al.*, 2000). Where data were available the authors also considered the secondary factors of emissions from candle usage and battery-charging stations.

Information Administration, 1998, and White House, 1998). If carbon were to trade at $50/tC, this would amount to a lifetime carbon credit of about $50, for each 50 Wp panel – worth about one quarter of current wholesale module prices. While hardly decisive, this would marginally boost the number and size of SHS installed since the technology is already cost-effective in this application.[7] Of course, the CDM can play a useful role only if the transaction costs involved in certifying emissions reductions for SHS are kept to an absolute minimum (Kaufman et al., 1999).

In addition to direct carbon displacement, as noted every MWp of SHS sold helps to lower the global market price for PV. As PV prices fall, sales in existing markets increase and new niche markets open up. For example, at present, residential grid-connected PV systems are not economic even in states with the most favourable combination of high insolation and expensive retail rates. If module costs fall to $1.50/Wp, then rooftop systems would become cost-effective in about one tenth of new single-family homes constructed in the United States, or an annual market of about 500 MWp, i.e twice the global level of PV module sales in 2000. At $1/Wp, the new home market increases by a factor of four and large residential PV retrofit markets also become viable (Duke et al., 2001). Similar distributed grid PV markets exist globally.

As noted above, increased near-term SHS sales could help to generate indirect demand effects in the global PV market. This substantially raises the carbon abatement value of SHS.

15.7 CARBON ABATEMENT IMPLICATIONS OF AFRICAN SHS MARKETS

As of 2002, there were approximately 840 million people in Africa, of which 690 million are in sub-Saharan Africa.[8] Precise estimates are unavailable, but assuming less than half of these people have access to grid electricity implies a potential SHS market of at least 60 million households.[9] A 50% penetration rate with average system size growing to 50 Wp (as prices fall and rural incomes increase) translates into about 1,500 MWp of total PV demand, or six times global PV sales of 250 MWp in the year 2000.

There are, however, a number of unique aspects to the African SHS market that must be considered. Most importantly, the majority of the rural population in Africa lives in extreme poverty. This means that substantial subsidies and aggressive measures

to reduce the 'first-cost' barrier are particularly important in the African context. The fee-for-service programmes emerging in South Africa are encouraging in this regard, but delays in disbursing promised subsidies threaten to undermine their impact (Anderson and Duke, 2001).

Moreover, especially in the South African context, there are important political issues related to SHS dissemination. Rural South Africans generally view SHS as a second-best option relative to heavily subsidized grid connections that would provide them with considerably better service for similar or lower monthly payments. As it proceeds with its SHS efforts, the government of South Africa must therefore balance fiscal constraints on increasingly expensive grid-based electrification with the risk of being perceived as perpetuating a history of second-class electricity service for black South Africans (Anderson and Duke, 2001).

15.8 RENEWABLES SCENARIOS FOR AFRICA

As of 1995, Africa derived 16% of its grid electricity from renewable sources, and hydropower accounted for over 99% of this total (World Resources Institute, 1998). As of 2002, roughly 100 MWp of off-grid PV had been installed in Africa.[10] SHS account for approximately one quarter of this, with the remainder in telecommunications and various government and donor projects (e.g. water pumping, schools, and health clinics).[11] This is equivalent to just 0.03% of total grid electricity generated in the continent.

Projecting forward to the year 2025, if off-grid PV sales increase at 15% annually, then off-grid PV provides 0.8% of total expected grid electricity generation by 2025 based on 3 GWp of installed capacity.[12] This is an aggressive projection since it is equivalent to providing a 50 Wp SHS for all 60 million currently unelectrified African households. Nonetheless, electrification efforts have often failed to keep pace with population growth in rural Africa, average system sizes should increase as SHS prices fall, and non-SHS markets may continue to drive overall off-grid sales. Moreover, as off-grid markets begin to saturate, grid-connected PV could

---

[7] Note that the CDM would probably recognise carbon benefits only as they accrue. In that event, the stream of carbon abatement benefits from each SHS would have to be discounted. At a 5% real discount rate, this lowers the present value to about US$25.

[8] www.prb.org

[9] Assumes an unelectrified population of 420 million and an average rural household size of seven.

[10] Maycock (1996) suggests that Africa accounted for 10–13% of the global PV market in 1995. Cumulative global PV sales reached about 2 GWp by the end of 2002 and roughly half of this was in off-grid installations (though the off-grid share has been diminishing rapidly as subsidized grid-connected markets in industrialized countries have taken off). Assuming Africa maintained a 10% share implies that there has been about 100 MWp of cumulative off-grid sales in the continent through 2002.

[11] This assumes there will be about 150,000 SHS averaging 25 Wp each in Kenya and another 500,000 SHS scattered throughout the rest of Africa, with an average size of 40 Wp, yielding total SHS installations of 24 MWp. The Kenyan estimates are extrapolated from van der Plas and Hankins (1998).

[12] This assumes electricity consumption increases at 2% annually based on the GDP growth rate for Africa during 1990–1997 and total off-grid PV installations in Africa were about 16 MWp in 2002 (about 10% of the global off-grid PV market).

grow to become a major factor in African energy markets (and some of the larger-scale companies involved with providing rural solar installations might transfer their expertise to grid-connected markets as they emerge).

In sum, PV has the potential to contribute to the African energy supply while providing critical development benefits to rural populations and improving both the local and global environment. Growth in SHS and subsequent grid-connected markets will likely prove modest, however, without sustained and aggressive public support. Assessing the available policy options, and the desirability of this goal relative to other public priorities, requires further analysis.

ACKNOWLEDGEMENTS

The authors wish to thank the anonymous reviewers for useful edits and suggestions.

REFERENCES

Anderson, J. and Duke, R. D. (2001) *Solar for the Powerless: Electrifying Rural Africa with the Sun* (video documentary viewable at www.princeton.edu/duke).

Argote, L. and Epple, D. (1990). Learning curves in manufacturing. *Science*, **247**, 920–924.

Arrow, K. J. (1962) The economic implications of learning by doing. *Rev. Econ. Stud.*, **29**, 166–170.

Badiru, A. B. (1992) Computational survey of univariate and multivariate learning curve models. *IEEE Trans. Eng. Manage.*, **39** (2), 176–188.

Banks, D. (1998) *Off-grid Electrification for the Poor: Constraints and Possibilities*, Energy & Development Research Centre.

Boston Consulting Group, Inc. (1972) *Perspectives on Experience*. Boston, The Boston Consulting Group, Inc.

Cody, G. D. and Tiedje, T. (1997) A learning curve approach to projecting cost and performance for photovoltaic technologies. Presented to Future Generation of Photovoltaic Technologies NREL Conference held in Denver, Colorado, March 24–26.

Colombier, M. and Menanteau, P. (1997) From energy labelling to performance standards: some methods of stimulating technical change to obtain greater energy efficiency. *Energy Policy*, **25** (4), 425–434.

Duke, R. D. (2002) Clean Energy Technology Buydowns: Economic Theory, Analytic Tools, and the Photovoltaic Case, Ph.D. Dissertation, Princeton University.

Duke, R. D. and Kammen, D. M. (1999a) The economics of energy market transformation programs. *Energy Journal*, **20** (4), 15–64.

—— (1999b) PV market transformation: the virtuous circle between experience and demand and the strategic advantage of targeting thin-film photovoltaics. Presented at the IEA Workshop On Experience Curves For Policymaking –The Case Of Energy Technologies, Stuttgart, 10–11 May.

Duke, R. D., Graham, S., Hankins, M. *et al.* (2000) *Field Performance Evaluation of Amorphous Silicon (a-Si) Photovoltaic Systems in Kenya: Methods and Measurements in Support of a Sustainable Commercial Solar Energy Industry*. World Bank.

Duke, R. D., Jacobson, A. and Kammen, D. M. (2002) Product quality in the Kenyan solar home industry. *Energy Policy*, **30** (6), 477–499.

Duke, R. D., Williams, R. H. and Payne, A. (2004) *Accelerating Residential PV Expansion: Demand Analysis for Competitive Electricity Markets. Energy Policy*, in press.

Dutton, J. M. and Thomas, A. (1984) Treating progress functions as a managerial opportunity. *Acad. Manage. Rev.*, **9** (2), 235–247.

Energy Information Administration (1998) *Impacts of the Kyoto Protocol on U.S. Energy Markets and Economic Activity*. Washingon, DC, US Department of Energy.

Hall, G. and Howell, S. (1985) The experience curve from the economist's perspective. *Strategic Manage. J.*, **6**, 197–212.

Harmon, C. (2000) *Experience Curves of Photovoltaic Technology*. Laxenburg, Austria, International Institute for Applied Systems Analysis.

Hirschman, W. B. (1964) Profit from the learning curve. *Harvard Bus. Rev.*, January–February, 125–139.

IIASA/WEC (1995) *Global Energy Perspectives to 2050 and Beyond*. Laxenburg, Austria and London.

Isoard, S. and Soria, A. (2001) Technical change dynamics: evidence from the emerging renewable energy technologies. *Energy Economics*, **23**, 619–636.

Johnson, R. (2002) Personal communication.

Kammen, D. M. (1999) Bringing power to the people: promoting appropriate energy technologies in the developing world. *Environment*, **41** (5), 10–15, 34–41.

Kaufman, S., Duke, R. D., Hansen, R., Rogers, J., Schwartz, R. and Trexler, M. (1999) *Rural Electrification with Solar Energy as a Climate Protection Strategy*. Renewable Energy Policy Project.

Lieberman, M. B. (1984) The learning curve and pricing in the chemical industries. *Rand J. Economics*, **15**, 213–228.

—— (1987) The learning curve, diffusion, and competitive strategy. *Strategic Manage. J.*, **8**, 441–452.

Marnay, C., Richey, R. C., Mahler, S. A. and Markel, R. J. (1997) *Estimating the Environmental and Economic Effects of Widespread Residential PV Adoption Using GIS and NEMS*. Lawrence Berkeley National Laboratory, LBNL-41030.

Maycock, P. (1996) *Photovoltaic Technology, Performance, Cost and Market Forecast: 1975–2010*. PV Energy Systems, Inc.

Neij, L. (1997) Use of experience curves to analyze the prospects for diffusion and adoption of renewable energy technology. *Energy Policy*, **25** (13), 1,099–1,107.

Nitsch, J. (1998) Probleme der Langfristkostenschätzung – Beispiel Regenerative Energien. Vortrag beim Workshop Energiesparen – Klimaschutz der sich rechnet, Rotenburg an der Fulda, 8–9 October.

Payne, A., Duke, R. D. and Williams, R. H. (2001) Accelerating residential PV expansion: supply analysis for competitive electricity markets. *Energy Policy*, **29** (10), 787–800.

Peterson, T. (1997) *Making Photovoltaics Economical for Electric Utilities*. Electric Power Research Institute Strategic R&D Report, Fall.

UPVG (1994) *PV Vision*. The Utility Photovoltaics Group, **2** (2).

Van der Plas, R. J. and Hankins, M. (1998) Solar electricity in Africa: a reality. *Energy Policy*, **26** (4), 295–305.

Von Hippel, E. (1988) *The Sources of Innovation*. New York, Oxford University Press.

Watanabe, C. (1999) Industrial dynamism and the creation of a "virtuous cycle" between R&D, market growth and price reduction: the case of photovoltaic power generation (PV) in Japan. Experience curves for policy making: the case of energy technologies. In C. O. Wene, and T. Fried (Eds.), *Proceedings of the IEA International Workshop at Stuttgart, Germany, 10–11 May, 1999*. IEA.

White House (1998) *The Kyoto Protocol and the President's Policies to Address Climate Change: Administration Economic Analysis*. White House, Washington, DC, July 1998.

Williams, R. and Terzian, G. (1993) *A Benefit/Cost Analysis of Accelerated Development of Photovoltaic Technology*. Princeton University Centre for Energy and Environmental Studies, Report No. 281.

World Bank Group (1996) *Photovoltaic Market Transformation Initiative Background Paper*. International Finance Corporation and World Bank Staff.

(1998) *Solar Development Corporation*. Project Brief, International Finance Corporation and World Bank Staff.

World Resources Institute (1998) *World Resources 1998–99*. Oxford, Oxford University Press.

Ybema, J. R., Cloin, J., Nieuwenhout, F. D. J., Hunt, A. C. and Kaufman, S. L. (2000) *Towards a Streamlined CMD Process for Solar Home Systems: Emission Reductions from Implemented Systems and Development of Standardised Baselines*. ECN-C-00-109.

16 Climate change in sub-Saharan Africa: assumptions, realities and future investments

CHRISTOPHER O. JUSTICE[1], DAVID WILKIE[2], FRANCIS E. PUTZ[3], AND JAKE BRUNNER[4],[*]

[1]*University of Maryland, USA*
[2]*Wildlife Conservation Society, New York, USA*
[3]*University of Florida, USA*
[4]*Conservation International, Washington, DC, USA*

Keywords
Africa, global change research, scientific equity, climate change impacts.

Abstract
Of all the regions in the World, sub-Saharan Africa is the least well equipped to respond to the issues associated with climate change. This paper discusses the significance of climate change to African nations and the related needs and opportunities. At the centre of the discussion are the important issues, often raised by African colleagues, of scientific equity and the urgent need for investment in African scientific infrastructure to help African scientists inform and advise African governments and decision makers on the likely impacts of climate change on their nations' economy and resource base.

In most African countries, funding for global change research is a low priority compared with more immediate and pressing societal issues. However, understanding the processes and predicting the impacts of climate change on African environment and economies necessitates a series of focused scientific initiatives within the region. These initiatives will require a combination of applied biological and social science and would benefit from being led by African scientists. Some initiatives are currently under way, supported in large part by the international community. African scientists need to be true partners in these scientific endeavours, from identifying the priority research questions, to designing and implementing the research and presenting the policy implications of the results. Now more than ever, support for African science and education must be part of the national and international development portfolios for the countries of the region.

*Formerly with World Resources Institute, Washington, DC.
This chapter was first submitted in 2000.

There is an urgent need to raise broader awareness in Africa concerning the issues of global change. Dialogue needs to be encouraged between the African science and policy communities on the current state of scientific understanding, how best to manage resources with the various uncertainties associated with climate change, and the possibility of financial opportunities to promote sustainable development through participation in carbon offset programmes under the Clean Development Mechanism (CDM) of the Kyoto Protocol. In contrast to concerns about the potential impacts of climate change and the lack of resilience of African economies to climate change and variability, the concept of carbon trading offers some economic opportunities, which, if carefully developed, could contribute to improved resource management and provide tangible benefits to African communities. Carbon trades that help African countries to achieve their goals of improved human welfare and conservation of natural resources will be made more likely and more favourable when African scientists and their business partners are well aware of the associated opportunities and potential pitfalls.

16.1 INTRODUCTION

Following the Third Session of the Conference of Parties to the United Nations Framework Convention on Climate Change (UNFCCC) held in December 1997 in Kyoto, governments began to determine how they could respond to the Kyoto Protocol to the UNFCCC, whilst maintaining economic growth. Developing countries clearly have an important role in ensuring that the Convention is equitable and can be implemented effectively and that it meets the needs for global emissions control. African countries are not currently included in the list of 39 Annex 1 countries to which emissions limits or reduction commitments will apply. However, 41 African countries have ratified the Convention and the door is open for greater African participation in the Convention, for example, through the Clean Development Mechanism

Climate Change and Africa, ed. Pak Sum Low. Published by Cambridge University Press. © Cambridge University Press 2005.

(CDM) (Article 12 of the Kyoto Protocol). This mechanism allows Annex 1 countries to meet their commitments by undertaking projects with non-Annex 1 countries. The specific details of the Kyoto Protocol and the various implementation mechanisms continue to evolve and to be debated, but investments by industrialized countries in the energy sector and land-use change carbon offset projects on forestry and energy are being made in several tropical and subtropical countries, e.g. Costa Rica, Bolivia and India, with Africa falling well behind in knowledge and revenue gained. There are admittedly complexities in the socio-economic structure of African countries that present obstacles to full participation in the process. However, African scientists, policy makers, and representatives of conservation and business interests need to be fully engaged in these discussions (Maya and Churie, 1996).

To be persuasive, as participants in the climate change debate and substantial beneficiaries of any funds made available through implementation of the CDM adopted by the Kyoto Protocol, African governments need to understand how their development strategies are likely to influence their nations' roles as net contributors to or mitigators of global warming. African leaders need to be cognizant of and benefit from the lessons learned by Latin American countries such as Costa Rica and Guatemala, which are taking the lead in terms of developing country roles with respect to climate change and carbon trading, by proactively developing carbon trading opportunities (Totten, 1999; Tucker, 2001). Equally important, governments need to understand the impacts of climate change on their nations and to be prepared to respond effectively to these impacts. Decision makers must feel confident in predictions of inter-annual and decadal changes in weather patterns. They also should recognize the likely impact of these fluctuations on the natural resource base that forms the foundation for most of their economies. In particular, they need to be aware of the probable impacts of climate change on the provision of ecological goods and services, such as water and food supplies. Similarly, given the role climate variability plays in natural resource productivity in sub-Saharan Africa, international agencies that are currently supporting and encouraging long-term economic growth in Africa need to recognize and incorporate climate change and its mitigation as factors in their development assistance portfolios. This chapter outlines some of the more pressing issues associated with climate change and Africa.

16.2 CLIMATE CHANGE IN AFRICA

There are four aspects of the climate change issue that need to be considered by African governments and the international community:

- The contribution that African countries make to climate change through their net emissions of greenhouse gases;

- The prediction of the likely changes in climate and the resulting response of natural and managed ecosystems and socio-economic systems;
- The actions that need to be taken to mitigate continued production of greenhouse gases and mitigate the adverse societal effects of climate change;
- How participation in internationally funded greenhouse gas emissions mitigation projects can contribute to achieving national development and conservation goals.

16.3 GREENHOUSE GASES: SOURCES AND SINKS

At Kyoto it was agreed that initially six greenhouse gases would be considered together (carbon dioxide, methane, nitrous oxide, hydrofluorocarbons, perfluorocarbons and sulphur hexafluoride) and that an equivalent global warming potential of the different gases would be developed. It was also made clear that calculation of annual net emissions would necessitate a better quantification of both sources and sinks at the national scale. This is particularly the case with respect to emissions from forestry and other land use. Africa is not a major global contributor to climate change in terms of industrial emissions, and given the present and projected state of the economies of sub-Saharan Africa, this is unlikely to change over the next 10–20 years. Yet reliable estimates of carbon sources and sinks in Africa at a national scale are important prerequisites if governments are to participate effectively in any future carbon markets.

Sources of greenhouse gases are starting to be quantified for Africa and the IPCC/OECD methodology has been applied to several African countries (Braatz *et al.*, 1995). The 1992 Southern African Fires and Atmosphere Research Initiative (SAFARI) provided an international focus on regional emissions of gases and particulates (van Wilgen *et al.*, 1997). Preliminary findings show, unsurprisingly, that estimates of emissions from land-use change, forestry and agriculture are clearly more important than those from fossil-fuel burning and waste decomposition. While in aggregate terms sub-Saharan Africa is not a major contributor to global warming, there are a few emission 'hot spots' within the region that warrant particular attention.

16.3.1 Energy production in South Africa and the Gulf of Guinea

South African energy production using high-sulphur coal contributes the overwhelming majority of the region's emissions and a significant fraction of the global emission of CO_2, SO_x, and NO_x. For example, it is estimated that in 1997 South Africa generated 2% of global CO_2 emissions and 3–7% of the global total emissions of NO$_x$. South Africa is the fifteenth largest industrial emitter of greenhouse gases globally and the seventh largest developing country emitter. These emissions are relatively significant from a global warming and regional public health perspective. However, South Africa is anomalous in the region because it can be viewed more like an emerging economy where industrial-technological mitigation is both possible and preferable. South Africa is exploring ways to reduce growth in greenhouse gas emissions and atmospheric pollutants. Emissions of sulphate aerosol by South Africa's coal-burning energy production sector are high (Held *et al.*, 1996). Claims have been made that the nation's High-veld region exhibits among the highest levels of sulphate aerosols in the world (Tyson *et al.*, 1988). Sulphate aerosols are entering the climate change debate, as their presence in the lower atmosphere contributes to acid deposition and net surface cooling. Currently, the radiative role of aerosols and their relationship to cloud physics is a subject for global change research.

Expansion of oil extraction in the Gulf of Guinea and in Angola may also become an important point source of emissions at some time in the future. Gas-flaring from petroleum and natural gas facilities, primarily in Nigeria, currently accounts for 21% of global gas-flaring. Within the next 5–10 years, other areas may become significant point sources of emissions. For example, the Copper Belt of Zambia, following recent privatization, may return to heightened production levels similar to the late 1960s or early 1970s. The Maputo corridor of Mozambique, where construction of the world's largest aluminium smelter is under way, and the Maputo harbour, which is undergoing industrialization, could also become large source areas. As with South Africa, the control of emissions from oil production and industry is a well understood technology transfer and capital investment issue, which is largely within the capacity of the production companies to address.

16.3.2 Forest management and deforestation in the humid tropical forests

Forests and land-use change became major negotiating items in the climate change debate following the Kyoto meeting. At present, the forests of the Congo Basin cover approximately 2.8 million km^2, an area about one third the size of the USA, and serve as a reservoir of over 40 Gt (Gt $= 10^9$ tonnes) of carbon, which is equivalent to approximately six years of total global emissions of CO$_2$. Preliminary calculations of biomass density show the highest levels for Equatorial Guinea and Gabon, but there is much that could be done to improve current estimates of carbon stocks (Brown and Gaston, 1995). Approximately 1% of the Congo Basin's dense forests are cleared each year, compared to an estimated 4% in West Africa. In Côte d'Ivoire, preliminary estimates by FAO indicate that 6.5% of the forest is currently cleared annually, and that less than half of the original forest cover remains.

The rate of deforestation in the Congo Basin is relatively low at present compared with tropical forests in other regions of the world. But in many areas of the Congo Basin forests are being rapidly fragmented and degraded. Forest fragmentation as a result of road construction increases the edge-to-surface area ratio of forest patches and may result in the biomass collapse seen in the Amazon. Loss of woody biomass and changes in species composition and abundance within fragmented forest is a concern from both climate change and biodiversity conservation perspectives. The major factor driving forest fragmentation and degradation, and thus the key to the near-term fates of moist forests of Africa, is road access. Many of the roads penetrating into the forest frontier and thus stimulating colonization and deforestation are built by logging companies. Logging activities themselves are probably much more destructive than necessary given the volumes of timber being extracted. Nevertheless, it is road building that is most closely linked to deforestation and biomass loss in the moist forests of Africa and in other parts of the tropics (Kaimowitz and Angelsen, 1998). Logging is particularly critical to protected area management as concessions bordering national parks provide essential habitat to wide-ranging species, and logging is the primary land use within corridors that connect isolated protected areas.

Though the present estimates of deforestation are rough, there is no doubt that the Congo Basin contains an enormous stockpile of carbon and could contribute significantly to global CO$_2$ emissions if the rate and extent of forest clearing, burning, and fragmentation follow the trajectory of West Africa. Continuing efforts to monitor the tropical forests of Central Africa using satellite data are supported by NASA and the European Community (Malingreau *et al.*, 1995; Skole *et al.*, 1997; Laporte *et al.*, 1998). Refinements to current estimates are expected through increased use of high-resolution satellite data and with completion of the FAO Tropical Forest Assessment 2000. A reliable forest monitoring system is critical both for effective national forest management and for participation in any of the climate change mechanisms. A new international initiative entitled *Observatoire Satellital des Forêt d'Afrique Centrale* (OSFAC), involving government and non-government organizations, is aimed at improving capacity within the region for forest monitoring (Mayaux *et al.*, 2000).

Whatever agreements are made as a result of the international climate change debate, forest management in Africa is likely to be affected. It is important that existing regional forest management and biodiversity conservation plans are considered in decisions concerning African country involvement in climate change mitigation mechanisms.

16.3.3 Woodland clearing, savannah burning and woodland degradation

Rates of woodland clearing in Africa are even less readily available than data on deforestation in the humid zone. Part of the

Table 16.1. Dense forest and forest clearing in the Congo Basin.

Country	Dense moist forest 1993 in thousand km^2	Dense moist forest 1993 ha/capita	Deforestation rate 1990–95%
Cameroon	180 (39%)	1.44	1.5
CAR	44 (7%)	1.38	1.2
Congo	237 (69%)	9.88	0.7
DRC	1,108 (47%)	2.69	0.6
Eq. Guinea	100 (60%)	5.54	0.8
Gabon	214 (81%)	17.83	1.3

Sources: Forest cover data from TREES (1997); deforestation rates from FAO (1997).

problem is that detecting canopy loss through remote sensing is more difficult in open canopied woodlands than in closed forests. Furthermore, tree cover in many savannah woodlands often decreases gradually and not in a spatially contiguous fashion in response to fuelwood harvesting, charcoal-making, selective logging, and intensive late dry season fires that damage canopy trees but do not kill them outright (Campbell, 1996). Gradual or not, the process has resulted in the conversion of millions of hectares of woodland in Africa into sparsely wooded savannahs or nearly treeless grasslands. It is also reported anecdotally that in some areas with changes in management practice, tree density and woody biomass are increasing. Improvements in our capacity to detect changes in forest and woodland structure more subtle than outright deforestation are needed.

Savannah burning is extensive in Africa north and south of the Equator and is an integral part of savannah ecology. While over an annual cycle, savannah burning does not lead to a net emission of CO_2, it does generate significant emissions of other trace gases (carbon monoxide, methane, oxides of nitrogen) and aerosol particulates (Scholes *et al.*, 1996a). Methane is part of the greenhouse gas bundle being considered as part of the Kyoto Protocol. Scholes *et al.* (1996b) estimated that the biomass consumed by vegetation fires annually in Africa south of the Equator averages 177 ± 87 TgDm/yr. These fires are concentrated between 5 and 20° south during the months of June to October. Unlike the savannah grassland fires, biomass burning from fuelwood and carbon production, woodland clearing, and ash-fertilization agriculture produces net annual emissions.

16.4 MITIGATION OPTIONS

Given the present and projected contribution of industrial emissions from sub-Saharan Africa to global warming, emission mitigation using technological 'fixes' should not be the major focus of African climate change investment.

Africa has a substantial potential for additional carbon sequestration in biotic sinks, as well as for reduction of emissions caused by clearing and degradation of forests and woodlands. Reasonable estimates have been made for potential reductions in net emissions that could be achieved through managing woodlands (Kundhlande *et al.*, 2000), adopting more efficient woodfuel use technologies (e.g. efficient stoves and charcoal kilns), or by zero-tillage agriculture. However, few institutions are prepared to develop and implement land-use change based carbon offset projects. Part of the reason for this lack of interest in carbon offsets through improved natural resources management is that the focus has been on energy sector options (Maya and Gupta, 1996; Rowlands, 1998). The major problem, however, appears to be a lack of awareness, information and expertise, especially on the marketing side.

There is an urgent need for research to better understand carbon sequestration rates associated with tropical forest and woodland management. With increasing understanding and means for verification, CDM projects could be expanded to include alternative mechanisms. For example, by increasing soil organic matter in agricultural systems through minimum tillage or agroforestry methods, Africa could provide both a significant increase in the terrestrial carbon pool and greatly benefit soil fertility and productivity (Woomer *et al.*, 1994). Improved methods for natural forest management, such as reduced-impact logging techniques, could also increase carbon sequestration and provide a number of environmental co-benefits, including biodiversity protection and increased yields of timber in the future (Pinard and Putz, 1997). Though the specific mechanisms and guidelines for carbon offsets between developed and developing countries have yet to be fully elaborated, African countries need the information and capacity to develop and analyse the proposals, to be able to participate in the design and implementation of procedures for accreditation, monitoring and enforcement. In keeping with Article 12 of the Kyoto Protocol, African nations should develop approaches to mitigation that also match their national objectives for sustainable development, including improved resource management, enhancement of social welfare, and biodiversity conservation.

As experience with biotic offsets accumulates globally, increasing concerns are being raised about their social and environmental impacts. At its worst, land management for maximized rates of carbon sequestration (e.g. 'carbon forestry') could be socially disruptive, damaging to biodiversity, and have numerous other unwanted environmental impacts (Frumhoff *et al.*, 1998). Despite the potentially dire effects, there are several examples where biotic offsets are being effectively used to promote rural development and protect the environment (e.g. Jong *et al.*, 1997). For example, a community-based carbon offset project involving fire management, subsidies for purchase of fuel-efficient wood stoves, and agroforestry could result in substantial carbon sequestration, protect biodiversity, and promote rural development (Lowore and Putz, 1999). Although proponents of such a project would face numerous problems and challenges (e.g. monitoring and verification, ownership, duration), none of the problems seems insurmountable. Not to diminish the importance of institution-building, but a more likely impediment to development of such projects is lack of access to information, both by the communities themselves and by their potential partners among environmental and social NGOs, the private sector, or government. Information on opportunities and approaches for developing carbon offset projects needs to be widely disseminated in Africa.

Although additional biophysical and socio-economic research would help in developing verifiable CDM projects in Africa based on biotic offsets, a bigger problem appears to be finding investors for the projects once they are developed. During this period of uncertainty before the Kyoto Protocol enters into force and the guidelines for the CDM are fully developed and elaborated, investors are particularly concerned about the co-benefits of carbon offset projects, particularly those related to environmental protection and poverty alleviation (Newcombe, 1998). If effectively brokered, biotic offset projects in Africa might attract funding due to concerns about the poverty prevalent over much of the continent. Even if early estimations of the amount of money that will be available for biotic offsets prove exaggerated, substantial funds might nonetheless become accessible to properly formulated and effectively brokered projects (Smith *et al.*, 2000).

16.5 RESPONSES TO CLIMATE CHANGE

The IPCC (1998) report on *Regional Impacts of Climate Change: Assessment of Vulnerability* was constrained by the ability of the global climate models to make reliable climate projections. However, despite these vagaries, the findings of the report are, on the whole, bleak. Sub-Saharan Africa contributes little to global warming, is powerless to prevent climate change, and thus is left merely to respond to it. Furthermore, the nations of this region are woefully ill-prepared to deal effectively with the adverse impacts of climate change. This potentially places over 500 million Africans in jeopardy, because more than any other region of the world, the human toll of global climate change is liable to be most severe in sub-Saharan Africa. This is likely to be true for the following reasons:

- Global warming will make weather events more extreme and harder to predict;
- Weather is a powerful force in African economies, which are unusually dependent on agriculture and the direct consumption of wild resources;
- The majority of Africans live in areas prone to droughts or flooding;
- Average per capita income in Africa is the lowest in the world, limiting the options that families have to "weather" adverse impacts of global climate change;
- African governments often lack the technical expertise, infrastructure and funds to predict, plan for, and respond to the adverse impacts of climate change.

Of the 19 countries in the world that are classified as water-stressed, more are in Africa than any other continent. In most African countries, farming depends entirely on the quality of the rainy season. Changes in mean winter temperatures would have different impacts in different regions, reducing production in some areas and reducing susceptibility to frost in others. Coastal regions already under stress from population pressure and conflicting use would be seriously impacted by sea-level rise. Human health in Africa, as has been vividly demonstrated in Kenya recently, will be particularly susceptible to changes in vector-borne diseases and reductions in food or water supply.

To help African governments better understand how weather patterns are likely to change in the future, considerable advances are being made by the global change research community to improve climate prediction. Southern Africa weather patterns exhibit a strong El Niño Southern Oscillation (ENSO) connection and the national meteorological, university research, and international communities are applying new observational and modelling techniques to provide improved predictions of inter-annual variability in rainfall patterns. Improvements are now needed in regional-scale (mesoscale) modelling of decadal to centennial climate change. Collaborative international research involving southern African scientists would help provide access to the computer technology necessary for complex simulations.

Prediction of the impacts of climate change is also based on modelling. Some models simulate biogeochemistry and ecosystem response to climate change. Others predict crop yields, hydrological budgets, disease vectors and land-use change. Hulme *et al.* (1996) in their study on regional model prediction emphasized the need for more detailed process models calibrated by extensive field data. A new class of models that integrate socio-economic and physical interactions with climate is being developed to assess the likely impacts of global warming (e.g. Leemans, 1995). If African nations are to develop strategies for responding to the impacts of climate change, it is critical that these models be designed and tuned for the African environment (Desanker and Justice, 2001).

16.5.1 Concern about climate change is not new to Africa

Though human-induced global climate change is a new phenomenon, the majority of Africans have always been concerned about and have had to deal with the risks associated with variable, unpredictable and often severe weather. Even those who live in wet forested regions have had to contend with inclement weather, where maybe surprisingly, too much rain is often the problem (Wilkie *et al.*, 1999). Concerns about weather and strategies to deal with the vagaries of weather are central to the economies of most African families, and over millennia, African farmers and pastoralists have developed strategies to minimize weather-related risks. Though population expansion and the progressive globalization of markets in the last 30 years have added new elements to economies of rural households, their strategies for addressing weather-related risks may still offer lessons for policy makers concerned about mitigating the expected impacts of global climate change.

African farmers and pastoralists are not the only groups of concern in regard to climate change. The extent of African policy makers' involvement in recent IPCC meetings demonstrates that climate change is also a political issue in Africa. Policy makers' specific concerns are:

- That at present they feel that they lack the technical knowledge to participate effectively in the climate change debate, and lack the political power to influence the direction of the debate;
- That as signatories of global climate change accords they will be required to adhere to levels of greenhouse gas emissions that may severely constrain their nations' rates of economic growth and social development;
- That their nations (i) lack the trained personnel to analyse the growing corpus of scientific information on short and mid-term changes in climate, to assess the socio-economic impacts on their countries and to derive appropriate policy interventions; and (ii) lack the infrastructure to implement effective mitigation efforts;
- That the likely impact of climate change on land cover, food production, human health and water resources means that strategies for economic development cannot be developed in isolation from strategies for mitigating the impacts of global warming.

16.6 GLOBAL CHANGE IN AFRICA

African societies and the environment that supports them are exposed to several drivers of change in addition to climate. The internal demographic, socio-economic and political forces that drive changes in land cover and land use are coupled with external climatic and global economic factors. These factors drive inter-annual variability and long-term trends in climate and fluctuations in commodity prices to affect food production, water availability and human health. The resultant changes in disturbance regimes, such as fire frequency, forest regeneration rates and insect infestations, directly impact managed and unmanaged ecosystem structure and function and the goods and services that the ecosystems provide. In addition to changes in terrestrial ecosystems, these drivers of change have transboundary impacts on atmospheric composition and chemistry that feedback to the terrestrial systems.

Effective environmental management requires a strong basis in applied science, adapted to local land-use management practices, which address the major drivers of land-use change and production. Understanding these changes and their feedbacks to sustainability requires fundamental research. To achieve this objective, a true partnership is needed between national and international scientists. The international research community has developed an agenda for global change research through the International Geosphere and Biosphere Programme (IGBP) and the World Climate Research Programme (WCRP). A set of regional global change research priorities are being developed through the System for Analysis Research and Training (START), which focuses on those aspects of global change research that are directly relevant to developing countries and regional natural resource management.

START http://www.start.org is helping to develop regional research networks within Africa. These include the IGBP/START Miombo Network (Justice *et al.*, 1994; Desanker *et al.*, 1997), the Kalahari Transect (Scholes and Parsons, 1997), the Sustainable Rangelands Programme (Odada *et al.*, 1996), the Central African Land-Use and Land-Cover Change Network and the Southern Africa Fire and Atmospheric Research Initiative (SAFARI) 2000. IGBP/START has made a concerted effort to identify the global change priorities for the African regions (START, 1995). What is needed is a strategy to enable African scientists to participate in global change research in a sustainable way with only limited outside support.

Several US agencies are contributing to these networks and the regional global change research agenda through such activities as the US Country Studies Programme (http://www.gcrio.org/CSP/ap.html), the NASA Land-Cover Land-Use Change Programme (http://lcluc.gsfc.nasa.gov/), the NOAA ENSO Applications Programme (NOAA, 1997), and the USAID Central African Regional Programme for the Environment (CARPE, http://carpe.umd.edu/). Of particular note is the NOAA/USAID African Climate Outlook Forum (http://www.ogp.noaa.gov/), which has provided a mechanism for coordination of regional forecasting amongst meteorological and other research communities.

Results from these various preliminary global change activities show that:

- The global change research community has dramatically improved the accuracy of short- to mid-term predications of changes in regional weather patterns for Africa, in part through the work of the regional climate forums for Africa;
- The international global change research community is building global change science capacity within Africa through mentorship and partnership activities;
- There is a strong interest among African scientists to participate in global change research, particularly in those areas of direct policy relevance to their countries. South Africa already has a strong science community and offers potential as a focal point for training other Africans;
- The global change community needs help in (i) developing climate projections, specifically for the African region and making linkages between: scientific predictions and national policies; and (ii) capacities to respond to these predictions;
- Electronic communication networks are critical to enhancing the global change research capacity of African scientists;
- Open access to and sharing of data is an integral part of scientific collaboration and will be essential for international mitigation agreements. In Central Africa, the USAID Central Africa Regional Project for the Environment (CARPE) is playing a key role in the compilation of regional environmental data sets. The IGBP/START Miombo Network has developed a CD of regional data. Compilation and distribution of data sets on CD-ROMs to address specific research questions can greatly facilitate global change research in Africa:
- Even small-budget pilot programmes can be very effective in using scientific predictions of climate change events to help regional scientists make the link to decision makers. For example, one NOAA-supported study helped the government of Zimbabwe prepare and implement plans to mitigate outbreaks of malaria associated with El Niño-driven changes in regional weather patterns;
- Increased access to affordable environmental satellite data by regional scientists and resource managers would provide a major contribution to improved monitoring of environmental change in Africa;
- Regional integrated assessments of the causes and consequences of change, combining physical, social science, economic and policy options will be a critical step in effective policy formulation and mitigation. Thus, there is a need to develop models that are suitable for application to African environments and societies.

The growing participation of African scientists has pushed the global change research community to make the results of its research more policy relevant, an issue that has always been the case for African scientists, for whom funding for basic research is particularly scarce. A further lesson learned is that decision makers are often more accessible to scientists and science findings in Africa than in the developed world. The clear communication of scientific results to the decision-making community, whether at the level of householder, farmer or government minister, is critical for the transformation of research results into policies and actions on the ground. The global change research community is keen that donors with specific strengths in policy and the institutional capacity-building arena build on this indigenous capacity by supporting these information exchange activities, thus minimizing the tendency of natural scientists to attempt ad hoc policy responses for areas and on topics beyond their expertise.

16.7 GLOBAL CHANGE AND INTERNATIONAL DEVELOPMENT: A US EXAMPLE

It is hard to imagine a development agenda that would not incorporate consideration of the forces and impacts of global change in Africa. Nevertheless, development agencies have been slow to recognize the importance of global change as part of a development portfolio. In general, African governments addressing immediate societal and infrastructure needs do not have the resources to develop global change research programmes. But in most African countries there are well-trained scientists, interested in and capable of addressing national and regional questions of global change.

The US Global Change Research Programme and the component research agencies contribute significantly to the international global change research programmes. However, they are often unable to support scientific capacity-building in Africa directly. The US Agency for International Development (USAID) allocates a significant amount of resources to Africa, particularly in the area of natural resources management, but traditionally has not supported global change research or scientific capacity-building in Africa. Over the last five years, USAID has supported studies to define its role in global climate change (e.g. BSP, 1992; Hausman *et* al., 1993; USAID, 1998).

In the USAID Climate Change Initiative 1998–2002 report (USAID, 1998), Administrator Brian Atwood stated that climate change is a public policy issue that simply cannot be ignored, and that critical to promoting long-term sustainable development will be a comprehensive programme to combat the threat of climate change. In this document, USAID identified three focus areas: decreasing the rate of growth in net emissions, increasing developing and transition participation in the UNFCCC, and decreasing developing country vulnerability. The document identifies various tools and techniques that could be used in a development programme, such as policy reform, private sector partnership, technology cooperation, education and outreach. USAID's current climate change focus in Africa is on industrial emissions from South Africa and potential emissions from the Congo Basin. There is clearly room for enhancement of the African component in this initiative.

The Kyoto Protocol stresses the importance of promoting and supporting the development of scientific equity with respect to the developing world. This goal can be approached, for example, by providing scientists from African countries with the means to participate actively in the global change debate and providing the necessary technical support to their governments with respect to the emerging conventions.

Researchers in the United States are currently constrained in how they can further international cooperation and obtain support for their research counterparts in Africa. A cooperative twinning programme between US and African universities and research institutions, shared by the US research and development agencies, would provide a strong basis for improving the current scientific inequity and furthering research into the causes and impacts of global change in Africa. Investment in developing sustainable global change research capacity in Africa would help provide the necessary scientific underpinning to sustainable development.

The following suggestions would help US agencies strengthen their regional development portfolios and the current global change research in Africa, and assist African policy- and decision-making.

• US agencies should help build global change research and socio-economic impact assessment capacity in Africa. The objective of this recommended focus is to enable African scientists to participate fully in the global change debate and to provide technical support to their governments on global change issues. The existing global change programmes within the region provide an excellent starting point for expanded US involvement.

• USAID should help African governments and the private sector assess and respond appropriately to the science community's predictions regarding global change. This should build on existing global change activities, and focus USAID investments where it has a comparative advantage (i.e. resource management, capacity-building, policy dialogue and donor coordination, economic growth, and fostering regional cooperation). The link between science and policy is an area where USAID can make a unique and critical contribution.

• USAID should expand its work in the natural resource management and agriculture sectors to help reduce household risks associated with climate change.

• US agencies should continue efforts to expand Internet access to African scientists, NGOs and governments. One example of the benefits of such investments is that Namibia is presently using its high-speed Internet connection to South Africa to run its weather prediction models on a South African supercomputer.

• USAID should ensure that proposed investments under a climate change rubric receive appropriate technical review and monitoring to ensure results and an appropriate level of integration with other global change activities.

Such support from US agencies needs to be matched by a receptive response from African governments, providing the necessary structures and encouragement for scientific equity and providing their own infrastructure and support to enable the process to be sustained.

In summary, we suggest that the time has come for global change to be incorporated in the African development agenda. Pilot activities have shown that a relatively small number of well-managed activities, focused on natural resource management, water and food security, and human health would have a significant pay-off. A new partnership is called for between the US science and development agencies to develop a programme in Africa that would build capacity to provide an improved understanding of the causes and impacts of climate and global change on the African environment and societies. This understanding will require an integration of biological and social science with a foundation in local knowledge. Implicit in this programme would be the forging of strong links between scientific understanding and decision- and policy-making in Africa from local to regional scales.

ACKNOWLEDGEMENTS

The authors wish to thank Dr Pauline Dube and the reviewers for their suggested improvements to this paper and Dr Paul Desanker for his encouragement with this work.

REFERENCES

Biodiversity Support Programme (1992) *Central Africa: Global Climate Change and Development.* Landover, MD, USA, Corporate Press.

Braatz, B. V., Brown, S., Ischei, A. O. *et al.* (1995) African greenhouse gas emission inventories and mitigation options: forestry, land-use-change and agriculture. *Environ. Monit. Assess.*, **38**, 109–126.

Brown, S. and Gaston, G. (1995) Use of forest inventories and GIS to estimate biomass density of tropical forests: application to tropical Africa. *Environ. Monit. Assess.*, **38**, 51–62.

Campbell, B. (Ed.) (1996) *The Miombo in Transition: Woodlands and Welfare in Africa.* Bogor, Indonesia, Centre for International Forestry Research.

de Jong, B. H. J., Soto-Pinto, L., Montoya-Gomez, G., Nelson, K., Taylor, J. and Tipper, R. (1997) Forestry and agroforestry land-use systems for carbon mitigation: an analysis in Chiapas, Mexico. In W. N. Adger, D. Peterella, and M. Whitby (Eds.), *Climate*

Change Mitigation and European Land Use Policies, pp. 269–284. New York, Oxford University Press.

Desanker, P. V. and Justice, C. O. (2001) Africa and global climate change: critical issues and suggestions for further research and integrated assessment modelling. *Climate Res.,* **19**, 93–103.

Desanker, P. V., Frost, P. G. H., Justice, C. O. and Scholes, R. J. (1997) The Miombo Network: framework for a terrestrial transect study of land-use and land cover change in the Miombo ecosystems of Central Africa. *IGBP Report # 41,* IGBP Secretariat, Stockholm, Sweden.

FAO (1997) *State of the World's Forests 1997.* Food and Agriculture Organization of the United Nations, Rome, Italy.

Frumhoff, P. C., Goetze, D. C. and Hardner, J. J. (1998) *Linking Solutions to Climate Change and Biodiversity Loss Through the Kyoto Protocol's Clean Development Mechanism.* UCS Reports, Briefing Papers from the Union of Concerned Scientists, 14pp.

Hausman, L., Pardo, R., Ross, B. and Sherwin, W. (1993) *Recommendations for Africa Bureau Global Climate Change Programming.* Report for USAID, IQC Contract # PDC 5517-I-00-0103-00, Chemonics, Washington, DC, USA.

Held, G., Gore, B. J., Surridge, A. D., Tosen, G. R., Turner, C. R. and Walmsley, R. D. (Eds.) (1996) *Air Pollution and its Impacts on the South African Highveld.* Environmental Scientific Association, Cleveland 2022, South Africa.

Hulme, M. (Ed.) (1996) *Climate Change and Southern Africa: An Exploration of some Potential Impacts and Implications for the SADC Region.* Climate Research Unit, East Anglia / World Wildlife Fund International Report. ISBN:2-88085-193-9.

IPCC (1998) *The Regional Impacts of Climate Change: An Assessment of Vulnerability.* R. T. Watson, M. C. Zinyowera and R. H. Moss (Eds.), A Special Report of IPCC Working Group II. Cambridge, Cambridge University Press, 517pp.

Justice, C. O., Scholes, R. and Frost, P. (1994) African savannahs and the global atmosphere research agenda 1994–98. In *Proceedings of a Joint IGBP START/IGAC/GCTE/GAIM/DIS Workshop on African Savannahs, Land-use and Global Change: Interactions of Climate Productivity and Emissions. Victoria Falls, Zimbabwe,* IGBP Report 31, IGBP Secretariat, Stockholm.

Kaimowitz, D. and Angelsen, A. (1998) *Economic Models of Tropical Deforestation. A Review.* Centre for International Forestry Research (CIFOR), Bogor, Indonesia.

Kundhlande, G., Adamowicz, V. L. and Mapaure, I. (2000) Valuing ecological services in a savannah ecosystem: a case study from Zimbabwe. *Ecol. Econ.,* **33**, 3, 401–412.

Laporte, N., Heinicke, M., Justice, C., Goetz, S. (1998) A new land cover map of Central Africa derived from multiresolution, multitemporal satellite data. *Int. J. Remote Sensing,* **19**, 18, 3,537–3,550.

Leemans, R. (1995) Determining the global significance of local and regional mitigation strategies: setting the scene with global integrated assessment models. *Environ. Monit. Assess.,* **38**, 99–110.

Lowore, J. and Putz, F. E. (1999) Carbon offsetting through woodland protection and sustainable management; capturing the carbon

market in Malawi. *Forest Research Institute of Malawi Newsletter,* **15**, 1–6.

Malingreau, J. P., De Grandi, G. F. and Leysen, M. (1995) TREES-ERS-1 study: significant results for over Central and West Africa. *ESA Earth Observation Quarterly,* **48**, 6–11.

Maya, R. S. and Churie, A. (1996) Critique of the African approach to negotiating JI – bargaining or posturing. In R. S. Maya and J. Gupta (Eds.), *Joint Implementation: Carbon Colonies or Business Opportunities (Weighing the Odds in an Information Vacuum).* Southern Centre for Energy and Environment Climate Change, Series 3, 31–41.

Maya, R. S. and Gupta, J. (Eds.) (1996) *Joint Implementation: Carbon Colonies or Business Opportunities (Weighing the Odds in an Information Vacuum).* Southern Centre for Energy and Environmental Climate Change Series, 3, 164.

Mayaux, P., Justice C. and Lumbuenamo, R. S. (2000) *Observation par satellite des forets d'Afrique Centrale: creation du reseau GOFC-OSFAC.* Commission Europeene, Centre Commun de Recherche, EUR 19585, 26pp.

Newcombe, J. (1998) *The Risks and Environmental Benefits of Investing in Climate Change Projects Under the Kyoto Protocol: An Investor Perspective.* Report prepared for CIFOR, Indonesia.

NOAA (1997). *Workshop on Reducing Climate-Related Vulnerability in Southern Africa Workshop Report, 1–4 October 1996, Victoria Falls, Zimbabwe.* NOAA OGP Silver Spring, MD, USA.

Odada, E., Totolo, O., Stafford Smith, M., Ingram, J. (1996) *Global Change and Subsistence Rangelands in Southern Africa: The Impact of Climatic Variability and Resource Access on Rural Livelihoods.* Report of a workshop on Southern African Rangelands today and tomorrow: social institutions for an ecologically sustainable future, Gaborone, 10–14 June 1996. GCTE Working Document No. 20. 99.

Pinard, M. A. and Putz, F. E. (1997) Monitoring carbon sequestration benefits associated with a reduced-impact logging project in Malaysia. *Mitigation and Adaptation Strategies for Global Change,* **2**, 203–215.

Rowlands, I. H. (Ed.) (1998) *Climate Change Cooperation in Southern Africa.* London, Earthscan Publications.

Scholes, R. J. and Parsons, D. A. B. (1997) *The Kalahari Transect: Research on Global Change and Sustainable Development in Southern Africa. IGBP Report # 42,* IGBP Secretariat, Stockholm, Sweden.

Scholes, R. J., Ward, D. and Justice, C. O. (1996a) Emissions of trace gases and aerosol particles due to vegetation burning in southern hemisphere Africa. *J. Geophys. Res.,* **101**, 23,677–23,682.

Scholes, R. J., Kendall, J. and Justice, C. O. (1996b) The quantity of biomass consumed in Southern Africa. *J. Geophys. Res.,* **101**, 23,667–23,676.

Skole, D. L., Justice, C. O., Townshend, J. R. G. and Janetos, A. C. (1997) A land cover change monitoring programme: strategy for an international effort. *Mitigation and Adaptation Strategies for Global Change,* **2**, 157–175.

Smith, J., Mulongoy, K., Persson, R. and Sayer, J. (2000) Harnessing carbon markets for tropical forest conservation: towards a more realistic assessment. *Environ. Conserv.*, **27** (3), 300–311.

START (1995) *Regional Workshop on Global Changes in Southern, Central and Eastern Africa.* Workshop report, Gaborone, Botswana, July 4–7, 1994.

TREES (1997) *Central Africa Forest Cover Database.* Tropical Ecosystem Environment Observations by Satellite (TREES), Joint Research Centre, Ispra, Italy.

Trexler and Associates, Inc. (1998) *Final Report of the Biotic Offsets Assessment Workshop, September 5–7, 1997.* Prepared for United States Environmental Protection Agency, Office of Policy, Planning and Evaluation, Washington, DC.

Totten, M. (1999) *Getting it Right: Emerging Markets for Storing Carbon in Forests.* World Resources Institute Report, WRI, Washington, DC, 48pp.

Tucker, M. (2001) Trading carbon tradable offsets under Kyoto's clean development mechanism: the economic advantages to buyers and sellers of using call options. *Ecol. Econ.* **37**, 173–182.

Tyson, P. D., Kruger, F. J. and Louw, C. W. (1988) *Atmospheric Pollution and its Implications in the Eastern Transvaal Highveld.* South African National Scientific Programmes, Report No. 150.

USAID (1998) *Climate Change Initiative 1998–2000.* USAID Global Environmental Centre, Washington, DC.

van Wilgen, B. W., Andreae, M. O., Goldammer, J. G. and Lindsey, J. A. (Eds.) (1997) *Fire in Southern African Savannahs: Ecological and Atmospheric Perspectives.* Johannesburg, South Africa, Witwatersrand University Press.

Wilkie, D. S., Morelli, G. A., Rotberg, F. and Shaw, E. (1999) Wetter isn't better: global warming and food security in the Congo Basin, *Global Environmental Change* **9** (4), 323–328.

Woomer, P. L., Martin, A., Albricht, A., Resck, D. V. S. and Scharpense, H. W. (1994) The importance and management of soil organic matter in the tropics. In P. J. Woomer and M. J. Swift (Eds.), *The Biological Management of Tropical Soils.* Chichester, UK, Wiley and Sons.

17 Climate-friendly energy policies for Egypt's sustainable development

Arabian Gulf University, Bahrain

Keywords

Energy/environment policy, GHG emissions, climate change, management of energy resources, renewable energy, energy efficiency, fuel switching to natural gas, energy prices reform, market barriers, Egypt.

Abstract

It is of crucial importance to Egypt and most developing countries to make fundamental changes in the energy policies, so as to comply with environmental requirements and ensure a sustainable path for development. Without environmentally sound energy policies, the production and consumption of energy could be a major source of carbon dioxide emissions, leading potentially to global climate change, in addition to other adverse local environmental impacts and negative effects on human health.

This paper outlines Egypt's energy/environment policy framework. It highlights the energy situation in Egypt in terms of supply and demand, as well as energy and economy linkages through the year 2010. It demonstrates a developing country's success story of better management of indigenous energy resources while striving to meet domestic energy demand and secure sufficient oil exports earnings that are needed to finance economic development. With 93.5% dependence on fossil fuels, the environmental impacts of the current energy systems are important. The paper describes the policy framework within which a large number of policy initiatives have been successfully implemented to mitigate such impacts.

Energy efficiency, switching from petroleum products to natural gas, promotion of renewable energy development and use, and energy pricing reform are the key elements of this strategic policy framework. The paper also discusses a set of market barriers that need to be overcome in order to secure a sustainable energy system.

17.1 ENERGY SITUATION OF EGYPT

17.1.1 National context

The Arab Republic of Egypt has a total land area of nearly one million square kilometres, and a population of about 66 million. The inhabited land is along the River Nile at the Delta and along the northern coast, representing about 5% of the country's total land area. The Gross Domestic Product (GDP) of Egypt in 2000/2001 was 296 billion (Egyptian Pound) LE. Figure 17.1 shows that the industry and mining sector contribute about 20.1% of the total GDP compared to 16.5% from agriculture, 6% from construction, 5.3% from petroleum production, 1.9% from electricity and nearly 50.2% from the services sectors.

17.1.2 Energy supply

Egypt's main energy resources are oil, natural gas, coal and hydropower, in addition to a good potential of renewable energy resources. Oil reserves are estimated to equal approximately 3.68 billion barrels (end of 2000–2001), most of which are located in the Gulf of Suez. Oil production during the same year accounted for about 32.3 million tonnes (Mt), of which nearly 73% were consumed domestically. Natural gas reserves equal nearly 10.64 billion barrels of oil equivalent. Most of the gas resources are located in the Mediterranean, Nile Delta and Western Desert. At the current annual production level of about 18.4 Mt, it is planned that natural gas will play an important role in the country's future fuel mix and commodity exports. Hydropower is the third major energy resource in Egypt. Most of the Nile's hydropower potential has already been exploited to generate about 13.7 Twh of electricity per annum. As shown in Figure 17.2, oil and gas account for about 95% of total commercial energy production, while hydropower represents the remaining 5%. In addition to oil, natural gas and hydropower, Egypt has limited coal reserves estimated at about 27 Mt.

Climate Change and Africa, ed. Pak Sum Low. Published by Cambridge University Press. © Cambridge University Press 2005.

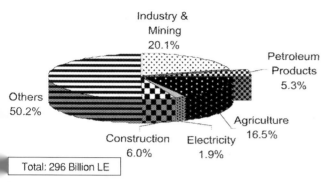

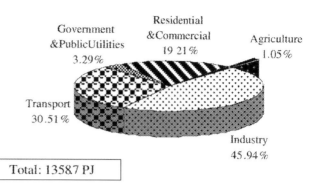

Figure 17.1. Egypt's total GDP composition in 2000/2001.

Figure 17.3. Egypt's energy consumption by sector, 2000/2001.

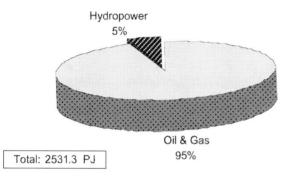

Figure 17.2. Commercial energy production, 2000/2001.

In addition to its commercial energy resources, Egypt has a good potential of renewable energy resources. These include solar, wind and biomass such as fuelwood, agriculture wastes and dried animal dung, which are used in rural areas as non-commercial energy resources to meet some energy demands. It is estimated that about 3.6 Mt of oil equivalent of biomass energy is consumed annually. Due to its geographic location, Egypt enjoys sunshine all year around with direct daily solar intensity ranging between 1,970 and 3,200 kWh/m². Meanwhile, wind speed assessments have shown that Egypt has locations of average annual wind speed of about 30 km/hr, suitable for producing power from commercial wind turbine farms.

According to 2000–2001 figures, primary commercial energy consumption totalled about 1930 PJ, of which oil represents about 50.6%, gas about 42.9% and hydropower about 6.5%. The per capita primary energy consumption is approximately 3.02 TJ, which is higher than the average of most developing countries.

17.1.3 Energy demand

Egypt has had a long history of rapidly growing energy demand, which grew by an average growth rate of about 8.6% per annum during the 1970s and 6.5% during the 1980s. During the same peri-

ods, the electricity sector was characterized by increasing reliance on oil as a basic source of fuel, due to limited availability of hydropower resources. Thermal power generation using petroleum products and natural gas currently represents nearly 82.3% of the total electricity generation; hydropower contributes 17.53%, while the balance is met by wind.

In addition to the energy transformation sectors, energy is consumed in four major end-use sectors: industry, transport, residential/commercial and agriculture. Figure 17.3 shows the energy consumption by sectors in 2000/2001, which reveals that industry and transport are the major energy-consuming sectors, accounting for about 76.5% of total final consumption.

Based on the Government's development plans, population growth, fuel prices, and using the income and price elasticities previously developed at the Organization for Energy Planning (OEP), the total and sectoral future energy demands during the period (2000/2001–2015/2016) have been determined using the Energy-Economic Model for Egypt. Table 17.1 indicates that total energy consumption is expected to increase from nearly 1,983 PJ in 2000/2001 to about 4,158 PJ in the year 2015/2016. The industry sector is expected to be one of the major consumers of petroleum energy in the year 2015/2016. Its demand is expected to increase from nearly 623 PJ in 2000/2001 to about 1,354 PJ in the year 2015/2016. The demand of the power sector is expected to increase from nearly 612 PJ in 2000/2001 to about 1,423 PJ in 2015/2016. The transport sector's total demand is expected to increase from nearly 415 PJ in 2000/2001, with a share of about 21% of total energy consumption, to about 712 PJ in 2015/2016, with a share of 17% of total energy consumption.

17.2 ENVIRONMENTAL IMPACT OF THE ENERGY SYSTEM

As is the case in many countries, the oil and gas industry is a major source of air pollution. This occurs during production, transport, refinement, distribution and the consumption of oil products.

Table 17.1. Projection of energy consumption (PJ) 2000/2001-2015/2016.

	Elec.	Petr.	Ind.	Res.	Trans.	Agri.	Total
2000/2001	612	56	623	263	415	14	1983
2005/2006	823	74	802	327	490	16	2531
2010/2011	1147	91	1082	420	604	19	3363
2015/2016	1423	105	1354	542	712	22	4158

Note: ELEC. Electricity; PETR. Petroleum; IND. Industry; RES. Residential; TRANS. Transport; AGRI. Agriculture.

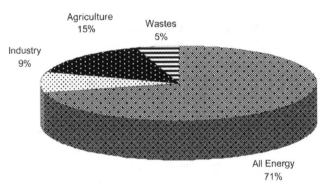

Figure 17.4. Contribution of greenhouse gases by sectors.

The main air pollutants are sulphur dioxide, particulates, hydrocarbons, carbon monoxide and nitrogen oxides. In addition to their combustion products, oil and gas operations are also responsible for significant fugitive emissions of carbon dioxide (CO_2), nitrous oxide (N_2O) and methane (CH_4). Another source of air pollution is thermal power generating units, which represent about 82.3% of the current electricity generation in Egypt. Oil and gas are the main sources for fuel, however, and as will be discussed below, a fuel substitution policy has been adopted so that 86% of thermal electricity generation has become natural gas based.

Another area of major concern in Egypt is the land-use conflicts between development projects, protected areas and the oil and gas industry, specifically along the Mediterranean and the Red Sea coastlines. In addition, most of the oil refineries and power stations exist in urban, heavily populated centres such as Cairo, Alexandria and Suez, causing adverse impacts on the urban environment.

The first greenhouse gases (GHG) inventory of Egypt for the year 1990 was undertaken in 1994, which has recently been refined based on the 1996 IPCC guidelines and default emission factors. The total net GHG emissions are estimated at 116.918 Gg of CO_2 equivalent using the 1995 Global Warming Potential (GWP). Figure 17.4 shows the GHG inventory by sector. With 93.5% dependence on fossil fuels, the energy sector is expected to be the major source of GHG emissions, contributing about 71% of the national total.

17.3 CLIMATE-FRIENDLY ENERGY POLICIES

The energy policies in Egypt have been developed and adopted to address many country-specific socio-economic and environmental considerations, in addition to some global economic and environmental concerns. Egypt is an oil and gas producing country. Its relatively limited hydrocarbon exports are, however, one of the main sources of the national income. On the other hand, the low land of the northern part of the Nile delta is reported by the Intergovernmental Panel on Climate Change (IPCC) to be vulnerable to the possible sea level rise due to global climate change. As the Nile Delta is one of the major sources of food production in Egypt, it became evident that assessments of the socio-economic implications of such environmental threats are badly needed. Being economically and environmentally vulnerable, Egypt formulated its position within the climate change negotiations to address these issues, and designed its energy and environmental policies to ensure sustainable development. Four major energy policies are outlined below.

17.3.1 Energy efficiency

Energy efficiency is a national necessity for Egypt's sustainable development. The numerous benefits of energy efficiency to the Egyptian economy include increasing hydrocarbon surplus available for exports, enhancing profitability of industry and competitiveness in the world market, creating new jobs, and improving environmental quality. Though the primary energy resources of Egypt are limited, they play a vital role in its economic and social developments. The energy intensity is currently equal to 24.7 GJ/000' 95 US$,[1] which is higher than the levels in most developing countries. Egypt took an early step to improve its energy efficiency in the early 1980s, when several studies were conducted to identify and better understand the main features of Egypt's energy system and its interrelated linkages with the national economy.

---

[1] This means 24.7 GJ primary energy is used to produce a 1000 US$ equivalent of GDP at the 1995 prices.

These studies have been a good vehicle to develop national professional capabilities. They have also proven that energy policies would have long-term implications on the performance of the Egyptian economy. Furthermore, these studies have identified a set of energy-related problems, which adversely affect most of the other economic sectors. One of the most alarming energy issues in Egypt has been the poor level of energy efficiency in the main consuming sectors, which leads to a demand growth that surpasses the economic and population growth rates. Such escalating demand has been always met with the typical supply-oriented approach. This accelerates the depletion of indigenous oil reserves, erodes export revenues, threatens the environment, and exhausts most of the capital needed for investments which, in turn, will make it difficult for Egypt to meet its future energy needs. However, well-balanced supply and demand strategies have been recently realised, and the policies of the energy sector have been developed to adopt energy efficiency as one of the strategies to enhance economic efficiency and protect the environment. There are several recent market drivers for energy efficiency in Egypt: (i) accelerating privatization of industrial state-owned enterprises has led to more rational energy use and more awareness of competition; (ii) growing domestic and international competition is creating pressure to improve the quality of products and reduce operating costs; (iii) availability of technologies is playing a significant role in driving energy efficiency market for products and services; (iv) due to removal of historical energy subsidies within the economic reform programme, the slowdown in energy demand growth has been obvious over the last few years; (v) economic pricing of energy products and services has emphasized the importance of energy efficiency to different energy consumers; and last but not least, (vi) the enforcement of environmental Law No. 4, which calls for specific levels of air and water quality, also creates an increasing demand for efficient and clean technologies. There have been a number of energy efficiency measures undertaken by many Egyptian institutions. These activities have clarified to the major energy users the extent of their wastage of energy and the potential of energy saving that could be achieved through investing in energy efficiency. Most of these efforts have focused on the industrial sector.

Industrial energy audits showed an average potential energy saving of about 25%, of which nearly 10% could be achieved by proper maintenance and housekeeping. These audits served as a basis for the Energy Conservation and Environmental Protection (ECEP) project funded by United States Agency for International Development (USAID). This project aimed at demonstrating nine energy efficiency technologies, these are: (i) cogeneration; (ii) power factor improvement; (iii) combustion control; (iv) energy management systems; (v) high efficiency lighting; (vi) waste heat recovery; (vii) process control; (viii) efficient motors and drives; and (ix) regenerative burners.

In parallel, and to improve energy efficiency of the power sector, the Egyptian Electricity Holding Company (EEHC) has embarked on a programme to rehabilitate the old power stations. During the past few years, rehabilitation has been carried out for eleven steam power plants of 1,977 MW total capacity. In addition, EEHC has planned to use more efficient, large, and new generating units and to expand the use of combined cycle plants as part of its expansion plans. Furthermore, EEHC is preparing to introduce a time of use (TOU) tariff to enhance its load management. These activities have led to the reduction of the average rate of fuel consumption for thermal electricity generation from 346 g/kWh in 1981/1982 to 221 g/kWh at present. EEHC has also upgraded its transmission and distribution system to reduce its losses from 19% in 1980/1981 to 14% at present. EEHC, in cooperation with other Egyptian institutions, is currently implementing another project supported by the Global Environment Facility (GEF) to reduce carbon dioxide emissions from the energy sector. This project, entitled Energy Efficiency Improvement and Greenhouse Gas Reduction Project, comprises three components, which are (i) loss reduction, load shifting and load management in the electrical network; (ii) energy efficiency market support; and (iii) cogeneration.

As part of Egypt's climate change national action plan, a national energy efficiency forum has been recently formulated to develop and oversee the implementation of a national energy efficiency strategy. The forum's membership includes the main players in the energy sector from the oil industry, power producers, major energy consumers, and the private sector energy services providers.

17.3.2 Fuel-switching policy

Natural gas is playing a key role in the country's energy policy. Given its economic and environmental advantages, natural gas will improve the overall energy efficiency and environmental quality of Egypt. As is well recognized, primary energy resources in Egypt have been nearly all hydrocarbons-based, in addition to some other limited resources, but proven natural gas reserves are more abundant than oil, so the energy policy of Egypt has been developed to promote the substitution of natural gas in various sectors. Strategies include: (i) developing gas infrastructure to expand gas markets and develop domestic gas demand – the market share of natural gas in the total hydrocarbon consumption has increased to about 45%; (ii) the substitution of heavy fuel oil with natural gas in electricity generation has made considerable reduction in air pollution; and (iii) promotion of unconventional applications such as Compressed Natural Gas (CNG) as a transport fuel is also under way. A number of private firms have been formed to participate in the construction of gas pipelines, building CNG fuelling stations and converting vehicles to use CNG. A total of 57 fuelling CNG stations have been built, and the number of vehicles converted to CNG has reached more than 38,800. Another initiative to use CNG as a fuel for two-stroke engines is under discussion between the Government of Egypt and the Government of Canada. At present, a Canadian technology provider is negotiating the transfer of this technology to the Egyptian two-stroke engine manufacturers to produce CNG fuelled motorcycles. The project would be a major step forward to reduce carbon dioxide emissions and improve air quality in Egypt.

17.3.3 Promotion of renewable energy resources

Egypt's energy situation and its future outlook necessitate urgent expansion of the development and use of renewable energy resources. Intensive efforts have been directed to achieve these goals. A renewable energy strategy was developed and has been incorporated as an integral element of national energy planning since 1980. The strategy targeted 3% of the electricity peak load in 2021/2022 from renewable energy. These are primarily solar, wind and biomass energy resources. In 1986 the New and Renewable Energy Authority (NREA) was established to provide the institutional framework for implementation of the strategy and act as a focal point for expanding efforts to develop and introduce renewable energy technologies to Egypt. Since then, NREA has been successful in assessing renewable energy resources. Solar and wind atlases were developed, and several studies on biomass resources in rural and urban areas have been undertaken. Studies show that solar radiation is highly available in all the Egyptian territories. The average annual national radiation varies between 1,900 and 2,600 kWh/m^2 per year (OEP, 2002). In addition, Egypt enjoys tremendous wind resources in the Gulf of Suez, where the annual average wind speed reaches 10 m/s at 30 m above sea level. Several solar technology options were assessed, field demonstrated and found to have a promising market potential. These include solar thermal low-temperature technologies, photovoltaic applications, and solar thermal electricity generation. NREA has already finalized the pre-feasibility studies for the first Integrated Solar Combined Cycle Systems (ISCCS) with a capacity of 127 MW. In addition, Egypt has already crossed the phase of demonstration and pilot projects of large-scale wind farms. Currently, a 5-MW grid-connected wind farm in Hurghada, on the Red Sea, is operating successfully. A wind energy technology centre was established to provide technical support to wind energy manufacturers and developers. Additionally, the first large-scale (60 MW) grid-connected wind farm, supported by the Danish government, is under construction at Zafarana on the Gulf of Suez. Discussions are under way between NREA and both the German and Japanese governments to add more capacity to this wind farm. The installed capacity of this wind farm is planned to reach 600 MW by the year 2010. Recently, NREA has developed a long-term vision for the priority applications of renewable energy technologies through the year 2017. This long-term vision envisages about 10% of the total industrial process heat by solar thermal technologies, about 9 TWh/year of solar electricity generation and 2,000 MW of installed capacity as wind farms. As a result, CO$_2$ emissions would be reduced by about 12 Mt annually.

To promote alternative energy technologies, the Egyptian Environmental Affairs Agency (EEAA), supported by the GEF, is currently leading two initiatives in the transport sector aimed at reducing carbon dioxide emissions and improving air quality. The first initiative was designed to study the feasibility of using hydrogen in fuel cell technology as an alternative fuel in Cairo buses. The demonstration project is to feature eight buses tested under normal operation for about three years.

The second initiative was designed to test the feasibility of operating electric buses in Cairo. Within the project, 22 electric buses will be demonstrated in selected sites in Egypt and a commercialization plan will be established for the promotion of electric vehicles in the Egyptian market. The final objective of the project is to integrate the electric and hybrid electric drive systems into Egyptian-made buses leading to cost reduction.

17.3.4 Energy pricing reform

Domestic energy prices have been characterized by a great deal of rigidity for a long period of time. The under-pricing of energy products has promoted unwise consumption behaviour and non-optimal investment strategies. For decades, the Egyptian economy has suffered from severely subsidized energy prices.

Furthermore, the artificially low domestic prices of energy have also impaired the economic viability of most of the efficiency projects. Within the economic reform programme, the government has decided to escalate the energy prices gradually to their economic levels. It should be noted that, although the market prices of petroleum products, natural gas and electricity have increased since 1985, their current real prices are still under the economic levels. However, the expected effect of the pricing policy on slowing down the growth in demand has been evident in the last few years. Energy intensity has also been improved by an average of 7.4% annually during the last five years.

17.4 BARRIERS TO IMPROVING ENVIRONMENTAL QUALITY

During the transition period to market economy, when most of the market distortions are not fully reformed and some market drivers are not fully operational and effective, it is expected that the energy efficiency and environmental quality market will be expanded slowly. There is a set of market barriers that inhibit energy efficiency and environmental market growth. These are lack of financing mechanisms for energy efficiency and environmental projects, poor cash flow of public sector industries, lack of government incentives to enhance environmental quality, and lack of application of the concept of life cycle costing. Investments in environmental quality improvement need to be placed on top of the country's portfolio. Within the context of the UNFCCC negotiations, and the evolution of the Clean Development Mechanism (CDM), it is evident that Egypt would be in a good position to participate in such a mechanism. Egypt's renewable energy resources and energy efficiency potential could be an excellent opportunity to demonstrate the viability of the emerging Kyoto mechanism. Egypt's climate policy welcomes the international cooperation and regards it as inevitable to achieve Kyoto targets as well as sustainable development in developing countries.

17.5 CONCLUSIONS

Though Egypt is energy self-sufficient, its primary energy resources are relatively limited, but its renewable energy resources are vast and should be better exploited. Natural gas is playing and is expected to play a key role in the country's energy profile for decades to come. A gas substitution policy is currently one of the main economically viable and environmentally benign options to achieve a sustainable energy path. The power sector is heavily reliant on natural gas. Evolution of the peak demand and continuing growth in electricity consumption has reversed the hydro-dominated fuel mix of the early 1970s to a current mix that is dominated by natural-gas-based thermal generation. Thermal efficiency in electricity generation has been remarkably improved with a consequent reduction in GHG emissions. Within the economic reform programme, the partial removal of the historical energy subsidies has led, together with other factors, to a noticeable slowdown in the growth in energy demand, as well as improvement in energy intensity. Energy intensity in Egypt is higher than the average of developing countries; however, its level has stabilised during the last few years. To achieve greater energy efficiency, a set of policy and market barriers needs to be overcome. Egypt's climate change policy is dictated by its socio-economic and environmental considerations. It promotes immediate action to reduce GHG emissions, fosters international cooperation, and calls for active participation in the post-Kyoto global initiatives to enhance the transfer of environmentally sound technologies from developed countries in order to achieve sustainable development.

ACKNOWLEDGEMENTS

The author is sincerely grateful to Pak Sum Low, Francis D. Yamba and Tom Hamlin for their efforts and valuable comments during the review process of this chapter. The assistance of Tawfik Azer of OEP on the preparation and updating of the energy statistics of Egypt is greatly appreciated.

REFERENCES

Abdel Gelil, I. (1997) Energy situation in Egypt, efficiency perspectives. In *Proceeding of the Fourth Solar Energy Conference, Cairo, Egypt.*

Abdel Gelil, I., Amin, A., El Quosy, D. *et al.* (1997) Egypt. In R. Benioff, E. Ness and J. Hirst (Eds.), *National Climate Change Action Plans: Interim Report for Developing and Transition Countries*, pp. 58–68. US Country Study Programme, USA.

Abdel Gelil, I., Korkor, H. A., Ragi, R. F. and Bedrous, M. A. (1998) *Environmentally Sound Energy Policies and Technologies for Egypt's Sustainable Development.* World Energy Council (WEC), 17th Conference, Houston, USA.

Bedewi, N. (1999) *Introduction of Viable Electric and Hybrid-Electric Bus Technology in Egypt and the Creation of an Export-oriented Manufacturing Base Leading to Job Creation and Sustainable Development.* Unpublished report. Egyptian Environmental Affairs Agency (EEAA), Egypt.

Chen, T. (1999) *Final Summary Report for Integrated System for Zero or Near Zero Emission Fuel Cell Bus Operation in Cairo.* Bechtel Corporation, USA.

Egyptian Electricity Holding Company (EEHC), *Annual Report (1996/97).* EEHC, Egypt.

IPCC (1994) *Guidelines for National Greenhouse Gas Inventories*, Intergovernmental Panel on Climate Change, WMO/UNEP. Cambridge and New York, Cambridge University Press.

Organization For Energy Planning (OEP) (2002) *Energy in Egypt, Annual Report (2000/2001).* Organization For Energy Planning, Cairo, Egypt.

PART III

Vulnerability and adaptation

18 Potential impacts of sea-level rise on populations and agriculture*

RENÉ GOMMES, JACQUES DU GUERNY, FREDDY O. NACHTERGAELE, AND ROBERT BRINKMAN

Food and Agriculture Organization of the United Nations, Rome, Italy

Keywords

Sea-level rise; extreme events; population density; scenarios

Abstract

The chapter discusses some issues related to the potential impacts of sea-level rise (SLR) on coastal populations and agriculture. This is a study mostly based on global data at the countrywide (national) scale. Indirect effects of SLR, as well as the potential impact of extreme events, may be more significant than direct effects in the future. In the absence of an accepted methodology for building long-term scenarios, two approaches are explored: an analysis of a large database of extreme events that have occurred over the last 100 years, and an analysis of population statistics in relation to a national vulnerability index based on physiographic features and population density. Recent historical data are examined with a view to identifying trends that could be extrapolated into the twenty-first century. Despite the limitations of the data sets, some trends do emerge, but they do not necessarily point in the direction of greater property and population losses in the future owing to sea-caused disasters. Rather, they seem to indicate that difficulties – independent of the global changes – will be relatively larger on land than along the coasts, and that the major component of life and property losses is associated with levels of economic development. The 'national vulnerability index' confirms that vulnerability – if considered at the scale of the globe – varies considerably, over several orders of magnitude. In addition, the index exhibits a marked positive skew. Combined with changes in population concentrations and the positive skew of many climatological elements (such as wind and rainfall), as well as of SLR itself, this indicates that relatively greater disasters are likely. The paper stresses the fact that both the impacted system (population and coastal agriculture) and the extreme physical factors have their own dynamics, and that those dynamics are independent. Some thought is given to the notions of 'shock waves', i.e. the repercussions at some distance from the seas.

18.1 INTRODUCTION

Just constructing dykes along all threatened coasts may, of course, not be a good idea

CZMS, 1990, page 86

In the 'worst case' scenario, global mean sea level is expected to rise 88 cm by the year 2100, with large local differences due to tides, wind and atmospheric pressure patterns, changes in ocean circulation, vertical movements of continents, etc.; the central value is 48 cm (IPCC, 2001). The relative movement of sea and land is the main factor of local sea-level change: some areas may experience sea-level drop in cases where land is rising faster than sea level. In addition, according to a study by Titus and Narayanan, quoted by CZMS (1992), the statistical distribution of future sea-level rise (SLR) exhibits a marked positive skew (i.e. many near-average values and some very large ones in only a few locations).

Regarding human settlements, Scott (1996) expresses the view that the impacts of SLR and extreme events are likely to be experienced indirectly through effects on other sectors – for instance changes in water supply, agricultural productivity (Brinkman, 1995) and human migration. In addition, it is uncertain whether extreme events associated with oceans will change in intensity and frequency (Ittekkot, 1996; Nicholls et al., 1996). It seems

* Disclaimer: The designations employed and the presentation of the material in this publication do not imply the expression of any opinion whatsoever on the part of the Food and Agriculture Organization of the United Nations concerning the legal status of any country, territory, city or area or of its authorities, or concerning the delimitation of its frontiers or boundaries. The designations "developed" and "developing" economies are intended for statistical convenience and do not necessarily express a judgement about the stage reached by a particular country, country territory or area in the development process. The views expressed herein are those of the authors and do not necessarily represent those of the Food and Agriculture Organization of the United Nations nor of their affiliated organization(s).

Climate Change and Africa, ed. Pak Sum Low. Published by Cambridge University Press. © Cambridge University Press 2005.

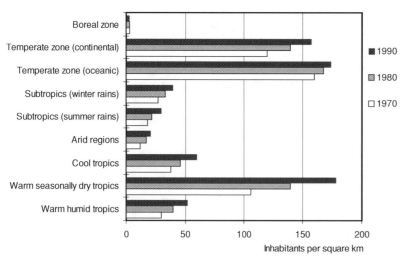

Figure 18.1. Recent population density changes according to major agro-climatic regions. Worldwide data; graph prepared using data from Tobler *et al.* (1995, 1997).

likely, therefore, that the internal dynamics of human demography, coupled with a series of indirect factors, including health factors (WHO, 1996), may eventually play a dominant part.[1]

According to Nicholls (1995), quoted by WHO (1996), the majority of the people that would be affected under the worst scenario live in China (72 million) and in Bangladesh (13 million). Between 0.3% (Venezuela) and 100% (Kiribati and the Marshall Islands) of the population would be affected. It is worth noting, however, that population per se receives relatively little attention in the literature as compared, for instance, to natural ecosystems or agriculture.

A disaster results from the impact of an extreme physical event on a vulnerable society or human activity (Susman *et al.*, 1983). Disasters can be quantified and predicted only insofar as the factors of the product 'extreme event × vulnerable system' are reasonably well known and quantified. In the specific case of SLR and population, only some terms of the equation are known. At the macro level, population growth is affected by the least error, but details of future population distribution, as well as the level of urbanization, are more open to debate – especially the question whether the future concentration of population will coincide with the area with large relative SLR. The vulnerable system itself is currently difficult to describe at the global level, for two reasons. First, sufficiently detailed digital maps of elevation, crops and population are not available. Second, although the dynamics of the response of coasts to SLR can be inferred with some confidence (Jelgersma *et al.*, 1993), the interaction with human activities and populations is largely open to debate. As to future impacts around years 2050 or 2100, we are not in a position either to describe with any level of accuracy and confidence what the impacted systems will be like. This is because, inter alia, both the coastal landscape and

buildings and infrastructure will adapt gradually in response to the changing environment and the socio-economic driving forces. The main weaknesses are thus in the SLR predictions, as well as the interactions with human activities.

18.2 THE GENERAL DEMOGRAPHIC, PHYSIOGRAPHIC AND SOCIO-ECONOMIC SETTING

Current population distribution varies markedly according to a number of factors. In particular, there are marked differences in population growth according to agro-climatic zones (Figure 18.1).

Population densities grow fastest in the seasonally dry tropics, while in temperate countries little change takes place. This will lead to the need to produce more food in the former zones, which have quite severe ecological and agricultural limits (Gommes, 1992).

Contrary to a common assertion[2] according to which 'it is estimated that 50–70% of the global human population lives in the coastal zone' (Bijlsma *et al.*, 1996), the population is rather land-bound, as illustrated in Table 18.1. The densities given are approximate in that they are based on an assumed total length of the coastline of 100,000 km and on 'large, round' continents. Global population density is about 39 persons per km^2. In spite of the gross approximations involved in the last column of Table 18.1, it is clear that population densities are far higher along the coasts than inland. Small (personal communication) indicates the percentages to be 37% within 100 km, and 66% within 400 km.

---

[1] Direct effects can be assumed to be more significant in the ecological sphere (see, for instance, Bijlsma *et al.*, 1996).

[2] The statement is ubiquitous, with many variants, including, 'By some estimates, nearly two-thirds of the world's population lives within 100 miles of an ocean, inland sea or major freshwater lake' (Engelman, 1997), or 'Six out of ten people live within 60 km of coastal waters' (IUCN, 1991) or 'At present, six out of 10 people live within 60 km of the coast' (FAO, 1997).

Table 18.1. Distribution of world population as a function of the distance from the nearest coastline.

Distance from the coast (km)	Population (million)	Accumulated population (million)	Accumulated percentage	Approximate density (persons/km^2)
Up to 30	1,147	1,147	20.6	382
>30 to 60	480	1,627	29.2	160
>60 to 90	327	1,954	35.0	
>90 to 120	251	2,205	39.5	
Beyond 120	33,395	5,567	100	

Based on the digital vector map by Tobler *et al.* (1995 and 1997), roughly 1:5 Million scale, population standardized to 1994.

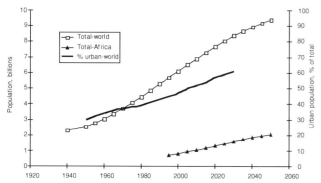

Figure 18.2. Recent and future population and urbanization trends (based on data from UN 1996a, 1996b, 1997).

There are, of course, large local differences. For instance, Sestini (1992; quoted by Zwick, 1997), writes that:

The importance of the Mediterranean seafront in relation to the rest of the country varies; as an example, it is relatively less so in Spain, France and Turkey than in Italy, Greece, Albania, Algeria, Israel. In Greece as much as 90% of the population lives within 50 km of the coast and all major industrial centres are coast-related as well as much of agriculture. In Egypt, the Nile delta north of Cairo represents 2.3% of the area of the country, but contains 46% of its total cultivated surface and 50% of its population; the [altitude] belt 0–3 m harbours about 20% of the population (with Alexandria 3.5 million, and Port Said 450,000 inhabitants), 40% of industry, 80% of port facilities, 60% of fish production.

World population, now at about 6.3 billion, is expected to begin levelling off around 2050 at about 9.3 to 9.4 billion people (UN, 1996a, 1996b), although in some continents, notably Africa, population will probably continue growing at a more sustained pace well into the twenty-first century. The urbanized population should exceed 60% of the total in 2030, from current values of around 50% (Figure 18.2).

It is well known that most of the current largest urban concentrations[3] are on the sea coasts (Engelman, 1997). The population in the world's 15 largest cities reached 277 million in the year 2002. It appears that overall urban population trends are not so obvious. While most coastal megacities do grow in size, their share of the total population of the country often remains stable (1% for Calcutta and Shanghai from 1950 to 2015), and sometimes decreases (from 7% to 6% for New York over the same period).

Also, the percentage of the urban population living in megacities often decreases (New York: 12% to 7% between 1950 and 2015; Cairo: 35% to 32%; Rio de Janeiro: 14% to 6%; Calcutta: 7% to 4%; Beijing: 6% to 2%; Jakarta: 15% to 10%, etc.), which points to the growth of other urban areas.

Also noteworthy is the fact that many cities suffer land subsidence due to groundwater withdrawal (Nicholls and Leatherman, 1995). This may be compounded by sea-level rise, the more so since current rates of subsidence may exceed the rate of sea-level rise between now and 2100.

Table 18.2 indicates how some socio-economic and physiographic indicators vary among landlocked countries, those with coastlines and the smallest islands, which are members of the Alliance of Small Island States (AOSIS). A list of these indicators for the African countries with a seacoast is given in Annex A. It is striking that the average Gross Domestic Product (GDP) per capita in landlocked countries is just above that of AOSIS members, and well below the global average. It is also worth noting that the population densities of the small island states are currently rather high and still projected to grow. It is unlikely that this will be accompanied by a marked increase in GDP per capita.

In order to examine in more detail the relation between some of the indicators in Table 18.2 and the 'insularity' of the respective countries, we define an 'Insularity Index' as the ratio between the

[3] They include, in decreasing order of the projected population in the year 2000, Tokyo, Mexico City, Bombay, Sao Paulo, New York, Shanghai, Lagos, Los Angeles, Calcutta, Buenos Aires, Seoul, Beijing, Karachi, Delhi and Dhaka (UN, 1997).

Table 18.2. Some general statistics about landlocked countries and territories and AOSIS member states.

	No.	Area '000 (km^2)	Arable land (%)	GDP/Capita US$ (1996)	Pop. million (1995)	Pop. density (1995)	Pop. density (2050)	Coast-line length (km)
Landlocked	44	378	13.8	4,711	348	21	52	0
AOSIS	32	9.3	17.2	4,518	37	122	198	1091
Others	154	83.4	13.0	7,829	5325	42	66	5144
Average		129	13.8	6,772				
Total/global	230				5710	39	65	

The data in the table are averages of country values, with the exception of the population densities, which are averages over all the countries covered in the three categories. 'Others' stands for countries that have direct access to the ocean but are not AOSIS members.
All data from *Factbook* (1997), except population statistics taken from UN (1996a, 1996b, 1997).

Table 18.3. The Insularity Index for various countries and territories grouped by 'continents'.

	America (N and S)	Asia and Australia	Europe and Africa	World
No. of countries	57	46	121	224
Average Insularity Index	0.411	0.611	0.0772	0.272
Median Insularity Index	0.0944	0.0226	0.00234	0.00770

The landlocked countries are included in the statistics (there are 2, 9 and 31 landlocked countries, respectively, in the America, Asia and Australia, and Europe and Africa groups). 'America' covers the area west of 30° western longitude (30° W) and 180° W; 'Europe and Africa' spans the longitudes from 30° W to 60° E, and 'Asia and Australia' goes from 60° E to 180° E. The Pacific Islands (roughly from 145° E to 109° W at equatorial latitudes) are partly included in 'America' and partly in 'Asia and Australia'.
The data used to compute the Insularity Index values are from *Factbook* (1997). The different numbers of countries given in different tables are due to missing data in the original data sets.
Refer to Annex A for details on Africa.

length of the coastline (km) and the total land area (km^2) that it encloses:

$$\text{Insularity index} = \frac{\text{Coastline (km)}}{\text{Land area (km}^2)}$$

The definition of the 'Insularity Index' is, of course, fraught with problems[4] linked with the actual shape of countries, the distribu-

tion and extent of low-lying areas within each country, as well as the fractal nature of coasts and therefore, the scale at which the coastline is determined. It is admittedly a crude index, but meaningful if a consistent method is used to estimate the length of the coastlines. Some interesting links with other variables can be found at the global level. Table 18.6 lists some typical values of the Insularity Index.

It is obvious, to start with, that Africa and Europe – which represent half of all countries and territories – are far less 'insular' than the other continents. It is also apparent that the Insularity Index has a strong positive skew and covers five to six orders of magnitude, so that it can be represented only on a logarithmic scale. The positive skew is clearly visible in Table 18.3, in which the large difference between median and average is due to the occurrence of a limited number of very high values.

Figure 18.3 shows the link between the insular character of countries and population density. Less insular countries are generally relatively less populated, which is linked to the fact that

---

[4] The length of the coastline can vary a lot according to the methodology used. For the total length, numbers oscillate between 46,185 km and 339,185 km if a multiplier is used to consider the complexity of low-lying coastlines (see Hoozemans *et al.* (1993) as well as Appendix D of the CZMS publication in 1990). For instance, the *Factbook* (1997) indicates 14,500 km for China (without Taiwan), but Du (1993), quoted by CGER (1996), uses 18,700 km (including Taiwan). India has 7,000 km in both CGER (1996, quoting Asian Development Bank) and *Factbook* (1997), while CZMS (1990) indicates 3,280 for the length of the coast 'as the crow flies' with a step of 50 to 100 km. The same source applies a multiplier of 10 to obtain 32,440 km of 'length of low coast', i.e. the actual length of coast that should be protected, taking into consideration its 'micro' structure.

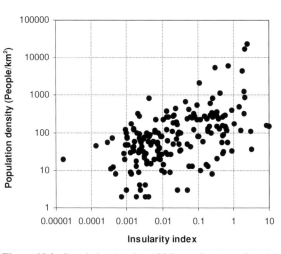

Figure 18.3. Population density (1995) as a function of the Insularity Index (based on 192 countries and territories for which the data are available in UN (1996a) and UN (1997) (population data) and in *Factbook* (1997) for the computation of the Insularity Index).

landlocked countries are mostly at higher elevations and latitudes (Eurasia) or in semi-arid areas (Africa), where productivity and population-supporting capacities tend to be low. In a similar way, and as a consequence of the situation described previously, GDP tends to decrease with insularity, mostly so above a 'threshold' of 0.01 (Figures 18.4a and 18.4b).

18.3 COASTAL AREAS AND CLIMATE

More than direct land loss due to seas rising, indirect factors are generally listed as the main difficulties associated with SLR. These include erosion patterns and damage to coastal infras-tructure, salinization of wells, sub-optimal functioning of the sewage systems of coastal cities with resulting health impacts (WHO, 1996), loss of littoral ecosystems and loss of biotic resources.

While it is possible, at any location, to define a 'climate com-plex', i.e. a set of interrelated climatic variables (Sombroek and Gommes, 1996), it should be noted in the current context that coastal climate complexes tend to be rather different from the cli-mate complexes at some distance from the sea (Figure 18.5). It is unknown how global climate change would alter these patterns, but it is obvious that zonal scenarios are unlikely to apply on the coast. Similarly, the sea- and land-breeze patterns are very likely to be altered. Since delta agriculture typically uses a combination of irrigated and rain-fed crops, a change in the relative behaviour of the local delta climate and the climate in the basin may lead to unprecedented difficulties. It might impose, for instance, a shift to more salt-tolerant crops.

In coastal areas, vulnerability to SLR will depend on the magni-tude of the projected relative SLR, on modified ocean circulation

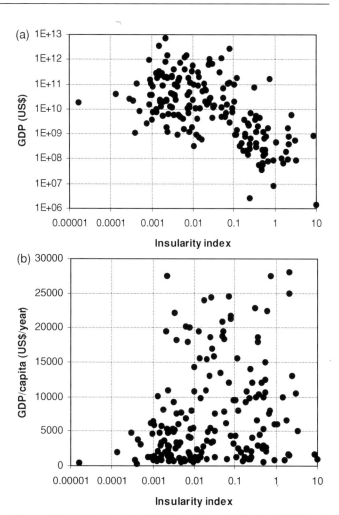

Figure 18.4. Variation of (a) 1996 Gross Domestic Product (GDP) and (b) GDP per capita, as a function of Insularity Index (based on data from *Factbook*, 1997).

patterns as they affect the building and erosion of the coast, on changes in the coastal climate and in the frequency of extreme events, as well as on climate change in the catchment basin. This constitutes a major difference with small islands, where remote land-based changes are likely to be of less importance.

Conversely, major disasters or changes in deltas and small is-lands could have repercussions over large areas. Given the size of populations potentially involved, this is more likely to seri-ously affect the major deltas, such as the Ganges-Brahmaputra, Mekong and Nile. In both the cases of deltas and small islands, a likely scenario could be out-migration when disasters due to SLR reach levels or frequencies considered unacceptable. It is at such thresholds that maximum damage and loss of life could be expected.

It may be useful to more closely examine the evolution of coastal lowlands during the period from about 18,000 to 6,000 years ago: then, the sea level rose faster than ever since, by some 120 m in

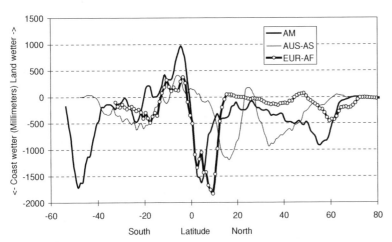

Figure 18.5. Difference between 'coastal' and 'continental' rainfall in the three conventional continents defined in Table 18.3 (based on the IIASA data set of Leemans and Cramer, 1990). Negative values indicate rainier conditions along the coasts, defined by 0.5° squares (up to 60 km from the coast at the equator), while land areas are all land found at the same latitude. The graph shows 5° moving averages.

12,000 years (Jelgersma and Tooley, 1995). In comparison, the sea level on the coast of the Netherlands rose by 4 m over the last 4,000 years.

Speculating on the impacts of that rapid and prolonged SLR on human society, we might note that near its beginning, some 9,000 years ago, arable agriculture began. Just after the end, some 6,000 to 5,500 years ago, there was the rise of a series of well-planned cities, with well-ordered agricultural areas surrounding them, in various coastal lowlands or lower river valleys, e.g. of the Euphrates-Tigris, Nile and Indus.

Where are the precursors of these cultures, so sophisticated in their spatial planning? Did they originate near the coast much earlier, at lower elevations, with those societies that survived and prospered being those that could best plan their successive evacuations to higher ground, re-establish their physical infrastructure and maintain their social organization? A re-reading of the several 'Flood' records in ancient literature worldwide (for example, George, 1999) might provide new insights. Also, underwater investigation of the shallow parts of the relevant undersea deltas — for example, by seismic means or possibly ground-penetrating radar — might yield new information on that intriguing period. We might learn something of use to present society, which is facing similarly great changes, albeit of a different kind.

As mentioned in the introduction, small islands and low-lying coastal areas have received much attention in the literature, including the recent IPCC assessments. Deltas fall into the same category of areas very vulnerable to SLR. But unlike deltas and other coastal areas, small islands have no hinterland to move to in the case of loss of land. In addition, their land resources are very limited.

18.4 DELTAS

According to Nicholls and Leatherman (1995), a SLR of one metre would affect 6 million people in Egypt, with 12% to 15% of agricultural land lost; 13 million in Bangladesh, with 16% of national rice production lost; and 72 million in China and 'tens of thousands' of hectares of agricultural land. These estimates do not take into account the natural adaptations of coastal areas to SLR or the human interference, both upstream and in deltas, with natural sediment supply.

Deltas facing open seas or oceans pose a set of more specific problems, as they are the areas where sea and land most closely interact. Their average elevation is usually very low, to the extent that tidal effects can be felt for several tens, and in some cases, hundreds of kilometres inland. Deltas in enclosed seas, such as the Nile and Danube, are subject to less tidal influence. This land/sea interaction results in very complex agricultural systems, where irrigation and rain-fed agriculture may be practised in alternate seasons, with attention to irrigation water quality (salinity) and to the washing out of salts by rains before planting crops.

The Mekong Delta provides a clear example of this complexity (see, for example, Jelgersma *et al.*, 1993). Siltation is usually very active during certain seasons of the year: deposition in the Mekong Delta continues to extend the Ca Mau peninsula to the south and west at a rate of 150 m/year in some places (Fedra *et al.*, 1991). The delta, therefore, is as much the result of conditions upstream as the result of local or coastal interactions. Sea-level rise, therefore, cannot be examined without some sound assumptions about climate and rainfall changes in the river catchment as a whole.

Contrary to many other, 'normal' coasts deltas are difficult and expensive to protect, because of the very dissected nature of the coastline. A measure of the difficulty is provided by the 'multiplier'[5] used by CZMS (1990) to evaluate protection costs. Multipliers for deltas usually range between 6 and 8, as for the Ganges-Brahmaputra-Mahanadi (7.5), but may reach higher values, as for the Magdalena Delta in Venezuela (10.4), Orinoco (11.3), Parana (26.7), or Mekong (9.8). CZMS (1990) stresses that the multiplier may be affected by large errors (100%), but the parameter gives at least a crude measure of the level of intricacy of the coastline in deltas.

Although deltas tend to be densely populated, it is not really relevant to assess exactly how many people live in them.[6] Due to their very high productivity (generally fertile soils, water availability, opportunities for multiple cropping, especially in tropical areas) they produce significantly more food than required for local consumption. In Vietnam, for instance, 50% of national rice production comes from the Mekong Delta in the south, while 20% is produced in the Red River Delta near Hanoi. Although much of this production is for export, a disaster in the deltas would have profound effects on the whole country. This indicates a fragile situation in which any major disturbance would result in economic and possibly political shock waves well beyond the delta proper.

It is also noteworthy that deltas, although not specifically fragile, are the result of a long evolution and that they are, in the words of Riebsame et al. (1995), tuned to the current climate. Any departure from the current equilibrium, even one that would a priori be positive (e.g. better rainfall/evaporation ratio), could result in a temporary disturbance of the ecology and productivity.

18.5 EXTREME EVENTS: LESSONS FROM THE RECENT PAST

The purpose of this section is to examine whether any trends can be recognized in disasters that have involved coasts and population over recent decades.

There are several databases on disasters, usually focusing on specific themes. A brief introduction to the subject is given by IFRCS (1996). For the current study, we adopted the database assembled by the US Office of Foreign Disaster Assistance, among

Table 18.4. Number of people killed per disaster and number per disaster per million world population.

	Sea	Land	Population	Other
No. of disasters	555	1,918	1,043	322
Dead/disaster				
Average	1,244	1,749	6,956	190
Median	555	12	54	13
Dead/disaster per million population				
Average	0.336	0.597	2.155	0.051
Median	0.007	0.003	0.011	0.003

1940–1995 data (derived from OFDA, 1996).

Disasters have been grouped in categories as follows: Sea (555 disasters), caused by extreme factors associated with the seas and oceans, mainly tropical cyclones, hurricanes, tsunamis and typhoons; Land (1,918 disasters), including avalanches, cold waves, drought, earthquakes, fires, floods, heat waves and landslides; Population (1,043 disasters), mostly events due to war and diseases, including internally displaced persons and refugees, epidemics, famine and food shortage; and Others (322 disasters), covering power shortages, storms, volcanic eruptions and unusual phenomena.

others, because all types of disasters are covered. Information is more complete on the events since 1964 (OFDA, 1996), and the records exclude disasters within the US and its territories. Basically it includes only 'major' disasters, based on losses and number of people affected, homeless or killed. The criteria are less strict for small island countries, with the consequence that they tend to be over-represented.[7]

Table 18.4 quantifies the loss of life associated with disasters in the twentieth century. It appears that man-made 'population' disasters tend to create more victims than both sea and land-bound disasters. As an example, wars and political disputes have, over the recent decades, replaced drought as one of the main causes of food shortages. Again, note the marked positive skew of the severity of all types of disasters.

The years since 1964 have witnessed an upward trend in the frequency of most disasters (Table 18.5).[8] While the increase is close to 10-fold for disasters associated with seas and land, the frequency of reported population-related disasters increased by a factor 38 between the 1960s and the 1990s. Sea-related disasters have shown a positive, but much less marked, trend as well, with a drop around 1990. Figure 18.6 shows the relative importance of the four categories defined in Table 18.4 over the years.

---

[5] The 'multiplier' expresses how many times longer is the coast – because of dissection – than would be estimated from the length measured by straight-line steps of 50 to 100 km 'as the crow flies'.

[6] The main deltas are the Ganges-Brahmaputra Delta in Bangladesh, the Nile in Egypt, and the Mahanadi and Ganges in West Bengal. There are about nine major deltas in the Americas (two flowing into the Glacial Arctic Ocean and seven in sub-tropical and tropical areas), three in Europe, one in glacial Asia, ten in tropical Asia, four in Australasia and five in Africa (Jelgersma et al., 1993).

[7] Volcanoes tend to occur along coasts, and many small islands are actually of volcanic origin, but there is no absolute association.

[8] The trends exist before 1964, but only the period after 1964 is covered because observations are more complete beginning around that year.

Table 18.5. Average frequency of disasters by category, year/number.

Year/No.	Sea	Land	Population	Others
1960s	12	30	6	5
1970s	11	41	15	9
1980s	18	67	38	11
1990s	105	393	228	62

(Based on data in OFDA, 1996).
Refer to Table 18.4 for the definition of categories.

Table 18.6. Some typical values of Insularity Index, Vulnerability Index and protection costs.

Country	Insularity Index	Vulnerability Index	Protection cost (% GDP)
Libya	0.0010	0.0030	0.08
Zaire	0.000016	0.031	0.12
United States	0.0022	0.063	0.02
Sweden	0.0078	0.17	0.14
Nicaragua	0.0076	0.24	0.35
Belgium	0.0021	0.70	0.01
Italy	0.026	4.9	0.04
Netherlands	0.013	5.0	0.03
Greece	0.10	8.3	0.10
United Kingdom	0.051	12	0.02
Jamaica	0.028	21	0.19
Mauritius	0.096	52	0.15
Anguilla	0.67	77	10
Gaza Strip	0.11	230	NA
Singapore	0.31	1,700	0.05
Tokelau	10.1	1,500	11
Maldives	2.1	1,800	34
Monaco	2.2	36,000	0.13

Data are from *Factbook* (1997), UN (1996a), UN (1997), median population scenario, as well as CZMS (1990) for the protection costs.

As noted, population-related disasters have been on the increase, while coastal areas have been losing importance as disaster areas, even though the population of coastal megacities has about doubled from 124 to 220 million (mid-sixties to mid-nineties). Cities evolve over time and may be expected to adjust to gradual change in SLR over the next century – of course up to certain limits, impossible to specify. If methodological problems and sampling problems can indeed be excluded (see, for instance Pielke and Landsea, 1997), and provided epidemics, famine, etc. are not concentrated in the main world urban centres, one may infer that cities, even coastal ones, constitute a rather safe environment.

In Figure 18.7, an attempt was made to extrapolate the total annual damage due to land- and sea-related disasters. Extrapolations

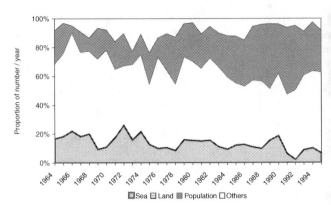

Figure 18.6. Proportions of each category of disasters, 1964–1994 (based on data in OFDA, 1996).

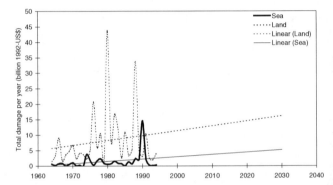

Figure 18.7. Total annual damage in constant US$ (base 1992) due to land- and sea-related disasters, together with their linear trends (based on data from OFDA (1996), and the implicit price deflator (Umich, 1997)).

of this kind have a large statistical variation, but they illustrate the fact that it is mainly the increase in wealth that leads to increased damage, rather than the increased frequency of disasters.

In addition, there is no indication as to whether the 'wealth structure' (wealth is now mainly concentrated in buildings and infrastructure, and to some extent in land) will not undergo significant qualitative changes (for instance, wealth could be in electronic equipment, or databases, knowledge, etc.).

18.6 VULNERABILITY INDEX (VI)

In order to assess the combined effect of insularity and population, we have used below a very simple 'vulnerability index' (or VI), defined as the product of the Insularity Index and population density:

$$\text{Vulnerability Index} = \text{Insularity Index} \times \text{population density}$$

The purpose, again, is not to develop a new indicator, but simply to allow a fast global statistical discussion of some of the

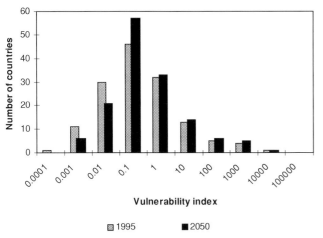

Figure 18.8. Frequency distribution of 143 national Vulnerability Indices for the countries and territories of the world in 1995 and 2050 (based on data in *Factbook* (1997), and UN (1997), medium population scenario).

population and SLR issues under consideration. There are other indices (see, for instance, the good discussion of the Global Vulnerability Assessment in WHO, 1996, and Nicholls, 1995). Most indices suffer from one or more of the following problems: they assume linear responses and a somehow uniform distribution of the target population. They are also difficult to use for projections as they relate only to some of the relevant factors. CZMS (1992) defined a Common Methodology for carrying out vulnerability assessments in different countries. Most of the criticisms of the Common Methodology stress difficulties related to the lack of even elementary data, the use of monetary value only to evaluate losses (which is of little relevance in subsistence economies), a lack of attention paid to the resilience and adaptation of certain systems to sea-level rise, and the assumption of linear responses (CZMS, 1992).[9]

The main weakness of the Vulnerability Index is that it does not take into consideration the spatial distribution of population within countries, nor the actual vulnerability of specific low-lying areas along the coasts. For example, for Libya the population density of 3 people per km^2 refers to the population density of the whole country. But if one considers the population in the coastal zone (that is the one that could be affected by SLR), the density rises to 105 in 1990 (cf. Hoozemans *et al.*, 1993). This will also become apparent at the conclusion of this section, where the distribution of arable land is shown as a function of the VI.

Like the Insularity Index, the VI varies over several orders of magnitude. Typical values are given and compared in Figure 18.8.

[9] An ideal 'Vulnerability Index' should be additive. For instance Tonga has 171 named islands, 36 of which are permanently inhabited, and Fiji has 320 (CGER, 1996), and it would be logical to expect that the total vulnerability should be the sum of the island values. The concept of vulnerability is a very complex one. For a more systematic approach to climatic risk and vulnerability, see, e.g. Downing (1991, 1992) or Gommes (1998).

The VI shows a good qualitative agreement with other indicators. Due to its close link with population density, we suggest that it can be used to evaluate some of the changes in vulnerability that may take place in the future. Figure 18.8 indicates a relative 'narrowing' of the distribution between 1995 and 2050, because of the relative decrease of population growth rates in a number of the countries and territories that are the most vulnerable by current standards.

For instance, in 1995, 46 (32%) countries displayed a Vulnerability Index between 0.01 and 0.1. The percentage may increase to 40% (57 countries) in 2050. This is also clear in Figure 18.8, which shows a relative decrease of low values, with a corresponding increase at intermediate and high VI values above 0.1.

If expressed in terms of population, it appears that more people will be living in vulnerable countries. The average of national Vulnerability Indices changes from about 5 to 6.6 between 1995 and 2050, but there could be a relative decrease of extremely high values when compared to current conditions.

Figure 18.9 shows an interesting relation between the much more complex assessment of CZMS (1990) and the Vulnerability Index, from which an upper limit of the cost associated with different vulnerabilities can be derived. CZMS assumes a population inertia (protection rather than other options) that may not occur. In practice, retreat, accommodation and protection will coexist.

Finally, turning to agriculture, a plot of the proportion of arable land as a percentage of total land against the Vulnerability Index on a logarithmic scale (Figure 18.10) shows a rather symmetric distribution (roughly log-normal). Some countries with high vulnerabilities and various proportions of arable land are indicated in the graph.

It is clear that the percentage of arable land is rather independent from the Vulnerability Index as defined here, except probably at high percentages. The most agriculturally oriented economies, as expressed by the percentage of arable land, are generally at moderate vulnerability levels, probably because the most vulnerable economies – those of the small islands – also rely on the ocean for their income and subsistence.

18.7 CONCLUSIONS

Several facts emerge from the literature, analyses and the data presented in the present paper. First, accelerated SLR seems to be one of the more probable consequences of global climate change, with a 'worst scenario' increase of 88 cm by 2100, while the population of the world will stabilize around 2050 with urbanization continuing to increase. Current population densities are generally highest – and the GDP lowest – in the more insular countries of the world.

Large local differences will be observed in SLR, resulting in large differences in impacts. In general, it is likely that the relative importance of coastal disasters will decrease, even if their number

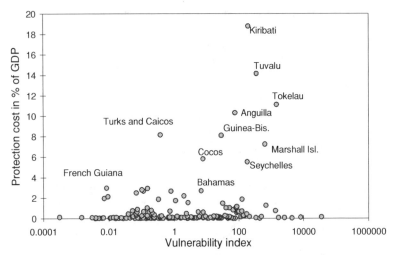

Figure 18.9. Cost of protection (from CZMS, 1990), percentage of GDP, versus Vulnerability Index for 182 countries and territories.

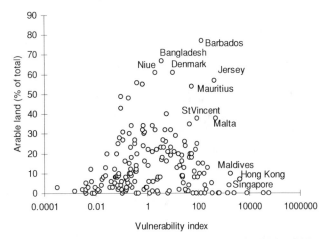

Figure 18.10. Arable land (percentage of total land) and Vulnerability Index.

and the associated economic losses are likely to increase due to general economic development.

Direct effects on the most vulnerable coastal areas, for instance deltas, are difficult to assess, in particular because the dynamic of deltas is determined by climatic conditions in the whole catchment basin and because coastal climates do not follow general patterns. It is clear that, given the gradual nature of the expected changes, populations and agriculture will gradually adapt or migrate. The cost of protection is relatively high in many small island states (5% of GDP and above); protective measures may be possible only where their cost remains low compared with GDP growth.

Current trends indicate that the proportion of population in areas of moderate vulnerability will increase at the expense of low-vulnerability areas. The trend towards greater urbanization may constitute a positive factor in the present context, as cities seem to provide a rather safe environment. Many variables examined display a positive skew; therefore, the distribution of severity of disasters will be very skewed, with many relatively small and very few, but extremely large and costly disasters.

Policy measures will have to include construction standards and other non-structural preventive measures (emergency plans, insurance) in addition to protection. Some countries should take action now to ensure that their agricultural production is not concentrated in areas that are likely to become more vulnerable with a rise in sea level.

It is clear that efforts should be made to identify possible run-away situations, and they should be kept under constant observation.

ACKNOWLEDGEMENTS

The authors are grateful to the editor of this volume and to Dr Isabelle Niang-Diop, Département de Géologie, Faculté des Sciences et Techniques, Université C. A. Diop, Dakar-Fann, Senegal, for a careful and thorough critique of their work. They readily acknowledge the resulting improvement in the paper and accept responsibility for any remaining shortcomings, including those that may derive from their not always following the commentators' views.

REFERENCES

Bijlsma, L., Ehler, C. N., Klein, R. J. T. *et al.* (1996) Coastal zones and small islands. Chapter 9, pp. 289–324. In IPCC (1996b).

Brinkman, R. (1995) Impact of climatic change on coastal agriculture. In D. Eisma (Ed.), *Climate Change Impact on Coastal Habitation*, Chapter 11, pp. 235–245. Boca Raton, FL, Lewis Publishers.

CGER (1996) *Data Book on Sea-Level Rise*. Center for Global Environmental Research, National Institute for Environmental Studies, Japan Environment Agency, Tsukuba, Japan. 87pp.

CZMS (Coastal Zone Management System) (1990) *Strategies for Adaptation to Sea-Level Rise*. Published for IPCC, Geneva, by Ministry of Transport, Public Works and Water Management, The Hague, Netherlands. 122pp.

 (1992) *Global Climate Change and the Rising Challenge of the Sea*. Published for IPCC, Geneva, by Ministry of Transport, Public Works and Water Management, The Hague, Netherlands. 35pp and 5 appendices.

Downing, T. E. (1991) *Assessing Socio-economic Vulnerability to Famine: Frameworks, Concepts and Applications*. FEWS Working Paper 2.1. USAID, Washington, DC, 102pp.

 (1992) *Climate Change and Vulnerable Places: Global Food Security and Country Studies in Zimbabwe, Kenya, Senegal and Chile*. Research Rep. No. 1, Environmental Change Unit, Oxford, UK. x + 54pp.

Du, B. (1993) The preliminary vulnerability assessment of the Chinese coastal zone due to sea level rise. In R. McLean and M. Mimura (Eds.), *Vulnerability Assessment to Sea-Level Rise and Coastal Zone Management*, pp. 177–188. Proc. of the IPCC Eastern Hemisphere Workshop.

Engelman, R. (1997) Why population matters. Population action international. Washington, DC, 56pp.

Factbook (1997) *The World Factbook 1996*: http://www.odci.gov/cia/publications/nsolo/wfb-all.htm

FAO (1997) *Seawater Intrusion in Coastal Aquifers. Guidelines for Study, Monitoring and Control*. FAP Water Reports No. 11, Rome, Italy. 152pp.

Fedra, K., Winkelbauer, L. and Pantalu, V. R. (1991) *Expert Systems for Environmental Screening. An Application in the Lower Mekong Basin*. IIASA report RR-91–19, Laxenburg, Austria, 169pp.

George, A. (1999) *The Epic of Gilgamesh. The Babylonian Epic Poem and Other Texts in Akkadian and Sumerian*. London, Allen Lane, Penguin Press, Lxii + 225pp.

Gommes, R. (1992) Current climate and populations constraints on agriculture. In H. M. Kaiser and T. E. Drennen (Eds.), *Agricultural Dimensions of Global Climatic Change*, Chapter 4, pp. 67–86. Delray Beach, FL, St. Lucie Press, 311pp.

 (1998) Some aspects of climate variability and food security in sub-Saharan Africa. In G. Demarée, J. Alexandre and M. De Papper (Eds.), *Proc. Internat. Conf. On Tropical Climatology, Meteorology and Hydrology*, Brussels, 22–24 May 1996, pp. 655–673. Brussels, Royal Meteorological Institute and Royal Academy of Overseas Science.

Hoozemans, F. M. J., Marchand, M., Pennekamp, H. A. (1993) *Sea-Level Rise. A Global Vulnerability Assessment*. The Hague, Delft Hydraulics, Delft / Rijkswaterstaat, 184pp.

IFRCS (1996) *World Disaster Report, 1996*. International Federation of Red Cross and Red Crescent Societies, Oxford, Oxford University Press, 178pp.

IPCC (1996a) *Climate Change 1995, The Science of Climate Change*. Contribution of Working Group I to the Second Assessment Report of the Intergovernmental Panel on Climate Change. J. T. Houghton, L. G. Meira Filho, B. A. Callander, N. Harris, A. Kattenberg and K. Maskell (Eds.). xii + 572pp.

 (1996b) *Climate Change 1995, Impacts, Adaptations and Mitigation of Climate Change, Scientific-Technical Analyses*. Contribution of Working Group II to the Second Assessment Report of the Intergovernmental Panel on Climate Change. R. T. Watson, M. C. Zinyowera and R. H. Moss (Eds.). x + 879pp.

 (1996c) *Climate Change 1995, Economic and Social Dimensions of Climate Change*. Contribution of Working Group III to the Second Assessment Report of the Intergovernmental Panel on Climate Change. J. Bruce, Hoesung Lee and E. F. Haites (Eds.). x + 448pp.

 (2001). *Third Assessment Report: Climate Change 2001, The Scientific Basis*. J. T. Houghton, Y. Ding, D. J. Griggs, M. Noguer, P. J. van Linden and D. Xiaosu (Eds.), Cambridge, Cambridge University Press, 944pp.

Ittekkot, V. (1996) Oceans. Chapter 8, pp. 269–288 in IPCC (1996b).

IUCN (1991) *Caring for the Earth, a Strategy for Sustainable Living*. Gland, Switzerland, IUCN, UNEP and WWF, IUCN, 228pp.

Jelgersma, S., and Tooley, M. J. (1995) Sea-level changes during the recent geological past. *J. Coastal Res.*, Fort Lauderdale, Sp. Issue N. 17, Holocene cycles: climate, sea levels and sedimentation, 123–129.

Jelgersma, S., vander Zijp, M. and Brinkman, R. (1993) Sea-level rise and the coastal lowlands in the developing world. *J. Coastal Res.*, 9 (4); 958–972.

Leemans, R. and Cramer, W. P. (1990) *The IIASA Database for Mean Monthly Values of Temperature, Precipitation and Cloudiness of a Global Terrestrial Grid*. Working Paper WP-90-41. International Institute of Applied Systems Analysis, Laxenburg, Austria.

Nicholls, N. (1995) Synthesis of vulnerability analysis studies. In *Preparing to Meet the Coastal Challenges of the 21st Century*, pp. 181–216. Report of the World Coast Conference, Noordwijk, 1–5 November 1993. Dutch Ministry of Transport, Public Works and Water management, Den Haag.

Nicholls, N., Gruza, G. V., Jouzel, J., Karl, T. R., Ogallo, L. A. and Parker, D. E. (1996) Observed climate variability and change. Chapter 3, pp. 137–192. In IPCC (1996a).

Nicholls, R. J. and Leatherman, S. P. (1995) Global sea-level rise. pp. 92–123. In Strzepek and Smith (1995).

OFDA (1996) *Disaster History, Significant Data on Major Disasters Worldwide, 1900–1995*. Office of Foreign Disaster Assistance, US Agency for International Development, Washington, 265pp.

Pielke, R. A. and Landsea, C. W. (1997) Normalized hurricane damages in the United States: 1925–1995. Paper prepared for a panel on social, economic and policy aspects of hurricanes at the 22nd Conference on Hurricanes and Tropical Meteorology,

19–23 May, AMS, Fort Collins (http://www.dir.ucar.edu/esig/HP_roger/hurr_norm.htm)

Riebsame, W. E., Strzepek, K. M., Wescoat Jr. J. L. *et al.* (1995) Complex river systems. pp. 57–91. In Strzepek and Smith (1995).

Scott, M. J. (1996) Human settlements in a changing climate: impacts and adaptation. Chapter 12, pp. 399–426. In IPCC (1996b).

Sombroek, W. S. and Gommes, R. (1996) The climate change–agriculture conundrum. In F. Bazzaz and W. Sombroek (Eds.), *Global Climate Change and Agricultural Production*, Rome, FAO pp. 1–14. and Chichester, UK, John Wiley, 345pp.

Strzepek, K. M. and Smith, J. B. (Eds.) (1995) *As Climate Changes: International Impacts and Implications.* Cambridge, Cambridge University Press, 213pp.

Susman, P., O'Keefe, P. and Wisner, B. (1983) Global disasters, a radical interpretation. In K. Hewitt (Ed.), *Interpretations of Calamity*, pp. 263–280. The Risks & Hazards Series: 1. Boston, London, Sydney, Allen & Unwin Inc., 304pp.

Tobler, W., Deichmann, U., Gottsegen, J. and Maloy, K. (1995) The Global Demography Project National Center for Geographic Information and Analysis, University of California, Technical Report TR-95-6, 75pp.

(1997) World population in a grid of spherical quadrilaterals. *Int. J. Population Geogr.*, 3, 203–225.

Umich (1997) 'Implicit price deflator' from Table 10.1, Gross domestic product and deflators used in the historical tables: 1940–2002, University of Michigan, http://asa.ugl.lib.umich.edu/libhome/Documents.center/steccpi.html

UN (1996a) *World Population Prospects, the 1996 Revision. Annex I: Demographic Indicators.* United Nations Secretariat, Population Division, New York, 253pp.

(1996b) *World Population Prospects 1950–2050 (The 1996 revision).* Data sets on magnetic tape and diskettes. Description of the IBM-compatible diskettes. United Nations Secretariat, Population Division, New York.

(1997) *World Urbanization Prospects, the 1996 Revision. Annex Tables.* United Nations Secretariat, Population Division, New York, 114pp.

WHO (1996) *Climate Change and Human Health. A. J. McMichael,* A Haines, R. Slooff and S. Kovats (Eds.). WHO, Geneva, Switzerland, 296pp.

Zwick, A. (1997) *Monitoring and Assessment of Recent Research Results on Global Climate Change with a Special Section on Sea Level Rise and Joint Implementation.* Climate change research and policy, a periodic survey for the Commission of the European Communities, Update N. 9, EU Joint Research Centre, Sevilla, Spain. EUR17303EN, 76pp.

Appendix A. Some indicators pertaining to coastal African countries

Country	Insularity Index	Vulnerability Index	Population density pers/km^2	Population growth rate[a] %/year	Arable land %	GDP/capita 1995 US$	Protection cost % of GNP[b]
Algeria	0.00042	0.01	12	2.30-0.56	3	3800	0.02
Angola	0.00128	0.01	9	3.20-1.22	2	700	0.07
Benin	0.00109	0.05	48	2.67-1.15	12	1380	0.74
Cameroon	0.00086	0.02	28	2.70-1.13	13	1200	0.07
Cape Verde	0.23945	22.99	96	2.34-0.71	9	1040	0.15
Comoros	0.15668	42.93	274	2.72-0.92	35	700	0.13
Congo	0.00049	0.00	8	2.79-1.30	2	3100	0.05
Congo D.R.	0.00002	0.00	19	2.57-1.35	3	400	0.12
Côte d'Ivoire	0.00162	0.07	42	1.78-0.82	9	1500	0.44
Djibouti	0.01429	0.37	26	1.19-0.94	0	1200	0.19
Egypt	0.00246	0.15	62	1.89-0.54	3	2760	0.09
Equ. Guinea	0.01055	0.15	14	2.50-1.11	8	800	1.02
Eritrea	0.00949	0.26	27	3.78-1.03	3	570	n.a.
Gabon	0.00343	0.01	4	2.59-1.06	1	5200	0.09
Gambia	0.00800	0.78	98	3.23-0.85	16	1100	2.64
Ghana	0.00234	0.17	73	2.71-1.13	5	1400	0.64
Guinea	0.00130	0.04	30	0.76-0.90	6	1020	0.43
Guinea-Bissau	0.01250	0.38	30	2.21-1.05	11	900	8.15
Kenya	0.00094	0.04	47	2.00-0.54	3	1300	0.06
Liberia	0.00601	0.11	19	8.23-1.27	1	770	2.66
Libya	0.00101	0.00	3	2.42-0.68	2	6510	0.08
Madagascar	0.00830	0.21	25	2.97-1.12	4	820	0.56
Mauritania	0.00073	0.00	2	2.73-1.08	1	1200	0.1
Mauritius	0.09568	52.34	547	0.78-(0.02)	54	9600	0.15
Morocco	0.00411	0.24	59	1.76-0.43	18	3000	0.11
Mozambique	0.00315	0.07	22	2.48-1.07	4	700	2.48
Namibia	0.00190	0.00	2	2.24-0.85	1	3600	0
Nigeria	0.00094	0.11	121	2.39-0.96	31	1300	0.04
Réunion	0.08040	21.87	272	1.31-0.14	16	4300	0.12
São Tomé	0.21771	32.66	150	n.a.	1	1000	1.46
Senegal	0.00277	0.12	42	2.59-1.09	27	1600	0.65
Sierra Leone	0.00561	0.33	58	2.95-0.98	25	960	1.86
Somalia	0.00482	0.07	15	4.16-1.27	2	500	0.62
South Africa	0.00229	0.08	34	1.49-0.35	10	4800	0.04
Sudan	0.00036	0.00	11	2.05-0.83	5	800	0.01
Tanzania	0.00161	0.05	32	2.27-1.09	5	800	0.16
Togo	0.00103	0.07	72	2.63-1.19	25	900	0.87
Tunisia	0.00739	0.41	55	1.39-0.40	20	4250	0.24
W. Sahara	0.00417	0.00	1	3.38-0.64	0	n.a.	0.02

[a]1995–2000 values, and projected 2045–2050 values; medium variant of UN (1996a) and UN (1996b) if available. Data not available from the mentioned sources were taken from *Factbook* (1997). Negative growth rates between brackets.
[b]From Annex D of CZMS (1990). Includes, cities, harbours, etc.
n.a., not available.
Data from *Factbook* (1997); UN (1996a, 1996b, 1997); CZMS (1990) and OFDA (1996).

19 Sea-level rise and coastal biodiversity in West Africa: a case study from Ghana

AYAA KOJO ARMAH[1], GEORGE WIAFE[1], AND DAVID G. KPELLE[2]

[1]University of Ghana, Legon, Ghana
[2]Department of Wildlife, Accra, Ghana

Keywords

Ghana; sea-level rise; biodiversity; coastal; ecosystem; erosion

Abstract

Available evidence points to an eventual rise of sea level for coastal areas as a result of global warming. Sea-level rise would introduce or aggravate existing threats to the continued survival of the biodiversity of low-lying coastal areas. In Ghana and the coastal states of West Africa, vulnerable habitats include the strand zone, lagoons, wetlands, and intertidal sandy and rocky areas. Physical and biological parameters of several of these habitats would change substantially as a result of submersion and increased salinity regimes. This would adversely affect, for example, the habitats of water birds, nesting beaches of sea turtles, and the brackish-water dependent fauna and flora of the estuaries and lagoons. Species extirpation leading to local loss of genetic diversity is envisaged to affect fauna like ghost crabs (*Ocypoda* spp.) and the fiddler crab (*Uca tangeri*). Plants in this category include five species of true mangroves (*Rhizophora racemosa*, *R. harrisonii*, *R. mangle*, *Avicennia germinans* and *Laguncularia racemosa*) and their associates (*Conocarpus erectus*, *Acrostichum aureum* and the uncommon creeper, *Phylloxerus vermicularis*). Strategies that would mitigate or protect biodiversity of the coastal zone from the anticipated effects of rising sea level are advanced. The potential role of coastal infrastructures in biodiversity conservation is addressed.

19.1 INTRODUCTION

The potential for future sea-level rise in coastal areas has been documented in many studies (e.g. Frassetto, 1991; Milliman and Ren, 1995; Leatherman *et al.*, 2000). Predictions are that many low-lying coastal areas will be flooded as sea levels rise up to about one metre over the next one hundred years (Raper *et al.*, 1996;

IPCC, 2001). Available evidence indicates a direct relationship between sea-level rise and coastal habitat loss in a number of coastal areas (Warrick and Ahmad, 1996; Zhang, 1998), with subsequent loss of biodiversity. The low-level tidal ranges of West Africa would mean that wetlands, for example, will be more quickly affected because vertical distribution of vegetation is related to local tide ranges (Oteng-Yeboah, 1999).

Coastal habitats and ecosystems consist of a wide variety of plant and animal assemblages living in the interface spanning terrestrial, semi-terrestrial and marine environments. Organisms found in these environments often are physiologically adapted in such a way that only a few are semi-terrestrial, that is to say, can neither survive entirely on land nor entirely submerged in water. Rising sea level and, in particular, the rate of rise would tend to dislocate the equilibrium and associations between species occupying such marginal habitats, particularly the semi-terrestrial forms.

In Ghana and most of the tropical world, coastal habitats are vitally important to communities inhabiting them, as they provide means of livelihood as well as the bulk of the required dietary protein. Additionally, they support the productivity of certain fish species exploited by both artisanal and industrial fleets, serve as buffers to flooding, and provide pollutant-trapping systems and recreational facilities. This situation is remarkably common to all of the coastal nations of western Africa.

The coastal zone of Ghana, which flanks a coastline of approximately 550 km, has been under heavy pressure because of high rates of influx of people from the hinterland, population growth and urbanization within the last three decades. The zone occupies less than 7% of the national land area but is the home of 25% of the populace (World Bank and EPA, 1997). This has imposed severe stress on resources and has led to the emergence of multifaceted problems, of which fisheries overexploitation, poor sanitation, wetland degradation and coastal erosion are key examples (Armah and Amlalo, 1998). These problems have serious implications for the functional integrity of the biodiversity, defined as the variability among living organisms and with the ecosystem

Climate Change and Africa, ed. Pak Sum Low. Published by Cambridge University Press. © Cambridge University Press 2005.

as a whole (CBD, 1992). Rising sea levels would worsen these growing pressures facing coastal habitats and their species. A good fore-knowledge of the status of coastal biodiversity would enable planners to anticipate likely problems and thus provide the much-needed opportunity for a response to protect species and their habitats, not only in the face of anthropogenic stresses, but also as sea level rises. From this perspective, biodiversity assessment should constitute a pivotal objective of coastal management policies.

19.1.1 Sea-level rise

Global temperatures and sea levels have undergone dramatic changes throughout geological history. For example, about 120,000 years before the present, global temperatures were higher than those at present and the seas were about 6 m higher. At the beginning of the Holocene period (approximately 15,000 years ago), the earth was much colder and the seas were about 100 m below today's level (Pirazolli, 1991).

Data gathered from the Permanent Service for Mean Sea Level (PSMSL) of the International Council for Scientific Unions (ICSU) confirms that there has been a gradual rise in sea level over the past century. The Intergovernmental Panel on Climate Change (IPCC) Third Assessment Report indicates a 10–20 cm rise since 1900 (IPCC, 2001). Tidal measurements from Ghana indicate rising trends in conformity with the IPCC global estimates (Woodworth, 1990). If the current rate of discharge of greenhouse gases into the atmosphere continues, it is projected that global mean sea level will rise by 9–88 cm between 1990 and 2100, for the full range of Special Report on Emissions Scenarios (IPCC, 2001).

Several portions of the coast of Ghana would undergo physical changes that potentially would threaten coastal biodiversity in the event of the sea level rising as predicted. The main physical changes envisaged would be the following:

• Establishment of permanent connections with the sea for most of the 90 lagoons within the coastal areas;
• Increased penetration of salt water inland into estuaries and open lagoons;
• Increased rates of coastal erosion, resulting in beach recession;
• Decrease in the depth of the water table in coastal areas (i.e. decrease in the thickness of the freshwater lens overlying saline water);
• Intrusion of salt water into coastal freshwater aquifers.

19.2 VULNERABLE COASTAL ECOSYSTEMS AND HABITATS

Erosion, submergence and salinity increase constitute perhaps the three foremost direct impacts of sea-level rise on coastal ecosystems. Erosion removes protective sandbars and catalyses both submergence and salinity increase. Progressive submergence causes reduced plant productivity and eventual death.

Sea-level rise may pose risks to many coastal habitats stretching from the terrestrial margins across the intervening brackish systems to the marine intertidal. Five main categories of coastal habitats and ecosystems in the states of western Africa are recognizable, namely: terrestrial, wetlands (lagoons and creeks), mangroves, estuaries, and the marine intertidal.

In Ghana, these habitats and ecosystems, with the exception of the terrestrial, constitute important habitats for migratory and resident water birds (avifauna). The water birds of the coastal zone have been categorized into seven feeding guilds by Piersma and Ntiamoa-Baidu (1995). These are: fishing terns, stalking herons/egrets, herbivorous tree ducks and fishing pelicans. Others are pelagic foraging waders, visual, and tactile surface foraging waders. A rising sea level would alter water depth and salinity, two factors that could be critical to the feeding success of these guilds.

19.2.1 The coastal terrestrial vegetation

The terrestrial vegetation of the coastal zone of Ghana consists of a rainforest portion to the west and coastal thicket savannah to the east. Approximately two-thirds of the coastline consists of the former type. Fringing the shoreline are the strand vegetation, and mangrove forests of wetlands and estuaries. Should the sea level rise as predicted, several species adapted to living at the land–sea interface stand the risk of being dislocated or disappearing from their habitats. This includes species that typically occur along the margins of wetlands, estuaries, and in the strand zone. Species found in these habitats, and which also occur at higher elevations beyond the influence of salt spray, would, naturally, not be threatened by the effects of sea-level rise. Angiosperm diversity in coastal wetlands of Ghana is very high. In a study of three coastal wetlands of Ghana (i.e. Densu, Sakumo and Muni-Pomadzi), Oteng-Yeboah (1999) reported that between two and nine plant families were restricted to specific wetlands and their floodplains. The maximum number of species recorded (both restricted and non-restricted) was 136 and the minimum was 114. Out of the above, about a quarter were represented by monocotyledon families (Table 19.1).

Strand zone

Much of the West African coastline consists of sandy beaches (Lawson, 1985). Above the high water mark is a characteristic strand vegetation. It has a relatively small number of species belonging to widely differing families, but with many common features. Stems are prostrate and lie either on the surface or just below the surface, rooting or throwing out tufts of leaves at intervals.

Table 19.1. Distribution of terrestrial plant taxa in three coastal wetlands in Ghana.

Category	Densu	Sakumo	Muni-Pomadze
Total number of species	136	114	133
Number of families	50	48	54
Dicotyledon families	40	39	45
Monocotyledon families	10	9	9
Dicotyledon species	94	87	86
Monocotyledon species	42	27	37
Restricted families	2	3	9

After Oteng-Yeboah, 1999.

Many have seeds that survive immersion in seawater. Examples include *Canavalia rosea, Ipomea pes-caprae, Sporobolus virginicus and Remirea maritima.* In certain parts of the country, the strand zone is extensively used for crop production. For example, the lagoon side of the sandbar that separates the Keta lagoon from the sea is an area of intensive commercial cropping of shallots (*Allium ascalonicum*). The industry depends on the amount of fresh water overlying a deeper saline layer. If sea level rises, two things are likely to occur to this fragile ecosystem. The first will be a rise in the salt water table, with a corresponding decrease in the overlying freshwater column. The second will be seawater intrusion into the lagoon through the creek that links up to the Volta River estuary. Should this happen, the water level of the lagoon would be much higher and average salinity would increase. The shallot farmlands on the shores of the lagoon would eventually become flooded with saline water from both ground and lagoon sources, leading to a total collapse of the industry (Armah *et al.,* 1997). Another commercial crop that would be threatened is the coconut palm, *Cocos nucifera,* which is cultivated in large plantations in the western area of the coast.

A rising sea level is expected to increase rates of erosion in areas where the shorelines or berms consist of unconsolidated material. This would lead to higher rates of beach recession, and the characteristic strand zone plant species would be at risk. The strand vegetation is very extensive and occurs on both consolidated and unconsolidated cliffs and berms. Erosion of consolidated beaches would be relatively minimal and the risk of vegetation loss would be more of human extirpation than loss due to sea-level rise.

19.2.2 Wetlands

Over 90 wetlands occur in Ghana and these have been classified into estuarine and lagoonal types. The latter are further divided into open, closed and 'depression' types (Kwei, 1977; Armah,

1993). The open lagoons maintain a permanent opening to the sea all year round, whereas the closed types are closed to the sea all or most of the year. The lagoons are typically very shallow with the average depth not exceeding 1 m. A good number of these lagoons are used for fishing (Table 19.2). Over 50 species of fin and shellfishes occur in the wetlands (Table 19.3). The algae in these brackish habitats are rather poor and are represented by only 10 species (John and Lawson, 1977).

The most vulnerable wetlands are those of the Volta River deltaic system (from Songor to Denu) and the Densu wetland (Sakumo I). The Volta River deltaic wetlands alone, in aggregate, constitute about 70% of the areal coverage of all the coastal wetlands of Ghana (Table 19.2). It is also the home of thousands of migratory birds. Small mammals, reptiles, fishes and invertebrates also form an important component of the biodiversity of the coastal wetlands.

Impacts of sea-level rise would be minimal if the wetlands could keep pace with the rate of rise. But this is likely only if the wetlands are healthy and if there is sufficient setback for wetland migration as well as adequate sediment inputs from the rivers to keep pace with the rate of rise. Should the wetlands fail to keep pace, most coastal lagoons and their associated fauna and flora are likely to undergo drastic changes. Many of the sand bars enclosing the lagoons would be broken and permanent connections established with the sea. First, the closed lagoons, invariably, would be transformed into open types, and the open types would have wider openings and become more marine than brackish. This situation may alter the overall diversity. A major shift towards marine types in lagoon fishery production, both finfish and shellfish, is very likely (Table 19.3). Increases in fishery production, however, would occur only if sufficient wetland nursery areas remain. Water birds (avifauna) would also be adversely affected, as submergence of tidal flats will reduce the area available for feeding, roosting and nesting.

Avifauna of lagoonal wetlands

The lagoonal wetlands are significant because of the occurrence of high avifaunal diversity. Species count exceeds 70, with total abundance exceeding 55,000 on occasions (EPA, 1990).

Eight areas have been identified as being sites of international importance for waterfowl habitats, of which seven are lagoonal wetlands. These include the Korle, Keta, Songor, Muni, Densu wetland (Sakumo I), Sakumo II and Elmina salt pans (Ntiamoa-Baidu and Gordon, 1991). Seven feeding guilds can be found in the lagoonal wetlands. The feeding habits include diving, siphoning, pecking, probing, ploughing, sweeping, stabbing and scooping (Piersma and Ntiamoah-Baidu, 1995). The diets range from weeds and seeds to a variety of small invertebrates and fish (Table 19.4).

Table 19.2. Examples of wetland types and their estimated sizes.

Type and name	Size (km²)	Uses
Estuarine wetlands		
Ankobra	2.8	Fishing, mangroves, wildlife
Pra	16.0	Fishing, mangroves, wildlife
Volta	70.2	Fishing, mangroves, Ramsar site, wildlife
Open lagoons		
Butre	3.5	Fishing, wildlife, mangroves
Hwin	2.0	Fishing, wildlife, mangroves, salt extraction
Eture	3.7	Fishing, wildlife, mangroves
Amisa	6.3	Fishing, wildlife, mangroves
Nakwa	3.6	Fishing, wildlife, mangroves
Closed lagoons		
Mukwe	0.5	Fishing
Songor	15.0	Fishing, salt extraction, wildlife, Ramsar site
Fosu	1.0	Fishing
Muni	2.7	Fishing, wildlife, Ramsar site, salt extraction
Depression lagoons		
Belibangara	0.5	Fishing, wildlife
Ndumakaka	0.0	Fishing, wildlife
Efasu	1.0	Wildlife
Ehunli	0.4	Fishing, wildlife

Armah and Amlalo, 1998.

19.2.3 Mangroves

Estuaries and lagoons provide suitable habitats for mangroves. The total area of land occupied by mangroves is estimated to be around 10,000 ha (Saenger and Bellan,1995). The stands of mangroves in most areas are of secondary and tertiary growth. Significant stands, however, occur along the coastline notably at Amansure, Oyibi, Korle and Eture lagoons as well as the Volta delta. Five species of true mangroves occur in the country and in the subregion. These are *Rhizophora racemosa, R. harrisonii, R. mangle, Avicennia germinans* and *Laguncularia racemosa. Avicennia germinans* is common on the higher ground and in the closed lagoons. The mudflats and associated shallow bodies provide suitable habitats for a great variety of aquatic birds, finfish and shellfish. The diet and feeding habits of the avifauna associated with mangroves are provided in Table 19.5.

Certain plant species associated with mangroves may be at risk from sea-level rise. These include *Conocarpus erectus, Thespesia populnea, Acrostichum aureum, Phoenix reclinata*, and *Sesuvium portulacrastrum* and *Phylloxerus vermicularis*. Faunal associations are generally unique and consist of several semi-terrestrial crab species, notably *Cardiosoma armatum, Sesarma* spp. and *Uca tangeri*. Others are the bivalves *Anadara senilis* and *Os-* *trea rhizophorae*, and the gastropods *Tympanotonus* spp. and *Pachymelania* spp. These fauna die when exposed to prolonged submergence (or exposure), as could happen as depths of wetlands increase with rising sea level.

19.2.4 Marine intertidal

Characteristic fauna and flora inhabit the marine intertidal sandy and rocky beaches. Over 116 species of algae occur in the marine rocky intertidal and about 80 species sub-tidally (John and Lawson, 1997). Their distribution is governed by the geology and geomorphology of the coast. Geomorphologically, the shoreline is divisible into three sections, namely western, central and eastern. The western shoreline is about 120 km long and stretches from the Côte d'Ivoire border to Axim. The eastern shoreline starts from Prampram (west of Tema harbour) to the Togo border and covers approximately 200 km. The western and eastern sections predominantly comprise sandy beaches. The central section consists of rocky shores and shorelines interspersed with sandy bays. It stretches between Axim and Prampram and covers a distance of 230 km. Together with the intervening sandy bays, sandy shores constitute about 70% of the entire

Table 19.3. Some fin- and shellfish species found in coastal wetlands of Ghana: (♦) fresh water;
(*) brackish; (♣) marine.

Eastern Zone	Central Zone	Western Zone
Fish		
* *Sarotherodon melanotheron*	♦ *Sarotherodon galilaeus*	* *Sarotherodon melanotheron*
♦ *S. galilaeus*	* *S. melanotheron*	♦* *Hemichromis fasciatus*
♦ *Tilapia zillii*	♦* *Hemichromis fasciatus*	♦* *H. bimaculatus*
* *T. guineensis*	♦ *Chrysicthys auratus*	♦ *Heterobranchus bidorsalis*
♦* *Hemichromis fasciatus*	♦ *Clarias anguillaris*	♦ *Clarias anguillaris*
♦* *H. bimaculatus*	♦ *Tilapia zillii*	♦ *Tilapia zillii*
* *Gobius* spp.	* *Ethmalosa fimbriata*	* *Ethmalosa fimbriata*
* *Psettias sebae*	* *Liza falcipanis*	* *Gerres melanopterus*
* *Gerres melanopterus*	♣ *Pegusa lascaris*	♣ *Psettodes belcheri*
♣ *Psettodes belcheri*	* *Gobioides ansorgii*	♣ *Lutjanus fulgens*
♣ *Lutjanus fulgens*	* *Porogobius schlegeli*	♣ *Trachinotus goreensis*
♣ *Trachinotus goreensis*	* *Citharicthys stampfli*	♣ *Ephinocephalus aeneus*
♣ *Ephinocephalus aeneus*	* *Periophthalmus papilio*	♣ *Caranx hippos*
♣ *Caranx hippos*	* *Gerres melanopterus*	♣ *Syacium micrurum*
♣ *Syacium micrurum*	♣ *Psettodes belcheri*	♣ *Pseudotolithus senegalensis*
♣ *Pseudotolithus senegalensis*	♣ *Lutjanus fulgens*	*♣ *Galeodactylus decadactylus*
*♣ *Galeodactylus decadactylus*	♣ *Trachinotus goreensis*	* *Mugil cephalus*
* *Mugil cephalus*	♣ *Ephinocephalus aeneus*	♣ *Pomadasys rogeri*
♣ *Pomadasys rogeri*	♣ *Caranx hippos*	♣ *Engraulis encrasicolus.*
♣ *Engraulis encrasicolus.*	♣ *Syacium micrurum*	♣ *Sardinella* spp.
♣ *Sardinella* spp.	♣ *Pseudotolithus senegalensis*	* *Elops senegalensis*
* *Elops senegalensis*	*♣ *Galeodactylus decadactylus*	* *Liza falcipinis*
	* *Mugil cephalus*	* *Pegusa lascaris*
	♣ *Pomadasys rogeri*	* *Gobioid* spp
	♣ *Engraulis encrasicolus.*	* *Periophthalmus papilio.*
	♣ *Sardinella* spp.	
	* *Elops senegalensis*	
Crustaceans		
* *Callinectes amnicola*	* *Callinectes amnicola*	* *Callinectes amnicola*
* *C. Pallidus*	* *Cardiosoma armata.*	* *C. pallidus*
* *Cardiosoma armata*	♦ *Macrobrachium* spp.	* *Uca tangeri*
♦ *Macrobrachium* spp.	* *Uca tangeri*	*♣ *Penaeus* spp.
* *Uca tangeri*	* *Sesarma huzardi*	
* *Penaeus* spp.	*♣ *Penaeus* spp.	
Molluscs		
* *Tympanostonus fuscatus*	* *Tympanostonus fuscatus*	* *Tympanostonus fuscatus*
* *Pachymelania* spp.	* *Crassostrea agar*	* *Pachymelania* spp.
* *Crassostrea agar*		* *Crassostrea agar*
♦* *Egeria radiata*		

After Entsua-Mensah, 1998.

shoreline. Organisms occupying the intertidal show zonation in response to the periodicity of emersion and immersion as the tides flood and ebb. Rising sea levels could disrupt or dislocate such zonation patterns in areas where beach recession is slow or impossible.

Sandy beaches

Beach recession as a result of sea-level rise would adversely affect the survival of intertidal organisms and others that utilize sandy beaches at one stage of their life cycle. Two examples are

Table 19.4. Avifauna of lagoonal wetlands and lagoonal depressions.

Name	Scientific name	Feeding habit	Diet
Common tern	*Sterna hirundo*	diving	fish
Little tern	*S. albifrons*	diving	fish
Roseate tern	*S. dougalli*	diving	fish
Sandwich tern	*S. sandviscensis*	diving	fish
Royal tern	*S. maxima*	diving	fish
Caspian tern	*S. caspia*	diving	fish
Whiskered tern	*S. hybrida*	diving	fish
Black tern	*S. nigra*	diving	fish
Gull-billed tern	*S. nilotica*	diving	fish
Fulvous tree-duck	*Dendrocygna bicolour*	siphoning	weeds, seeds
White-faced tree duck	*D. viduata*	siphoning	weeds, seeds
Kittlitz's plover	*Charadrius pecuarius*	pecking	invertebrates
Ringed plover	*C. hiaticula*	pecking	invertebrates
Grey plover	*Pluvialis squatarola*	pecking	worms
Common sandpiper	*Tringa hypoleucos*	pecking	arthropods
Marsh sandpiper	*T. stagnatilis*	pecking, ploughing	fish
Wood sandpiper	*T. glareola*	pecking, ploughing	arthropods, worms
Redshank	*T. tetanus*	pecking, ploughing	invert, fish
Greenshank	*T. nebualria*	pecking, ploughing	fish
Spotted redshank	*T. erythropus*	pecking, ploughing	fish
Turnstone	*Arenaria interpres*	pecking, ploughing	invertebrates
Whimbrel	*Numenius phaeopus*	pecking, ploughing	crabs
Curlew	*N. arquata*	pecking, ploughing	crabs, worms
Black-tailed godwit	*Limosa lomosa*	probing	molluscs
Bar-tailed godwit	*L. lapponica*	probing	worms
Curlew sandpiper	*Calidris ferruginea*	pecking, probing	invert, seeds
Knot	*C. canutus*	probing	molluscs, seeds
Sanderling	*C. alba*	peck, probe	invert
Black-wing stilt	*Himantopus himantopus*	probe, peck, sweep	fish, invert
Avocet	*Recurvirostra avosetta*	sweeping	fish
Little egret	*Egretta garzetta*	stabbing	fish
Reef heron	*E. gularis*	stabbing	fish
Great white egret	*E. alba*	stabbing	fish
Grey heron	*Ardea cinerea*	stabbing	fish
Purple heron	*A. purpurea*	stabbing	fish
Black-headed heron	*A. melanocephala*	stabbing	fish
Pink-backed pelican	*Pelecanus rufescens*	scooping	fish
Long-tailed shag	*Phacrocorax africanus*	under-water diving	fish
White-breasted cormorant	*P. carbo*	under-water diving	fish
Pied kingfisher	*Ceryle rudis*	diving	fish
Pigmy kingfisher	*Ceyx picta*	diving	fish
Blue-breasted kingfisher	*Halcyon malimbicus*	diving	fish
African darter	*Anhinga rufa*	diving	fish
Green-backed heron	*Butorides striatus*	stabbing	fish

Table 19.5. The avifauna of mangrove swamps.

Name	Scientific name	Feeding habit	Diet
Little egret	*Egretta garzetta*	stabbing	fish
Great white egret	*E. alba*	stabbing	fish
Reef heron	*E. gularis*	stabbing	fish
Grey heron	*Ardea cinerea*	stabbing	fish
Squacco heron	*Ardeola ralloides*	stabbing	fish
Purple heron	*Ardea purpurea*	stabbing	fish
Little bittern	*Ixobrychus minutus*	stabbing	fish
Tiger bittern	*Tigrirnis leucolopha*	stabbing	fish
Night heron	*Nyctocorax nycticorax*	stabbing	fish
Pied kingfisher	*Ceryle rudis*	diving	fish
Giant kingfisher	*Ceryle maxima*	diving	fish
Blue-breasted kingfisher	*Halcyon senegalis*	diving	fish
Pigmy kingfisher	*Ceyx picta*	diving	fish
Black-tailed godwit	*Limosa limosa*	probing	molluscs
Bar-tailed godwit	*L. lapponica*	probing	worms
Curlew	*Numenius arquata*	probing	crabs, worms
Whimbrel	*N. phaeopus*	pecking	crabs
Common sandpiper	*Tringa hypoleucos*	pecking	arthropods
Wood sandpiper	*Tringa glareola*	pecking	arthropods

sea turtles and the burrowing semi-terrestrial ghost crabs, *Ocypoda cursor* and *O. africana*. Sea turtles nest on sandy beaches beyond or at the high tide mark. The prime sea turtle nesting sites in Ghana are located within the rapidly eroding sites between Prampram and the Volta estuary. Five species occur in Ghana: the leatherback (*Dermochelys coriacea*), the hawksbill (*Erectmochelys imbricata*), green turtle (*Chelonia mydas*), the loggerhead (*Caretta caretta*) and the most abundant of all, the olive ridley (*Lepidochelys olivacea*). The hawksbill is very rare and only occasionally seen.

Ghost crabs normally live in the sediments of the upper shore of sandy beaches. As beach sand is eroded away, the underlying rocks become exposed and render the beaches unsafe for habitation by ghost crabs, and also unsuitable for nesting by turtles. Of the two species of ghost crabs, *O. africana* can withstand more exposure and survive on the supra-littoral among the strand vegetation. Thus, with the loss of the preferred intertidal sandy habitat, many areas will have lowered densities of ghost crabs, with *O. africana* taking refuge in the less benign terrestrial strand habitat.

Avifauna of sandy shores, mainly visual foraging waders, favour the sandy shorelines. These are Kittlitz's sand plover (*Charadrius pecarius*), white-fronted sand plover (*C. marginatus*), ringed plover (*C. hiaticula*), grey plover (Pluvialis squatarola) and common sandpiper (*Tringa hypoleucos*). The others are the whimbrel (*Numenius phaeopus*), curlew (*N. arquata*), turnstone (*Arenaria interpres*), wood sandpiper (*T. glareola*) and redshank

(*T. totanus*). The most common are the crab-eating waders such as curlew (*N. arquata*) and whimbrel (*Numenius phaeopus*). Loss of sandy beaches could result in reduction in numbers and possibly avifaunal biodiversity especially along the Essiama beach in the western region of Ghana.

Rocky beaches

Rocky beaches and coasts comprise about 30% of the coastline. In such areas with high consolidated cliffs (e.g. Cape Three Points, and at Ussher Fort in Accra), semi-aquatic invertebrates like the pulmonate (air-breathing) gastropod, *Siphonaria pectinata* (a limpet), would be subjected to increased periods of submergence beyond their physiological tolerance limits. A rise in the sea level might dislocate such species from the rocky intertidal and reduce their densities if no new rocky areas are exposed through beach recession or shoreline erosion. The avifauna found on rocky beaches include the common sandpiper (*T. hypoleucos*), turnstone (*A. interpres*), ringed plover (*C. hiaticula*), curlew (*N. arquata*) and whimbrel (*N. phaeopus*). These species would not be adversely affected by sea-level rise, as they are not restricted to the rocky intertidal habitat.

19.2.5 Estuaries

Deeper and longer duration of saline water is likely to occur in estuaries as sea level rises. In Ghana, this could have positive

Table 19.6. Avifauna of estuarine wetlands of the coastal zone of Ghana.

Name	Scientific name	Feeding habit	Diet
Fulvous tree duck	*Dendrocygna bicolor*	grazing	weeds, seeds
White-faced tree duck	*D. vidua*	grazing	weeds, seeds
Black-tailed godwit	*Limosa limosa*	probing	molluscs
Bar-tailed godwit	*L. lapponica*	probing	worms
Curlew	*Numenius arquata*	probing	crabs, worms
Whimbrel	*N. phaeopus*	probing	crabs
Knot	*Calidris canutus*	probing	molluscs, seeds
Sanderling	*C. alba*	peck, probing	invertebrates
Little stint	*C. minuta*	peck, probing	invertebrates
Curlew sandpiper	*C. ferruginea*	peck, probing	inverts, seeds
Kittlitz's sand plover	*Charadrius pecuarius*	pecking	invertebrates
White-fronted plover	*C. marginatus*	pecking	invertebrates
Ringed plover	*C. hiaticula*	pecking	invertebrates
Grey plover	*Pluvialis squatarola*	pecking	invertebrates
Common sandpiper	*Tringa hypoleucos*	pecking	arthropods
Wood sandpiper	*T. glareola*	pecking	arthropods
Redshank	*T. tetanus*	pecking	invertebrates, fish
Greenshank	*T. nebularia*	pecking	invertebrates, fish
Turnstone	*Arenaria interpres*	pecking	invertebrates
Little egret	*Egretta garzetta*	stabbing	fish
Reef heron	*E. gularis*	stabbing	fish
Great white egret	*E. alba*	stabbing	fish
Grey heron	*Ardea cinerea*	stabbing	fish
Pigmy kingfisher	*Ceyx picta*	diving	fish
Pied kingfisher	*Ceryle rudis*	diving	fish
Blue-breasted kingfisher	*Halcyon malimbicus*	diving	fish

impacts on fisheries in the three major estuaries of the Ankobra, Pra and Volta rivers. Deeper penetration of saline water could lead to expansion in the size of potential nursery areas for shrimps and other estuarine species. On the Volta River, for example, an important fishery for the clam *Egeria egeria,* which collapsed due to reduction in the extent of saline water intrusion (as a result of the Volta Dam), would stand a good chance of recovery (Pople and Rygoyska, 1969).

The avifauna of these areas consist of herbivorous ducks that feed in the freshwater marshes, as well as visual and tactile surface foraging waders feeding on both dry and wet mudflats. Stalking herons can also be found in this habitat, where they feed in the dry or wet mudflats and in the shallow waters. The feeding habits and diets of bird species occurring in this habitat are shown in Table 19.6.

19.3 RELATIVE VULNERABILITY OF COASTAL HABITAT TYPES

A reclassification of Ghanaian coastal habitats into three broad types – wetlands, beaches and mangroves – is instructive as it

reveals their relative importance and vulnerability to accelerated sea-level rise. Each may be subdivided into sub-habitats and their vulnerability assessed, from 'low vulnerability' to 'extremely high vulnerability', based on their anticipated response to impacts of erosion, flooding, seawater intrusion and aquifer salinization (Table 19.7).

Physical changes to coastline and habitat vulnerability vary from one coastal area to another. The degree of vulnerability also depends on the level of anthropogenic degradation that has already taken place. For example, wetland areas with severe degradation of mangrove forests through over-cutting stand very little chance of autonomous recovery.

19.4 BIODIVERSITY AND COASTAL INFRASTRUCTURES

Certain developmental infrastructures and activities along the coast may have the potential of altering, either positively or negatively, coastal biodiversity as sea level rises. These include salt industries, shoreline protection, port development, oil and gas development, and thermal power plants.

Table 19.7. Physical changes and vulnerability of coastal habitat types of Ghana.

Ecosystem/Habitat	Sub-habitat	Type	Physical changes and remarks
Beaches	Rocky	—	Low vulnerability. Only few organisms likely to be lost; e.g. Prampram and Cape Coast.
Beaches	Sandy	—	Moderately vulnerable. Berm and beach would be eroded. Fauna and terrestrial vegetation, e.g. *Sesuvium* sp., *Ipomea* sp., coconuts, at risk; e.g. East Prampram to Ada.
Mangrove	Lagoon	Open	Low to moderate vulnerability, depending on ability to re-colonize new areas; e.g. Oyibi lagoon.
Mangrove	Lagoon	Closed	High vulnerability, as salinity would approach sea values; e.g. Fosu lagoon.
Mangrove	Lagoon	Depression	Extremely high vulnerability, as salinity would approach sea values or higher; e.g. Ehunli lagoon.
Mangrove	Estuarine	—	Low vulnerability, as mangroves could retreat upstream; e.g. Pra and Ankobra rivers.
Mangrove	Deltaic	—	Low vulnerability, as mangroves could retreat upstream depending on elevation above mean sea level; e.g. Volta delta.
Wetland	Estuarine	—	Low vulnerability, as habitat already has continuous exchange with the sea. Wetland area often at appreciable elevation from mean sea level; e.g. Ankobra and Pra estuaries.
Wetland	Deltaic	—	Very high vulnerability, as most areas are low-lying and easily inundated, erodable and prone to seawater intrusion and salinization of aquifers; e.g. Volta delta and Keta lagoon wetland zones.
Wetland	Lagoon	Open	Moderate vulnerability because of continuous exchange with the sea. Freshwater inflows are often appreciable to retain brackish nature as sea level rises; e.g. Amisa, Oyibi and Butre lagoons.
Wetland	Lagoon	Closed	High vulnerability. Low freshwater inflows. Salinity would approach sea values when sandbar breaks permanently; e.g. Mukwe, Fosu and Muni lagoons.
Wetland	Lagoon	Depression	Extremely high vulnerability. No freshwater inflows. Habitat would turn essentially marine; e.g. Ehunli, Ndumakaka and Belibangara lagoons.

19.4.1 Salt industry

Significant salt production occurs within or on the edges of several of the coastal lagoons lying in the dry coastal savannah vegetational zone. Total national production is estimated at 150,000 tonnes. The major production areas are at the Songor lagoon and Densu wetlands (Sakumo I). Others are at the Keta, Djange (Gyankai), Laloi (Laiwi), Nyanya, Apabaka, Eture and the Ahwin lagoonal wetlands. The conversion of portions of these wetlands for salt production has led to substantial habitat loss and, in some cases, has significantly reduced local biodiversity. Good examples are evident at the Nyanya, Songor and Densu (Sakumo I) lagoonal wetlands.

A rising sea level may have some positive impacts on the biodiversity of lagoons where the saltpans could become inundated as a result of higher inflow regimes. In areas where the hydrology of the wetland has been modified for salt production, the expectation could be the return of some of the former biodiversity.

19.4.2 Shoreline protection

Beach erosion would increase as sea level rises. In areas where the shoreline is prone to erosion, shoreline recession would be expected to be on the ascendancy. Two main methods of shoreline protection have been employed in Ghana: revetments and groynes. Revetments refer to the method where the protecting elements are aligned parallel to the shoreline. In the case of groynes, they are aligned perpendicularly to the shoreline. In either situation, the original habitat is impacted. Revetments usually lead to loss of beach space and the existing beach biodiversity, e.g. Sakumo II beach at Tema and Sekondi-Nkontompo beach at Sekondi. Groynes often protect the beach and usually would affect the original biodiversity to a much lesser extent. Examples of groyne-protected areas are at Keta, Ada, Dzita and Accra. The protecting elements, which invariably comprise armour rocks and/or gabions, may offer habitats for colonization by existing species or others new to the area. Arguably, shoreline protection cannot lead to loss of biodiversity, because as some species disappear, new opportunistic colonizers would emerge. The real question is how beneficial the final assemblage of organisms would be in terms of biodiversity conservation.

19.4.3 Port development

Port development impacts on biodiversity and its relation to sea-level rise can be visualized from two viewpoints: (i) impacts arising from the construction of new ports; and (ii) those relating to existing ones. Two main ports are in existence at Tema and Takoradi, and minor fishing ports are at Elmina and Sekondi. In several ways, impacts of sea-level rise on biodiversity in relation to ports are similar to those associated with shoreline protection. This is because both involve loss of original habitats and production of new habitats, primarily in the form of breakwaters, either parallel or perpendicular to the shoreline.

A rising sea level would increase the mean height of waters surrounding breakwaters. Impacts will be both positive and negative. There would be new habitats for colonization by sub-tidal organisms, as well as a decrease in the area available for roosting by sea birds. Development of new ports would create new habitats, but there would be loss of the original floor space and its biodiversity.

19.4.4 Energy development: thermal plants

Development of thermal plants to augment hydroelectric power generation began in earnest a few years ago when poor rainfall in the catchment areas for the Akosombo dam created an energy crisis in the country. At present, there are two thermal plants at Aboadze and Efasu in the western region. Both plants employ seawater for cooling and the hot water is discharged into the nearby sea. The hot effluents will have a localized negative impact on biodiversity at their points of discharge. A rising sea level might increase the area influenced by the hot effluents and correspondingly affect a larger proportion of organisms. This situation will affect, predominantly, the sessile or infaunal organisms because of their limited mobility.

19.4.5 Oil and gas development

Oil and gas development are in the exploratory stages and hence pose no immediate threat to biodiversity. The only factor perhaps worth mentioning is the disused offshore platform at Saltpond, which has created habitats for organisms to colonize. The above notwithstanding, active offshore oil exploration is ongoing and the potential for erection of more platforms, spillage and blowouts in the near future are realistic projections.

19.5 DISCUSSION

Biodiversity loss as a result of sea-level rise could be attributed to three critical problems: erosion, submergence and salinity increase. Submergence occurs when vertical accretion of wetlands cannot keep pace with the rate of sea-level rise. Two major factors, sediment input and biomass production by vegetation, contribute to the rate of vertical accretion and the resilience of coastal wetlands.

Relative sea-level rise in the Mississippi delta has been about 1 cm per year or greater. Yet large areas have maintained their elevation due to significant inputs of sediments (Baumann et al., 1984). This has the effect of slowing down the impacts for several decades before widespread losses are evident (Day and Templet, 1989). Response and adaptation strategies would, therefore, depend on the resilience of habitats and ecosystems to the rate of sea-level rise. They would also be determined by the economic importance of the habitats/ecosystems and by their biodiversity status.

19.5.1 Autonomous adjustment

The autonomous adjustment of 'do nothing' and allowing the system to take care of itself may not be an acceptable option in situations where coastal biodiversity is of major national and international significance. Under the IPCC (2001) scenario of about 90 cm rise in 100 years, this option would certainly lead to dramatic loss in biodiversity of water birds from all five Ramsar wetlands if their connections with the sea are sufficiently widened to cause significant hydrological changes in the wetlands. How successful autonomous adjustment would be, however, depends on the health of the habitat or ecosystem, which, in turn, is a function of its ability to cope with the rate of change.

Areas that are fully built-up, as at Dansoman, near Accra, would naturally hamper autonomous adjustment of the Densu wetlands, as there would be no space for the wetland to migrate. Table 19.8 indicates the projected autonomous adjustment for some of the key wetlands.

19.5.2 Adaptation options

A range of options is available to decision makers on coastal development (and coastal ecosystems). These options typically have been described in three ways: accommodation, retreat, and protection (IPCC, 1990). By definition, the accommodation and protection options do not require physical relocation of habitats and infrastructure. This would imply that coastal habitats and ecosystems, with the exception of estuarine wetlands, might not qualify to be considered under the retreat option. The only exception could

Table 19.8. Response of wetlands under autonomous adjustment (do nothing) scenario.

Habitat/Ecosystem (wetlands and lagoons)	Ability for autonomous adjustment (under 100-cm rise in 100 years scenario)
Western Zone	
Balinbangara, Ndumakaka, Efasu, Ehunli	Extremely poor. Very limited self-adjustment possible, as wetlands lie in depressions with no watercourses. Habitat loss could exceed 75% of original size. Ehunli lagoon might become a small bay and lose all brackish biodiversity.
Amunsure, Aladoa, Anankwari	Moderate. Some landward areas along the streams and flat valleys would permit wetland migration. Habitat loss about 40%.
Butre, Ankobra, Kpani-Nyila	Good. Some landward migration possible along the watercourses. Habitat loss is estimated at 20% of present size.
Central Zone	
Abrobeano, Apabaka, Muni, Oyibi	Moderate. Landward migration possible to some extent. Habitat loss is estimated at 40%.
Densu, Fosu, Korle, Kpeshie, Sakumo	Extremely poor. Very limited self-adjustment possible as wetlands are in built-up areas with little or no space for migration. Habitat loss is estimated to exceed 70% of present size. Extensive salt production in the Densu wetland could be a problem.
Pra, Ahwin, Eture, Amisa, Nakwa	Good. Landward migration possible along the watercourses. Habitat loss is estimated to be minimal, about 20% of present size.
Eastern Zone	
Songor, Keta	Extremely poor. Self-adjustment is restricted and rather poor for the Keta lagoon because built-up, and inadequate space for wetland migration. Habitat loss is estimated to exceed 60%.
Volta delta, creeks and wetlands	Good. Self-adjustment very good if there is sufficient sediment from the Volta River to sustain mangroves. Habitat loss is estimated at 20%.

be the transfer of species from one threatened habitat to another or the artificial creation of new habitats.

Clearly, the options above will not suitably apply when dealing with coastal habitats and their biodiversity. From this viewpoint, the three options are modified as follows:

- Accommodation: implies the continued existence of the original habitat or ecosystem, allowing for natural 'adjustment';
- Retreat: implies colonization of new habitats and the original habitat lost or marginally available;
- Protection: implies no physical change since the original habitat is kept intact or only marginally lost.

Based on the above adaptation options, a projection is made on the potential changes of species diversity in Ghanaian coastal habitats and ecosystems (Table 19.9).

In summary, the protection option, where feasible, would be the most favoured strategy in conserving and protecting biodiversity of the coastal zone, since changes in species diversity would, in most cases, be minimal. On the other hand, the accommodation and retreat options would lead to reduction in species diversity.

19.5.3 Protection of habitats

Four specific areas would require protection from the adverse effects of sea-level rise. These are the Ramsar sites (Muni, Densu wetland or Sakumo I, Sakumo II, Songor and Keta), Eture wetlands and Essiama beach. The Ramsar sites and the Essiama beach are important because of their immense use as bird habitats. In addition, the Ramsar sites are prime sea turtle nesting beaches, which begin from Prampram and decrease in intensity eastwards to Ada. The Eture wetland is important for its mangrove diversity. It is the only wetland where all five species of true mangroves occur in Ghana.

As shown in Table 19.8, all the Ramsar sites are incapable of retreating, either because physical space for expansion is restricted, or because in some cases, watercourses flowing into them are too small or dry up; and hence these sites are unable to sustain the brackishness of the system. The options available, therefore, are accommodation and protection. The former would lead to substantial habitat loss and the latter would provide minimal loss of habitat and biodiversity. The preferred option, therefore, would be protection of these habitats.

Table 19.9. Adaptation options of key coastal habitats in Ghana: (−) not physically feasible; (+) change in species diversity low; (++) change in species diversity moderate; (+++) change in species diversity high.

Wetland/ Habitat	Accommodation	Retreat	Protection	Remarks
Belibangara	+++	—	+	Closed/semi-closed lagoonal wetlands. Protection
Ndumakaka	+++	—	+	and accommodation are possible but species
Efasu	+++	—	+	composition is likely to change significantly in
Ehunli	—	—	+	response to new hydrological conditions. For
				Ehunli, protection is the only option; otherwise
				lagoon would become a bay, losing all of its
				brackish character.
Butre	++	++	—	Estuarine wetlands fed by medium-to-large
Ankobra	++	++	—	watercourses that flow all year round.
Hwin	++	++	—	Accommodation and retreat possible. Original
Kpani-Nyila	++	++	—	species composition would not vary
Pra	++	++	—	significantly.
Ahwin	++	++	—	
Eture	++	++	—	
Amisa	++	++	—	
Nakwa	++	++	—	
Oyibi	++	++	—	
Amansure	++	++	+	Coastal habitats fed by small watercourses. All
Aladoa	+++	++	+	three options are possible.
Anankwari	++	++	+	
Songor	++	++	+	
Abrobeano	++	++	+	
Apabaka	+++	++	+	
Muni	++	++	+	
Densu	++	—	+	Retreat option not possible because of
wetland	+++	—	+	surrounding built-up areas.
(Sakumo I)	++	—	+	
Fosu	+++	—	+	
Korle	++	—	+	
Kpeshie	++	—	+	
Sakumo II	++	—	+	
Keta	++	—	+	
Sandy and rocky intertidal	—	++	+	Protection by revetment not suitable for sandy shore biodiversity, as beaches would be lost.

Protection of the most vulnerable part of the Keta lagoon stretching over a distance of 7 km (from Kedzi to Havedzi) began in the year 2000. The Muni and Songor by their bigger sizes has lost much habitat in relation to the Densu wetlands (Sakumo I). The seaward portion of Sakumo II and Eture wetlands are flanked by major coastal roads, which, as stated, are under protection, and thus confer automatic protection to the wetlands as well. Bird biodiversity is threatened in the Keta, Songor, Densu wetlands and Muni, should these lagoons become deeper. Creation of

islands with sand from other parts of the wetland might resolve the problem. Beach erosion between Prampram and the Volta estuary would require protection with the use of groynes to preserve the beaches for turtles.

19.5.4 Community involvement

The ability of habitats to cope with or adapt to any of the options identified or recommended also depends to a large extent on the active involvement of the local communities. Sand extraction from beaches for the construction of houses is a widespread practice. This worsens the rate of erosion as sea level rises, and even where physical protection is provided, it will seriously undermine the effort. Community education and the provision of alternative material for building needs to be an integral part of management plans for the coastal zone.

Recent studies conducted on the mangroves of the Volta delta indicate a loss of 60% during the 12-year period from 1973/1974 to 1986 (Adu-Prah et al., 1977). Mangroves are heavily disturbed through over-cutting for fuelwood or conversion of the areas for solar-salt production. Evidence available from studies in the Caribbean and the Indo-Pacific (Ellison, 1989) suggests that mangroves, if undisturbed, are capable of accumulating enough sediment to keep pace with rising sea levels at the projected global rate of 5 mm per year. Again, community involvement is important if mangroves are to remain healthy. Replanting of degraded areas and provision of alternative woodlots, coupled with awareness creation through government and non-governmental agencies, are essential ingredients for a successful coastal habitat and biodiversity conservation programme.

19.5.5 Conclusion

Biodiversity loss can be negligible if comprehensive, integrated, long-term planning is put into place to take care of the problem of sea-level rise. This should include zoning of critical areas, allowing reasonable and adequate buffer zones in new areas, and formulating new policies and legislation with regard to the coastal zone.

Given the fact that the coastal ecosystem types and characteristics as described for Ghana are representative of the western African coastal zone, this area could potentially serve as a large sentinel zone for the detection of early signs of the impacts of sea-level rise on low-lying tropical coastal wetlands worldwide.

ACKNOWLEDGEMENTS

The study was supported by the Environmental Protection Agency of the Ministry of Environment and Science, Ghana. The authors wish to acknowledge the useful comments of Dr David Duthie, who reviewed the manuscript.

REFERENCES

Adu-Prah, S., Gyamfi-Aidoo, J. and Agyepong, G. T. (1997) Lower Volta Mangrove Project – Remote Sensing Component. *Lower Volta Mangrove Project Technical Report* No. 5. Volta Basin Research Project, University of Ghana, Legon, Ghana.

Armah, A. K. (1993) Coastal wetlands of Ghana. *Coastal Zone*, **93**, 313–322.

Armah, A. K. and Amlalo, D. S. (1998) *Coastal Zone Profile of Ghana*. Accra, Ghana, Royal Crown Press Ltd.

Armah, A. K., Awumbila, M., Clark, S. *et al.* (1997) Coping responses and strategies in the coastal zone of Ghana: a case study in the Anloga area. In S. M. Evans, C. J. Vanderpuye and A. K. Armah (Eds.). *The Coastal Zone of West Africa: Problems and Management, pp. 17–27.* UK, Penshaw Press.

Baumann, R. H., Day, J. W. and Miller, C. A. (1984) Mississippi Deltaic wetland survival: sedimentation versus coastal submergence. *Science*, **224**, 1,093–95.

CBD (1992) Convention on Biological Diversity. http://www.wcmc.org.uk/igcmc/convent/cbd/cn_cbd.html

Day, J. W. and Templet, P. H. (1989) Consequences of sea-level rise: implications from the Mississippi delta. *Coastal Management*, **17**, 241–257.

Ellison, J. C. (1989) Possible effects of sea-level rise on Mangrove ecosystem. *Small States Conference on Sea-level Rise*, Male, Maldives. 14–18 November, 1989.

Entsua-Mensah, R. E. M. (1998) *Comparative Studies of the Dynamics and Management of Fish Populations in Open and Closed Lagoons in Ghana*. Ph.D. Thesis, University of Ghana, Legon, Ghana.

EPA (1990) *Coastal Zone Indicative Management Plan*. Environmental Protection Council, Accra, Ghana.

Frassetto, R. (1991) *Impact of Sea Rise on Cities and Regions*. Centro Internacional Citta d'Acqua, Venice, Italy.

IPCC (1990) *Strategies for Adaptation to Sea-Level Rise*. R. Misdorp, J. Donkersand and J. R. Spradley (Eds.), Intergovernmental Panel on Climate Change Coastal Zone Management sub-group. Ministry of Transport and Public Works, The Hague.

    (2001) *Intergovernmental Panel on Climate Change Working Group 1 Third Assessment Report*. WMO/UNEP.

John, D. M. and Lawson, G. (1997) Seaweed biodiversity in West Africa: a criterion for designating marine protected areas. In S. M. Evans, C. J. Vanderpuye and A. K. Armah (Eds.), *The Coastal Zone of West Africa: Problems and Management*, pp.111–123. UK, Penshaw Press.

Kwei, E. A. (1977) Biological, chemical and hydrological characters of coastal lagoons of Ghana, West Africa. *Hydrobiologia*, **56**, 157–174.

Lawson, G. W. (1985) *Plant Life in West Africa. Ghana*. Accra, Ghana, Universities Press.

Leatherman, S. P., Zhang, K. and Douglas, B. C. (2000) Sea-level rise shown to drive coastal erosion. *EOS Trans. AGU*, **81**, 55–57.

Milliman, J. D. and Ren, M. (1995) River flux to the sea: impact on human intervention on river systems and adjacent coastal areas. In E. Eisma (Ed.), *Climate Change: Impact on Coastal Habitation*, pp. 57–83. Boca Raton, FL, CRC Press.

Ntiamoa-Baidu, Y. and Gordon, C. (1991) *Coastal Wetlands Management Plans: Ghana.* World Bank and Environmental Protection Council Report. Government of Ghana.

Oteng-Yeboah, A. (1999) Biodiversity studies in three coastal wetlands in Ghana. *J. Ghana Sci. Assoc.*, **1**, 147–149.

Piersma, T. and Ntiamoa-Baidu, Y. (1995) *Waterbird Ecology and the Management of Coastal Wetlands in Ghana.* Ghana Coastal Wetlands Management Project/Netherlands Institute for Sea Research (NIOZ-Report 1995–6)/Ghana Wildlife Report.

Pirazolli, P. A. (1991) *World Atlas of Holocene Sea-Level Changes.* Amsterdam, Elsevier.

Pople, W. and Rogoyska, M. (1969) The effect of the Volta River hydroelectric project on the salinity of the Lower Volta River. *Ghana J. Sci.* **9**, 9–20.

Raper, S. C. B, Wigley, T. M. L. and Warrick, R. A. (1996) Global sea-level rise: past and future. In J. D. Milliman and U. H. Bilal (Eds.), *Sea-Level Rise and Coastal Subsidence*, pp. 11–45. Dordrecht, The Netherlands, Kluwer Academic Publishers.

Saenger, P. and Bellan, M. F. (1995) *The Mangrove Vegetation of the Atlantic Coast of West Africa – A Review.* Laboratoire d'Ecologie Terrestre (UMR 1964), Université de Toulouse, France.

Warrick, R. A. and Ahmad, Q. K. (1996) *The Implications of Climate and Sea-Level Change for Bangladesh.* Dordrecht, The Netherlands, Kluwer Academic Publishers.

Woodworth, P. (1990) Measuring and predicting long-term sea level changes. *NERC News*, No. 15, NERC, Swindon, UK.

World Bank and EPA (1997) *Towards an Integrated Coastal Zone Management Strategy for Ghana.* The World Bank/Environmental Protection Agency, Accra, Ghana.

Zhang, K. (1998) *Twentieth Century Storm Activity and Sea-Level Rise Along the US East Coast and Their Impact on Shoreline Position.* Ph.D. Dissertation, University of Maryland, College Park, USA.

20 The impacts of ENSO in Africa

GODWIN O. P. OBASI[*]

World Meteorological Organization, Geneva, Switzerland

Keywords

El Niño; La Niña; ENSO; Southern Oscillation Index;
African rainfall climatology; vulnerability; impacts; Africa

Abstract

El Niño and La Niña phenomena are simply referred to
as the warm and cold ENSO phases respectively. ENSO
events generally last from 3 to 6 seasons, sometimes as long
as 24 months, and tend to recur every 3 to 7 years. The
warming/cooling (El Niño/La Niña) of the eastern and central
equatorial Pacific Ocean are known to lead to worldwide
anomalies in sea surface temperatures (SSTs) and the circulation of the ocean currents. The most significant influence is
found in the tropics, but such influences have been found to
vary significantly from place to place and from season to season, as well as with the evolution pattern of the ENSO phases.
Although ENSO impacts are strongest in the Pacific Ocean
region, past records in Africa show that some severe droughts
and floods that have been observed over parts of the continent
have been associated with ENSO events. As elucidated in
this paper, the impacts of some of these extreme droughts
and floods have seriously affected the social and economic
development of various countries in the African continent.

20.1 BACKGROUND

The El Niño and La Niña phenomena have, in recent years,
become widely known by the public at large around the
world. These phenomena have a long historical base in Latin
America. For many generations, the occasional extensive warming or cooling of the eastern Pacific Ocean was known by the
local fishermen to be associated with extreme climate anomalies

[*]Professor Godwin O. P. Obasi submitted this paper before he retired from the
World Meterological Organization as Secretary-General at the end of 2003.

and adverse socio-economic impacts of the ensuing phase in their
area. El Niño and La Niña phenomena are of global dimension
and have probably existed for thousands of years, but it is only
in the last two decades that they have been properly understood
(Ropelewski and Halpert, 1987; Obasi, 1999). How is this knowledge being disseminated to the world to enable appropriate actions
to be taken? The impacts of the severe 1997–1998 El Niño event,
for example, mobilized governments, the media, social scientists
and international intergovernmental and non-governmental organizations to address the many negative impacts on several countries or societies around the world.

This paper reviews the impacts of both El Niño and La Niña
and the associated ENSO events in Africa. Five major sections
are discussed. These are the definitions of El Niño, La Niña and
ENSO; the climatology of the African continent; the global influence of ENSO and the influence of ENSO in Africa, including its
impacts on socio-economic activities.

20.1.1 What are El Niño, La Niña and ENSO?

The term El Niño (Spanish for 'the boy-child' or 'the Christ-child'
because of its timing) refers to the periodic building up of a large
pool of unusually warm waters in large parts of the eastern and
central equatorial Pacific Ocean (Figure 20.1(b)). On the other
hand, the opposite, which is known as La Niña ('the girl-child'),
is used to describe the periodic building up of unusually cold
waters in large parts of the same ocean basin (Figure 20.1(c)).

The atmosphere and the neighbouring oceans respond to the
warming and cooling of the equatorial eastern Pacific Ocean region (El Niño and La Niña events) in various ways. The warming
and cooling of the equatorial Pacific Ocean are associated with
changes in atmospheric pressure known as the Southern Oscillation (SO). This is a 'sea-saw' in the atmospheric pressure between
the western and eastern sides of the equatorial Pacific Ocean. The
centres of action are represented by measurements of atmospheric
pressure from Darwin and Tahiti. One index that measures the

Climate Change and Africa, ed. Pak Sum Low. Published by Cambridge University Press. © Cambridge University Press 2005.

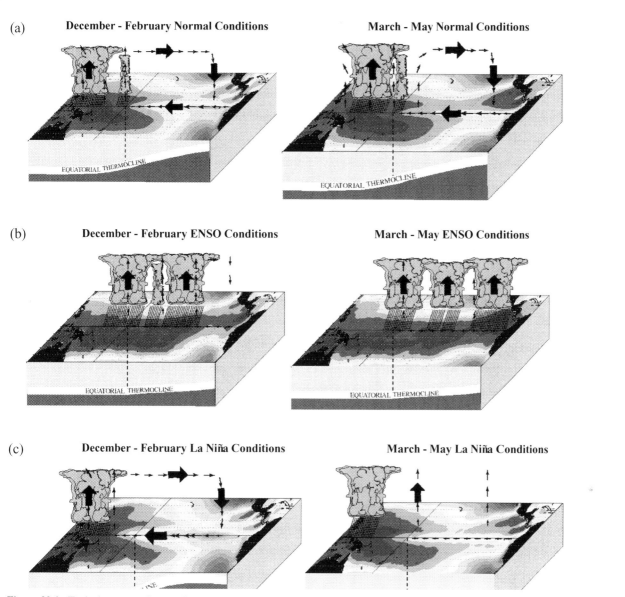

Figure 20.1. Typical patterns of sea surface temperatures and associated convection during (a) normal conditions; (b) El Niño conditions; and (c) La Niña conditions in December–February and in *March–May* (*Source:* http://www.pmel.noaa.gov/tao/elnino/nino-home.html).

magnitude of the Southern Oscillation is the Southern Oscillation Index (SOI). It is basically the difference in standardized pressure between Darwin and Tahiti. Since the SO is closely linked to El Niño episodes, they are collectively referred to as El Niño Southern Oscillation (ENSO) (WMO, 1984). Following such logic, El Niño and La Niña are simply referred to as warm and cold ENSO phases, respectively.

ENSO events generally last from 3 to 6 seasons (9 to 18 months), sometimes as long as 24 months, and tend to recur periodically every 3 to 7 years. It should be noted that El Niño events are, at times, immediately followed by La Niña episodes and vice versa. However, the evolution characteristics of any of the events and the associated impacts have shown that there are

no two events that are perfectly identical. They are just one aspect of the extremes of climate variability and signal a major departure from more average conditions.

The 1997–1998 El Niño event has been considered as perhaps the strongest of the twentieth century, comparable to or even surpassing the famous 1982–1983 event (Obasi, 1999). This El Niño began in the first half of 1997 and ended in the middle of 1998. In the second half of 1998, a rapid transition from the strong El Niño to strong La Niña conditions occurred. The cold episode conditions then persisted through 1999 into 2000. The strong 1997–1998 El Niño and the 1998–2000 La Niña events imposed considerable socio-economic stress and set back development in many parts of the globe. Notable global impacts include the severe droughts and forest fires of Indonesia and north-east Brazil, and the catastrophic floods of tropical east Africa. Indeed, as highlighted in this paper, the impacts of past El Niño events have been considerable in Africa. However, before examining those impacts, it is necessary to place the impacts in context through a general look at the physical setting and climates of Africa.

20.1.2 Climatology of Africa

Topography

The continent of Africa covers an area of about 30 million square km. It stretches for about 7,200 km from east to west and roughly 8,000 km from south to north. It is almost symmetrically astride the equator running from latitude 35° S to 37° N. The continent is a giant plateau, which is very stable geologically. Volcanic activities in the continent in the mid-Pleistocene epoch and earlier, led to the formation of the Great Rift Valley with a track of 6,400 km. The Rift Valley contains most of the great lakes in the continent. Throughout much of the continent there is a relative lack of pronounced topography. The highest peak is Mount Kilimanjaro, which stands at 5,894 m, while the lowest point is the Qattara Depression, which reaches 132 m below mean level (Griffiths, 1972).

The vast plateau is broken by only a few mountain ranges such as the Atlas in the north-west, Ahaggar, Cameroons, Tibesti, Ethiopian, East African and Southern Highlands. In East Africa, in addition to Mount Kilimanjaro, there are also Mount Kenya (5,199 m), Ruwenzori (5,120 m) and Elgon (4,321 m). Lake Victoria is the largest freshwater lake on the continent and the second largest in the world.

Rainfall climatology

There are several major factors that determine climate: the balance of solar and terrestrial radiation at any place; the temperatures of the surface and of the atmosphere, which result partly from the radiation balance and partly from the heat transported into and out of the area by winds and ocean currents; the horizontal and vertical motion of the air; and the moisture cycle. These factors have been used as a basis for describing in detail the climates of Africa by Obasi (1991).

In all, Africa is affected by pressure and associated wind systems of the equatorial, subtropical and mid-latitude regions. The rainfall regimes in Africa are quite complex both in their nature and causes. A continental annual mean of 686 mm arises from precipitation totals as low as 0 mm in the desert regimes to as high as 5,000 mm in the tropical rain forests. Because of the nature of the rain-producing systems, the rainfall pattern exhibits very strong seasonality.

The poleward parts of the continent experience winter rainfall associated with the passage of mid-latitude phenomena such as frontal systems, embedded in the westerly wind regimes. Moving towards the Equator, vast tracts of the subtropical latitudes are dominated by arid conditions virtually throughout the year because of subsiding air in the global-scale Hadley cell circulation. This represents a dominance of quasi-permanent high pressure cells, descending air mass motion, and consequent desert-like conditions with very limited mechanisms for cloud and rainfall formation, apart from some locations that are near large water bodies and windward-facing highlands. Good examples of such arid climate regions are found in the Sahara and Kalahari deserts.

Again moving equatorward, and in contrast to the arid belt, moderate to heavy precipitation characterizes equatorial and tropical areas. The rains are associated with the convergence of easterly trade winds from the northern hemisphere (north-east trades) and the southern hemisphere (south-east trades) in a zone of low pressure. This, the near equatorial trough, is often referred to as the Inter-Tropical Convergence Zone (ITCZ) or the Inter-Tropical Discontinuity (ITD) in western Africa. The region of large-scale convergence of the inter-hemispheric tropical monsoonal wind systems is in harmony with the apparent movement of the overhead sun, giving rise to peak summer rains in northern and southern parts of sub-Saharan Africa during June to August and December to February respectively (Figure 20.2(a)). Two rainfall peaks are concentrated close to the equator because the ITCZ/ITD passes overhead twice each year, giving rise to a bimodal system with rain during March to May and September to November (Figure 20.2(b)).

The classical climate patterns described above are further modified by the presence of large variations in topography over some areas and the existence of large lakes in some parts of the continent. These create their own small-scale circulation patterns that interact with the large-scale flow to modify the climate patterns of some of the areas. Indeed, some locations near the large water bodies receive substantial rainfall throughout the whole year.

Major climate systems that are known to influence rainfall over the continent are complex and include tropical storms, squall lines, easterly/westerly wave perturbations, jet streams, extratropical weather systems, and tropical cyclones (Goddard and Graham, 1999; Nicholson and Seleto, 2000). Other systems include interactions between mesoscale circulations and the large-scale monsoonal flows, teleconnections with global-scale climate anomalies like those associated with sea surface temperatures (SSTs), Indian Ocean Dipole patterns (Webster et al., 1999; Reason, 2001), the Quasi-Biennial Oscillation (QBO) in the equatorial lower stratospheric zonal wind (Reason et al., 2000; Loschnigg et al., 2003), inter-seasonal, 30–60 day, Madden Julian wave, solar and lunar forcing (Ogallo ,1988, 1989; Anyamba, 1992; Mukabana and Pielke, 1996; Indeje and Semazzi, 2000).

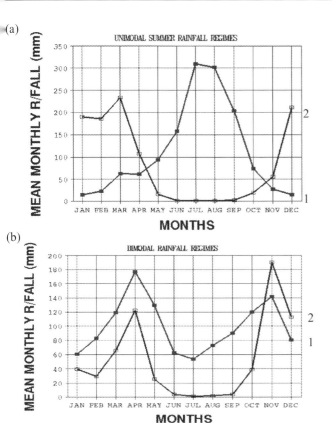

Figure 20.2. The annual variation of mean monthly rainfall indicating (a) unimodal and (b) bimodal rainfall regimes in Africa. (a) Peak summer rains for northern (curve 1) and southern (curve 2) regions of Africa. (b) Peak rainfall in March to May and in September to November/December, typical in tropical eastern Africa. (*Source*: Ambenje, 2002).

20.2 TYPICAL WORLDWIDE INFLUENCE OF ENSO EVENTS ON CLIMATE

The warming/cooling of the eastern and central equatorial Pacific Ocean during El Niño/La Niña events are known to lead to worldwide anomalies in sea surface temperatures (SSTs) and the circulation of the ocean currents. The most significant influence is found in the tropics, but such influences have been found to vary significantly from season to season and with the evolution pattern of the ENSO phases. Within the tropics, there are well-defined east–west circulation cells that make up the Walker circulation with well-defined centres of action. During a strong warm/cold ENSO event, it has been observed that there are displacements of the warm and cold air masses (WMO, 1984; Coghlan 2002), with consequent changes in patterns of air convergence/divergence. In a strong ENSO event, such displacements cause anomalous and enhanced convective activities over the central eastern Pacific (Figures 20.1(b) and (c)), over the central western equatorial

Indian Ocean, along the coast of eastern Africa and off/near the Atlantic equatorial coast of Africa, north-western South America, the northern part of the Greater Horn of Africa (GHA) subregion and among many other parts of the tropics. As a result, abundant rainfall tends to occur in these areas. In contrast, over Indonesia, Australia, India, south-east Africa and some other parts of the tropics, drought conditions are usually experienced (see Figures 20.3(a) and (b), Ropelewski and Halpert, 1987).

The observed impacts have a large degree of variability both in time and space (Ogallo, 1988), i.e. they are location- and season-dependent. They also depend on the phase of the El Niño/La Niña event. Note that El Niño/La Niña development phases may be classified as the onset, peak and withdrawal. The onset/withdrawal phases of an El Niño/ La Niña correspond to the months of onset/cessation of the specific El Niño/La Niña event, while peak phase refers to the months when the strength of the El Niño/La Niña event is at its maximum.

20.3 OVERVIEW OF ENSO INFLUENCE ON THE CLIMATE OF AFRICA

Extreme climate events such as droughts and floods are very common in some regions of Africa. Although ENSO impacts are strongest in the Pacific Ocean region, past records from Africa show that some severe droughts and floods that have been observed over parts of the continent have been associated with ENSO events. At the same time, recent studies from the region have shown that although ENSO signals are discernible, both the Atlantic and Indian Oceans play significant roles in the determination of the regional climate extremes in the Sahelian, eastern and southern Africa subregions. In addition, the large inland lakes, and the complex inland topography, including the Great Rift Valley that runs across the region from north to south with attendant chains of mountains on both sides, also play a significant role in modulating the regional climate anomalies.

This section highlights some of the major rainfall anomalies that are associated with ENSO events in Africa. Because of the complex nature of rainfall over the continent and its high variability both in time and space, the presentation is made on a region-to-region basis. The regions considered are the Sahel and West Africa, eastern Africa and southern Africa. It should be noted that the ENSO is the most dominant perturbation responsible for inter-annual climate variability over eastern and southern Africa (Nicholson and Entekhapi, 1986). ENSO linkages are however known to vary significantly from not only location to location and season to season, but also with the evolution pattern of the ENSO phases (Figure 20.4). As mentioned in the previous sections, its influence on rainfall is however dependent on the evolution phase (onset, peak and withdrawal) of the event as well as the sea surface

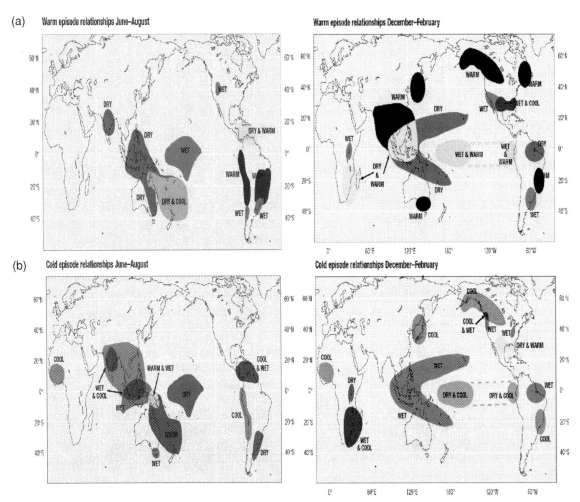

Figure 20.3. Typical worldwide impacts of (a) warm and (b) cold ENSO conditions, June to August and December to February. (*Source*: Ropelewski and Halpert, 1987).

temperature evolutions of the neighbouring Atlantic and Indian Oceans.

20.3.1 Sahel and West Africa

The rainfall variability in the Sahel and similar regions of West Africa is quite complex, as is its response to ENSO events. Global-scale studies by Ropelewski and Halpert (1987, 1989) showed only a weak association between ENSO and Sahelian precipitation, but a number of subsequent studies show stronger relationships (Janicot et al., 1996; Ward, 1998; Thiaw and Kumar, 2001). In these studies, the Sahel and the Guinea coast are generally treated as separate climatic zones.

Janicot et al. (1996) demonstrated that correlations between summer (July–September) rainfall in the Sahel and the Southern Oscillation Index increased over the 1945–1993 period, and were statistically significant for the most recent 25 years of that analysis. Prior to 1970, rainfall in the Sahel

and the Guinea coast regions was highly negatively correlated and corresponded to the latitudinal variability of the ITCZ (i.e. when the ITCZ was located northward, the Sahel rainfall anomaly was positive and that of the Guinea coast was negative, and vice versa).

Ward (1998) found a clear ENSO connection in the Western Sahel and Guinea coast regions. In his analysis, data were divided into years with dipole behaviour (where the rainfall anomalies in the Sahel and Guinea coast regions were of opposite sign) and non-dipole behaviour (when the anomalies were of the same sign). In non-dipole years and in a warm event phase, a SST pattern associated with ENSO is also associated with reduced Sahelian (and Indian) precipitation in the JAS (summer) season, and wetter than normal conditions in the Sahel in June and October. Ward noted a complex interaction between a direct connection from the eastern Pacific and the Atlantic sector, and a connection with ENSO via the Indian sector and concluded that to successfully model the teleconnectivity that gives rise to the SST-forced component of

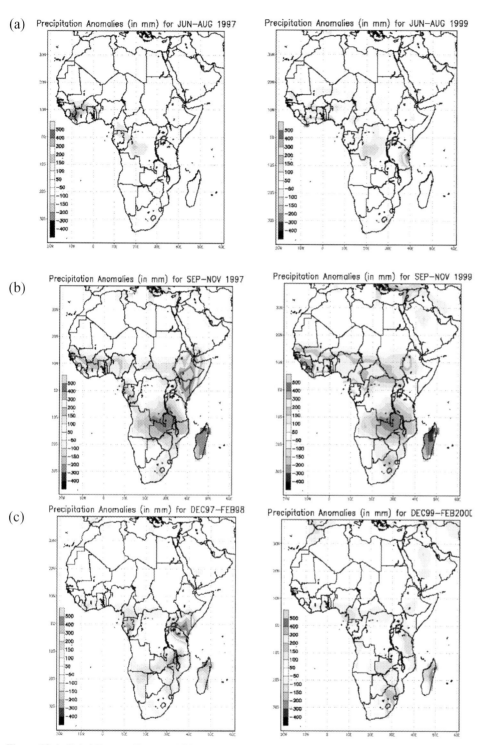

Figure 20.4. Rainfall anomalies over Africa during the warm and cold ENSO of 1997–1998 and 1999–2000 (*Source*: Climate Prediction Centre, USA).

the African rainfall anomalies, GCMs should include the Indian monsoon.

Thiaw and Kumar (2001) noted that the predictability of Sahel rainfall inter-annual variability is primarily driven by large-scale ocean-atmosphere forcing through changes in the global

monsoon circulation partially induced by ENSO. They further stated that the response of the Sahel rainfall to this large-scale forcing is likely influenced by Atlantic SSTs and by land-surface conditions. Using a different approach in which tropical convection is inferred from Outgoing Longwave Radiation (OLR), Thiaw

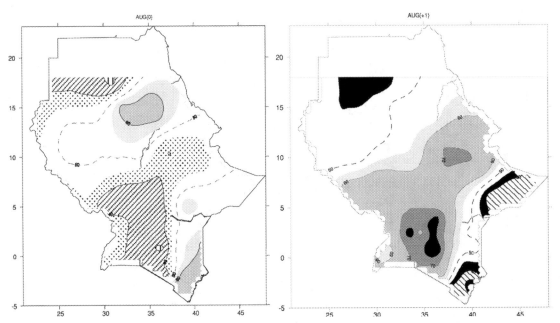

Figure 20.5. (a) Warm and (b) cold ENSO-related rainfall anomalies (percentile ranks) over equatorial parts of eastern Africa for August during ENSO events. The smooth shaded areas indicate enhanced rainfall while the hatched areas indicate depressed rainfall. The vertical and horizontal scales are latitude and longitude, respectively (*Source*: Drought Monitoring Centre-Harare (DMCH), Drought Monitoring Centre-Nairobi (DMCN) and Majugu (2002)).

and Kumar (2001) found that in strong El Niño or La Niña years, OLR anomalies in the central Africa–Gulf of Guinea region provide a useful technique for rainfall prediction. OLR reflects the large-scale monsoon systems in the region, which are influenced by the changes in atmospheric circulation during ENSO episodes.

20.3.2 Eastern and southern Africa

The typical rainfall anomaly associated with ENSO over eastern and southern Africa is a dipole rainfall pattern. Eastern Africa is in phase with warm ENSO episodes, whereas southern Africa is negatively correlated with these events (Nicholson and Kim, 1997) during the October to December period.

Eastern Africa

Following the discussion in Section 20.1.2, the eastern Africa subregion can be divided into three sectors based on the rainfall regimes.

Northern sector areas of eastern Africa have peak rainfall during the northern hemisphere summer months of June to September. On the other hand, the southern sector of eastern Africa has peak rainfall concentrated mainly within the southern hemisphere summer months of December to February. Various studies have shown that the onset of the warm/cold ENSO events are often, but not al-

ways, associated with below/above average rainfall amounts over most parts of the northern sector as well as over the southern parts of the eastern sector (Figure 20.5 and Figure 20.6) during their respective summer months.

Seasonal rainfall characteristics over the equatorial parts of eastern Africa are quite complex. Locations near the large water bodies receive rainfall throughout the year, while parts of the western and coastal regions also receive substantial amounts of rainfall during the months of June to August. Recent diagnostic studies have shown a high degree of relationship between ENSO events and seasonal rainfall anomalies over the region. The linkages vary significantly from season to season, and also with the specific ENSO phases. In general, above/below-average rainfall conditions are common in March to May and October to December during the onset of the warm/cold ENSO events. On the other hand, below/above-average rainfall conditions dominate many areas of the sector during June to September period of the onset of the warm/cold ENSO event (Figure 20.4).

Southern Africa

One of the stronger relationships of ENSO is between the warm onset phase of the ENSO cycle (El Niño) and drought in southern Africa. Warm episodes beginning in April or May have a particularly strong relationship with drought in this area in the following summer months of December to February. During most warm

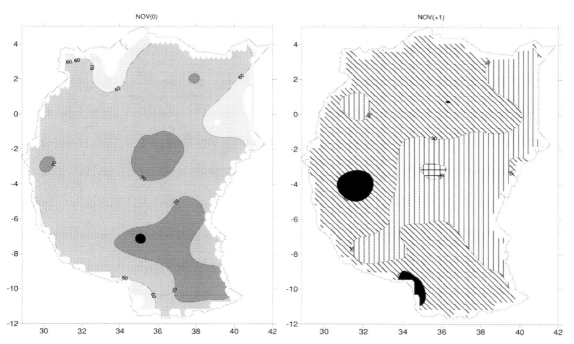

Figure 20.6. (a) Warm and (b) cold ENSO related rainfall anomalies (percentile ranks) over equatorial parts of eastern Africa for November during ENSO events. The smooth shaded areas indicate enhanced rainfall, while the hatched areas indicate depressed rainfall. The vertical and horizontal scales are latitude and longitude, respectively (*Source*: Drought Monitoring Centre-Harare (DMCH), Drought Monitoring Centre-Nairobi (DMCN) and Majugu (2002)).

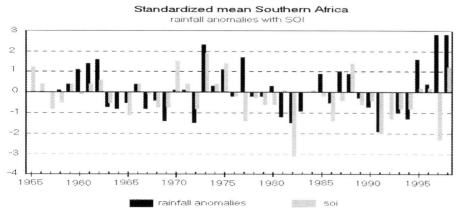

Figure 20.7. Standardized rainfall anomalies over southern Africa juxtaposed with Southern Oscillation Index (*Source*: Garanganga, (2003)).

ENSO episodes, the bulk of southern Africa experiences considerable rainfall deficits. This is well depicted in the time series of seasonal rainfall (October to March) over southern Africa and the SOI (Figure 20.7). It is evident from the time series that negative seasonal rainfall anomalies tend to coincide with negative SOI values (warm episode). This is reversed during the cold onset phase of ENSO, with a tendency for the subregion to receive enhanced rainfall as was experienced in 1999–2000.

The strength of the ENSO signal is not uniform throughout the subcontinent of southern Africa. Figure 20.8 shows the spatial patterns of rainfall during a warm and a cold episode in January, which is the peak rainfall month over most of the area.

It has been noted that in both the eastern and southern parts of Africa, Indian and Atlantic Ocean SST anomaly patterns, and anomalies in the other regional circulation patterns have significant influence on the region's ENSO impacts.

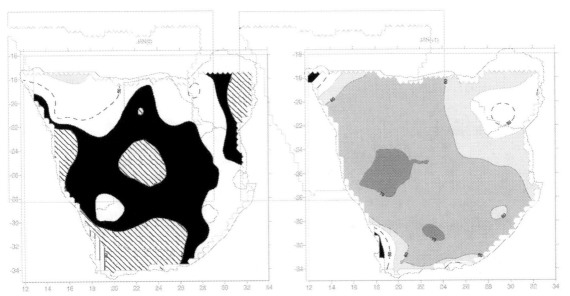

Figure 20.8. (a) Warm and (b) cold ENSO-related rainfall anomalies (percentile ranks) over southern Africa for January following a mature ENSO event. The smooth shaded areas indicate enhanced rainfall, while the hatched areas indicate depressed rainfall. The vertical and horizontal scales are latitude and longitude, respectively
(*Source*: Drought Monitoring Centre-Harare (DMCH), Drought Monitoring Centre-Nairobi (DMCN) and Majugu (2002)).

20.4 OVERVIEW OF THE IMPACTS OF ENSO ON SOCIO-ECONOMIC ACTIVITIES IN AFRICA

20.4.1 Vulnerability of Africa to variations in climate

Many socio-economic activities in Africa heavily depend on climate and specifically on rainfall. The formal and informal economies of most African countries are strongly based on natural resources: agriculture, pastoralism, logging, ecotourism and mining are dominant. Consequently, climatic variations that alter the production systems of these activities have very high leverage on the economy. It is estimated that on a global scale, about 75% of the natural disasters are associated with extreme weather and climate events such as floods, droughts, heat waves, etc. In Africa, the vulnerability of society to such events is much higher. The continent contains a large number of the poorest and least-developed nations of the world. Per capita Gross Domestic Product (GDP), life expectancy, infant mortality, and adult literacy are all in the bottom quartile globally when averaged across Africa, although individual nations may perform somewhat better on one or more of these indices.

20.4.2 Impacts of ENSO on socio-economic activities

Extreme climate events such as droughts and floods have far-reaching impacts on many socio-economic activities. They often lead to loss of life and property, destruction of infrastructure and

large losses to the economy. They also have harsh negative impacts on agriculture, livestock, wildlife, tourism, water resources and hydroelectric power generation, and on many other socio-economic sectors. The impacts are often beyond the coping capacity of many developing countries, especially those that often have to rely on support from international communities. The impacts of extreme weather and climate events also affect the welfare of the communities and tend to increase and aggravate poverty, especially in regions where rain-fed agriculture as well as livestock and hydroelectric power form major sources of food and energy, respectively. Destruction of crops by floods and low yields that result from drought reduce the economic status of most of the rural communities, especially affecting women, who form most of the active farmers in Africa. Similarly, lack of adequate hydroelectric power, due to prolonged drought, results in loss of employment and also reduces the economic status of the people.

During drought periods, lack of food for humans and pasture for animals (domesticated and wild) often leads to mass migration of both humans and animals in search of the limited water and food resources. This can lead to conflicts between humans and also between humans and animals.

20.4.3 The African experience with ENSO: selected key examples of impacts from recent ENSO events

The 1997–1998 warm ENSO episode is the best-recorded and studied ENSO event, and is sometimes called the 'El Niño of the Twentieth Century'. Glantz (2001) used lessons learned from this event to identify problems in coping with the many impacts of El Niño, such as jurisdictional disputes among government agencies, forecast reliability, lack of education and training about the El Niño phenomenon, political and economic conditions (or crises) existing during the event, lack of resources to cope in a preventive or mitigative way, lack of donor sensitivity to the local needs, poor communication, lag time between forecast and impacts and between impacts and responses, responses and reconstruction, and so on. Many of these issues are not exclusive to coping with ENSO events, but apply to all manner of natural hazards. Amongst the conclusions in his report is that these weaknesses must be addressed, and that it will take social, economic and political change, not just sophisticated technological solutions such as early warning systems.

Sixteen countries participated in case studies of impacts and responses to the major 1997–1998 El Niño event, including Ethiopia, Kenya and Mozambique (Glantz, 2001). Highlighted are some excerpts of the observed impacts on various socio-economic sectors/activities in Ethiopia and Kenya.

Subsequent to the 1997–1998 El Niño, a major La Niña developed and held sway from 1998–2000. Impacts in Africa from that event are also included.

Ethiopia

The June to September 1997 seasonal rainfall totals in Ethiopia at 20 observation sites were 20% less than that those received in 1996. Almost all parts of Ethiopia had dry spells in the Kiremt rainy season months of July and August 1997. Out of 33 administrative zones in Ethiopia, the onset of the rain in 18 zones was delayed, affecting land preparation and sowing. The 15 zones that had a good start in rainfall were affected by dry spells in the peak rainfall months of August and September, which adversely affected the maturation stage of the crops.

Erratic rainfall reduced the area cultivated by 9% from that of 1996, caused by oxen with low energy levels due to sparse pasture. Poor farmers could not rent or borrow oxen at the right time because the owners gave priority to their own plots. The replanting of the land several times following the coming and retreating of rainfall depleted the seed reserves of farmers. Yields were low because of reduced land preparation, and poor and early cessation of rainfall. Lack of fodder reduced the price of cattle and some animals died, especially in the Raya region of northern Ethiopia. Coffee, the main cash crop of the country, was reduced in production because 'coffee berries ready to be picked from the trees have been falling on the ground due to heavy rains' that came later in October and November. Food production declined after two years of good harvest in 1995–1996 and 1996–1997. Total output in the Meher rainy season in 1997–1998 was reduced by 24% from the 1996–1997 output. Prices of agricultural commodities also increased by 13–53% from those of 1996.

Kenya

The heavy rains associated with the 1997–1998 warm ENSO event had severe impacts on various socio-economic sectors and activities. These are highlighted below.

Water resources sector

The water resources sector was both negatively and positively affected by the 1997–1998 El Niño event. Negative impacts included widespread flooding that led to the destruction of property in several sections of the country, increased soil erosion in areas with poor land use and management practices, and increased frequency of mud and landslides, especially in the hilly areas with loose soil types. Other negative impacts included surface and groundwater pollution, destruction of small storage earth dams, and the increased sedimentation and siltation in the rivers and streams that led to the sedimentation and siltation of the major water storage reservoirs. The general cost of these negative impacts amounted to about US$9 million. However, this sector also benefited from the excess rainfall during this period. Pollution loads were reduced through the washout effect of the rainfall, soil moisture for agricultural production was enhanced, and the water reservoirs were adequately recharged boosting the output levels of the hydroelectric dams.

Agricultural sector

The agricultural sector was also negatively and positively affected by the phenomenon. The abundance of rainfall resulted in increased plant and animal diseases that affected the livestock and crop production in several regions in the country. Flooding also affected the farms through water logging, leading to further reduction in yields and destruction of livestock water facilities. Several cases of deaths of animals through drowning were also reported. The estimated combined loss suffered by this sector reached US$236 million.

However, in the arid and semi-arid areas the rains were a welcome relief from the perennial dry situation, leading to development of good pasture and the resultant improved livestock performance. Agricultural production in some areas increased due to the enhanced availability of moisture for the crops. The rains enhanced and prolonged the time of moisture availability for the biological soil and water conservation structures to take up. Tree planting and survival rates were generally increased to nearly 100%.

Transport and communication sector

The El Niño rains devastated the transport sector. Accompanying floods and landslides wreaked havoc on the roads and transportation infrastructure throughout the country. Several bridges and an estimated 100,000 km of both rural and urban roads were destroyed, leading to a general paralysis of the transportation system in most parts of the country. The estimated cost of these damages was US$670 million. Aviation and shipping industries were also disrupted through the flooding of the facilities. Scheduled and chartered flights were disrupted due to poor visibility and the submergence of the navigational equipment and runways by floodwaters. Docking facilities at the shipping ports were submerged in floodwaters, making it impossible to off-load merchandise from the ships. Telecommunications were severely affected by falling trees that destroyed the communications lines. The underground cable channels were also flooded, causing a disruption in services. Interruptions of electric energy supply were experienced as some equipment was destroyed by floodwaters, falling trees, and collapsing buildings. However, a positive effect of the event was experienced by the energy sector, with the complete recharging of the hydroelectric dams, and hence the enhancement of the production of electricity.

Health

During the 1997–1998 El Niño, Kenya's health resources were pushed beyond their limits. Impacts included the destruction of several health facilities, contaminated drinking water, increases in the number of stagnant ponds, blockage and overflow of sewers and open drains, and an increase in breeding of flies as a result of the decomposition of refuse. These all led to an upsurge in disease epidemics and an increase in morbidity and mortality rates.

The heavy rains also saw the re-emergence of diseases that had disappeared for some time. Notable among these was the Rift Valley Fever, which affected livestock in the marginal areas.

The impacts on Africa of the 1998–2000 La Niña

The western Africa (Sahel) rainy season was wetter than normal for 1999 but experienced deficits the following year. The western part of the Guinea region was especially dry during the year 2000 rainy season. During 1998–2000, southern Africa (including northern Mozambique and northern Madagascar) received significantly below-normal precipitation, while the islands in the southwest Indian Ocean experienced their 'drought of the century'. In some areas, this drought saw the proliferation of the insect *Cinara cupressi* that, in search of moisture, attacked and decimated many tropical conifers of the *Cypress cupressus* family. The Gross Domestic Product of countries in this region was affected by as much as 3.5% during this period. Rainfall deficits continued into 2001 along the east coast from north-eastern South Africa up into Mozambique.

Failure of the rainy season in the Greater Horn of Africa (GHA) region in 2000, following two years of erratic rainfall, triggered food shortages and losses of livestock not seen since the early 1980s. Widespread drought affected northern Kenya and southern Ethiopia most severely, but was serious as well in Sudan, Somalia, the United Republic of Tanzania and Eritrea. At a time of civil strife and drought, an estimated 20 million people faced food shortages in the GHA, 10 million of them in Ethiopia alone. The drought in parts of Kenya, Somalia, Mozambique and the United Republic of Tanzania continued into 2001.

In contrast, the 1998–2000 La Niña brought devastating floods to parts of Africa. Heavy rains in Sudan in 1999 damaged or destroyed more than 2,000 homes, while in Mozambique, some of the worst flooding in 40 years cost dozens of lives and massive property losses. The Sahel is normally dry in the July–October period, but unseasonable heavy rains and floods in that period cost hundreds of lives. Flooding recurred in the Mozambique area in 2000, partly due to the La Niña conditions, but exacerbated by the landfall of cyclone Connie in early February. Even worse disaster struck shortly afterwards with the landfall of cyclone Leon-Eline. The region was already saturated, and the additional rainfall led to great loss of life.

20.5 CONCLUSION

Socio-economic losses cannot be entirely eliminated, but timely and appropriate mitigation measures can certainly reduce the impacts. In fact, timely and advanced warnings of El Niño episodes allow for advanced national planning, with considerable advantages in many sectors of the economy, such as in water resources management, tourism, and fisheries and agricultural production (Obasi, 1996). For example, in the case of the 1997–1998 El Niño event, advances in El Niño-related science and in monitoring the sea surface temperatures in the Pacific Ocean, enabled scientists in the National Meteorological and Hydrological Services to predict its formation much earlier than for all previous similar events. With recent developments in communication technology, including the use of the Internet, information on the El Niño grew rapidly and was disseminated throughout the world in a timely manner. This enabled many governments to take appropriate measures, and stimulated international cooperation and integrated efforts to address the associated impacts.

Following experience gained from the major El Niño event in 1997–1998, many African countries reflected on their vulnerabilities and proposed plans of action for their governments, their climate experts and their citizens and media groups. Countries highlighted the benefits of, for example:

- Better coordination between the various agencies concerned with early warnings;
- Enhanced levels of knowledge of ENSO, its characteristics and impacts amongst government and other agencies and society in general;

Investment in monitoring networks and in strengthening forecast capacity;

Taking preventative actions, when and where possible, based on knowledge of climatology and local conditions and using the increasingly efficient seasonal predictions that are available.

To illustrate the benefits of recent progress, the 1999–2000 drought in parts of eastern Africa was the most severe since 1961, climatologically (worse than 1984), but its impacts were not as severe as those observed in the 1984 drought. This was mainly because governments of the affected nations applied the forecast information provided by the National Meteorological and Hydrological Services to put mitigation measures in place to address the associated impacts.

The after-effects of a major climate-related event can undo years of development efforts, and so it is hoped that investments in education, communications, monitoring and prediction will continue to help to mitigate the effects of future ENSO events on the nations of Africa. This is particularly important now that Africa also faces the threat of long-term climate change as a result of human activities.

ACKNOWLEDGEMENTS

The authors would like to acknowledge the useful peer reviews by Dr Wassila Thiaw, Dr Mick Kelly and Dr Graham Farmer and the kind assistance of Dr Yinka Adebayo.

REFERENCES

Ambenje, P. G. (2002) Climate of the Greater Horn of Africa. Paper presented at the Tenth Climate Outlook Forum for the Greater Horn of Africa, 26–28 August 2002, Nairobi, Kenya.

Anyamba, E. K. (1992) Some properties of the 20–30 day oscillation in tropical convection. *J. Afr. Met. Soc.*, **1**, 1–19.

Coghlan, C. (2002) El Niño – causes, consequences and solutions. *Weather*, **57**, 209–215.

DMCH, DMCN and Majugu, A. (2002) In *Drought Monitoring Centre Nairobi – Great Horn of Africa Climate Atlas*, DMCN 2002.

Garanganga, B. J. (2003) *Circulations Patterns Associated with Droughts Over Southern Africa*, DMC, Harare.

Glantz, M. (2001) *Once Burned, Twice Shy? Lessons Learned From the 1997–98 El Niño*. The United Nations University.

Goddard, L. and Graham, N. E. (1999) Importance of the Indian Ocean for simulating rainfall anomalies over eastern and southern Africa. *J. Geophys. Res.*, **104**, D16, 19,099–19,116.

Griffiths, J. F. (1972) Climates of Africa. In H. E.Landsberg (Ed.), *World Survey of Climatology*, Vol. 10. Amsterdam, Elsevier.

Indeje, M. and Semazzi, F. (2000) Relationships between QBO in the lower equatorial stratospheric zonal winds and East African seasonal rainfall. *Meteorol. Atmos. Phys.*, **73**, 227–244.

Janicot, S., Moron, V. and Fontaine, B. (1996) Sahel droughts and ENSO dynamics. *Geophys. Res. Lett.*, **23**, 515–518.

Loschnigg, J., Meehl, G. A., Webster, P. J., Arblaster, J. M. and Compo, G. P. (2003) The Asian Monsoon, the Tropospheric Biennial Oscillation, and the Indian Ocean Zonal Mode in the NCAR CSM. *J. Climate*, **16**, 1617–1642.

Mukabana, J. R. and Pielke, R.A. (1996) Investigating the influence of synoptic-scale monsoonal winds and mesoscale circulations and diurnal weather patterns over Kenya using a mesoscale numerical model. *Mon. Weather Rev.*, **124**, 224–243.

Nicholson, S. E. and Entekhabi, D. (1986) The quasi-periodic behaviour of rainfall variability in Africa and its relationship to the Southern Oscillation. *Arch. Meteorol. Geophys. Biolklimatol. Ser. A*, **34**, 311.

Nicholson, S.E. and Kim, J. (1997) The relationship of the El Niño-Southern Oscillation to African rainfall. *Int. J. Climatol.*, **17**, 117–136.

Nicholson, S. and Selato, J. C. (2000) The influence of La Niña on African rainfall. *Int. J. Climatol.*, **20**, 1761–1776.

Obasi, G. O .P. (1991) Climatic resources of Africa: problems and potentials in their management for increased agricultural productivity and sustainable development. Lecture presented at the First United Nations University (UNU) Programme on Natural Resources in Africa, Orientation/Training Course, Nairobi, 16 August 1991.

(1996) *Climate, Climate Change, Variability and Predictability.* Rajiv Gandhi Institute for Contemporary Studies (RGICS). RGICS Paper No. 36. New Delhi, 35pp.

(1999) The role of WMO in addressing the El Niño phenomenon. Lecture at the Evaluation Seminar on the Impact of the 1997–98 El Niño Phenomenon for Directors of National Meteorological and Hydrological Services in WMO Regions III and IV. Lima, Peru, 15 March 1999. SG/91.

Ogallo, L. J. (1988) Relationship between seasonal rainfall in East Africa and the Southern Oscillation Index. *J. Climatol.*, **8**, 31–43.

(1989) The spatial and temporal patterns of the East African seasonal rainfall derived from principal component analysis. *Int. J. Climatol.*, **9**, 145–167.

Reason, C. J. C. (2001) Subtropical Indian Ocean SST dipole events and southern African rainfall. *Geophys. Res. Lett.*, **28**, 2225–2227.

Reason, C. J. C., Allan, R. J., Lindesay, J. A. and Ansell, T. A. (2000) ENSO and climatic signals across the Indian Ocean basin in the global context. *Int. J. Climatol.*, **20**, 1285–1327.

Ropelewski, C. F. and Halpert, M. S. (1987) Global and regional precipitation patterns associated with the El Niño/Southern Oscillation. *Mon. Weather Rev.*, **115**, 1606–1626.

(1989) Precipitation patterns associated with the high index phase of the Southern Oscillation. *J. Climate*, **2**, 268–284.

Thiaw, W. M. and Kumar, V. (2001) Effects of tropical convection on the predictability of the Sahel summer rainfall interannual variability. *Geophys. Res. Lett.*, **24**, 4627–4630.

Ward, M. N. (1998) Diagnosis and short-lead time prediction of summer rainfall in tropical North Africa at interannual and multidecadal timescales. *J. Climate*, **11**, 3167–3191.

Webster, P. J., Moore, A. M., Loschnigg, J. P. and Lebden, R. R. (1999) Coupled ocean-atmospheres dynamics in the Indian Ocean during 1997–98. *Nature*, **401**, 356–360.

WMO (1984) *The Global Climate System: A Critical Review of the Climate System during 1982 to 1984*. World Meteorological Organization, Geneva.

(2003) *The Global Climate System Review, June 1996–December 2001*. World Climate Data and Monitoring Programme, WMO-No. 950, 144pp.

21 Climate policy implications of the recent ENSO events in a small island context

ROLPH ANTOINE PAYET

Ministry of Environment & Natural Resources, Mahe, Seychelles

Keywords

Climate policy; El Niño; La Niña; ENSO; small island states; vulnerability and adaptation; the Seychelles

Abstract

El Niño and La Niña are the warm and cold phases of El Niño-Southern Oscillation (ENSO) that develop in the eastern and central equatorial Pacific. The strong El Niño in 1997–1998 was rapidly followed by a strong La Niña in 1998–2000. Both events caused severe economic and human losses in many parts of the world, including small island states. In the Seychelles, the El Niño event brought extreme rainfall causing flooding and landslides; and due to the elevated sea surface temperatures, severe coral bleaching occurred. On the other hand, the La Niña event brought an extended drought, resulting in acute water shortages. Both events caused profound impacts that led to economic and infrastructure losses, especially in the fisheries and tourism sectors, as well as the infrastructure on the coastal areas.

This chapter highlights the vulnerability of small island states to the extreme climatic events caused by ENSO and to the climate change, which is projected to increase the frequency of such extreme climatic events. Without appropriate observing systems and an adaptive policy framework, adaptation to such extreme climate events, which is already an extra burden on those states, will be difficult. A case study on the impacts of the 1997–1998 El Niño and 1998–2000 La Niña on the Seychelles and the resulting policy responses are discussed, with an emphasis on the potential of using the response to ENSO events as a policy window to develop adaptive capacity to extreme events as a result of long-term climate change.

21.1 INTRODUCTION

El Niño and La Niña are the warm and cold phases of El Niño-Southern Oscillation (ENSO) that develop in the eastern and cen- tral equatorial Pacific (Obasi, 2005). However, sea surface tem- perature (SST) anomalous warming and cooling in the eastern Pacific and south-western Indian Ocean have been found to show significant correlation, although the SST anomalous warming is much larger in the eastern Pacific (Tourre and White, 1995). Nicholson (1997), among others, has also shown strong cor- relations between the Indian Ocean and the Pacific Ocean for almost every ENSO event that dates back to 1946, indicating that the two systems may somehow be linked or influence each other.

Figure 21.1 shows the SST anomalies in the Pacific (equatorial) and the Indian Ocean (the Seychelles region) from January 1997 to December 2001. It is clear that the SST anomalies in both oceans are strongly correlated. The onset of the El Niño event in the Pacific Ocean started around April/May 1997 and ended in July 1998. Immediately after, a rapid transition from the strong El Niño to the moderately strong La Niña conditions occurred. The cold episode conditions persisted throughout 1999 and well into 2000 (Obasi, 2004).

The impacts of both the El Niño 1997–1998 and the La Niña 1998–2000 events were felt around the globe (WMO, 1999; Obasi, 2005). It has been estimated that economic losses attributed to global-scale disasters, ranging from flooding to droughts as a result of ENSO and ENSO-related events, amounted to US\$76 billion in 1998 (WMO, 1998). However, the actual losses due to affected economic sectors and decrease in ecosystem function are likely to be much higher.

This chapter highlights the impacts of the El Niño event in 1997–1998 and the La Niña event in 1998–2000 in the Seychelles as a small island state. The events have clearly led to a reduction in adaptive capacity of the Seychelles to ENSO, as well as to fu- ture climate change, which is projected to increase the frequency of extreme climatic events (IPCC, 2001). The effects of climate change threaten many small island states, and therefore it is criti- cal that governments and regional and international organizations push forward the process of developing adequate local policy and

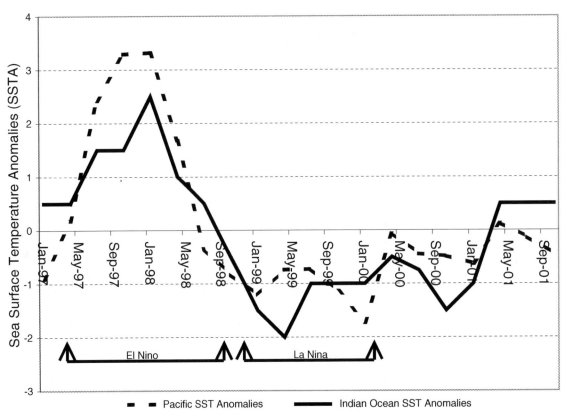

Figure 21.1. Sea surface temperature anomalies in the Pacific (equatorial) and Indian Ocean (Seychelles region). The El Niño in 1997–1998 and the La Niña 1998–2000 are indicated in the figure (*Source*: NOAA Climate Diagnostics Center).

adaptive response strategies, as described in Pernetta and Hughes (1990).

In addition, the escalating economic costs associated with ENSO further emphasize the need for both government and the international community to improve monitoring and sharing of information on weather and climate, as well as transfer of knowledge and technology to mitigate and adapt. Risks associated with ENSO and other extreme climatic events should therefore be reflected in policies focused on building capacity for adaptation, including early warning and preparedness.

Thus, the impacts of ENSO provide the opportunity and the basis for the emergence of policy frameworks, which may either focus on measures to increase adaptive capacity to climate change or approaches aimed at technological advances, institutional arrangements, sustainable financing, reduction in greenhouse gas (GHG) emissions, and information exchange (IPCC, 1995). In this context, consideration needs to be given to the relevance of this policy window to drive home feasible adaptive options in small island states with low coastal areas, ahead of any future impacts of climate change.

21.2 THE IMPACTS OF ENSO IN THE SEYCHELLES

21.2.1 The impacts of El Niño in 1997–1998

During the El Niño event in 1997–1998, extreme rainfall conditions in the Seychelles led to floods and landslides, which had last been experienced in 1868, more than 100 years earlier (Payet, 1998). The disaster of 1868, which claimed a third of the area on which the capital city stands today, caused extensive damage to infrastructure and more than 70 deaths.

The floods and landslides in 1997 caused damage to more than 500 houses at an estimated cost of US$2 million, and almost 40% of the public roads, with temporary repair estimated to cost US$1.5 million. Numerous homes on the coastal plateaux were severely affected as a result of the flooding, and communications were paralysed for several days. Abnormally high tides, probably due to thermal expansion of the ocean as a result of higher sea-surface temperatures, spewed seawater into coastal houses and hotels all over the island. There was also damage to the islands' coastal infrastructure, such as roads and airstrips, hampering communication between the main island and the outer islands. Agricultural losses were estimated to be about US$1.5 million (Moustache, 1997).

The industrial sector also suffered losses as a result of the intensive rainfall and landslides during the event, which caused damage to construction materials and manufacturing equipment to the value of an estimated US$1.25 million, some of which were covered by insurance. The indirect cost to tourism has not properly been accounted for.

The impact of the elevated sea surface temperatures due to El Niño on coral reefs within the Indian Ocean was severe, causing up to 95% mortality in some areas in the Seychelles (Wilkinson *et al.*, 1999). The extent and severity of secondary impacts, such as decreases in reef fish and diving tourism, are to be quantitatively evaluated, as the reefs cease to play an active role as one of the most diverse and productive ecosystems and a natural coastal defence. Although conclusive data have not been published yet, it is expected that coastal fisheries and dependent communities may be severely affected as a result of the coral-bleaching event. More information on the likely impacts and costs will emerge as more research is undertaken on the coral-bleaching event in 1997–1998. However, the long-term economic consequences of potential impact on an accelerated destabilization of the coastline and other coastal infrastructure, as well as the design life of existing infrastructures, such as roads, water supply and foundations, would be difficult to estimate (UNESCO, 1991).

Indeed, the disasters brought by the very strong El Niño event in 1997–1998 in many parts of the world led the First Intergovernmental Meeting of Experts on El Niño, held in November 1998 in Guayaquil, Ecuador, to declare that it has 'imposed continuing poverty on people and set back development in so many parts of the globe' (UNEP, 1999). In particular, those without the resources and capabilities to adjust have increasingly become more vulnerable. A number of small island states fall within this category.

21.2.2 The impacts of La Niña in 1998–2000

During the La Niña period (1998–2000), drought conditions persisted in the Seychelles, especially during the last few months of 1998 and the first few months of 1999. This caused acute freshwater shortages in the country, resulting in some establishments shutting down operations, and the agricultural sector, tourism industry and other related services really suffered. Although the Seychelles expects on average about 3,000 mm of rainfall a year, the limited capacity for water storage further exacerbated the situation. The demand for water was so intense that many companies decided to make heavy investments in storage and acquire small desalinators. The value of these investments is not readily available. Growth in population and tourism will continue to in-

crease the national deficit in water. With increasing pressures for preservation of catchment areas, construction of dams has become unacceptable and very costly.

The La Niña event is believed to have caused an estimated loss of 30,000 tonnes in average fishery catch (SFA, 1999). As a result, gross earnings from the fisheries sector dropped by 20% in 1998. Other unaccounted impacts include loss of revenue from the tuna-canning factory and for fishing vessels utilizing port services, which reportedly moved to other locations.

An increase in the frequency of forest fires was observed due to the persistent drought, placing at risk numerous endemic species and unique habitats. In total, forest fires caused about US$0.5 million in losses to the fire control services, but the loss of biodiversity-use values is yet to be accounted for (GoS, 2000a). The cost of reforestation and fire prevention equipment has been estimated to exceed US$1 million per year (GoS, 2000a).

The indirect impacts during the La Niña period in 1998-2000 included health and fisheries. Although there were some disease outbreaks, mainly diarrhoea, in the Seychelles during this period, no studies have since been undertaken to establish such a link. The impact on health is critical for small island states, where an entire population can easily be threatened with mass epidemics. Current research elsewhere shows increasing evidence linking epidemic outbreaks to the drought condition (Diaz and McCabe 1999).

21.2.3 The economic impacts of the 1997–1998 El Niño and 1998–2000 La Niña

The 1997–1998 El Niño and the 1998–2000 La Niña events have had profound impacts on the economy of the Seychelles. Figure 21.2 shows the estimated combined losses in monetary terms caused by the events. Fisheries suffered the greatest loss, accounting for 45% of the total. This was followed by agriculture (28%), tourism (12%), industry (7%), construction (5%) and forestry (3%). The services providing support to these sectors were also affected.

Other impacts included damage to coastal infrastructure such as roads and airstrips. Because of the topography of the main granitic islands of the Seychelles, about 85–90% of the population currently live on the coast (Micock, 1998). As a result, there is increased pressure for land for both development and conservation, resulting in high property costs. Increased recession of the coastline may push property costs further up. The vulnerability of critical infrastructure necessary to support the economy and human welfare, such as houses, hospitals, power stations, electricity, communication, emergency response centres, water supplies and schools, all located in the low-lying areas, is therefore very high.

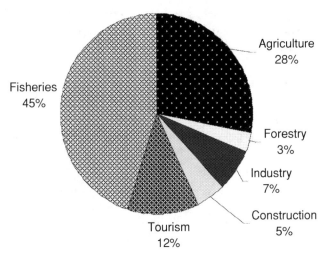

Figure 21.2. Estimated combined losses in monetary terms arising from the 1997–1998 El Niño and 1998–2000 La Niña events in the Seychelles (*Sources*: Gos, 2000b; Moustache,1997).

There could also be impacts on markets for goods and services, and impacts on natural resources on which economic activity depends. It is estimated that the loss of reef value as a result of the 1998 mass bleaching event could easily top US$5–7 billion per year, dwarfing any of the impacts of direct human interference.

The ENSO events may also have significant implications for climate change, as industries adapt and put in place water desalinators and air-conditioning units, both energy-consuming units (Payet, 1998).

The future economic losses from tourism and fisheries, and the adaptation costs to such events are indeed, in many instances, not assigned any monetary value. This may have implications for the actual assessment of the economic losses, and how adaptation from an economic perspective may be addressed in the context of future ENSO events and the long-term climate change.

21.3 FORECASTING ENSO AT THE NATIONAL LEVEL

Over the last two decades, various approaches and instrument facilities have been developed to track the development and impact of ENSO (Halpern, 1987; Stockdale *et al.*, 1998; McPhaden *et al.*, 1998; Ji *et al.*, 2000) and climate change (IPCC, 1996). In both cases, the advent of better monitoring networks, advances in computing and communications, and an increasing body of knowledge of the global climate system and its processes have been key factors for enhancing the understanding of the climate over time and spatial scales.

The ability to predict El Niño events up to one year before the event has been demonstrated (Cane *et al.*, 1986; National Research Council, 1995). Clearly, such opportunities, if consistent

and further developed, may provide governments, the private sector, farmers and individuals with the time necessary to mobilize resources to engage in disaster preparedness and undertake response measures and adaptive adjustments (Adams *et al.*, 1995).

The Seychelles, being a group of 115 small islands sprawling over an Exclusive Economic Zone of 1.3 million square kilometres, is a typical example of an area where global forecasting efforts appear not to be useful at local scales. There are significant gaps at all spatial levels. At the national level, there is a lack of adequately located land- and sea-based meteorological monitoring stations. There is currently a network of over 40 rainfall gauge networks and a few automatic weather stations, with over 95% located in the three main islands within a radius of 50 km. There is no ocean observing system, though some voluntary observing ship data may be available. Access to deployed buoy data is limited.

On a regional level, the islands in the Indian Ocean are far apart, but due to improved satellite information and the Internet, there is improved cooperation and exchange of information. The Regional Tropical Cyclone Committee for the Indian Ocean, which was established with the support of the World Meteorological Organization (WMO), provides this essential platform for the development of early warning systems and extreme weather preparedness. However, there is a need for more active support in terms of predicting ENSO and climate change-related events.

The Seychelles National Meteorological Services (SNMS), with assistance from a voluntary contributions programme (VCP) through WMO, has evolved and progressed over the last 25 years, though with a strong focus on aviation meteorological services. However, there are still many areas that need to be implemented, strengthened or considered. These include linking meteorological services to end-users and sectors, forecasting of extreme weather, and enabling the country to adopt early adaptive measures to ENSO-related events and future climate change (Zillman, 1999).

In countries subject to frequent and extreme weather hazards, appropriate weather services have been developed over time, driven by the need to reduce the loss of human life, economy and infrastructure. Unfortunately for the Seychelles and a number of other small island states, which have had relatively 'ideal' or paradise-like weather conditions over several hundred years, the increased threats posed by ENSO events and future climate change have led to a significant burden and need for re-adjustment of policy vis-à-vis the role and operations of the meteorological, hydrological and oceanographic services.

As a consequence, there is a policy drive to improve climate and ocean data collection, monitoring, hazard detection and prediction, community education, long-range forecasting, and to build a fully functional meteorological warning system (Payet, 1999a).

Furthermore, if the cost implications are considered, and budget spending is compared with a host of other countries, a whole new picture emerges, as shown in Figure 21.3. Assuming we

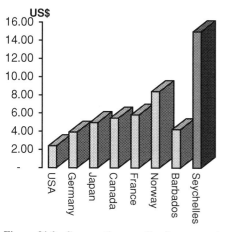

Figure 21.3. Comparative spending for meteorological services in selected countries on a per capita basis in 1998 (*Sources*: GoS 1998; WMO, 1999).

consider an anticipated increase of about 10% every year due to inflation and cost of consumables, and include the additional cost of response as a result of ENSO, the implications for the economies of small island states are phenomenal.

The impact on small island economies, which spend US$15 per capita per year for meteorological services, is far-reaching (as compared with the USA, where per capita spending is about US$2 per capita; see Figure 21.3). It may be argued that this cost may be offset by the reduction of losses because of early warning, but the issue that remains is that at present there are few mechanisms to allow for the cost-effective transfer of such advanced approaches and technologies to small island states. With pressure from international organizations for the Seychelles to meet all costs nationally and through commercial means, there is the danger that in the long-term the meteorological services will probably suffer, as the costs of operations escalate and government budget allocation diminishes.

21.4 IMMEDIATE POLICY RESPONSES

21.4.1 Policy measures as a consequence of the ENSO events in 1997–2000

The government of the Seychelles attempted to revise its policies in response to the ENSO events in 1997–2000. The first move involved the setting up of a National Disaster Preparedness Committee, a multidisciplinary team chaired by the Vice-President of the country. There was, fortunately, a rapid response from both civil and military services during the August 1997 disaster brought about by El Niño, but the lack of adequate preparedness and resources was evident and the extent of the losses could have been reduced with a proper national disaster response plan in place. This led to the development of a project to address the need for a

comprehensive national disaster plan, which is now in place but requires updating and the procurement of support equipment.

Second, to engage in disaster recovery and restoration, the government had to solicit the approval of the National Assembly for an extraordinary budget, and requested assistance from international organizations and governments. The extraordinary budget was also used for the repair of roads and the restoration of basic services such as electricity, communication and water. Such unplanned demands on the budget resulted in an increased budget deficit, which led to a number of problems and delays in the government's development plans. Although the Seychelles also benefited from external assistance to rebuild some of the damaged infrastructure, it still could not fully restore other aspects of the economy in order to reduce future risks since it is considered by many organizations to be a rich country. This consequently reduces the country's ability to develop adequate adaptive capacity to ENSO and further constrains efforts towards sustainable development.

The desire to reduce economic vulnerability and losses due to climate-related events has also compelled the Seychelles government to review the existing institutional structure and seek more effective approaches within all sectors. The government created a new body, under the Ministry of Environment, to look after issues of climate, drainage and coastal stability in a national context. Another new body to study the effects of landslides and to identify mitigation measures was also proposed Abbey (1998), but due to the lack of technical capacity, funds and human resources, this body has yet to be set up.

Policies for future land use and development are being developed to reduce impacts on housing and infrastructure. For example, the number of houses being constructed along the unstable slopes of the granitic islands has decreased, thanks to the adoption of stricter guidelines. Numerous roads are now being diverted further inland, and coastal setback lines and buffer zones are being introduced by hotel developers. The accentuated topography of the islands and the lack of land resources have forced reclamation activities over the last 30 years. Without reclamation, the Seychelles would have had no space for an international airport and a commercial and cruise ship port, and the economy would have been far less developed than it is now. It is also believed that economic diversification and reclamation of the coastline have permitted the country to continue to devote more than 45% of its territory as protected reserves. With the increased threat to coral reefs and coastal zones, this policy has been brought into question; however, the paradigm remains – without economic diversification, there is no stable revenue, and without stable revenue, the country cannot continue to develop long-term sustainable actions and build its adaptive capacity to climate change.

As a result of increased forest fires induced by the drought conditions during the La Niña event, the government is also revisiting various aspects of its policy on lighting of fires and forestry management. However, technical and financial resources remain the major constraints in implementing that particular policy in the short term.

Due to the economic and social impacts of the extended drought and lack of water resources, the government embarked on a programme to extract water from small rivulets on the islands. However, in late 1999, the government decided to purchase water desalination plants to meet demand for water during the drought period.

On the other hand, impacts of the climate on coral reefs and monitoring programmes have increased significantly over the last few years in the western Indian Ocean. Worth noting is the work undertaken by Coral Reef Degradation in the Indian Ocean (CORDIO) (Souter et al., 2000), the International Coral Reef Action Network (ICRAN), and a Global Environment Facility (GEF)-funded project to conduct detailed surveys of the extent and impact of the coral bleaching and develop approaches to reduce impacts that may hamper recovery in the Seychelles.

The management of the fisheries resource is extremely complex, and the Seychelles has very clear policies on fisheries exploitation, ensuring that there is no over-fishing and that natural fish nurturing areas are protected. However, with decreased revenue from fisheries as a result of sea surface temperature variability, the pressure for increasing fishing efforts may likely intensify. In many ways, revenue from the fishing industry is also partly allocated to resource monitoring and support, so a decrease in revenue will also affect this important decision support and the enforcement activities.

21.4.2 Addressing early warning of ENSO

The need to develop better forecasting tools for ENSO impacts in other regions provides a window whereby policy makers can muster local and global support for further commitment of resources towards developing a range of better quality meteorological services. This also includes building capacity at a local level, and partnerships with key international climate research organizations and regional climate centres.

Adams et al. (1995) analysed the economic benefits from improvements in forecasting the El Niño phenomenon for crop yields, and showed that there were substantial losses that could have been prevented. This relationship further emphasizes the financial attention that governments must provide to improve forecasting. Since most island states are characterized by small landmass and large sea boundaries (CSD, 1999), special consideration is required when planning forecasting needs and requirements.

Better forecasting capacity could decrease vulnerability and improve planned adjustments; however, in real cost terms, an average island state (on a per capita basis) will have to invest much

more to engage in the same level of meteorological services as in developed countries. The requirement for the Clean Development Mechanism or any other mechanism that may be adopted, to mitigate and adapt to climate change, should endeavour to assist island states in developing better forecasting and disaster management systems, as these are directly related to adaptive capacity (Goldemberg, 1998).

Since contributions to climate change by small island states are insignificant, the actual cost for improved forecasting should not be passed on exclusively to such states. Rather, a global fund mechanism should evolve whereby voluntary contributions become more formal and which would allow the distribution of the Global Climate Observing System (GCOS) to adequately cover the Indian Ocean (UNESCO, 2001).

It was, therefore, as a result of this critical need that the Indian Ocean Global Ocean Observing System (IOGOOS) was launched in November 2002 in Mauritius, the membership of which is open to countries within and outside the region. This system comprises three basic components: (a) a data collection subsystem; (b) a data and information management subsystem; and (c) a modelling and applications subsystem for the development of products and services, parts of which could be national or regional. This arrangement will, in the long term, allow the active participation of local institutions within global science initiatives.

Research and institutional capacity-building policies aimed at strengthening and expanding both island-based stations, as well as ocean-monitoring buoys, should provide the basis for improved long-range forecasting and better preparedness for ENSO events and climate change. Access to satellite data such as sea surface temperatures will also be useful. Such developments, however, should depend not only upon acquisition of high technology, but also training and human resource development within the sector. It is expected that much support will therefore be given to IOGOOS by the international community.

The integrated coastal zone management (ICZM) approach offers a good platform for small island states to develop capacity at all levels of society. At the 1993 World Coast Conference, participants exhibited clear agreement that the continued degradation of the coastal areas would exacerbate the potential impacts of climate change. They further concluded that countries would need to manage the use of coastal resources now in order to reduce vulnerability to climate change in the future (IPCC, 1994). The recent ENSO impacts, therefore, provide opportunities for countries to address climate change issues now directly to the community and stakeholders.

21.5 FUTURE POLICY RESPONSES

The special considerations related to environmental vulnerability have been discussed in numerous forums, including the United

Nations Conference on Environment and Development (UNCED) (Halpern, 1992), and included in Chapter 17G of Agenda 21. This paved the way for the integration of sustainable development mechanisms for states that are vulnerable in economic terms, as it is the performance of island economies that will, in many ways, drive many of the environmental options and any adaptation options to climate change. In the Seychelles, the per capita budget per year allocated for direct environmental management and protection is about US$100 (GoS, 1998). The pressure for increasing the present allocation will likely escalate in view of decreasing international funding possibilities and increased costs as a result of the ENSO events and future climate change.

Therefore, long-term policy responses are needed to address a wide range of issues pertinent to the specificities of small island states. These include mainstreaming land-use planning, enhancing monitoring, and building institutional and human capacity (Payet, 1999b). Unfortunately, many of these actions would not be possible without significant financial and technical support. Even small island states with high GDP per capita will not be able to adapt effectively to climate change (Jepma and Munasinghe, 1998). For example, it has been estimated that the marginal cost per capita to protect small island states from a 1-metre sea-level rise is over 10 times more than that for developed countries (IPCC, 1990). Furthermore, budget allocations for meteorological and environmental services are already considerably high, compared with those in most countries, and the costs of technology and expertise transfer will, in many cases, surpass those in countries with higher economies of scale and complementary resources.

The further development of coastal zone management to improve the adaptive capacity of coastal habitats is an equally vital policy option. The ICZM provides governments, the private sector and communities with the necessary tools to incorporate adaptation options in a sustainable manner. The challenge to integrate climate change issues in land-use planning, construction and coastal development should be facilitated through improved assessment tools and methodologies that can be easily transplanted to small island states.

Policies aimed at addressing the effects and the socio-economic implications of the impacts of climate change and sea-level rise on small island states has yet to be integrated within the concept of environmental cost. There is also considerable lack of information on the socio-economic implications of climate change, especially within the context of small island states (IPCC, 2001).

Small island developing states, being particularly vulnerable to global climate change, climate variability and sea-level rise, cannot develop adequate adaptive capacity should such events continue to delay development and cause major siphoning of funds into disaster rehabilitation (CSD, 1999). Almost 90% of island populations, agricultural lands and infrastructure are located virtually at sea level. Destruction of coastal habitats, inundation of outlying islands and loss of land could result in loss of economic value and exclusive economic rights, as well as the destruction of existing economic infrastructure. As indicated in a report, global climate change may cause irreparable damage to coral reefs, alter the distribution of zones of upwelling, and may consequently affect both subsistence and commercial fishery production (Hoegh-Gulberg, 1999). Indeed, as discussed above, during the El Niño event in 1997–1998, irreparable damage to coral reefs had already occurred, with 90% coral cover bleached in some areas. More damage is likely to occur if future severe El Niño events occur too frequently for ecosystem recovery to take place between the intervals.

There is, therefore, a need for the United Nations Framework Convention on Climate Change (UNFCCC) to continue to facilitate the development of mechanisms to implement its equity principles on the basis of national responsibility, and, in order to ensure the survival of all nations, concrete actions need to be taken to reduce GHG emissions.

The acute water shortage during the La Niña event has caused the government to examine again its water resource management capacity, with a view to installing desalination plants. However, this type of reactive adjustment is not cost-effective in the long term, but rather a quick and perhaps unplanned approach to adaptation. Conversely, the ENSO phenomenon should provide the international community with a basis for such preventive and adaptive policies to be developed and put in place, with the option of steering policy actions to reduce the vulnerability of small island states to climate change in the most cost-effective manner. Several policy re-evaluations may need to be undertaken:

- The current approach, which considers only the cost-effectiveness of abatement costs but not the costs of adaptation in affected countries, falls short of Article 3.1 of the UNFCCC;
- Lack of a fully transparent adaptation component to the Clean Development Mechanism that would seek to discourage and reduce GHG emissions whilst placing more emphasis on meeting the costs of adaptation;
- A policy mechanism that attempts to engage further research and development in non-GHG-emitting technologies, push renewable energy costs down, and promote rapid transfer of such technologies throughout the world.

For example, Bernow et al. (1999) showed that, should the United States adopt a carbon emissions reduction target, even below that of the Kyoto target, there would still be net economic savings, equal to an average of US$400 per American household per year. Although this study is based upon models, one issue is clear: shifting to policies that encourage technology transfer, innovation and capacity-building within all sectors of the economy will yield long-term economic benefits, not only to the

United States, but also to small island states, and hence reduce the liabilities associated with climate change.

There should be more active support for small island developing states to develop the ICZM plans, including the integration of adaptive response measures to climate change and sea-level rise. Although there have been a number of initiatives on the ICZM in several island states, there is a need for these to catalyse the transfer of expertise and technology in planning and coastal resource management. Efforts to guide this process into a more practicable and tangible approach are also required. For example, cost-effective options for coastal erosion control and coastal resources management should be seen as a result of the ICZM approaches and not the other way round. The ICZM should provide the opportunities for transfer of expertise and technology, sustainable development and a forum for negotiations on matters of similar interest.

21.6 CONCLUDING REMARKS

Whilst this chapter has examined the impacts of ENSO on small island states, it is evident that in many cases poor access to financial and technical resources for monitoring the changes in the climate, for assessment of the impacts and for the timely development and implementation of adaptation strategies may significantly contribute to the increase in vulnerability of island states to climate change. The 1997–1998 El Niño and 1998–2000 La Niña events provide the opportunity for both island state governments and the international community to consider long-term policy options, which aim primarily at optimizing adaptive capacity in vulnerable regions of the world. The chapter has attempted to address this opportunity in an effort to develop an active dialogue between island states and other nations, and put in place some key adaptation measures in anticipation of climate change.

Recognizing the specific vulnerabilities and disproportionate cost borne by small island developing states, emphasis on the need to negotiate more substantial contributions through existing and future financial mechanisms should not be ignored.

ACKNOWLEDGEMENTS

The author acknowledges the Government of the Seychelles, in particular, Wills Agricole and Michel Vielle for providing access to the data and publications cited. Thanks to the peer reviewers Professor Abdelkrim Ben Mohamed, Dr David Hastings, Antoine Moustache and an anonymous expert reviewer for their constructive comments and revisions to the document; Kai Yin Low for kindly proofreading the chapter and finally to Dr Pak Sum Low for providing guidance. In 2001, the author was awarded a AIACC/START research grant (http://www.aiaccproject.org), which contributed towards the preparation of this paper.

REFERENCES

Abbey, P. (1998) *Guidelines for Stormwater Management in the Seychelles.* Land Transport Division, Government of the Seychelles.

Adams, R. M., Bryant, K. S., McCarl, B. A., Degler, D. M., O'Brien, J., Solow, A. and Weiher, R. (1995) Value of improved long-range weather information. *Contemp. Econ. Policy*, **13**, 10.

Bernow, S., Cory, K., Dougherty, W., Duckworth, M., Kartha, S. and Ruth, M. (1999) *America's Global Warming Solutions.* World Wildlife Fund and Energy Foundation.

Cane, M. A., Zebiak, S. E. and Dolan, S. C. (1986) Experimental forecasts of El Niño. *Nature*, **321**, 827–832.

CSD (1999) Progress in the implementation of the Programme of Action for the Sustainable Development of Small Island Developing States. *Report to the UN Secretary General. Commission on Sustainable Development*, Seventh Session, 19–30 April 1999.

Diaz, H. F. and McCabe, G. J. (1999) A possible connection between the 1878 yellow fever epidemic in the southern United States and the 1877–78 El Niño episode. *Bull. Am. Meteorol. Soc.*, **80**, 21–27.

Goldemberg, J. (1998) *Issues and Options: The Clean Development Mechanism.* United Nations Environment Programme, Nairobi, Kenya.

GoS (1998) *Approved 1998 Budget.* Government of the Seychelles publication.

(2000a) Forest fire damage in the Seychelles in 1998–1999. Unpublished report, Government of the Seychelles.

(2000b) *National Budget Supplement 1997–1999.* Government of the Seychelles Publication.

Halpern, D. (1987) Observations of annual and El Niño flow variations at 0°, 110° W and 0°, 95° W during 1980–85. *J. Geophys. Res.*, **92**, 8,197–8,212.

Halpern, S. (1992) *United Nations Conference on Environment and Development: Process and Documentation.* Providence, RI, Academic Council for the United Nations System (ACUNS). UN publication.

Hoegh-Gulberg, O. (1999) *Climate Change, Coral Bleaching and the Future of the World's Coral Reefs.* Australia, Greenpeace publication.

IPCC (1990) *Strategies for Adaptation to Sea-level Rise.* Report of the Coastal Zone Management Subgroup. IPCC, November 1990. Intergovernmental Panel on Climate Change.

(1994) *Preparing to Meet the Coastal Challenges of the 21ˢ Century.* Conference Report, World Coast Conference, 1993. IPCC.

(1995) *Climate Change 1995: Impacts, Adaptations, and Mitigation.* Summary for Policymakers. IPCC/WMO/UNEP.

(1996) *Climate Change 1995: The Science of Climate Change.* Contribution of Working Group I to the Second Assessment Report of the Intergovernmental Panel on Climate Change. WMO/UNEP, Cambridge University Press.

(2001) *Impacts, Adaptations, and Vulnerability.* Contribution of Working Group II to the Third Scientific Assessment Report of the Intergovernmental Panel on Climate Change. WMO/UNEP, Cambridge University Press.

Jepma, C. J. and Munasinghe, M. (1998) *Climate Change Policy: Facts, Issues and Analyses.* Cambridge University Press.

Ji, M., Reynolds, R. W. and Behringer, D. W. (2000) Use of TOPEX/POSEIDON sea level data of ocean analyses and ENSO prediction: some early results. *J. Climate*, **13**, 216–231.

McPhaden, M. J., Busalacchi, A. J., Cheney, R. *et al.* (1998) The Tropical Ocean-Global Atmosphere (TOGA) observing system: a decade of progress. *J. Geophys. Res.*, **103**, 14,169–14, 240.

Micock, B. C. J. (1998) *Urban Planning in the Seychelles: A GIS Approach.* Dissertation towards an M.Sc. in Geographical Information for Development, University of Durham, UK.

Moustache, A. (1997) *A Report on the Consequences of the Natural Disaster of the 12th–17th August, 1997 on the Agricultural Sector of the Republic of the Seychelles.* Submitted to the Food and Agriculture Organization (FAO). Ministry of Agriculture and Marine Resources.

National Research Council (1995) *A Review of the US Global Change Research Programme and NASA's Mission to Planet Earth/Earth Observing System.* Washington, DC, National Academy Press.

Nicholson, S. E. (1997) An analysis of the ENSO signal in the tropical Atlantic and western Indian Oceans. *Int. J. Climatol.*, **17**, 345–375.

Obasi, G. O. P (2005) The impacts of ENSO in Africa. This volume.

Payet, R. A. (ed.) (1998) *Vulnerability Assessment of the Republic of the Seychelles.* Government of the Seychelles/ENVIRO.

Payet, R. A. (1999a) *Operational Action Plan and Strategy for the National Meteorological Services, 1999–2002.* Government of the Seychelles.

(1999b) *Adaptation Options to Climate Change for the Republic of the Seychelles*, Government of the Seychelles.

Pernetta, J. C. and Hughes, P. J. (eds.) (1990) Implications of expected climate changes in the South Pacific region: an overview. *UNEP Regional Seas Reports and Studies* No. 128, UNEP/ASPEI/SREP.

SFA (1999) *1998 Annual Report.* Seychelles Fishing Authority, Government of the Seychelles.

Souter, D., Obura, D. and Linden O. (2000) *Coral Reef Degradation in the Indian Ocean: Status Report 2000.*CORDIO, Sweden.

Stockdale, T. N., Busalacchi, A. J., Harrison, D. E. and Seager, R. (1998) Ocean modelling for ENSO. *J. Geophys. Res.*, **103**, 14,325–14,355.

Tourre, Y. M. and White, W. B. (1995) Does the Indian Ocean have an El Niño of its own? *Eos*, **75**, 585–6.

UNEP (1999) *Declaration of Guayaquil: First Intergovernmental Meeting of Experts on El Niño, Ecuador, 13 November 1998.* UN System-wide Earthwatch Coordination. http://www.unep.ch/earthw/declguay.htm

UNESCO (1991) *IOC-SAREC-KMFRI Workshop Report No. 77.* Regional workshop on causes and consequences of sea-level changes on the western Indian Ocean coasts and islands, Mombassa, Kenya, 24–28 June 1991.

(2001) *An Indian Ocean Observing Strategy.* Fifth Session of the IOC-WMO-UNEP Committee for the Global Ocean Observing System (GOOS) held in Paris, 28–30 June 2001. Document No. IOC-WMO-UNEP/I-GOOS-V/10A.

Wilkinson, C., Linden, O., Cesar, H., Hodgson, G., Rubens, J. and Strong, A. E. (1999) Ecological and socio-economic impacts of 1998 coral mortality in the Indian Ocean: an ENSO impact and a warning of future change. *Ambio*, **28**. No. 2, March 1999.

WMO (1998) *Annual WMO Statement on Global Climate.* http://www.wmo.ch/web/wcp/wcdmp/wcdmp.html

WMO (1999) *World Climate News.* No. 15, June 1999.

Zillman, J. W. (1999) The national meteorological service. *WMO Bulletin*, **48**, No. 2, 129.

22 El Niño causes dramatic outbreak of *Paederus* dermatitis in East Africa

INGEBORG M.C.J. VAN SCHAYK[1,*], RUBEN O. AGWANDA[2] (deceased),
JOHN I. GITHURE[1,†], JOHN C. BEIER[3], AND BART G. J. KNOLS[4,‡]

[1]*International Centre of Insect Physiology and Ecology, Nairobi, Kenya*
[2]*Kenya Medical Research Institute, Nairobi, Kenya*
[3]*Tulane University, New Orleans, USA*
[4]*Wageningen University Research Centre, the Netherlands*

Keywords

Paederus sabaeus; Nairobi Fly; outbreak; El Niño; *Paederus* dermatitis; conjunctivitis; Nairobi; Kenya

Abstract

An outbreak of *Paederus sabaeus* rove beetles in Kenya during the 1997–1998 El Niño resulted in a dramatic increase of vesicular dermatitis in its capital Nairobi. The beetle, popularly called 'Nairobi Fly', contains a potent toxic fluid that causes epidermolysis and acute conjunctivitis. A cross-sectional epidemiological study involving 1,208 Nairobi residents was conducted to determine the health impact of this outbreak. The results showed that one-third of the Nairobi population were infected during this period. The majority of the respondents reported lesions on exposed body parts above the shoulders. Disfiguring, painful blisters and skin rashes in and around the facial area had a strong personal and social impact. Policy makers and public health specialists need to recognize that outbreaks of insects of medical importance resulting from global climatic events require urgent remedial action.

22.1 INTRODUCTION

Modern human activities have an obvious negative impact on the environment and contribute to the irreversible alteration of the global climate. Evidence shows that successions of abnormal climatic events have a disturbing effect on the world's ecosystems (Epstein *et al.*, 1998). Rise of temperatures and more frequent occurrence of extreme weather conditions alter the flora and affect the development of competitive insect species in certain parts of the world, causing unexpected insect explosions with consequences for human health (Dukes and Mooney, 1999).

*Currently consultant to the US National Library of Medicine, Trautmannsdorf/Leitha, Austria
†Currently at University of Miami School of Medicine, FL, USA
‡Currently with International Atomic Energy Agency (IAEA), Seibersdorf, Austria

During an average rainy season in Kenya, the 'Nairobi Fly' appears in moderate to low numbers with negligible effects on the human population. One of the strongest El Niño warmings of the Pacific Ocean in history, however, dramatically changed the scene (Kerr, 1998a). Grasses, plants and insect breeding sites developed uncontrollably between October 1997 and March 1998 as a result of the high temperatures and prolonged heavy rainfall. Insect plagues involving mosquitoes, grasshoppers and rove beetles spread at a bewildering tempo all over the country. Millions of Kenyans suffered from the related human diseases malaria, dengue fever, rift valley fever and *Paederus* dermatitis. Many suffered and an excessive social and economic burden, increased poverty and insecurity.

In Kenya's capital, Nairobi, the most severe impact on human health was caused by the occurrence of millions of *Paederus sabaeus* rove beetles. Apparently many citizens suffered from vesicular dermatitis despite regular media dissemination of information on possible ways to prevent infection.

This chapter reports on the consequences of the 1997–1998 El Niño event on the rapid development of *Paederus* rove beetles and their impact on human health and well-being. The theoretical reflection is illustrated by a socio-epidemiological study on the prevalence and impact of infection among Nairobi residents.

22.2 *PAEDERUS SABAEUS*, THE TOXIN AND CLINICAL EFFECTS

Beetles belonging to the genus *Paederus* include over 600 species. At least 30 species have been reported to cause vesicular dermatitis and conjunctivitis in humans and other mammals. *Paederus* beetles occur in all temperate and tropical regions of the world. *Paederi* contain a potent toxic fluid called pederin that causes epidermolysis upon contact with the human skin (Okiwelu *et al.*, 1996). Both female and male beetles possess this toxic fluid, but females have been noticed to cause more severe dermatitis

Climate Change and Africa, ed. Pak Sum Low. Published by Cambridge University Press. © Cambridge University Press 2005.

Figure 22.1. Paederus sabaeus (Photo courtesy of B. Copeland).

because the toxin is related to their reproductive system (Frank and Kanamitsu, 1987).

The insect exists in moist habitats and abundance is related to rainfall and temperature. In Kenya and throughout parts of East Africa, the beetle is well known as 'Nairobi Fly'. This name dates back to 1916 when Ross reported *Paederus* beetles in Nairobi – this being the first official record from East Africa (McCrae and Visser, 1975).

Paederus sabaeus (Coleoptera: Staphylinidae) is a small wasp-like rove beetle with a length of 5–7 mm. It is clearly recognizable by its distinctive appearance: bright orange parts alternate the black head, elytra and posterior end of the abdomen. The cover wings have a metallic, dark green colour. Their bright orange coloration is associated with their defence mechanism – the possession of a potent toxic fluid called pederin.

It has been observed that *Paederus* beetles, in captivity, defend themselves by curvature of the abdomen like a scorpion tail. Some of these rove beetle species are known to possess retractile anal vesicles producing a defensive fluid that is different from pederin (Frank and Kanamitsu, 1987). Although this would be a plausible explanation for the curvature of the abdomen, there is no available proof that this is true for *Paederus sabaeus*.

The *Paederus sabaeus* beetle (Figure 22.1) lives in sandy and damp soils close to water. It feeds mainly on decomposing organic material and predates on insect larvae (Frank and Kanamitsu, 1987). Adult beetles emerge during the rainy season. The insects are immediately attracted by artificial light and prefer to rest on ceilings and walls (Fox, 1993; Veraldi and Süss, 1994). *Paederi* frequently drop down and unexpectedly land on people. They do not bite, sting or attack and do not cause any harm when left undisturbed. It is only when the beetle is crushed that it re-

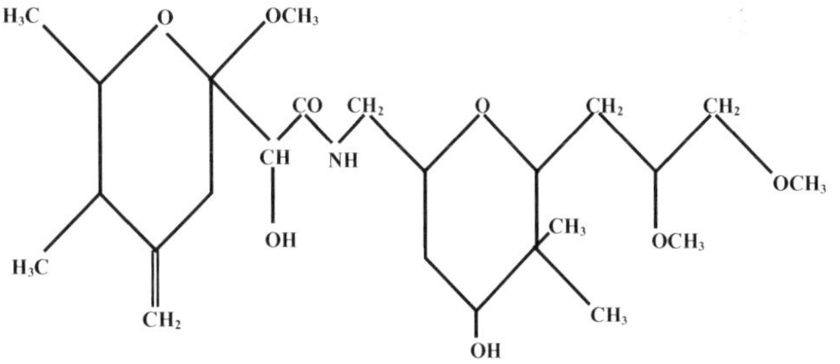

Figure 22.2. The molecular structure of pederin.

leases pederin, a toxic fluid present in its hemolymph that causes acute lesions of the skin and inflammation of the eye (Gelmetti and Grimalt, 1993).

Pederin (Figure 22.2) is one of the most powerful toxic substances of animal origin known. This amide is one of the most complex toxins found in beetles (Frank and Kanamitsu, 1987). It is known to be at least 15 times more potent than cobra venom. Tests have shown that the toxin is not destroyed when boiled for one hour in water and that it retains its toxicity for a minimum of 10 months. During experiments with mice it was found that 0.14 mg/kg of pederin injected intraperitoneally was sufficient to kill (Kellner and Dettner, 1995).

Although it would be expected that the primary function of pederin would be a defence mechanism against predators like frogs, toads, lizards and birds, it has been mainly observed to affect mammals. The toxin is not known to be distasteful to vertebrate predators and does not introduce a negative association with the beetle (Frank and Kanamitsu, 1987). The function of pederin seems to be an opportunistic defence mechanism against any form of attack.

When released on human skin, pederin resembles a drop of boiling water on skin causing irritation and eventual inflammation of the affected area.

22.2.1 Paederus dermatitis

The rove beetle *Paederus sabaeus* is a very small and lightweight insect that does not cause any pain when making contact with a person. Often unaware of its presence, people inattentively brush it away. During this action the beetle may be crushed and produces the toxin. Subsequently *Paederus* dermatitis develops a severe skin reaction that turns into ugly blisters and scars. The dermatitis can be severe and may persist for months (Kamaladasa *et al.*, 1977; Frank and Kanamitsu, 1987). Upon contact with the eyes, the fluid may cause acute conjunctivitis, locally known as "Nairobi Eye" (McCrae and Visser, 1975). Both inflammations may be complicated by secondary infections.

In general, the first symptoms will appear 12–36 hours after the toxin has affected the skin. In the case of *Paederus* dermatitis, the victim starts to develop a red spot after 1–2 days accompanied by a feeling that varies from a burning sensation to severe pain. A few days later the sensitive spot turns into a blister, often filled with mucus. The blister easily breaks, revealing sensitive, red skin. Eventually the dead skin will peel off, leaving a transitory pigmentation. The whole process usually takes 7–21 days, but a remaining scar will be visible for months. Secondary bacterial infections may prolong the healing process, causing severe pain, headaches, fever and general malaise (McCrae and Visser, 1975; Fox, 1993). Upon contact with the skin, pederin usually causes a skin lesion corresponding with the area affected by the toxin. However, when the fluid is pressed between skin surfaces (e.g. between the upper arm and the forearm or between the thighs), the toxin may have a 'mirror effect' and damage the skin with two similar sized blisters. This appearance is known as 'kissing lesions' (Veraldi and Süss, 1994).

The human eye is very sensitive to pederin. Conjunctivitis is usually induced when the toxic fluid is being rubbed into the eyes. On very few occasions the insect crawls or flies into the eye and releases its toxin when the victim tries to remove it. The infected area around the eye swells up and may produce a yellow discharge. It may be impossible to open the eye for a few days. During the period of infection, the eye becomes very sensitive to (day) light and the inflammation is often very painful. Temporary blindness may occur (McCrae and Visser, 1975; Frank and Kanamitsu, 1987).

There is no effective treatment for *Paederus* dermatitis. References mention the ineffectiveness of cortisone, antihistamines and penicillin. Antibiotics only seemed to work against secondary bacterial infections. When someone realizes at the time of occurrence that the beetle has been crushed on the skin, the most effective remedy would be to wash the affected area with water and soap. Additionally, it is known that pederin can be destroyed by iodine tincture, but this needs to be applied immediately, before the toxin has a chance to damage the skin. In the case of acute conjunctivitis, the infected eye immediately needs to be washed with clean water. If the eye itself is inflamed, immediate medical treatment with antibiotics, corticosteroids and atropine is required to prevent long-term damage of the eye (Frank and Kanamitsu, 1987).

22.3 *PAEDERUS SABAEUS* OUTBREAK IN NAIROBI

The El Niño phenomenon of 1997–1998 surprised East Africa with heavy, prolonged rainfall responsible for many water related problems throughout the region. In Kenya, this contributed to yet another human discomfort: the occurrence of millions of *Paederus* rove beetles all over Kenya's capital Nairobi. Thousands of *Paederi* were reported on ceilings around fluorescent tube lights. Various media channels frequently informed Nairobi residents on the appearance of the insect and possible ways to prevent infection. A full-page article called 'The Flying Menace' was published in the daily newspaper the *East African Standard* on 20 January 1998. It included a large picture of the Nairobi fly and provided detailed information on the insect and its behaviour. To a melody of Don Mclean, a new text was composed and the song *Bye, bye Nairobi Fly* was played over the radio throughout. The Nairobi Fly became the talk of the town. Despite all these efforts *Paederi* infected thousands of people and caused distress among the citizens of Nairobi.

A washerwoman hides her mottled face;
a band leader nurses an ugly patch of blisters on his neck;
a small boy scratches his cheek raw.

The 'Nairobi fly' got' em.

Dispatch online, Monday, January 26, 1998.

Inspired by this situation, the International Centre of Insect Physiology and Ecology (ICIPE) and the Kenya Medical Research Institute (KEMRI) recognized the need to jointly assess the social and public health impact of this outbreak of *Paederus sabaeus* in Nairobi as a result of extreme weather conditions. It was decided to conduct a Nairobi-wide epidemiological study on the public's knowledge of, and experience with, the inconveniences caused by the Nairobi Fly.

22.4 STUDY DESIGN

A pilot study was designed to assess the prevalence of Nairobi Eye infections on a small scale. These data were used to determine the minimum number of respondents required to develop a

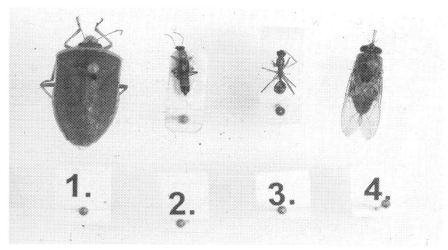

Figure 22.3. Box used during the survey with 1, True bug (Pentatomidae); 2, Nairobi fly beetle (Staphylinidae); 3, Ant (Formicidae); 4, Tsetse fly (Glossinidae) (Photo courtesy B. Copeland).

representative sample for the entire Nairobi population. Additionally, this information provided valuable insight for a large-scale survey.

Pilot study. During the month of March 1998, 74 interviews were carried out using a structured questionnaire with closed and open questions. The respondents were randomly selected from among employees of ICIPE and KEMRI. The study covered 10% of the total employees of both institutes based in Nairobi, with 25 respondents from ICIPE and 49 from KEMRI.

The pilot study showed an infection rate of 23% (13–33% with 95% confidence interval). 17 out of 74 people had experienced an infection by the Nairobi Fly between December 1997 and March 1998. Among the affected respondents, skin lesions were most common. 94% of these victims suffered from *Paederus* dermatitis and only 1.4% experienced conjunctivitis. In the majority of the cases, it took 7–14 days for the wound(s) to heal with exceptions taking up to 25 days. Most interviewees reported that the scars resulting from the blisters had not disappeared. With respect to peoples' knowledge about the insect, 95.9% mentioned they had heard about the Nairobi Fly, mainly through other people.

Based on this prevalence figure of 23% and the fact that this pilot study was a stratified random sampling with proportional allocation, the optimum sample size for the entire city was calculated using $n =** (W_i p_i q_i)/V$, where $i = 1,2$; $W_1 = 25/250$; $W_2 = 49/490$; $p_1 = p_2 = 0.1$; $q_1 = q_2 = 0.9$; and $V =$ desired variance. The sampling variance for the pilot study was 0.00122. Because the target group for the whole of Nairobi would be much larger than that for the pilot study, a smaller variance was desired. V was set at 0.0001 and the resulting sample size (n) became 900. Using a finite population correction of 1.1, n eventually turned out to be 820. Hence, in the large-scale study, a minimum of 820 interviews would need to be conducted in order to represent the entire population of Nairobi.

The pilot study did not indicate any link between infection rate and sex, age or education. It was therefore decided that respondents of both sexes, all ages and educational levels would have to be included. Since the occurrence of the Nairobi Fly may be related to specific areas or places (for instance with presence of electricity), it was concluded that the Nairobi-wide survey would need to cover different areas and locations of the city.

Nairobi survey. The development and implementation of the survey has been a close collaboration between ICIPE and KEMRI. The questionnaire was designed to cover data on peoples' knowledge and experience in relation to the Nairobi Fly and the related infection. Both multiple choice and open questions were included. A visual aid being a small box with four prepared insects containing a tsetse fly (Glossinidae), an ant (Formicidae), a true bug (Pentatomidae) and a Nairobi Fly beetle (Staphylinidae) was used to test if people were familiar with the appearance of the insect (Figure 22.3).

Before implementation of the study, six Kenyan interviewers (three women and three men) were recruited through KEMRI. A training session was organized for the interviewers in which the questionnaire was explained and practised. The supervising team closely monitored the first interviews and assistance was provided where needed. Not all areas of Nairobi are safe for strangers and many people do not welcome unfamiliar faces in their private premises. To provide the best security for the interviewers, it was agreed that the survey would be conducted in public places throughout Nairobi.

Paederus sabaeus lives in moist soil and adult beetles mainly appear in areas around artificial lights, preferably fluorescent tube lights. To address the fact that the abundant occurrence of the beetle might be related to specific locations, all areas of the city were included covering parts of the town both with and without electricity.

A cross-sectional socio-epidemiological survey was implemented for a period of 3.5 weeks during the months of April–May 1998, covering a total of 1,208 respondents.

22.5 RESULTS

An effort was made to avoid any bias related to gender, education or age, by addressing all categories. Every interviewer was instructed to interview more or less equal numbers of men and women. Female respondents did not always welcome female interviewers, but male interviewers did not seem to experience any problems. The eventual respondent men-to-women ratio was 1:0.9. Although people of different ages were addressed, the majority of the respondents (80.6%) were between 16 and 40 years old. Because the interviewees were chosen randomly but at different public locations (e.g. market places, schools and shops) and in geographically spread areas, people with varying educational backgrounds (between 0 and 12+ years) were interviewed. A total of 82.1% had experienced at least 9 years of formal education and 43.3% had been to school for more than 12 years. Results from this study do not indicate a significant difference in Nairobi Fly infection rates between sexes, ages or educational background.

Occurrence. The survey (Table 22.1) shows that between August 1997 and May 1998, a total of 30.2% ($n = 1,208$) of Nairobi residents were affected by the Nairobi Fly. Of these victims, 79.5% ($n = 365$) suffered from infections on exposed areas above the shoulders, with blisters and scars on the neck, throat and facial areas. The symptoms of *Paederus* dermatitis and conjunctivitis are obvious and people could easily indicate if they had been infected and what symptoms they had experienced. Through examination of preserved specimens of the Nairobi Fly, the insect responsible for the epidemic was identified as *Paederus sabaeus*.

Symptoms. Skin rashes, blisters and itching skin were most commonly listed as being the major symptoms of *Paederus* dermatitis. Of all affected, 8.5% experienced acute conjunctivitis. Fever, headaches and general malaise were mentioned occasionally. Although a burning sensation is experienced in most cases, few respondents mentioned really serious infections. In some cases, people had suffered from multiple infections between August 1997 and April 1998. The study mainly focused on peoples' own experiences with the Nairobi Fly. Additionally, a question was included about other victims known to the interviewee. It occurred that 82.7% of all respondents knew at least one other affected person and/or had observed several cases in Nairobi.

Treatment. People seem to be aware that *Paederus* dermatitis is a self-healing wound. In general the symptoms are disturbing but not experienced as extremely painful. About half of the infected people (51.5%) did not seek any medical advice or consult anybody else on the symptoms of dermatitis. A relatively large group (32.6%) referred to a friend, a relative or a pharmacist for advice, while only 15.3% visited a physician. The actual treatment of

Table 22.1. Knowledge of, experience with and action taken regarding Nairobi Eye infection by 1,208 Nairobi residents during a 1997–1998 outbreak of *Paederus* dermatitis.

Knowledge	Heard about the Nairobi Fly	98.3% ($n = 1,208$)
	Know how it effects people	75.2% ($n = 950$)
	Recognized appearance of *Paederus sabaeus*	95.4% ($n = 1,187$)
	Relate occurrence of beetle to rainy season	88.1% ($n = 1,149$)
Experience	Have been infected by Nairobi Fly	30.2% ($n = 1,208$)
	Have been infected above the shoulders	79.5% ($n = 365$)
	Suffered from vesicular dermatitis	93.7% ($n = 365$)
	Suffered from conjunctivitis	8.5% ($n = 365$)
Action	Consulted a physician	15.3% ($n = 365$)
	Consulted relative/friend/pharmacy	32.6% ($n = 365$)
	Did not consult anybody	51.5% ($n = 365$)
Treatment	Nothing	24.7% ($n = 365$)
	Rinse with water	15.3% ($n = 365$)
	Herbs	0.8% ($n = 365$)
	Medicine	23.3% ($n = 365$)
	Other	38.1% ($n = 365$)
Healing Time	1–6 days	30.1% ($n = 365$)
	7–15 days	55.9% ($n = 365$)
	Not known	14.0% ($n = 365$)

the wounds varied. A quarter of the affected people did not find it necessary to treat the wound at all. Upon medical advice, most people used medicated creams (antibiotics or antihistamines) and/or analgesics. Most creative were respondents who applied home remedies. Popular was rinsing with water, application of Vaseline, Dettol, Vicks and various other creams. Some tried brake fluid (4.5%), toothpaste, paraffin, shoe polish, battery acid and even gun fluid. The effectiveness of these remedies remains unknown. The healing period was usually 1–2 weeks, but a pigmented spot was reported to remain. Since this study was implemented towards the end of the outbreak, no data were provided on the time needed for the scars to completely disappear.

Discomfort caused by the infection led in 12% of the reported cases to absenteeism from work or school. Private and public functions (weddings, receptions, sport events, meetings, etc.) had to be cancelled. In some schools, the Nairobi Fly appeared in such a large numbers and infected so many children that all educational activities were temporarily cancelled.

Identification and knowledge. The study indicates that the majority of the respondents (98.3%) had heard about the Nairobi Fly and knew that the insect can cause harm to people. Virtually everybody named the recognizable symptoms of *Paederus* dermatitis and conjunctivitis correctly. Of all people interviewed, 59.1% ($n = 1,208$) knew that the beetle needed to be crushed in order to harm people. Others believed that the insect bites (6.5%), stings (2.6%) or 'attacks' (0.01%).

Upon showing the small box with prepared insects, respondents were asked to point out the Nairobi Fly. Apparently people were familiar with the appearance of the insect, as 95.4% of the respondents who had heard about the Nairobi Fly ($n = 1,187$) pointed out the *Paederus sabaeus* beetle correctly.

The Nairobi Fly had been noticed in different locations (indoors on ceilings and floors, outdoors on the grass and in bushes) by many people. Over 80% of the respondents related general occurrence of the Nairobi Fly with the rainy season, and 79.5% ($n = 1,208$) blamed the heavy rainfall for causing the outbreak of the *Paederus* beetle. Although the majority of the people were obviously familiar with the Nairobi Fly and its behaviour, personal protection against becoming infected by its toxic fluid still seemed almost impossible.

Location. The south of the city centre in the area around Kenyatta hospital was chosen as the starting study area because of its vicinity to KEMRI, allowing intensive supervision. The interviewers shifted to a new area at least once a day, moving clockwise around the city. In some areas more interviews were conducted than in other areas, depending on the population density and availability of cooperative respondents in specific parts of town. Many areas of Nairobi were covered to ensure that the results of this study are representative for the entire Nairobi population (Figure 22.4). Public places like markets, hospital areas, schools, shopping

centres and individual shops were visited to include a wide variety of people with respect to age, educational background, employment and social status.

Large areas of Nairobi do have electricity supply. Hence, the majority of people interviewed live in areas and houses where electricity is available. Since the Nairobi Fly is attracted to artificial lights, the effect of electricity had to be taken into consideration. Indeed, a significant difference in prevalence between people who lived in houses with and without electricity was found: 17.1% ($n = 207$) of all respondents with no electricity at home showed an infection rate of 19.0%, while of the 1,001 persons living in houses with electricity supply, 30.5% had been infected ($\chi^2 = 11.4$, $p < 0.001$).

Although this study focused on the capital city of Kenya, the outbreak was not restricted to Nairobi itself. Cases were reported throughout Kenya, Uganda and Tanzania, and might have extended into other neighbouring countries.

22.6 DISCUSSION

The epidemiological study indicates that there was indeed a dramatic epidemic of *Paederus* dermatitis in Nairobi, caused by an outbreak of *Paederus sabaeus* or Nairobi Fly, during the prolonged heavy rainfall of 1997–1998 related to the El Niño phenomenon. According to information released by the Kenya Ministry of Health, rainfall was 500% above normal. The beetle caused severe distress and discomfort among the Nairobi residents by affecting one in three people.

In metropolitan cities like Nairobi, electricity is available in most areas. Because *Paederus sabaeus* is attracted to artificial light, the beetle seemed to have appeared in all areas of the city. Many people were affected in public places and without noticing the appearance of the insect, as it also appears on grass and other vegetation.

The most effective way to kill the insect is by using an insecticide spray. Although this was commonly known in Nairobi, it might not have been among the daily survival priorities of the majority of the people.

Frank and Kanamitsu (1987) claim that public awareness on the *Paederus* beetle, its behaviour and harmful effects is the key to prevention. The Nairobi survey, however, indicates that prevention of infection is a far more complex process. It involves not only attitudinal and economical factors, but is principally influenced by global climatic events that go beyond the control of the individual. However, human-induced global warming seems to have affected the global climatic events, including the frequency, persistence and magnitude of El Niño.

Paederus beetles occur on different continents all over the world and outbreaks have been reported from Africa, south-east Asia, Japan, Taiwan, to Australia, the Middle East and Europe

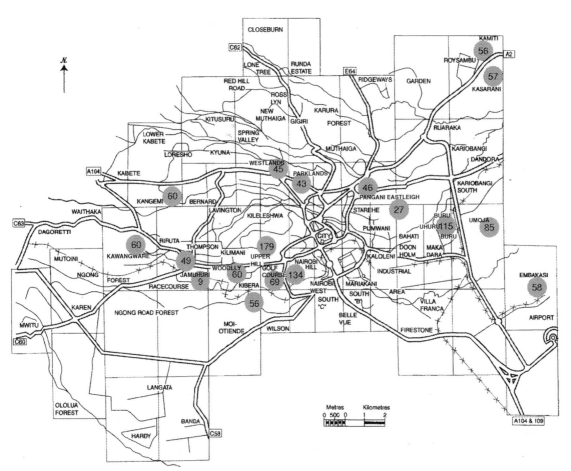

Figure 22.4. City map of Nairobi with number of interviews by area.

(Okiwelu *et al.*, 1996). The climatic and environmental factors that trigger outbreaks of *Paederus* rove beetles remain unspecified and may vary in different regions. In the case of the Nairobi outbreak, it was obvious that the abnormal heavy rainfall contributed to the abundance of the insects.

Despite its harmful effect on the human skin and eye, the toxic fluid pederin has important implications for health research. Pederin has been known to the western medical world only since the beginning of the twentieth century, but it has been used in Chinese medicine for more than 1,000 years. In experimental mice and rats, pederin had an obvious suppressing effect on cancerous tumours. Additionally, evidence shows that it can stimulate the regeneration of damaged tissues and has a healing effect on chronic, necrotic lesions in geriatric patients (Frank and Kanamitsu, 1987).

Nevertheless, important as the *Paederus* rove beetle may be for medical research, global climatic events obviously induce unexpected outbreaks of arthropods. Uncontrollable abundance of these insects will have unknown consequences for human well-being and deteriorate the ecosystems of the world.

Increasingly we are confronted with the consequences of human destruction to our living environment. Global climate change is a problem of natural complexity. We therefore need to look beyond a single factor like changes in temperature or the effect on a single species. Climatic changes contribute to altering of ecosystems and affect the distribution of and the interaction between species (Davis *et al.*, 1998; Harrington *et al.*, 1999). Records show that the 1997–1998 El Niño was exceptional. It was one of the two strongest El Niño events of the last 120 years, and it contributed to a rise in global temperature that made 1998 to be the warmest year of the millennium. If no immediate action is undertaken, we can expect that more extreme weather conditions with unknown consequences will occur in the future (Kerr, 1998b; Vogel and Lawler, 1998).

ACKNOWLEDGEMENTS

The International Centre of Insect Physiology and Ecology, Nairobi, Kenya, financed the survey described in this chapter. We are indebted to Mary Masawi, Susan Mbatha, Jessica Nyokil, Stephen Kwamanga, Victor Omondi and Ismael Owiti for their assistance with the implementation of the study. We wish to thank Dr Scott Miller for assistance with the identification of the rove beetle. The authors owe profound debt to the Director Generals of ICIPE and KEMRI, for their cooperation and continuous support. Special thanks are due to all Nairobi residents who dedicated their time, knowledge and experiences to this study. The constructive comments of Dr Hiremagalur Gopalan on the earlier draft are also gratefully acknowledged.

REFERENCES

Davis, A. J., Jenkinson, L. S., Lawton, J. H., Shorrocks, B. and Wood, S. (1998) Making mistakes when predicting shifts in species range in response to global warming. *Nature*, **391**, 783–786.

Dukes, J. S. and Mooney, H. A. (1999) Does global change increase the success of biological invaders? *Trends Ecol. Evol.*, **14**, 135–139.

Epstein, P. R., Diaz, H. F., Elias, S. *et al.* (1998) Biological and physical signs of climatic change: focus on mosquito-borne diseases. *Bull. Am. Meteorol. Soc.*, **79** (3), 409–417.

Fox, R. (1993) *Paederus* (Nairobi Fly) vesicular dermatitis in Tanzania. *Tropical Doctor*, **23** (1), 17–19.

Frank, J. H. and Kanamitsu, K. (1987) *Paederus*, sensu lato (coleoptera: staphylinidae): Natural history and medical importance. *J. Med. Entomol.*, **24** (2), 155–191.

Gelmetti, C. and Grimalt, R. (1993) *Paederus* dermatitis: an easy diagnosable but misdiagnosed eruption. *Eur. J. Pediatrics*, **152**, 6–8.

Harrington, R., Woiwod, I. and Sparks, T. (1999) Climate change and trophic interactions. *Trends Ecol. Evol.*, **14**, 146–151.

Kamaladasa, S. D., Perera, W. D. H. and Weeratunge, L. (1977) An outbreak of *paederus* dermatitis in a suburban hospital in Sri Lanka. *Int. J. Dermatol.*, **36** (1), 34–36.

Kellner, R. L. L. and Dettner, K. (1995) Allocation of pederin during lifetime of *Paederus* rove beetles (Coleoptera: Staphylinidae): Evidence for polymorphism of hemolymph toxin. *J. Chem. Ecol.*, **21** (11), 1,719–1,733.

Kerr, R. A. (1998a) The hottest year, by a hair. *Science*, **279**, 315–316. (1998b) Models win big in forecasting El Niño. *Science*, **280**, 522–523.

McCrae, A. W. R. and Visser, S. A. (1975) *Paederus* (Coleoptera: Staphylinidae) in Uganda. 1: Outbreaks, clinical effects, extraction and bioassay of the vesicating toxin. *Ann. Trop. Med. Parasitol.*, **69** (1), 109–120.

Okiwelu, S. N., Umeozor, O. O. and Akpan, A. J. (1996) An outbreak of the vesicating beetle *Paederus sabaeus* Er. (Coleoptera: Staphylinidae) in Rivers State, Nigeria. *Ann. Trop. Med. Parasitol.*, **90** (3), 345–346.

Veraldi, S. and Süss, L. (1994) Dermatitis caused by *Paederus fuscipes* Curt. *Int. J. Dermatol.*, **33** (4), 277–278.

Vogel, G. and Lawler, A. (1998) Hot year, but cool responses in Congress. *Science*, **280**, 1,684.

23 The role of indigenous plants in household adaptation to climate change: the Kenyan experience

SIRI ERIKSEN

University of Oslo, Norway, and Centre for International Climate and Environmental Research, Oslo, Norway

Keywords

Climate variability; drought; coping mechanisms/strategies; biodiversity; indigenous plants; adaptation policies; farmer households; drylands; Kitui; Kenya

Abstract

This chapter discusses household coping mechanisms during the 1996 drought in Kitui, a dryland agricultural area in Kenya. The 1996 drought and its effects on farmer households are described. Most households exhausted their food stocks several months before the next harvest was due in 1997. The alternative sources of food and income to which most households turned are discussed. Broadly speaking, there are two types of coping mechanisms: first, sources of food or income that are rated highly by interviewee households but to which few households have access; and second, sources of food or income to which most households resort, but which are not highly rated. The latter are often informal activities. Indigenous plants are input raw material in most household coping mechanisms, in particular in informal activities. Therefore, they are an important resource for the majority of households that do not have access to more formal sources of food and income during drought. There are, however, several constraints to indigenous plant-based activities. Unreliable sales and low profits contribute to their low status. Decreasing natural vegetation may limit the currently relatively cheap and easy access to indigenous plant resources. The chapter concludes that policies aimed at enhancing local indigenous plants and household capacity to cope with climatic variability can improve local welfare. At the same time, such policies can contribute to the implementation of international conventions on climate change and biodiversity, in terms of preparing for climate change adaptation and conserving local biodiversity and knowledge.

23.1 INTRODUCTION

Climate change may have dramatic effects on human activities, in particular in developing countries that depend on natural resources. As global climate change manifests itself as diverse local changes, the environmental parameters within which human activities operate also change. The United Nations Framework Convention on Climate Change (UNFCCC) recognizes the need to adapt human activities to climate-induced changes. The Convention commits Parties to prepare for such adaptation; in particular, policies regarding the environment and social and economic development are to take climate change into account to minimize adverse effects (UNEP/WMO, 1992, Article 4.1 (e), (f) and (h)).

The challenge of implementing adaptation policies for future changes that are uncertain and variable is compounded by the lack of resources in developing countries to instigate costly programmes. Many rural households are, however, already adapting to variability inherent in the local environment, both climatic variability such as drought, or economic changes following liberalization (see Mbithi and Wisner, 1973; Scoones *et al.*, 1996; Rocheleau *et al.*, 1995). Vulnerability theory, which discusses why some are more likely than others to experience adversity in the face of environmental change, proposes that exposure to change (or risk of a particular event) and capability of groups or individuals are two important factors (Chambers, 1989; Downing *et al.*, 1995). Capability refers to the ability to cope, or alter practices, in order to maintain a living standard when exposed to different forms of environmental change.

Given the difficulty in addressing exposure to future changes about which knowledge is still scarce, the capacity of individuals or groups to cope and adapt represents an appropriate focus for adaptation policies. Moreover, present climatic variability provides a good proxy for household ability to respond to future climate uncertainty. The household, as well as individuals in the household, are the economic and social units that ultimately feel

Climate Change and Africa, ed. Pak Sum Low. Published by Cambridge University Press. © Cambridge University Press 2005.

the effects of climate change. Studying household adaptation to current climate variability provides insight into the kind of policies that are likely to enhance household ability to adapt to future climate change. Such a bottom-up approach ensures that the interests and needs of different types of households may be taken into account in adaptation efforts.

Households represent a level of convergence between several global environmental issues. Rural households make up 80% of the population in Kenya (Republic of Kenya, 1996), and enhancing their social and economic viability is crucial for the welfare of a large section of the population. Rural households are the managers of local resources, and their activities are critical for, for example, conservation of biological diversity. There is an increasing realization that the local flora and habitats are important, in addition to the conventional reserves and national parks (see, for example, Biodiversity Support Programme, 1993). It is estimated that the majority of wildlife in Kenya is contained outside the areas set aside for conservation (Eriksen *et al.*, 1996). The UN Convention on Biological Diversity encourages indigenous knowledge and practices where these are linked to conservation and sustainable use of biological diversity (UNEP, 1992). Indigenous plant species are one aspect of biological diversity and an important component of the local resource base. How indigenous plants are used in coping mechanisms to climatic variability is vital both to local capacity in adaptation, and in biodiversity conservation. The link between local biological resources and climate change has been observed by Kelly and Adger (1999) in Vietnam, where the conservation of mangrove forests is important both for local livelihoods and for storm defences, a form of climate change impact mitigation.

This chapter illustrates the convergence between biodiversity and local capacity in adaptation through a case study of drought use of indigenous plants in Mbitini, a dryland farming area in Kenya. Though specifics of local coping strategies cannot be generalized, some of the issues arising from the case study are relevant in the formulation of adaptation and biodiversity policies aimed at strengthening local capacity. It is argued that many households do not have access to activities in the formal sphere during drought, and they rely on activities characterized by informal exchange and the input of indigenous plants. It is critical, therefore, that structures put in place to enhance informal activities are combined with efforts to ensure continued household access to these activities.

The following section introduces the 1996 drought and its effects on local households. Next, coping mechanisms in which households engage during drought are discussed. The role of indigenous plants in household coping strategies is assessed. In order to provide food and income when harvests fail, farmers intensify certain activities, such as casual labour, sale of livestock, production of bricks, charcoal and handicrafts, and collection of fruits. The relative advantages and constraints of various drought activi-

ties are identified. The final section discusses how policies aimed at adaptation can strengthen household capacity to cope with climatic variability.

23.2 THE 1996 DROUGHT IN MBITINI

23.2.1 Rainfall and drought

The selected case study site, Mbitini Location (henceforward referred to as Mbitini), is situated in Chuluni Division, Kitui District, in eastern Kenya (see Figure 23.1) and covers 129 square kilometres, with about 17,000 inhabitants (DIDC, 1997). It is located in hills sloping away from Kitui Central at between 800 and 1,000 metres above sea level. Mbitini is in an intermediary zone in terms of rainfall, where smallholder mixed farming and grazing exist. There are also areas of uncultivated land, and several indigenous plant species are utilized by farmer households as part of the agro-ecosystem.

Mbitini displays considerable climatic variability. Rainfall records for nearby stations, obtained from the Kenya Meteorological Department, illustrate this. The most complete long-term monthly rainfall records (1904–1996) were obtained for Kitui Agricultural Station, which is situated approximately 25 kilometres from Mbitini. The general topography at the site of Kitui Agricultural Station is similar to that of Mbitini, although the altitude is slightly higher at 1,100 metres above sea level. Comparison of monthly rainfall data for Kitui Agricultural Station and Kisasi Chief's Camp (located at the north-west boundary of the case study site) reveals that Kitui Agricultural Station data represents fairly well the variation in rainfall between months and years in the case study site. The total rainfall for the rainy season months March–May and October–December have correlation coefficients of 0.54 and 0.79 respectively for the years 1975–1996 for which data are available for both stations. Kitui Agricultural Station has slightly higher rainfall than Kisasi Chief's Camp, with mean total rainfall for the March–May and October–December rainy seasons of 412.4 mm and 573.9 mm respectively, compared to 257.8 mm and 371.0 mm respectively in Kisasi (in the time period from 1975–1996). The standard deviation for Kitui Agricultural Station for the two rainy seasons is 232.8 mm and 254.4 mm, compared to 215.9 mm and 158.6 mm at Kisasi Chief's Camp.

Agriculture in the area depends on the two rainy seasons: the so-called 'long' rains from March to May, and the 'short' rains from October–December. Unlike other parts of Kenya, the 'short' rains are actually more reliable than the 'long' rains in Kitui. The mean total rainfall (1961–1990) for Kitui Agricultural Station was 449 mm for the 'long' rains and 595 mm for the 'short' rains. Figures 23.2 and 23.3 show inter-annual variability in rainfall for the two seasons in terms of standard deviations from the mean. The figures reveal that there have been a few very wet incidences,

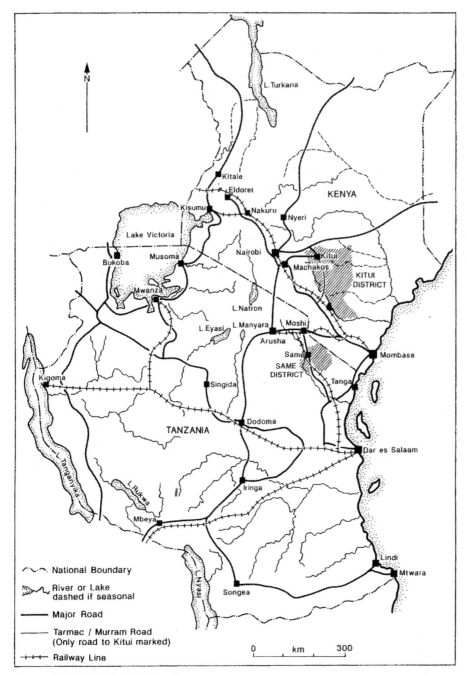

Figure 23.1. Map of Kitui District, Kenya.

and relatively more regular, and less extreme, dry events. The standard deviation is 224 mm for the 'short' rains and 265 mm for the 'long' rains. Therefore, 'long' season rains with rainfall one standard deviation less than the mean (such as in 1980) totalled 225 mm of rainfall. As can be seen from the figures, such low rainfall takes place fairly regularly, in the example of the 'long' rains, approximately once in five years.

Drought can be perceived in several different ways. According to Glantz (1994a), meteorological drought refers to a lack of rainfall compared to an expected amount over a certain time period. Agricultural drought, on the other hand, refers to insufficient rainfall to support agricultural activities prevalent in the area. Economic drought incorporates the requirements of productive activities that sustain the local economy as criteria of sufficient water availability and drought (Benson and Clay, 1998). Previous studies (O'Leary, 1980) and key informants in Mbitini recount that harvests in the Kitui area have failed for a number of reasons in the past, both climatic (insufficient or excessive quantity or

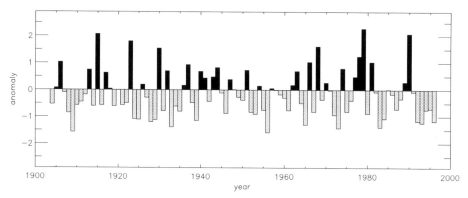

Figure 23.2. Rainfall anomalies, 'long' rains (March–May), Kitui Agriculture Station. The mean rainfall for the 1961–1990 period is 449 mm and the standard deviation is 224 mm.

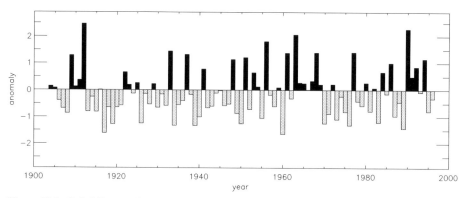

Figure 23.3. Rainfall anomalies, 'short' rains (November–December), Kitui Agriculture Station. The mean rainfall and standard deviation for the time period 1961–1990 is 486 mm and 211 mm, respectively. October rainfall is omitted from the plot, despite representing the start of the rainy season, due to several years of missing data for this month. Comparison of time series of rainfall anomalies of November–December and existing data for October–December confirms that omitting October does not change the overall pattern of the plot. Mean rainfall for October–December 1961–1990 is 595 mm, and the standard deviation is 265 mm.

unsuitable distribution of rainfall), environmental (such as locusts or other pests) and economic (such as prices of inputs or outputs). For this discussion, drought refers to harvest failure induced by poor rainfall, and specifically a harvest that yields so little that most households cannot rely on their own food stocks for most of the time period until the next harvest.

This chapter focuses on the drought that took place in 1996, following poor rainfall both during the October–December 1995 and the March–May 1996 rainy seasons. According to local farmers and administrators, Mbitini experienced droughts both in 1991 and 1992, in 1994 and in 1996. Figure 23.3 shows that the 'long' rains were consistently below normal (below the 1961–1990 mean) from 1991 to 1996. According to Figure 23.3, during the 1990s, the 1993, 1995, and 1996 'short' rain months had rainfall below normal. Examination of daily rainfall data for Kisasi confirms the poor distribution of that rainfall both in 1995 and 1996. 1996 was selected for study due to the poor rainfall, general perception of

the local population and administration of this as a bad drought year, as well as being a sufficiently recent incident for respondents to be able to recall their experiences.

Fifty-two randomly selected households in Mbitini were interviewed about their harvest and household activities after food stocks ran out in June 1996 until August 1998 (when interviews were carried out). During the time between the July 1996 harvest, which was poor for most households, and the next harvest in February 1997, most households ran out of food from their own farm. Of the 52 households interviewed in the case study site, almost a third produced no harvest of maize (the staple). Only two farmers had a crop that lasted them until the next harvest. Figure 23.4 shows that, by August 1996, fewer than half the households still had maize left from their June/July harvest. Very few had substantial amounts of other foods such as cassava or sorghum.

Attempts to distinguish a non-drought reference period illustrate the problem of determining a 'normal' year in an area of

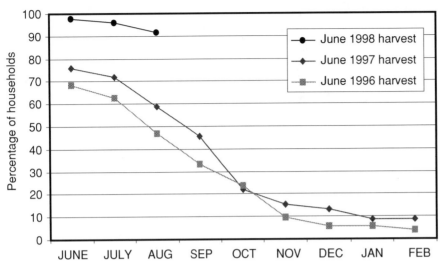

Figure 23.4. Duration of maize crop, after June/July harvest (named 'June harvest') in 1996, 1997 and 1998, Mbitini. Total number of households for which there were data was 51 in 1996, 46 in 1997 and 51 in 1998. Total number of households interviewed was 52.

high inter-annual variability in rainfall, and many below mean rainfall years. This is compounded by the fact that a poor harvest interacts with a number of other factors leading to households exhausting their food stocks. Due to the need for cash to pay for school fees, medical treatment and other household expenses, farmers are forced to sell part of their produce and make up for the shortfall by buying food later. 1997 was a slightly better year than 1996 in terms of the June/July harvest, but even then, many farmers ran out of maize food stocks well before the next harvest due in February. Rainfall was unusually heavy from October 1997 through to May 1998 and this period became locally known as the *El Niño rains*. While inducing drought in southern Africa, El Niño appears to be associated with unusually heavy rains in east Africa (Tyson *et al.*, 2002). Some areas of Kenya suffered flooding due to the rain, but some of the dryland areas in eastern Kenya, including Mbitini, enjoyed bumper harvests.

Most interviewees still had food stocks from the February/March harvest left when interviewed in August 1998, and many had continued to harvest maize until June/July. It therefore required the heaviest rain in Kisasi since records began in 1980 (combined with good distribution of rainfall) to produce a harvest where nearly all households harvested something in June/July, and over 90% still had food stocks left in August. One could argue that, due to economic marginalization of farmers, this area experiences 'economic drought' in all but exceptional years. The period from the June/July 1997 harvest until August 1998 was selected as the non-drought reference period. Interviewees were asked specifically about the months within this time period when they still had food stocks, in order to compare economic activities with the 1996 time period when they had no food stocks.

Figure 23.5 illustrates how prices of food reinforce the effects of poor harvests. As more and more households are forced to buy food, the prices of maize and other forms of food tend to rise unless the supply of food from other regions increases. As farmers' food stocks were exhausted in 1996, maize prices in markets near Mbitini rose. The rains improved and farmers harvested and started selling their crop at the end of 1997 and in 1998. The price of maize plummeted in less than two months, from nearly 20 Kenya Shillings per kilo to less than 10 Kenya Shillings per kilo.

23.2.2 Drought coping mechanisms

For most of the time period from July 1996 until February 1997, interviewee households in Mbitini had to look for sources of food other than their own food stocks. Alternative activities in which households engage in order to secure food or income during drought are referred to as 'coping mechanisms'. Farmer households engage in a diversity of activities or mechanisms. Such activities are not restricted to drought episodes, but are intensified during such instances. Some of the main sources of food and income on which households depend during drought include:

- Doing skilled work, such as carpentry, building or tailoring;
- Selling land;
- Collecting honey for consumption and for sale;
- Making bricks for sale;
- Engaging in group activities that result in food or money that is shared between group members;
- Business, such as selling snacks locally;
- Burning charcoal for sale;

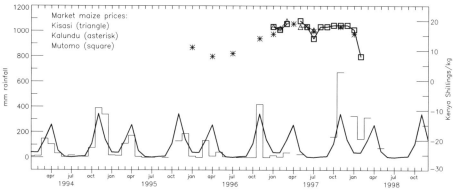

Figure 23.5. Monthly rainfall Kitui Agriculture Station and market maize prices, Kisasi Market (very near the field study site) and Kalundu Market and Mutomo Market (near field study site). Mean monthly rainfall 1961–1990 is plotted as a thick line, and observed rainfall January 1994 until May 1998 is plotted as histogram where data exist. Source of market prices: Kitui Quarterly Market Bulletin (Kitui Agricultural Project, (1998)).

- Receiving the salary of employed household members or remittances from spouse working elsewhere;
- Producing handicrafts for sale, such as ropes and baskets, tools, chairs and carving;
- Selling or consuming exotic fruits growing on the farm;
- Receiving credit from the local shop to be repaid later;
- Receiving food or money from family or relatives;
- Borrowing food or money from neighbours (sometimes received freely);
- Engaging in casual labour, usually for other farmers in the neighbourhood;
- Selling cattle, goats or chicken;
- Collecting indigenous fruit for own consumption or sale;
- Receiving food aid from the government or other organizations.

Many of these activities are informal in character. The informal sector has been defined as comprising activities that do not have formal systems of regulation, remuneration or control (Goodall, 1987; Todaro, 1994; Knox and Agnew, 1998). Lack of recognition by government agencies is another characteristic of a sector that 'includes cash and barter transactions that are part of everyday life in most places but go unrecorded by government agencies' (Hanink, 1997). Informal income generating activities, such as extraction of natural resources and crafts, as well as services and trading are important in rural livelihoods in East Africa (Seppälä, 1996).

Very few activities can be categorized as purely formal: salaries of resident household members or remittances, food aid, and sale of land. The rest of the listed coping mechanisms fall within the informal sector or partly in both sectors. Receiving food or money from family, relatives and neighbours, engaging in casual labour, and doing group activities where food or money is shared between members are mechanisms based mainly on informal exchange (though some group activities are assisted and regulated by government or aid organizations). The degree of formality of skilled work or conducting business depends on the agreement under which skilled work is done, and whether the business is of such a scale that it pays tax. Sale of products can be classed as informal or formal activities depending on whether these products are sold in the market, to shops, to business people from within or outside the area, or to neighbours. Products that are sold through several of these channels include honey, handicrafts, exotic fruit, livestock and poultry, charcoal, bricks and indigenous fruit (though the latter mostly locally). The above channels differ vastly in scale (sale of charcoal to a businessman buying truckloads to take to Nairobi compared to the sale of a papaw to a neighbour), and none of them guarantee registration or remuneration. Activities that gain some degree of formality in terms of government recognition, through inclusion in agricultural surveys and reports, include the sale of livestock and poultry, exotic fruit and honey.

There are several ways of viewing the relative importance of these different coping mechanisms. One way of measuring importance is whether or not many households engage in a particular mechanism. Another measure of importance is whether or not households rank the activity highly (from first, second, third until last compared to other activities).

The coping mechanisms in which sample households in Mbitini engage are plotted in Figure 23.6 according to these two measures of importance. Interviewed households described their activities in the time period from when the harvest ran out in the second half of 1996 until the next harvesting season in February 1997. The further right on the box-plot that the mechanism is situated, the higher number of households engage in that activity. The plot shows

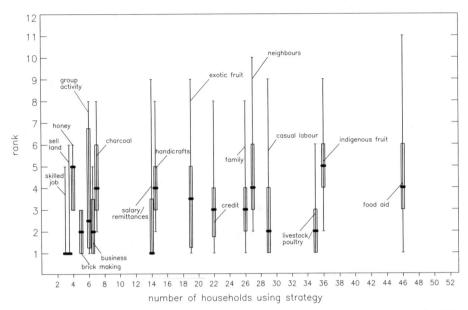

Figure 23.6. Household activities during the 1996 drought (end of July harvest to beginning of 1997), Mbitini. Value 1 on the Y-axis indicates the highest ranking and therefore the activity rated most important by the households in the given time period. The X-axis shows the number of households (the total is 52) that engaged in each activity during the same period.

ranking of an activity in terms of the maximum and minimum rank (thin line) given by the different interviewee households, the 25 and 75 percentile ranks (box), and the median rank (thick line). The number one means that an activity is ranked first, or most important. The lower the activity is placed in the plot, therefore, the better the rank. The lower right hand corner of the plot

signifies the highest importance in terms of a combination of many households engaging in the activity and ranking it highly.

This can be compared to Figure 23.7, which represents activities in the time period after the July 1997 harvest and early 1998 when sample households had sufficient harvest (and did not have to buy maize).

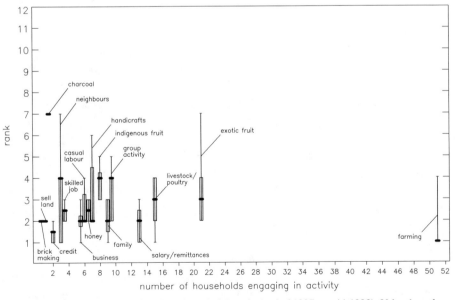

Figure 23.7. Household activities after the end of drought (end of 1997 to mid-1998). Value 1 on the Y-axis indicates the highest ranking and therefore the activity rated most important by the households in the given time period. The X-axis shows the number of households (the total is 52) that engaged in each activity during the same period.

One of the most conspicuous features in Figures 23.6 and 23.7 is the diversity of activities. Each household averages about six activities during drought, twice as many as during the non-drought reference period. There are many activities, however, which are more popular (in terms of number of households that engage in them) than others. The following pattern emerges.

First, there are certain activities that rank highly, but to which not all households have access. Activities that are ranked well, but in which less than half the households engage, include skilled employment (labelled 'skilled job'), selling land, making bricks, engaging in business, salary/remittances and obtaining credit. These activities typically necessitate either a particular skill and/or capital investment by the household. The exception is the sale of land. Only the poorest households with little labour or access to money resort to such depletion of their production capital. Among the interviewee households, sale of land was restricted to households headed by women 60 years of age or over, where the husband was deceased or not attached to the household. Salary/remittances indicate fairly reliable access to resources through employment or business. Credit also implies that the household has such a channel of access, as credit at the shop can only be obtained if the shopkeeper can be sure that one can pay back, usually through receiving a salary or remittance. Also among the activities that necessitate skill or investment are honey collection, charcoal burning and handicrafts. Exotic fruit further necessitate skill and investing in implements, required in planting and tending exotic trees.

Second, there are certain activities where households are not constrained by skill or investment needs, and which often operate through more informal channels. Receiving help in terms of a bag of maize or other food or money from family members; borrowing food from neighbours; doing casual labour, usually in the neighbourhood; and collecting indigenous fruit, are such informal mechanisms that do not require particular skills or investment beyond normal labour. More than half of the interviewed households engage in these activities during drought, although only casual labour ranks relatively highly because, for many households, this is the only option that can bring an income without having to sell off capital. Though open to most households, these informal mechanisms can be unreliable. Some interviewees claimed that during the very hard times, neighbours and family members no longer assist each other. Casual labour becomes very hard to obtain, and pay is so low that it may provide food only for one day if the family is large. Competition is fierce for casual labour, and reliance on this time-consuming mechanism may lead to neglect of the household's own land preparation.

Third, coping mechanisms are interconnected. This is illustrated by livestock and poultry sale, which is highly valued because of its versatility. Sale brings money immediately (though prices vary), which can be invested in a petty business such as the sale of porridge, or which can be used to buy food, send a sick household member to the hospital, or pay school fees. If one coping mechanism fails, another can be tried. One interviewee explained that if the drought were to persist, he would resort to charcoal burning since he had sold all his livestock.

As noted earlier, drought represents only one of the situations when the household needs money or food that cannot be provided by their harvest. The above activities are not confined to periods of poor rainfall. The pattern of activities nevertheless differs between a drought period, such as in 1996, and a period when harvests were fairly good. Farming is the most important activity, in that 51 of 52 interviewee households engage in farming, and almost all rank it first during non-drought periods. Farming is excluded from the Figure 23.6 drought activities since drought refers to the time period when the household has no crop left. Certain activities are abandoned when food is relatively plentiful. Informal mechanisms that households draw upon in drought, including help from neighbours and family, casual labour and collections of indigenous fruit, are more than halved in frequency. Sale of livestock and poultry is also reduced. Activities that are more readily recognized as part of the farming system or economy, such as salary/remittances, skilled employment and business remain almost unaltered in frequency.

The implication of the above pattern is that apart from livestock sale and food aid, there are few activities within the more formally recognized economy to which most households have access. Households typically resort to activities that are not supported by the formal infrastructure. These activities are often marginal in terms of income or food accrued. Sale of indigenous fruit or handicraft to neighbours or in the local market, for example, mostly yields low amounts of money (often less than 10 Kenya Shillings or the equivalent of 10 pence for a pile of fruit or for a piece of handicraft).

23.2.3 Indigenous plants

Indigenous plants play an important role in coping mechanisms. The term 'indigenous plant' refers to trees, shrubs, herbs, grasses and other floral species that grow naturally on the land. These are distinguished from exotic species which have been introduced from other parts of the world in the past few decades, and which are usually planted. Indigenous fruit are, for example, found growing wild in uncultivated areas.

The importance of indigenous plants derives from their role as raw material or input resource in many drought activities. Indigenous plants contribute directly to five of the 17 listed types of drought activities: as trees for making and hanging beehives as well as a source of nectar; as fuelwood for burning bricks and for burning charcoal; as reeds, fibres and wood for handicrafts such as making mats or tool handles; and as a source of fruit, vegetables and tubers for food and sale. Indigenous plants contribute indirectly to another five activities. Many group activities rely on indigenous plants, for example, collection of grass for sale. Business, such as selling snacks, may require fuelwood for cooking. Further, casual labour frequently involves tasks such as fencing using indigenous trees or shrubs. In addition, cattle, goats and chickens feed on indigenous foliage, grasses, pods and fruit found in the bush and left on the farm. Finally, skilled work, such as building, sometimes involves using materials from indigenous trees due to their superiority in strength and resistance to termites compared to exotic timber. Other activities that are not mentioned among the 17 types due to their low frequency in the interviewed sample include the sale of firewood, timber and pods for fodder.

There exist a number of other non-commercial household uses of indigenous plants, such as firewood for own use, household implements and tools, and medicinal plants. Non-commercial uses of indigenous plants are clearly important for household consumption, and without them the household would have to substitute with goods bought from the shops or market. These are not specifically discussed in this chapter because they are not deemed to directly contribute to food or income during drought.

The analysis above shows that most types of drought activities rely to a certain extent on indigenous plants. Exceptions include those activities that involve direct exchange rather than productive activities, such as food aid, help from neighbours or family, and receiving salary or remittances from outside. Producing exotic fruits for food and sale is a coping mechanism that does not involve indigenous plants. Most exotic species are, however, sensitive to drought, and yields are often low at the time when the household is struggling to cope with drought.

A second aspect of the importance of indigenous plants in coping mechanisms is that some of the most common activities (as indicated by the high number of households that engage in them in Figure 23.6) rely on indigenous plant resources. The most common activities include livestock and poultry sale, indigenous fruit for food and sale, and casual labour. Their high frequencies indicate that most households have access to these activities, in terms of having the required skill, labour and resources. Indeed, easier access to indigenous plants due to informal ownership arrangements is one of the advantages of indigenous plant-based activities.

The relationship that the household has to the owner of the land on which the indigenous plants are found, and the planned use of the products, determine the mode of acquiring indigenous plant produce, such as fuelwood or fodder. Indigenous fruits are a case in point: children can generally pick these without asking permission from the landowner; however, permission generally has to be asked if the fruits are for sale. Similarly, people may be able to collect firewood for free (with or without asking permission first) on the land of a relative or neighbour, while someone burning charcoal may buy a tree or acquire it for free from a relative.

In general, the higher the income potential, the more likely it is that someone has to pay for acquiring indigenous plant products from someone else's land. Informal relationships of access are, therefore, one of the main characteristics of coping mechanisms to which most households have access. Indigenous plant resources and coping mechanisms based on informal exchange mechanisms are particularly important for households that do not have special skills or particular access to resources through, for example, business, salary or remittances.

Despite the relatively easier access to indigenous plant-based activities compared to other drought sources of food and income, some of the indigenous plant-based activities are ranked relatively poorly by households in terms of importance during drought. The opportunities in the formal sector appear to be preferable according to favourable ranking by households, for example, of remittances of spouse/salary, skilled job, and business. When a household cannot sufficiently engage in one of the above-preferred formal mechanisms, household members will revert to more informal options.

For example, a significant association was found between a household not receiving remittances and engaging in casual labour and/or receiving assistance in the form of food or money from neighbours or relatives ($N = 51$, Phi coefficient $= -0.409$, Chi-square $= 0.011$). Indigenous plant-based activities, such as brick making and honey collection, were also common among households with no formal sources of drought income. In addition, such activities provide a supplement source of income when remittances are unreliable or irregular. This is a common situation for women whose husbands work elsewhere. These women often resort to producing handicrafts for sale or engaging in petty business. A respondent confirmed that there are longer time periods during which households have to find sources of income independently. The respondent engaged in a number of activities, including charcoal burning and handicraft sale while waiting for remittances from her truck driver husband. If one activity did not do well, she would switch to another.

Reasons for the different rankings given by interviewee households reveal several advantages and constraints. Some respondents rank collection of indigenous fruit highly because these fruits can be eaten by children; however, most respondents assign collection of indigenous fruits a low status as they are seldom seen as food for adults or substantial food at all beyond a snack. Households that process the fruit for eating by adults and selling give occasional higher ranking. Easier sale appears to be one of the reasons for respondents preferring exotic fruits to indigenous fruits. Exotic fruit trees are also promoted by government and aid agency extension and often valued as more progressive farming, even if a tree yields nothing or very little due to it being poorly suited to local conditions. The importance of reliability in sale was similarly reflected in ranking of other activities. In addition, the ability to generate a considerable amount of money, such as producing several sacks of charcoal, affects ranking positively.

An expressed advantage with certain activities based on indigenous plant resources is the relatively low labour requirement associated with, for example, collecting fruits or producing handicrafts. This is particularly important to households with single elderly or ill heads of household who are unable to do heavy work such as casual labour. Some handicrafts, such as making ropes or baskets, appear to be an important cash source for female-headed households with no reliable remittance from distant husbands or children.

An emerging constraint to indigenous plant-based activities is the increasing scarcity of such plants. Most of the households sampled in this study experienced a decrease in the number of indigenous trees and other plants in the last 10 years, both on their land and in the community in general. This decrease has forced some households to abandon certain activities, such as charcoal burning, to buy plant resources from neighbours, or to go further distances in order to find sufficient plant material. Muok and Owuor (1997) observed the trend of decreasing natural vegetation in a study in the same area. According to O'Leary (1980) and Pagiola (1995), processes such as the clearing of land for cultivation, and the cutting of trees for fuelwood and other purposes, are contributing to the decreasing availability of these resources.

23.3 CONCLUSIONS AND POLICY IMPLICATIONS

Regular droughts, as observed in Mbitini and Kitui, are a common phenomenon in large parts of Africa, including large areas of East Africa and the Sahel (Campbell, 1994; Glantz, 1994b). This study has illustrated the importance of household ability to engage in various activities in coping with climatic uncertainty in a smallholder rural setting. For example, the number of alternative non-crop activities to generate income or food doubles during drought. These drought activities are, however, subject to constraints. This study focused particularly on the crucial role of indigenous plant resources in drought coping at the household level. Results presented in this chapter suggest that initiating measures specifically directed at household capability is one way of strengthening the present policy framework regarding drought and climate change.

Policies aimed at improving household capacity to cope with climatic variability and drought can be targeted at making opportunities in the formal system easier to access. The analysis in this chapter points out that the importance of reliable sale of products and that the need for cash are factors that shape household drought coping decisions. Salary or sale of a product within the organized market appears to generate cash more easily in general than many of the more informal household-based activities.

Another aspect of enhancing household capabilities is developing measures to support household activities that are now largely informal and marginal in income. Many households are limited from participating in the more formally recognized activities mentioned above, due to cash investment and skill requirements. Activities characterized by informal transactions and inputs of local resources, such as indigenous plant materials, are more easily accessed by most households. These activities comprise economic diversification. Few activities generate a lot of money or food, but they complement each other in satisfying household needs. Cash investments are usually few, and the required skills, for example knowledge about local plants, are found in the informal sphere of the family, relatives and neighbours, rather than acquired through formal education or training.

Given the dependence of many households, particularly the poor, on informal activities, there is a need to improve their viability. Support of local skills and knowledge and improved marketing opportunities are likely to improve how people value these activities. An issue that would have to be addressed simultaneously, however, is that of ensuring access. There is a paradox in that it appears to be precisely the low profits associated with poor ranking that contribute to access regimes remaining informal and, therefore, also allows certain activities to remain open to most households. Improving economic potential and cash generation of informal activities could lead to increased cash investment and skill requirements, effectively reducing access by many households. Such reduction of access would have to be carefully avoided in order to sustain the diversity and complementarity of activities.

First, the local availability of indigenous plant resources is an important aspect of access to be addressed by policy makers. As resources become scarce, increasingly commercialized, or cash-based, access systems could be closed off to the poorest. Presently, many agricultural and forestry extension services focus on exotic farmed plants and trees rather than the production of indigenous plants. Extension services do emphasize the role of indigenous plants in watershed and soil conservation. Recognition by government and development agencies of the productive value of indigenous plants in economic and subsistence activities, however, is integral to reducing the problem of resource scarcity. For example, studies suggest that research and extension on the management and planting of indigenous plants on farm could yield huge improvements in productivity (Maghembe, 1995; Mwamba, 1995).

Second, social determinants of access to resources and particular drought activities have to be taken into account in formulating policies and developing specific measures. Many of the vulnerable households are female-headed with no male adult attached. Since some activities (such as carving, honey collection and to some extent charcoal burning) are undertaken almost exclusively by men, a gender-sensitive balance of activities should be targeted for support.

A third aspect of access is that of skills. In supporting the generation and development of knowledge in engaging in coping mechanisms, and in particular in processing indigenous plant materials, it is important that most households, including the poorest, gain access to this knowledge.

Enhancing household coping with short-term climate variability is, of course, only one of several approaches to climate change adaptation. The need for structural changes has been pointed out (Zinyowera et al., 1998), such as promoting economic activities that are less susceptible to climate perturbations; improving infrastructure, including implementing strategic food reserves and drought early warning systems; and adjusting agricultural practices, through introducing drought resistant crops and improving water supply systems. Farmer household strategies alone may not be sufficient in avoiding potential disruption to agricultural production, local economic development and social welfare, resulting from long-term climate change or extreme hazards. In this chapter, the focus has been on responses to drought-induced harvest failure rather than long-term changes. Several of the 17 identified types of drought activities may not be viable during permanently or long-term altered conditions, for example a drought of 5 or 10 years duration. Activities based on local markets and informal systems of exchange may fail as farmer assets are depleted, and indigenous plant resources may be exhausted.

Nevertheless, household capability plays a crucial role in future planning. First, improving the viability of household-based economic activities may reduce poverty in general and vulnerability to both economic and environmental events. Such reduction in short-term vulnerability may be particularly important, given the uncertainty about future climate change and its manifestations at the local level. Second, improving household capability represents a resource-efficient and fairly immediate measure. Larger scale structural changes, for example introducing drought resistant crops and constructing water supply systems, have proved expensive and difficult to implement in the past (Therkildsen, 1988; Moris, 1989; Republic of Kenya, 1989). Economic resources and planning capacity are already stretched in many African countries. In contrast, most of the measures proposed above to enhance household capability can be accommodated within current development policies and within the structure of government agencies, development agencies, and non-governmental organizations (NGOs) projects and activities.

In this chapter, a strong link between household capacity to cope with drought and the use of local indigenous plants was revealed. Household drought activities exemplify a point of convergence between biodiversity and adaptation to climate change.

Some of the measures suggested above may contribute to addressing both these issues simultaneously. This interconnectedness on a local level between the global issues of climate change and biodiversity loss suggest that various international environmental treaties, to which many developing countries are Parties, should be implemented in a coordinated manner. Household economic activities are a useful starting point for formulating policies for climate change impact mitigation and biodiversity conservation at both the national and sub-national (local and district) level.

ACKNOWLEDGEMENTS

This study was carried out as part of my doctoral research at the Climatic Research Unit, University of East Anglia, under the supervision of Dr Mick Kelly, Dr Kate Brown and Dr Jean Palutikof and as a visiting researcher at the African Centre for Technology Studies, Nairobi, under the supervision of Dr John Mugabe. Different stages of the research were funded by the Research Council of Norway, the British Foreign and Commonwealth Office (Chevening Scholarship), the Nansen's Fund, Christian Michelsen's Fund, the Dudley Stamp Memorial Trust, Lise and Arnfinn Heje's Fund, and Pastor Harald Kallevig's and Professor Fredrik Petersen's Fund. The research was assisted by the Kenya Forestry Research Institute in Muguga and Kitui, in particular by Bernard Owuor, Bernard Muok, Joshua Cheboiwo and Nelson Kavoi, and by the local authorities, including District Administration and Daniel Muasya, Chief of Mbitini Location, Peter Kyenze, Divisional Forest Extension Officer, as well as assistant chiefs and village elders. Many thanks to Lucy Kanini Mutunga, who assisted the interviews and to two anonymous reviewers of this chapter who provided useful feedback.

REFERENCES

Benson, C. and Clay, E. (1998) The Impact of Drought on Sub-Saharan African Economies. A Preliminary Examination, World Bank Technical Paper No. 401, World Bank, Washington, DC.

Biodiversity Support Programme (1993) African Biodiversity: Foundation for the Future. A Framework for Integrating Biodiversity Conservation and Sustainable Development, USAID/WWF/Nature Conservancy/World Resources Institute, Washington, DC.

Campbell, D. (1994) The dry regions of Kenya. In M. H. Glantz (Ed.), Drought Follows the Plow: Cultivating Marginal Areas, pp. 77–89. Cambridge, Cambridge University Press.

Chambers, R. (1989) Vulnerability, coping and policy. IDS Bulletin, 20 (2), 1–7.

DIDC (1997) Data at the District Information and Documentation Centre, Kitui, Kenya.

Downing, T. E., Watts, M. J . and Bohle, H. (1995) *Climate Change and Food Insecurity: Toward a Sociology and Geography of Vulnerability.* In T. E. Downing (Ed.), *Climate Change and World Food Security*, pp. 183–206. Berlin, Springer.

Eriksen, S., Ouko, E. and Marekia, N. (1996) Land tenure and wildlife management. In C. Juma and J. B. Ojwang (Eds.), *In Land We Trust: Environment, Private Property and Constitutional Change*, pp. 199–227. Nairobi/London, Initiatives Publishers/Zed Books.

Glantz, M. H. (1994a) Drought, desertification and food production. In M. H. Glantz (Ed.), *Drought Follows the Plow: Cultivating Marginal Areas*, pp. 7–30. Cambridge, Cambridge University Press.

——(1994b) The West African Sahel. In M. H. Glantz (Ed.), *Drought Follows the Plow: Cultivating Marginal Areas*, pp. 33–43. Cambridge, Cambridge University Press.

Goodall, B. (1987) *The Penguin Dictionary of Human Geography.* London, Penguin Books.

Hanink, D. M. (1997) *Principles and Applications of Economic Geography.* New York, John Wiley and Sons.

Kelly, P. M. and Adger, W. N. (1999) *Assessing Vulnerability of Climate Change and Facilitating Adaptation.* CSERGE Working Paper GEC 99–07, Centre for Social and Economic Research on the Global Environment, University of East Anglia, Norwich, and University College London, UK.

Kitui Agricultural Project (1998) *Kitui Quarterly Market Bulletin.* A Market Information Bulletin, funded by Kitui Agricultural Project, Issue No. 4 (March–June 1997) – Issue No. 7 (March 1998), Kenya.

Knox, P. and Agnew, J. (1998) *The Geography of the World Economy*, 3rd edition. London, Arnold.

Maghembe, J. A. (1995) Achievements in the establishments of indigenous fruit trees of the miombo woodlands of southern Africa. In J. A. Maghembe, Y. Ntupanyama and P. W. Chirwa (Eds.), *Improvement of Indigenous Fruit Trees of the Miombo Woodlands of Southern Africa*. Proceedings of a conference held on 23–27 January 1994 at Club Makokola, Mangochi, Malawi. ICRAF, Nairobi, Kenya; 39–49.

Mbithi, P. M. and Wisner, B. (1973) Drought and famine in Kenya: magnitude and attempted solutions. *J. Eastern African Res. Dev.*, **3** (2), 113–143.

Moris, J. R. (1989) Indigenous versus introduced solutions to food stress in Africa. In D. E. Sahn (Ed.), *Seasonal Variability in Third World Agriculture: The Consequences for Food Security*, pp. 209–234. International Food Policy Research Institute/Johns Hopkins University Press, Baltimore, USA.

Muok, B. and Owuor, B. (1997) *Report of Indigenous Fruit/Food Tree Species Survey in Kitui District*. KEFRI, Kitui and Nairobi.

Mwamba, C. K. (1995) Variations in fruit of *Uapaca kirkiana* and

effects of *in situ* silvicultural treatments on fruit parameters. In J. A. Maghembe, Y. Ntupanyama and P. W. Chirwa (Eds.), *Improvement of Indigenous Fruit Trees of the Miombo Woodlands of Southern Africa*. Proceedings of a conference held on 23–27 January 1994 at Club Makokola, Mangochi, Malawi. ICRAF, Nairobi, Kenya; 27–38.

O'Leary, M. (1980) Responses to drought in Kitui District, Kenya. *Disasters*, **4**, 315–327.

Pagiola, S. (1995) Kitui: resource pressures in a low-potential, small farm district. In S. Pearson *et al.* (Eds.), *Agricultural Policy in Kenya: Applications of the Policy Analysis Matrix*, pp. 214–248. Ithaca and London, Cornell University Press.

Republic of Kenya (1989) *Kitui District Development Plan 1989–93.* Office of the Vice President and Ministry of Planning and National Development, Nairobi, Kenya.

——(1996) *Economic Survey 1996, Central Bureau of Statistics.* Office of the Vice President and Ministry of Planning and National Development, Nairobi, Kenya.

Rocheleau, D., Steinberg, P. E. and Benjamin, P. A. (1995) Environment, development, crisis, and crusade: Ukambani, Kenya, 1890–1990, *World Dev.*, **23**, 1037–1051.

Scoones, I. with Chibudu, C., Chikura, S. *et al.* (1996) *Hazard and Opportunities: Farming Livelihoods in Dryland Africa: Lessons from Zimbabwe.* London, IIED.

Seppälä, P. (1996) The Politics of economic diversification: reconceptualizing the rural informal sector in south-east Tanzania. *Dev. Change*, **27**, 557–578.

Therkildsen, O. (1988) *Watering White Elephants? Lessons from Donor Funded Planning and Implementation of Rural Water Supplies in Tanzania.* Uppsala, Sweden, Scandinavian Institute of African Studies.

Todaro, M. P. (1994) *Economic Development*, 5th edition. Singapore, Longman.

Tyson, P., Odada, E., Schulze, R. and Vogel, C. (2002) Regional-global linkages: southern Africa. In P. Tyson *et al.* (Eds.), *Global-Regional Linkages in the Earth System*, pp. 3–73. Berlin, Heidelberg, New York, Springer.

UNEP (1992) *Convention on Biological Diversity.* Environmental Law and Institutions Programme Activity Centre, UNEP, Nairobi, Kenya.

UNEP/WMO (1992) *United Nations Framework Convention on Climate Change.* UNEP/WMO Information Unit on Climate Change, Geneva, Switzerland.

Zinyowera, M. C., Jallow, B. P., Maya, R. S. and Okoth-Ogendo, H. W. O. (1998) Africa. In R. T. Watson, M. C. Zinyowera, R. H. Moss and D. J. Dokken (Eds.), *The Regional Impacts of Climate Change, An Assessment of Vulnerability, A Special Report of IPCC Working Group II*, pp. 29–84. Cambridge, Cambridge University Press.

24 Requirements for integrated assessment modelling at the regional and national levels in Africa to address climate change

PAUL V. DESANKER[1], CHRISTOPHER O. JUSTICE[2], GRAY MUNTHALI[3],
AND KENNEDY MASAMVU[4]

[1] *Pennsylvania State University, USA*
[2] *University of Maryland, USA*
[3] *Department of Meteorological Services, Chileka, Malawi*
[4] *SADC Regional Remote Sensing Unit, Gaborone, Botswana*

Keywords

African policy maker needs; end-to-end model; IMAGE 2; integrated assessment model; impact assessment; national integrated model; SADC

Abstract

Integrated Assessment Models (IAMs) are important tools for assessing impacts and adaptation options for climate change at the global level. This chapter discusses some requirements for integrated models at the national and regional level in Africa. It is based on two meetings in southern Africa in April 1999, and subsequent discussions. The first meeting was held in Harare, Zimbabwe, and included representatives from different agencies and institutions doing work in the southern African region; the second was held in Lilongwe, Malawi, and involved representatives from different government departments and university scientists within Malawi. There is great potential for application of IAMs in African countries to help understand problems of water and food insecurity in the face of climate change. Alternative development futures will need to take into account how climate variability and extremes interact with health, social and economic factors such as HIV/AIDS and poverty in impacting upon development. In this chapter we outline important elements of an integrated model suitable for application at the national to regional level, and discuss important next steps for Africa.

24.1 INTRODUCTION

Integrated Assessment Models (IAMs) are models that represent within one integrated numerical model the physical, chemical and biological processes that control the concentration of greenhouse gases in the atmosphere, the physical processes that determine the effect of changing greenhouse gas concentrations on climate and sea level, and biology and ecology of ecosystems (natural and managed), the physical and human impacts of climate change and socio-economics of adaptation to and mitigation of climate change (IPCC, 1995). IAMs implement this integration from end-to-end, and include the ability to evaluate the effects of different policies and management options (either through optimization or simulation).

Integrated assessment modelling is a growing field, and promises to bridge the long existing gap between the science and policy communities. It involves addressing all aspects of a problem, from the physical science to the social and economic, as well as societal responses and policy. An interesting analogy in recent times was the 'Y2K bug' issue. An issue was posed (the potential catastrophic disruption of many systems that rely on computers) and the case made about likely impacts and potential responses to minimize the risks. Adaptation strategies were developed and implemented, and most of the public actively participated in the process. Potential risks were weighed against potential impacts, and people invested accordingly in safeguarding their computer-based data systems and many systems that rely on computers. Some of these measures were implemented in Africa too. Not much disruption happened, as it turned out, and it could be because everyone at risk acted to mitigate any adverse effects. This is an excellent example of an integrated project. For the climate change problem, the timescales are open-ended, and the scale even bigger and the processes highly complex. The need for the public's understanding of the problem and likely impacts, and degree of acceptance before mitigation is implemented, is fundamental to any risk. For the climate change issue, there is need for an intimate understanding of the issues, and likely risks, before the public (in Africa and elsewhere), can demand action.

IAMs have arisen out of the need to translate the physical impacts of global climate change (GCC) into economic terms. IAMs are 'end-to-end' models that focus on addressing multiple facets of a single issue and that incorporate physical science, social science and economic processes in an integrated analysis package (Figure 24.1). The aim of IAMs is to provide the interconnection

Climate Change and Africa, ed. Pak Sum Low. Published by Cambridge University Press. © Cambridge University Press 2005.

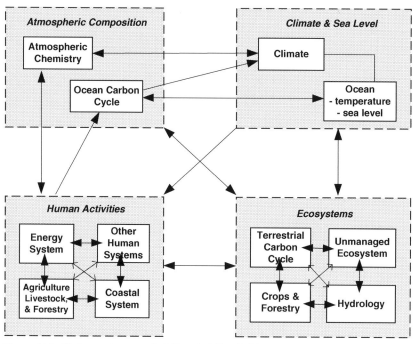

Figure 24.1. End-to-end nature of a full-scale IAM (IPCC, 1995).

of these multisector processes and to represent the response and feedback between the sectors.

Several IAMs have been built, and among the earlier and more widely used is IMAGE (Integrated Model to Assess the Greenhouse Effect). IMAGE has been documented widely (Alcamo *et al.*, 1994) and in this paper we will use it as an example. The potential exists to build a nationally specific GCC IAM for African countries, within economic subregions, that addresses:

- Implications of different global emission scenarios on climate change with detailed representation of short-term climate variability and extreme events at the regional scale, down to country level;
- The assessment of impacts of such scenarios on national sectors and commodities, especially issues related to food and water security, health and sustainable development;
- The regional implications of current policies to limit greenhouse gas concentrations (Kyoto Protocol) and to stabilize atmospheric concentrations of greenhouse gases (Article 2 of the United Nations Framework Convention on Climate Change (UNFCCC));
- The relative advantages of early action in adaptation and the exploration of synergies between climate change and other environmental agreements (biodiversity conservation; land degradation); and
- An explicit representation of the great diversity between countries and regions of Africa, with land-use systems that are reasonably accurate.

Developing a GCC IAM for African countries is particularly important for a number of reasons. To begin with, they are ill-prepared to deal with the adverse effects of GCC, and the human toll of GCC is likely to be great in this region, as evidenced by recent extreme events. The model results of a regionally focused GCC IAM would provide a first step in informing policy makers and leaders in the region as to the potential socio-economic impact of climate change and would allow them to evaluate the utility of establishing infrastructure to alleviate adverse socio-economic and human health effects of GCC. The process of deriving such a model would require interaction of all stakeholders, and Africa-specific characteristics would be built into the model. Africa would have intimate knowledge of how the model works, and the model would then become a practical tool for African countries in dealing with climate change issues, including international negotiations.

Immediate concern within the southern African region is with the socio-economic impacts of climate variability and extreme events, given the flood disasters in the region in recent years (Limpopo River and others in Mozambique in 1999/2000, and Zambezi River and its feeder basins in 2000). Developing an integrated assessment model sensitive to inter-annual variability would provide a strong foundation for moving planning from a few years to decadal time scales. Given that the natural resource sector is highly sensitive to climate variability, it is crucial to integrate realistic climate forecasting into predictive models that aim at integrating the physical, social, and economic consequences of climate variability. Regional climate models are being implemented in Africa to improve weather prediction. Global climate

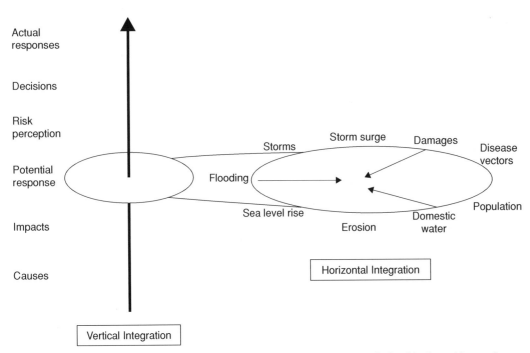

Figure 24.2. Differing dimensions and tasks of integrated assessment showing vertical and horizontal integration (after Lancaster, 1996).

models are improving in their representation of climate variability and El Niño Southern Oscillation (ENSO), and providing climate scenarios that can be used in impact assessments. Experiences with recent ENSO and extreme events provide useful case studies of how communities and ecosystems are impacted and how they respond to adverse effects of climate.

24.2 IAMS AND REGIONAL CLIMATE CHANGE

24.2.1 Horizontal and vertical integration

Lancaster (1996) defines IA to encompass vertical and horizontal integration (Figure 24.2). Vertically, IA begins by looking at basic causes of the climate change problem and moves up through physical impacts, societal impacts, potential responses, and management issues. The need to include an analysis of potential responses (adaptation) adds great complexity to the modelling. Horizontal integration, on the other hand, involves impact assessments within a geographic region. In other words, horizontally (or within a region), we have an integrated model that captures the biophysical and socio-economic processes and parameters. The vertical axis would then be the policy-management interface that incorporates possible decisions and their potential consequences in order to adapt and respond to given adverse effects and 'results' of the physical climate system. This conceptualization of IAMs sets them apart from mere integrated models that may combine a few sectors or disciplines. IAMs provide the end-to-end integration,

both vertically and horizontally. This enables analysis of causal links between specific emissions and their impacts on climate, and subsequent impacts on the biophysical and socio-economic components, and vice-versa. The inclusion of the energy sector in IAMs has made them particularly complete in terms of analysing emissions as a cause of greenhouse warming.

Extensive descriptions of IAMs and some sample outputs can be found at http://sedac.ciesin.org/MVA. The IPCC Working Group III Second Assessment report is another useful reference (IPCC, 1995). Some of these models are shown in Table 24.1.

24.3 EVALUATION OF IAMS WITH RESPECT TO DEVELOPING REGIONS

An IPCC workshop on IAMs for the Asia-Pacific region, held in 1997 in Japan, discussed at length the ins and outs of IAMs in relation to the regional level and, more importantly, in relation to developing countries (see Cameron *et al.*, 1997). The Proceedings of this workshop reflect the thinking of a diverse expert group on integrated assessment modelling. While Africa is not discussed explicitly, the issues raised apply to Africa equally well. Some of our own observations of IAMs with respect to Africa follow. It is clear that IAMs are a useful tool for capturing what is important in relation to the problem, and force one to think through all the linkages and implications. They stretch the limits of our understanding of individual sectors and regions (horizontal integration),

Table 24.1. Integrated assessment and modelling projects – a summary of more than 20 projects in integrated assessment and modelling of climate change, all of which attempt some level of end-to-end integration.

IAM	Remarks
CETA	A simple model, developed by the Electric Power Research Institute, to explore optimal combinations of abatement and adaptation policies (Peck and Teisberg, 1992)
The DICE Model	Dynamic integrated model of climate change (Nordhaus, 1992)
ICAM	Integrated Climate Assessment Model (ICAM) was developed by Carnegie Mellon University and several related research and assessment projects (Dowlatabadi and Morgan, 1993)
MERGE	A dynamic integrated model supported by the Electric Power Research Institute, based on the general equilibrium model Global 2200 (Manne *et al.*, 1993)
Asian Integrated Model (AIM)	A model developed at the Japanese National Institute for Environmental Studies to study the impacts of mitigation and adaptation scenarios on the Asian-Pacific region (Morita *et al.*, 1994)
CSERGE Model	The Centre for Social and Economic Research on the Global Environment (CSERGE) Model developed by University College, London (Maddison, 1994)
ESCAPE	An extension of IMAGE 1.0, intended to assess European policies and impacts, developed by the University of East Anglia and Oxford University (Hulme *et al.*, 1994)
MiniCAM and PGCAM	Developed by Battelle Pacific North-west Laboratories; the State of the Art Report (SOAR) on social science and global change (Edmonds *et al.*, 1994)
IMAGE 2.0	The first integrated climate assessment model to represent impacts and land-use at fine spatial scale, developed by RIVM (Alcamo *et al.*, 1994)
The CONNECTICUT Model	An integrated model that combines the approach of the DICE model with a probabilistic scenario analysis (Yohe and Wallace, 1995)
The CFUND Model	A nine-region dynamic model that permits exploration of policies based on inter-regional financial transfers and damage compensation (Tol, 1995)
TARGETS	A RIVM project designed to study issues surrounding global change and sustainable development (Rotmans and de Vries, 1997)
IMAGE 2.1	An updated version of IMAGE (Alcamo *et al.*, 1998)
Massachusetts Institute of Technology Model	A large integrated model with associated activities (Prinn *et al.*, 1999)

Source: CIESIN, 1995.

and force us to include the societal relevance through the vertical integration.

Critical to analysis of impacts is, of course, the input climate change scenario (including concentrations of CO_2 and other greenhouse gases). Approaches for creating these climate scenarios vary greatly, and each method has its advantages and disadvantages. IAMs overcome the limitations of using historical analogues and statistical downscaling by including simplified formulations of important components of the climate system. Simulated emissions from land-use change and energy use (both are functions of regional socio-economic, demographic and technological developments) are used to predict climate change and its impacts. Feedbacks and biogeochemical cycles are modelled, and thus give an integrated overview of the transient response of the climate–earth system (Leemans, 1997). Improvements are possible by linking the relatively simple constructions in the IAM to three-dimensional Atmospheric Global Circulation Models. The time-dependent emission and concentration path

combined with a consistent climate change makes IAMs suitable for impact assessments (assuming that they are performing satisfactorily).

It is clear that IA modelling is the way of the future in terms of performing assessments; the issue then becomes one of deciding what such a model should look like and do. Let us consider the IMAGE 2 (Alcamo *et al.*, 1994) and its three major components:

- Terrestrial Environment System (agriculture, land use, land cover and terrestrial carbon);
- Energy–Industry System (emissions from industry and energy, linked to the economy);
- Atmosphere–Ocean System.

Embedded across these systems are economic models and policy, and feedbacks. The IMAGE 2 IAM divides the world into 13 regions, with Africa as one of these regions. Economic and policy simulations are based on regions, while some calculations are carried out on a half-degree grid. See Figure 24.3 for the structure

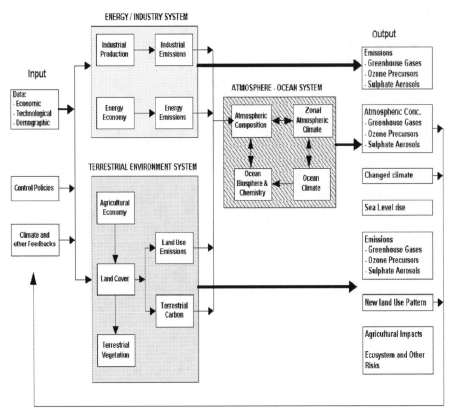

Figure 24.3. Structure of the IMAGE 2 Model (Alcamo *et al.*, 1994).

of IMAGE 2. As given by Leemans (1997), the scientific goals were:

- To provide insight into the relative importance of difference linkages, interactions and feedbacks in the society–biosphere–climate system;
- To estimate the most important sources of uncertainty in such a linked system.

And the policy-related goals of the model were:

- To link important scientific and policy aspects of global environmental change in a geographically explicit manner in order to assist decision-making;
- To provide a dynamic and long-term (50–100 years) perspective about the consequences of environmental change;
- To provide insight into the cross-linkages in the system and the side effects of various policy measures;
- To provide a quantitative basis for analysing the costs and benefits of various measures (including preventative and adaptive) to address environmental change. The model consists of three fully linked sub-systems of models: the Energy–Industry System (EIS); the Terrestrial Environment System (TES); and the Atmosphere–Ocean System (AOS).

EIS computes the emissions of greenhouse gases from world regions as a function of energy consumption and industrial production (de Vries *et al.*, 1994). The EIS models are designed to investigate the effectiveness of improved energy efficiency and technological development on future emissions in each region, and can be used to assess the consequences of different policies and socio-economic trends on future emissions.

AOS computes the build-up of greenhouse gases in the atmosphere and the resulting zonal-average temperature and precipitation patterns.

TES simulates the changes in global land cover on a grid-scale based on climatic, soil, demographic and economic factors. The role of land cover and other factors are then taken into account to compute the flux of CO_2 and other greenhouse gases from the biosphere to the atmosphere.

Most of the criticisms of IAMs like IMAGE 2 stem from the fact that a huge region like Africa is assumed to have common fuel-use patterns, socio-economic structures, land-use and tenure, and so on; and the land to satisfy 'African' demand for agriculture is basically distributed all across Africa. Most of these deficiencies relate to components of the model that can easily be corrected to provide better regional balance if detailed country-level data are included. For example, a land-use change for Africa is definitely possible at the aggregation of countries or better, as well as energy-use patterns, without the need for a totally inter-active model across the AOS–EIS–TES interfaces at that scale. In fact, this could be an entry point for a regional model, where the processes and properties that seem to matter locally can be mod-elled in more detail. Data exist to provide adequate baselines for most African countries for the land-use change, energy use and socio-economics. This would make the results more geographi-cally relevant.

There are many processes that are simply not well understood for Africa, for which assumptions are made to complete global models. The more obvious of these include those processes that vary with the geography of Africa (social value of different com-modities, e.g. cows are not simply hunks of beef, but a sign of wealth; significant role of access to ports as an indicator of eco-nomic potential; different adaptations to climate regimes across the diverse climates of Africa from rainforest to desert; different land-use systems; etc.). From a modelling point of view, model projections are very sensitive to technology projections and, for Africa, this is even harder, given experiences thus far in adapting new technologies, the failure to maintain a steady level of growth and, in some cases, because of the potential for rather drastic changes over relatively brief periods of time.

Given Africa's size and diversity in just about any variable or process, there is no justification for aggregating Africa into one region. For example, southern Africa forms a fairly distinct climate–atmosphere system, with atmospheric circulations form-ing a gyre, and this has been the focus of a major science ex-periment called Southern Africa Fire and Atmospheric Research program (SAFARI; see http://safari.gecp.virginia.edu). The eco-nomic integration and major river watersheds clearly carve up Africa into southern Africa (Southern African Development Com-munity or SADC), West Africa, East and Horn of Africa, central Africa, and northern Africa. These regions fall into relatively uni-form climate zones, within which land use and livelihoods are adapted accordingly. While the use of potential vegetation as a proxy for land cover attempts to capture the role of climate across Africa in some of the global models, there is in general very poor agreement between potential biomes and actual land cover, except on very coarse scales – land use in most of Africa is on very fine scales, given the rather small land holdings for the majority of Africa's subsistence farmers.

The large informal market is widely cited as another charac-teristic of Africa that does not conform to modern economics represented in IAMs. This is an area to which African economists can contribute greatly.

Another critical factor that has been analysed widely is that of economic growth/recovery and sustainable development. Theo-ries and potential solutions are numerous, yet these have not helped

many African countries overcome their slow growth. Given that the rate of economic growth has a significant impact on emis-sions and social welfare, future scenarios of socio-economic de-velopment for African countries are a critical component of any modelling on how the future might evolve.

24.4 POTENTIAL IAM APPLICATIONS FOR AFRICA

Any IAMs developed for Africa or its subregions and countries will need to include important elements of emissions, technology, energy and costing, while addressing the following critical issues for Africa:

• Country, regional to global economic linkages especially eco-nomic and political cooperation in regional groupings such as SADC;
• The role of multiple stresses such as impact of HIV/AIDS on the economy and on population dynamics, poverty and governance;
• The importance of climate variability and extremes besides long-term climatic change, given the high dependence on rain-fed agriculture and low economic capacity to cope with adverse climate; and,
• The great diversity in land uses by country and region, and by prevailing bioclimatic conditions.

Some important areas for which IAMs would be particularly use-ful in Africa are described below.

24.4.1 Food security

The problem of food insecurity for African countries is per-haps the most challenging, despite numerous assessments and initiatives. The analysis of food security goes beyond produc-tion, with socio-political factors playing pivotal roles after the onset of food shortages initiated by adverse weather. There are several existing programmes that assess food security for Africa, notably the United States Agency for International De-velopment (USAID) Famine Early Warning Systems (FEWS) programme (http://www.fews.net), the International Food Pol-icy Research Institute (IFPRI) studies (http://www.ifpri.org), the Food and Agriculture Organization of the United Nations (FAO) (http://www.fao.org) and the World Food Programme (WFP) (http://www.wfp.org). These programmes conduct inte-grated analyses in their own right to define food security, often with emphasis on short-term climate variability and prediction, but they are not tightly coupled to global processes in an 'end-to-end' fashion. The potential exists to incorporate these efforts into a fully coupled IAM at the national and regional level in order to provide projections of food security under a changing climate.

Such an assessment would include the production systems (driven by land use, climate, technology, inputs such as use of fertilizer and irrigation), issues of distribution through infrastructure, and ability to pay for food, as well as interacting factors of competing demands for financial resources at all levels (household to national levels). The ability of such a model to support decision analysis in the short term (seasonal to several years) would be particularly useful.

24.4.2 Water security

Food and water are closely related, and there are many projects related to these issues. While these often independent or piecemeal efforts are achieving their targets in some cases, they are unlikely to achieve maximum impact unless there is extensive integration across sectors and across different projects. There is a need to go beyond run-off analyses for major rivers of Africa (such as the Zambezi, Nile, Congo and Niger Rivers) to assess vulnerability of Africa's basic water needs. Most rural people rely on underground water sources, while major river systems are mainly important for major irrigation and hydroelectric power generation. The cross-calibration of flow rates with land use, climate and other factors that determine the flow and use of water is lacking. Available data on catchment and water bodies such as maps, river flow rates, climatic drivers and land use are adequate for modelling water budgets at district level for most countries in Africa.

24.4.3 Sustainable economic development

There is extensive literature and discussion of issues of sustainable development for Africa: why African countries continue to experience slow or negative growth, and what Africa needs to do to develop. There is no shortage of ideas. Perhaps one of the missing components is how these proposals fit into existing frameworks and how to go from current settings to the proposed targets. Big economic programmes implemented by the International Monetary Fund (IMF) and the World Bank have had very mixed results. Regional integration, structural adjustments, etc., are complex, and need to be devised and implemented in a fully integrated fashion. IAMs would provide a forum for integrating different proposals and ideas about Africa, especially in the context of climate change; and if the process involved all stakeholders, it would be possible to derive new ideas about how the future might be constructed. Social and environmental concerns would need to be included in discussions of economic development, and the models can be constructed in such a way that they capture important economic, social and environmental dimensions for the country in question. The duality of the economy in most developing countries between the market and informal sectors, activities being carried out to eradicate poverty, and the trade-offs between

the environment and development would be important aspects of the modelling and assessment.

24.4.4 Energy and health

The energy sector is a key driver of industrial development and is closely linked to economic development and social welfare. In urban areas, and in heavy industrial regions, emissions are a major concern for health. The possible spread of vector-borne diseases related to climate change is a major cause for concern; and there is increasing interest to relate diseases such as malaria to economic development. The important impacts of the HIV/AIDS epidemic on the population, on the economy and on factors that determine household production and coping would need to be included in integrated models for African countries and regions. Ideally, IAMs would be the tool of choice in evaluating possible impacts of, and adaptation strategies to, climate change, including the social dimensions and economic costs and benefits (basically the work of the IPCC assessments). Previous assessments have attempted to mimic this integration by synthesizing the literature by sector and theme, and trying to provide answers that are consistent with the 'outputs' or scenarios of other sectors. For example, impacts on food and water security would be analysed with respect to specific scenarios of climate change (temperature, rainfall, emissions, etc.), as well as socio-economics. This has proved challenging, at least for Africa, where extensive analyses simply do not exist in this format, partially because the tools or models for analysing these questions do not exist.

The core of a model for southern Africa is shown in Figure 24.4. Such a model would include a regional climate model nested within a global climate model, a land-use change model, an economic model that represents both formal and informal markets (or upper and lower circuits), a demographic model and an energy model. There are enough building blocks and adequate understanding of the issues for Africa to allow implementation of such a model without too much difficulty.

24.5 DATA REQUIRED FOR IA MODELLING

Lack of adequate data is one of the often-cited reasons for not including more detail for Africa. While this is true for some data sets, there is a wealth of data at the national level, most of which are archived and distributed by such international organizations as the World Resources Institute (WRI), the World Bank, FAO and so on. Within individual countries, national statistical centres are a useful source of sub-national data, and some of these are now available via the Internet (see http://www.geohive.com for links).

In preparation of a regional integrated modelling activity for southern Africa, we have started to compile a database of various

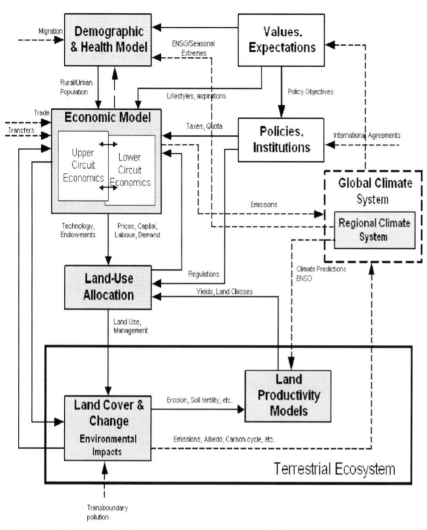

Figure 24.4. Proposed IAM for subregions and countries of Africa.

data layers and to store them in a relational database format for easy access and easy import into models or GIS for further processing. As part of the regional discussions on IAMs in Africa, a review of data compilation efforts was made (see Desanker, 1999 and http://www.miombo.org/climatecd) – several data bundles have been created by different groups, and data are distributed via CD-ROMs as well as through online data servers. Many of them are digital maps (i.e. paper maps converted to digital form). What is required now are digital spatial databases that can be used for modelling, ideally derived from a few basic inputs and, where possible, linked to remote-sensing products.

GIS-derived surfaces have the advantage of easy updates when better inputs become available. We have attempted to collate all basic layers such as boundaries and the best interpolated climate ·surfaces from which we can derive most other required indices and parameters, such as agricultural zones and land potential. It has been our experience that those databases that are designed with

modelling in mind will have data layers that are presented most effectively in formats suitable for use (such as gridded formats, and with straightforward legends). Data sets that are bundled into digital atlases with various interfaces, formats and overlays are very difficult to use for modelling and new applications because of the extra processing required. Standard GIS data formats exist to facilitate data sharing. Data preparation remains the most time-consuming activity in modelling. Data sharing and use of common standards for storage and distribution will facilitate modelling in the long run.

24.6 DISCUSSION

IAMs are highly complex and are unlikely to reach a level of acceptance that simpler models can achieve. The fact that there are so many global models under development means there will not be a shortage of sources of variation in predictions. Few models are documented well, and many do not provide ideas on how they can be implemented, since source code is rarely available. Open models would make testing and modifications for specific regions easy; however, this is unlikely for some of the bigger and more interesting models. IAMs are no longer individual models that can be ignored if so desired. They are being manoeuvred into the UNFCCC to act as baseline predictors of the future and yardsticks for implementation of mitigation and adaptation strategies. Africa must act for itself and develop credible IAMs that Africa can use in international climate negotiations and, equally important, apply to help solve environmental and development problems in Africa.

Some more observations on IA modelling follow:

- IA modelling is a long-term process. It must be made relevant for the policy-making and decision-making community. This means that policy makers must be brought in from the beginning and their input used within the overall design of the process. Existing regional integrations would make sense as organizing concepts. For southern Africa, a regional model within the SADC countries would be appropriate.
- Perhaps one criticism of IA modelling is its heavy reliance on consensus, versus fundamentals – while there is value in reaching consensus about what is important and true, there is great potential for being vulnerable to human folly and political agendas, especially where knowledge is limited and some views are not well represented. It is important that decisions and model structures be based on sound logic and a scientific basis, rather than ad hoc conceptualization by groups.
- The focus of IA must be on the end product. In the design of IA, time must be given to discussing and defining the principles and the requirements of the process(es) to be used.
- It must be kept in mind that IA modelling is a difficult process and hence it is important to take a realistic approach. Consider the US example, where historical data are easily accessible and technology is readily available. Even in this case, where all the factors are favourable for conducting IA, considerable challenges have presented themselves. It is difficult to make IA relevant for the policy community.
- Regional specificity will be key to the overall success of IA modelling. African scientists need to stress the characteristics of Africa that make it special. These need to go beyond representing national identities in such a model. Relationships (quantitative where possible) and interactions should be elucidated. New studies should address major gaps.
- It is desirable for individual nations to have the ability to analyse and synthesize their position in relation to the climate change issues, while taking into account other national priorities of food and water security, economic development, health, and sustainable natural resources management.

- The African scientific community does need to consider its predictive capacity. Are African scientists/institutions capable and ready in this respect? There is significant capacity-building that is required, and organizations such as START that have extensive expertise in capacity-building in developing countries must embrace the spirit of (regional) IAMs and facilitate the formation of IA modelling teams across Africa. The involvement of formal regional organizations such as SADC will be critical to success for these efforts.

IAMs are undoubtedly useful and there is great value in motivating their development for as many subregions as possible. The following random thoughts will serve as conclusions:

- An IAM should be useful to a country that needs to apply it – aggregations over large expanses (such as sub-Saharan Africa) are only useful to agencies outside of Africa that need to make policies for all of Africa in their planning. For anyone in Africa, continental assessment is of limited value, if any. While such aggregations are necessary from a modelling point of view for those processes that can be aggregated, it should be possible to disaggregate those processes that require country-specific detail. For example, when analysing global economics, Africa can be lumped, given its minimal driving force at that level. However, there should be efforts to link such large-scale processes to each nation, by using a nested or other approach, to make results directly useful to policies at the national level. Such downscaling should not require extensive data to implement, at least not much beyond what is routinely available.
- The end-to-end linkage of climate with the biophysical and socio-economic systems is a necessary and important feature of IAMs. It is critical that feedbacks and linkages are all properly understood, and that major areas of uncertainty are highlighted when results are applied. The reality is that as models get large and complex, the footnotes and warnings about weak linkages will get lost in the detail, even though the results are often applied as they are.
- For models that have to aggregate Africa, the subsystems that distribute predictions/projections spatially become the most critical. Land-use change, for example, is one component that can be improved greatly in big models such as IMAGE 2.1 without great loss of generality, to provide better detail over Africa.

Given Africa's diversity of issues and potential players in an IAM, it is unlikely that any one team can develop a model suitable or acceptable to everyone. For the future, the greatest number of efforts is encouraged and, in fact, needed to advance this field. We suggest a structure that would create a team of scientists from the component disciplines who would construct models suitable for specific targets (food and water security issues, energy, health, etc.), and then a central coordinating body (such as START)

ACKNOWLEDGEMENTS

This paper was developed as a result of workshops and projects funded by the European Community (EC DG XI through grant B7-8110/97/753/D4 EH to START), NOAA (Office of Global Programmes Order Number 40AANA403092), and NASA (Grant Number NAG5-6384). Participants at two regional workshops in Zimbabwe and Malawi (see http://miombo.org/climatecd) during April 1999 contributed to the ideas in this paper. Useful comments from anonymous reviewers and the editor are gratefully acknowledged.

REFERENCES

Alcamo, J., Kreileman, G. J. J., Krol, M. S. and Zuidema, G. (1994) Modelling the global society-biosphere-climate system: Part 1: Model description and testing. In J. Alcamo (Ed.), *IMAGE 2.0: Integrated Modelling of Global Climate Change*, pp. 1–36. Dordrecht, The Netherlands, Kluwer Academic Publishers.

Alcamo, J., Leemans, R. and Kreileman, E. (1998) *Global Change Scenarios of the 21st Century. Results from the IMAGE 2.1 Model*. London, Pergamon and Elsevier Science, 296pp.

Cameron, O. K., Fukuwatari, K. and Morita, T. (1997) Climate change and Integrated Assessment Models [IAMS] – bridging the gaps. *Proceedings of the IPCC Asia-Pacific Workshop on Integrated Assessment Models*, United Nations University, Japan, March 10–12, 1997. Center for Global Environmental Research (CGER), National Institute of Environmental Studies, Environmental Agency of Japan.

Consortium for International Earth Science Information Network (CIESIN) (1995) *Thematic Guide to Integrated Assessment Modelling of Climate Change* [online]. University Centre, Mich. CIESIN, http://sedac.ciesin.org/mva/iamcc.tg/TGHP.html, accessed on December 1, 2002.

de Vries, H. J. M., Olivier, J. G. J., van den Wijngaart, R. A., Kreileman, G. J. J. and Toet, A. M. C. (1994) Modelling the global society-biosphere-climate system: Part 2: Computed scenarios. In J. Alcamo (Ed.), *IMAGE 2.0: Integrated Modelling of Global Climate Change*, pp. 79–132. Dordrecht, The Netherlands, Kluwer Academic Publishers.

Desanker, P. V. (1999) *Miombo Data, Spatial and Integrated Modelling Workshops, Southern Africa, Workshop Report*. START Report No. 5, 1999. http://www.start.org, accessed on December 1, 2002.

Dowlatabadi, H. and Morgan, M. G. (1993) A model framework for integrated studies of the climate problem. *Energy Policy*, **21**, 209–221.

Edmonds, J. A., Wise, M. A. and MacCracken, C. (1994) *Advanced Energy Technologies and Climate Change: An Analysis Using the Global Change Assessment Model (GCAM)*. PNL-9798, UC-402, Pacific North-west Laboratory, Richland, WA, USA.

Hulme, M., Raper, S. C. B. and Wigley, T. M. L. (1994) An integrated framework to address climate change (ESCAPE) and further development of the global and regional climate modules (MAGICC). In N. Nakicenovic, W. D. Nordhaus, R. Richels and F. L. Toth (Eds.), *Integrative Assessment of Mitigation, Impacts, and Adaptation to Climate Change*. International Institute for Applied Systems Analysis (IIASA) Collaborative Paper CP-94-9. Laxenburg, Austria.

Intergovernmental Panel on Climate Change (IPCC) (1995). *Climate Change 1995: Economic and Social Dimensions of Climate Change*. Contribution of Working Group III to the Second Assessment of the Intergovernmental Panel on Climate Change. J. P. Bruce, H. Lee and E. F. Haites (Eds.), Cambridge, Cambridge University Press, 448pp.

Lancaster, J. (1996) Integrated assessment and interdisciplinary research. In S. J. Ghan, E. Rykiel, W. T. Pennell, M. J. Scott, K.L. Peterson and L. W. Vail (Eds.), *Regional Impacts of Global Climate Change: Assessing Change and Response at the Scales that Matter*, pp. 107–115. Hanford Symposium on Health and the Environment (32nd 1993: Richland, WA, USA; Columbus, OH, USA, Battelle Memorial Press.

Leemans, R. (1997) State-of-the-art computer models related to climate change impact models and their uncertainties. In Cameron *et al.* (1997), pp. 55–77.

Maddison, D. (1994) The shadow price of greenhouse gases and aerosols. In N. Nakicenovic, W. D. Nordhaus, R. Richels and F. L. Toth (Eds.), *Integrative Assessment of Mitigation, Impacts, and Adaptation to Climate Change*. International Institute for Applied Systems Analysis (IIASA) Collaborative Paper CP-94-9. IIASA, Laxenburg, Austria.

Manne, A. S., Mendelsohn, R. and Richels, R. J. (1993) MERGE: A Model for Evaluating Regional and Global Effects of GHG reduction policies. In N. Nakicenovic, W. D. Nordhaus, R. Richels and F. L. Toth (Eds.), *Integrative Assessment of Mitigation, Impacts, and Adaptation to Climate Change*. International Institute for Applied Systems Analysis (IIASA) Collaborative Paper CP-94-9. IIASA, Laxenburg, Austria.

Morita, T., Kainuma, M., Harasawa, H., Kai, K. and Matsuoka, Y. (1994) *Asian-Pacific Integrated Model for Evaluating Policy Options to Reduce Greenhouse Gas Emissions and Global Warming Impacts*. Japan Environment Agency, National Institute for Environmental Studies, Tsukuba, Japan.

Nordhaus, W. D. (1992) *The 'DICE' Model: Background and Structure of a Dynamic Integrated Climate-Economy Model of the Economics of Global Warming*. Cowles Foundation Discussion Paper No. 1009, Cowles Foundation for Research in Economics, New Haven, CT, USA.

would coordinate these activities and broker specific applications with user groups (agencies such as USAID, the World Bank, regional groups such as SADC, specific sectors with countries, etc.). START would facilitate the dialogues between the different groups.

Peck, S. C. and Teisberg, T. J. (1992) CETA: A model for Carbon Emissions Trajectory Assessment. *Energy J.*, **13**, 55–77.

Prinn, R., Jacoby, H., Sokolov, A. *et al.* (1999) Integrated global system model for climate policy assessment: feedbacks and sensitivity studies. *Climatic Change*, **41**, 469–546.

Rotmans, J. and de Vries, B. (Eds.) (1997) *Perspectives on Global Change: the TARGETS approach.* Cambridge, Cambridge University Press, 463pp.

Tol, R. S. J. (1995) *The Climate Fund: Sensitivity, Uncertainty, and Robustness Analysis.* Institute for Environmental Studies W95/02. Free University, Amsterdam, The Netherlands.

Yohe, G. and Wallace, R. (1995) *Near Term Mitigation Policy for Global Change Under Uncertainty: Minimizing the Expected Cost of Meeting Unknown Concentration Thresholds.* Department of Economics, Wesleyan University, Middletown, CT, USA.

25 Climate and disaster risk reduction in Africa

REID BASHER[*] AND SÁLVANO BRICEÑO

United Nations Secretariat of the International Strategy for Disaster Reduction, Geneva, Switzerland

Keywords

Africa; disaster risk; weather; climate change; disaster risk reduction

Abstract

Droughts, floods and storms affect millions of people in Africa every year, frequently with devastating impacts. The multiple stresses on African countries, particularly from poverty, infectious disease, fragile environments, limited institutional capacities and unsustainable development, mean that even modest fluctuations in weather or climate conditions can lead to severe consequences. However, there is sufficient understanding of how these vulnerabilities arise and compound each other, and therefore of how to act to reduce the risks. Disaster risk reduction encompasses three main areas of activity – assessing the risks, practices to reduce and manage risk, and policies and institutions to lead and support these activities. Climate change from rising greenhouse gas concentrations is an additional serious long-term threat, though there is little scientific (IPCC) evidence so far of material changes in the frequency or intensity of disaster-producing floods, droughts or storms, contrary to popular belief. Instead, the increased numbers of climate-related disasters over the last few decades appear to be mostly due to growing vulnerability and closer awareness and reporting of events. Attention therefore must remain focused on the vulnerabilities and risks associated with existing climate variability. Nevertheless, climate adaptation initiatives provide a welcome opportunity to advance the reduction of disaster risk. Conversely, disaster risk reduction provides a potent means to advance the adaptation agenda. Africa has several regional institutions that provide mechanisms for disaster risk reduction activities, and the United Nations International Strategy for Disaster Reduction (UN/ISDR) provides an international framework for linking the continent's needs and experience to international programmes.

25.1 INTRODUCTION

Worldwide, extreme climate events deliver enormous impacts, with thousands killed, millions of livelihoods destroyed or disrupted, and billions of dollars lost by economies. In Africa, drought and flooding are all too common, often resulting in devastating effects on crops, food security, housing and infrastructure and sometimes leading to starvation, migration, epidemics, conflict and environmental destruction. Disasters are a serious handicap in the struggle to develop sustainable livelihoods and economies and to reduce poverty.

In this chapter we review the issue of climate-related disasters primarily from a policy perspective rather than a research perspective. This includes a general outline of the nature and causes of disasters in Africa, a close examination of the relationship of disasters to climate variability and the climate change issue, and a summary of the main institutions and initiatives that are seeking to address the problems of weather and climate related disasters in Africa.

Disasters are defined as 'a serious disruption of the functioning of a community or a society causing widespread human, material, economic or environmental losses which exceed the ability of the affected community or society to cope using its own resources' (IDNDR, 1992; ISDR, 2002a). The term *disaster risk* means the probability of a disaster occurring, while *disaster risk management* means the systematic, proactive process of identifying, analysing and responding to risk, mainly to minimize the probability and consequences of adverse events. The concept of risk is very useful in disaster policy, as it offers a fundamental measure of the conditions that need to be managed.

The concept of *disaster risk reduction* encompasses all of the policies and practices that contribute to lower levels of risk. Since disaster risk depends as much on the human circumstances of vulnerability as on the originating hazard, a primary focus of disaster risk reduction must be on the human factors that raise or compound the risks, including the broad context of environmental and economic circumstances.

[*]Platform for the Promotion of Early Warning, Bonn, Germany

A key concept of disaster risk reduction is the idea of *living with risk* (ISDR, 2002a). Underlying this is the recognition that the climate naturally fluctuates in a variety of often unpredictable ways and is best coped with, now and in the future, by developing a culture of resilience in livelihoods, infrastructure and economic policy. By reducing individual and societal economic losses, disaster risk reduction helps conserve community resources and assets needed for sustainable development and future resilience. Disaster risk reduction is intimately connected to a range of development issues.

25.2 CLIMATE-RELATED DISASTER STATISTICS FOR AFRICA

The above definition of disasters is essentially a qualitative concept, which results in considerable difficulties in establishing quantitative data sets and providing reliable analyses of patterns and trends in disasters. The nature of a disaster situation will appear very differently when viewed through the eyes of a local farmer, a health official, an emergency manager, an insurer, or a meteorologist. The reporting of disasters by governments or researchers is voluntary and there is no standard protocol for disaster reporting. There are particular difficulties in defining drought, owing to its slow onset and uncertain termination, and its often patchy geographical extent.

The largest publicly available data set is the EM-DAT data set managed by the Centre for Research on the Epidemiology of Disasters (CRED), at the *Université Catholique de Louvain*, Belgium. In order for a disaster to be entered into the CRED database, one of the following criteria must be fulfilled: 10 or more people reported killed, 100 people reported affected, a call for international assistance, or a declaration of a state of emergency. Disaster numbers therefore must be treated with caution – a single large disaster may be much more significant than a hundred small ones. Summaries of EM-DAT data also may be found in the annual World Disasters Report, whose 2002 issue (IFRC, 2002) has a special focus on disaster risk reduction. A detailed assessment of the impacts of extreme weather events in 2002 is provided by Cornford (2003).

Table 25.1 shows that the numbers of reported disasters have grown rapidly over the last three decades, and that Africa accounts for about 20% of the global total. Globally, most of the growth has occurred in the category of climate-related events. There has been a very substantial growth in insurance payouts over the last three decades (Munich Re, 2003; Swiss Re, 2003). These trends are believed to be due largely to higher populations and assets at risk in more vulnerable circumstances, and to more effective capture and entry of climate-related events to the database.

Table 25.2 shows that climate-related disasters form about half of the reported disasters of natural origin for Africa, flooding

Table 25.1. Reported number of natural and technological disasters for Africa over last three decades.

	1973–1982	1983–1992	1993–2002
Africa	235	600	1403
World	1564	3721	5967
As percentage	15%	16%	23%

Data source: EM-DAT, http://www.cred.be

Table 25.2. Reported number of different types of disasters for Africa over period 1973–2002.

	Type[a]	Reports	Share
Climatic (53%)	Flood	356	21%
	Drought/famine	346	20%
	Windstorms	168	10%
	Wild fire	18	1%
	Extreme temperature	8	<1%
Geological (3%)	Earthquakes	28	2%
	Avalanches/landslides	19	1%
	Volcanic eruptions	11	<1%
Biological (44%)	Epidemics	698	41%
	Insect infestation	53	3%
Total		1705	

[a]Does not include AIDS pandemic or other chronic public health disasters.
Data source: EM-DAT, http://www.cred.be

and drought being the most common types. These are closely followed by biological disasters, in particular epidemics, which form the most numerous category overall. Epidemics have complex sources, largely rooted in poverty, but also in some cases arising from climate-related disasters. Disasters of geological origin are relatively few in Africa.

Table 25.3 presents the regional counts of reported climatic disasters for Africa. All of the regions are affected to some extent by all of the main hazards, but it is not clear that the data are of sufficient quality to draw more detailed conclusions about comparative vulnerabilities.

For example, the table does not reflect the large impact of drought in southern Africa. Drought is a recurrent feature in most parts of the region, with five recent major episodes, in 1980–1983, 1987–1988, 1991–1992, 1994–1995, 1997–1998 and 2002. Many of these events were regional in scale, with the 1991–1992

Table 25.3. Reported number of different types of climatic disasters for the main regions of Africa over period 1973–2002.

Subregion[a]	Drought	Flood	Windstorm	Total
East Africa	120	136	68	324
West Africa	102	87	18	207
North Africa	21	66	8	95
Central Africa	30	39	4	73
Southern Africa	35	28	20	73
Total	308	356	118	782

[a]Following EM-DAT criteria (predominantly based on groupings used by the United Nations Statistics Division). No affiliations or associations between countries contained within a regional grouping are implied.
Data source: EM-DAT, http://www.cred.be

Table 25.4. Reported number of people killed and affected by different types of climate-related disasters for Africa over decade 1993–2002.

Type	Killed	Affected
Flood	9,642	19,939,000
Drought/famine	4,453	110,956,000
Windstorms	1,335	5,687,000
Extreme temperature	147	—
Wild fire	136	8,000
Total	15,713	136,590,000

Data source: EM-DAT, http://www.cred.be

drought considered the worst in living memory, placing more than 20 million people at risk. The famine of 2002, whose origins were less attributable to drought alone, put 14 million people in need of food aid (see Section 25.4)

Table 25.4 summarizes the reported numbers of people killed and affected by climate-related disasters for the most recent decade. This shows that floods account for the most deaths, about twice as many as reported for droughts, but that droughts affected over a hundred million people over the decade, about five times the number of flood-affected people. This is consistent with the limited geographical extent and high short-term dangers of floods, and the extensive spatial reach and duration of droughts. Improved drought monitoring and food security assessment, coupled with targeted food aid, has markedly assisted survival rates in droughts. Windstorms are a lesser but still important hazard. Geological hazards together accounted for 781 reported deaths over the period. Again, one should be careful not to read too much

detail from these figures, owing to the many limitations of the raw data.

One encouraging feature is the sharp reduction in the number of people reported killed by disasters over the last two decades, from 579,452 reported deaths over 1983–1992 to 43,078 reported deaths over 1993–2002. This represents a real reduction in death rates, as the numbers reported affected in these two decades were similar, being 149 million and 137 million respectively. Some African countries have experienced extraordinary losses from drought in the past, the data for 1980–2001 showing 300,000 reported deaths for Ethiopia, 150,000 for Sudan, and 100,000 in Mozambique.

The estimates of economic losses for disasters in Africa are not likely to be very accurate, but they give some idea of the sizeable economic impacts on the continent's developing countries. The 1993–2002 totals derived from the EM-DAT database are US$424 million for droughts and famine, US$1,109 million for floods, and US$841 million for windstorms.

25.3 CLIMATE VARIABILITY AND CHANGE

Disasters can arise from either short-term weather events, or longer-term episodes of climate extremes. By 'weather' we mean the phenomena of wind, rain, sunshine, cloud and temperature that occur on a day-to-day basis, while by 'climate' we mean the overall or average weather conditions that prevail over longer periods of months and years. The predominant weather and climate hazards are:

(i) Prolonged deficiency of rainfall, which causes droughts, crop failures and water shortages, and may contribute to widespread fires and smoke haze;
(ii) Periods of high rainfall, which cause floods and rain-induced landslides, and sometimes may lead to epidemics and pest outbreaks;
(iii) Strong winds, which damage buildings, trees and crops and may cause coastal wave damage, and
(iv) Extreme heat or cold that can stress and kill humans, animals and crops.

Tropical cyclones (also termed hurricanes or typhoons), which affect eastern parts of southern Africa, can be particularly damaging owing to their combination of very high winds, high rainfall, and pressure-driven ocean surges. Some health conditions, such as respiratory diseases and meningitis, are known to be associated with climatic conditions.

The characteristics of Africa's average climate can be determined from long historical measurements of weather parameters, such as shown in Hulme *et al.* (2005). The marked regional patterns include the arid or semi-arid subtropical regions, the equatorial rainy zones, and the south-east region affected by summertime tropical cyclones. Over the year there are typically either one or two rainy seasons. Eastern and southern Africa are affected by the state of the El Niño Southern Oscillation (ENSO) phenomenon, and by the temperature of the Indian Ocean, while West Africa is more affected by conditions in the Atlantic Ocean. Weather data sets also can be used to estimate the risks of damaging extreme conditions. The spatial and temporal patterns of Africa's climate are now quite well understood in terms of the global climate system's behaviour and can be modelled by computer-based models of the global climate, also known as general circulation models (GCMs).

Historically, it was usually assumed that a locality's climatic patterns and its risks of extreme events were essentially unchanging over the long term and hence these risks could be adequately determined if enough years of data were in hand. However, it is now known that the climate exhibits a range of behaviours and fluctuations of natural origin, including slow long-term changes. The widely publicized El Niño phenomenon, which can bring large year-to-year changes for some parts of Africa, is driven by interactions between the Pacific tropical ocean and the global atmosphere and can be partially predicted one or two seasons ahead using GCMs (Goddard *et al.*, 2001). Multi-year and decadal timescale fluctuations, most notably demonstrated in the Sahel rainfall variations over the last century (see Figure 3.2, Hulme *et al.*, 2005), appear to be linked to slow oceanic fluctuations on decadal timescales.

Although some of these fluctuations and phenomena can be explained in a physical sense, it is also recognized that chaotic, random behaviour is a natural part of the climate system and that this sets fundamental limits on our ability to predict climatic conditions in advance, whether of a tropical storm a few days ahead or an El Niño fluctuation a season ahead. Nevertheless, advances in monitoring, modelling and forecasting techniques have increasingly enabled the generation of usable early warning information. For example, flood risk can be estimated through a combination of observed soil water conditions and river levels, catchment models, and forecast rainfalls. Drought risk can be estimated from observations of soil moisture and plant conditions coupled with GCM seasonal climate predictions.

On top of these natural processes, rising concentrations of greenhouse gases appear set to provide another source of long-term climate change. One of the most compelling concerns about climate change is the fear that it will lead to an increase in damaging weather events and disasters, especially for the least-developed countries, many of which are in Africa, and the low-lying small island states. Of course, the increased concentration of greenhouse gases does not directly cause the hazards – these are already an intrinsic feature of the climate system – but rather it nudges the existing patterns of the climate system, and thereby alters the frequencies and intensities of the hazardous events.

The comprehensive assessments of the Intergovernmental Panel on Climate Change (IPCC) (http://www.ipcc.ch/index.html) provide an authoritative benchmark on the scientific evidence concerning climate change. The IPCC Third Assessment Report establishes that climate change is a serious long-term threat for the coming centuries. The detailed assessments of past changes and projected changes are presented in the report of IPCC Working Group 1 (IPCC, 2001a), while the assessment of the likely impacts of the projected changes are provided in the report of Working Group 2 (IPCC, 2001b). These and other IPCC reports can be downloaded from the IPCC website noted above. A review of climate change projections for Africa is provided by Hulme *et al.* (2005).

Some projected changes, particularly rising temperatures and rising sea level, are more certain than others. By 2100, the global mean temperature is very likely to increase by 1.4 °C to 5.8 °C over the period 1990 to 2100, while sea level is very likely to rise by 9 cm to 88 cm (IPCC, 2001a). The wide ranges in these projections mainly arise from differences between the different climate models used and uncertainties in the future growth of greenhouse gas concentrations.

The projections are much less certain for phenomena that are more commonly associated with disaster risk. An important general conclusion of the IPCC is that it is likely that atmospheric water vapour and surface evaporation will be higher in a warmer world, and that this would tend to result in increased precipitation, greater intensities of rainfall, and more extensive droughts.

However, there is considerable uncertainty in the rainfall projections for specific regions and seasons. The results of the available climate models often differ significantly. For example, the models' projected rainfall changes at the end of the twenty-first century for West Africa range from 0% to +40% for the December–January–February season, and from −20% to +20% for the June–July–August season (IPCC, 2001a). Such projections provide valuable broad-brush indications of the future possibilities for Africa's climate but not the regionally specific and certain predictions desired by disaster risk managers and other decision makers.

Higher moisture levels also would likely result in some increases in the wind speeds and rainfall intensities of tropical cyclones, and in the rainfalls and drought intensities associated with El Niño events. However, IPCC reports that the model results for tropical cyclones, other storms and the El Niño over the next hundred years are varied and as yet do not show a consistent pattern of projected change in the intrinsic behaviour of these phenomena.

Looking backward, the analysis of historical measurements of climate provides a means to test for the emergence of the projected long-term changes. The IPCC Third Assessment does indeed confirm a consistent picture of a warming world over the last century, with rises in global average temperature, increasing frequency of high temperatures, and associated rises in sea levels and depletion of glaciers. The rate of observed temperature rise has noticeably accelerated over the last decade. The IPCC assessment concludes 'most of the observed warming over the last 50 years is likely to have been due to the increases in greenhouse gases... and that it is very likely that the twentieth century warming has contributed significantly to the observed sea-level rise.' (IPCC, 2001a). Africa is not well endowed with the long data sets needed to independently assess its temperature trends, but the IPCC report shows that the available data for Africa is generally consistent with the global patterns of temperature trends.

The changes in observed precipitation are less clear-cut. IPCC (2001a) reports that it is likely that precipitation increases of 0.2% to 0.3% per decade have occurred over tropical ($10°$ S $– 0°$ N) land areas over the last century, though this trend has weakened over the last decade. Higher increases are reported for most mid- and high-latitude regions of the northern hemisphere. Significantly, it is also reported that where precipitation has increased, it is very likely that there have been even more pronounced increases in heavy and extreme precipitation. Unfortunately precipitation trends for Africa cannot be established well, except in a few areas, owing to the limited data sets and to the complicating effects of the significant natural variations over decadal timescales (see Figure 3.2 Hulme *et al.*, 2005).

Only small or inconclusive changes are reported for the observed large-scale weather phenomena that are dominant in disasters, such as extended anticyclonic (drought) conditions, tropical cyclones, other storms and El Niño events. Trends for severe drought over 1900–1995 are reported by IPCC to be relatively small. Trend research results for tropical and extra-tropical storms are either not significant or are conflicting.

The frequency and intensity of the El Niño phenomenon has been unusual since the mid-1970s compared with the previous 100 years, with more frequent, persistent or intense El Niño (warm events), and therefore a somewhat enhanced pattern of impacts on those countries affected by El Niño. However, this shift may be simply a natural decadal-scale fluctuation, rather than any long-term trend. This would be consistent with IPCC projections which indicate only relatively small changes in El Niño amplitudes over the next 100 years.

Responsibility within the IPCC for assessing the likely impacts of the projected changes in climate on downstream geophysical, ecological and socio-economic systems, and possible adaptations, falls to Working Group 2 (IPCC, 2001b). The chapter on Africa (Desanker *et al.*, 2001) provides a highly informative background to the region and assessment of possible sectoral impacts. The impacts in several critical areas like water and agriculture are likely to become significant as the century unfolds. The chapter particularly notes the existing pervasive vulnerability to stress, including stresses arising from current climate fluctuations and extremes.

In summary, the current evidence indicates that changes in drought occurrence, heavy rainfall and storminess are likely to develop slowly over the coming century. Over the next decade, small changes may occur though these are unlikely to be distinguishable from the natural high variability of the present day climate, and therefore are unlikely to have a clearly discernible effect on disaster occurrence. The large uncertainty in the projections is itself a major issue that must be factored into long-term decisions and policies. The trends in temperatures and sea levels are likely to continue and may exacerbate stresses and possibly disasters in areas where heat waves and coastal inundations are already a problem.

The case for taking action on the root cause of climate change, i.e. to control greenhouse gas emissions, is very clear and urgent. Most governments recognize that this is the priority and are actively participating in relevant processes under the United Nations Framework Convention on Climate Change (UNFCCC) (http://www.unfccc.int).

Policy makers are also paying attention to the issue of adaptation to future climate conditions. There is considerable concern that climate change will disproportionately affect developing countries and that it will be a serious impediment to the achievement of the core development challenge of poverty reduction (AfDB *et al.*, 2003).

Important initiatives include the UNFCCC national adaptation plans of action (UNFCCC, 2002), adaptation projects being developed under support of the Global Environment Facility, and the mainstreaming of adaptation measures in development projects. Given the inexorable growth in greenhouse gas concentrations and the high uncertainty about future climatic conditions, it is especially prudent to consider 'no-regrets' adaptation steps, and to seek to develop greater *adaptive capacity* to climatic risks and shocks (Adger *et al.*, 2003). An important principle is that our capacities to cope with future climate variability and risks must start with the capability to manage existing variability and risks (Bruce, 1999).

The climate change issue thus brings a welcome weight of interest to the fundamental and pressing issues for Africa of poverty, climatic vulnerability, and disaster risk reduction, and there could be many synergies to be obtained from adaptation activities. However, as we have seen, the popular idea that greenhouse gas-induced climate change is already visiting increased disasters upon us and requires urgent adaptation action is not supported by the IPCC assessments. Adaptation initiatives need to be developed on a frank assessment of the real problems faced in Africa and should be targeted at decision processes that have long-term implications, for example, in respect to land use legislation, urban planning and public infrastructure.

Disaster risk reduction has a key role to play in the climate change agenda. It provides a ready-made set of concepts, approaches, tools and information to advance the adaptation agenda. It could achieve rapid progress with activities that are both effective and meaningful to the communities at risk to climate events. Equally, the adaptation agenda provides a new and potent means to advance the disaster community's agenda to reduce disaster risks.

25.4 DISASTER RISK AND RISK REDUCTION CONCEPTS

Climate hazards on their own do not necessarily lead to disasters; it is the exposure of a community that cannot cope with the hazard that leads to a disaster (UNDP, 2003). However, in common thinking there remains a substantial perception that disasters are 'natural' and unavoidable. For many, the belief is that, apart from some preparation, the main policy response should be to move swiftly to rescue those affected, whether from drowning, starvation or disease, and to clean up and rebuild as necessary. This is reflected in the substantially greater financial resources that are applied to the emergency phases of disasters compared to long-term investment in disaster risk reduction.

The real nature of disasters gets obscured by the preoccupation with the external nature of hazard threats, the vividness and drama of the rescuer images (via television), and the use of language that describes the event in terms of the external source (rain, flood, drought, etc.) rather than the internal conditions that result in vulnerability to the hazard. The various parties involved, whether individuals or governments, are treated as innocent victims without any role in the creation of the disaster.

As noted earlier (Table 25.1), there has been a substantial growth in the impacts of weather-related disasters over the past three decades. Global average economic losses from extreme weather events over the 1990s were six times greater than in the 1960s (Munich Re, 2003) and the economic impacts for developing countries were about five times higher per unit of GDP than for the rich countries. Since the observed changes to date in the characteristics of weather and climate hazards are quite small, this rapid growth in disasters must be due to increases in the underlying vulnerability to climate hazards. This is the consensus view of disaster risk reduction professionals (UNDP, 2003).

By the term *vulnerable* we mean the relative lack of capacity to resist, accommodate and recover from a stress event. Many African countries are particularly vulnerable to natural hazards, owing to an often-prevailing syndrome of endemic poverty, limited livelihood options, fragile and degraded soils and environments, and loss of productivity from HIV/AIDS and other infectious diseases. Additional factors are the widespread dependence on subsistence agriculture, limited transportation and markets, increasing urbanization, minimal social services, governance shortcomings, military conflict, population displacements and the

powerlessness of the poor to better their situation. In many cases, a government perspective does not extend far beyond the cities, leaving rural areas to suffer without services or visibility. About 30 million African adults and children are HIV/AIDS sufferers, accounting for 70% of the global cases of the disease (UNAIDS, 2003).

In such circumstances, countries and individuals have little capacity and few resources to resist and respond to the normal patterns of climatic fluctuation, much less the severe extremes. A quite modest event may result in devastation and even further decline. When drought occurs, the impact often lingers and is widespread and severe. Floods have been very destructive.

The growth in vulnerability is occurring in both developing and developed countries, through the concentration of growing populations in hazardous situations, the stripping of environmental capacities to withstand hazards, and the creation of social and economic vulnerabilities from migration, urban development, and economic growth. To some extent, the growing exposure to hazards is an outcome of development activity. Through myriad decisions being made every day at local, national and international levels, the risk burdens of countries are being modified and inadvertently compounded. The combination of development and poverty is leading to the accumulation of large amounts of latent risk, which can be subsequently exposed by hazard events, often with unexpectedly high levels of impacts. Disasters are thus a manifestation of *unsustainability*.

The specific circumstances of particular disaster events can be complex. Holloway (2003) points out that the famine conditions in southern Africa in 2002, while exacerbated by rainfall deficiency, were largely due to socio-economic factors, specifically widespread disruptions to food availability, failure of governance and extreme levels of prevailing poverty. Government policies in Zimbabwe were a critical factor in the substantially reduced regional food supplies. The severity of the crisis, at both household and macro-economic levels, was also partly attributable to the impacts of HIV/AIDS. Fourteen million people across several countries were in need of food aid in June 2002. The event was regional not only in its impacts but also in its causes. Clearly such events can only be avoided and managed through regional-scale approaches.

An analysis by Devereux (2002) of the famine in Malawi pointed to a complex set of factors involving misreading of food supply information, the sale of the national Strategic Grain Reserve, logistics problems, market inadequacies and possible profiteering, the neglect and depletion of rural and social capital, and the pervasive effects of growing poverty and HIV/AIDS. The climate hazard involved is barely mentioned by Devereux; in fact the rainfall deficiency was less marked than the 1992–1993 El Niño period when sizeable reductions of regional food production did not lead to famine.

Holloway (2003) argues that efforts to reduce disaster vulnerability in the region have not received priority, despite the long-understood linkages between disaster risks and development, partly because generous international relief provided over the years has resulted in an 'externalization' of responsibility for food insecurity. Holloway also argues that disaster risk reduction is handicapped by the mismatch in perceptions and organizational responsibility between the humanitarian disaster management community, which focuses on an operational, event-based preparedness and response cycle, and the development community, which takes a long-term, process oriented, poverty-focused view.

This is equally true at international levels and in bilateral relationships, where donor policies and practices generally continue this traditional separation and the related under-investment in development-oriented risk reduction. 'While international best practice in disaster reduction repeatedly underlines the need to reduce chronic risk factors, international humanitarian support continues to follow the established "African" pattern – that is, provide food relief.' (Holloway, 2003).

The frustration and dissatisfaction reflected in the above two references is understandable and merited. Nevertheless the recognition of the linkage of disasters to development in Africa is growing and there is increasing action at national, regional and international levels to address the many problems and to introduce risk reduction activities, as we shall see in later sections.

25.5 DISASTER RISK REDUCTION

Disaster risk reduction, also known as *disaster risk management*, focuses on the areas known to be critical to risk amelioration, which can be grouped into three key areas: (i) assessment of the risk factors present; (ii) tools and practices to reduce the risks; and (iii) institutional mechanisms to support both risk assessment and risk reduction.

Risk assessment concerns the identification and analysis of hazards (natural and otherwise), and the analysis of the environmental changes and degradation and the socio-economic vulnerabilities that exacerbate risk. It also includes the integration and compounding of multiple risks, and the analysis and summary of the spatial patterns and the temporal trends in hazards and other risk factors. Risk information is needed in readily usable forms, both in hardcopy maps and geographically referenced digital data.

Tools and practices to reduce risk encompass event management, context management and capacity development, as identified below. Public involvement is essential in these activities.

Event management includes the ongoing monitoring of all elements of risks (not just the hazards), early warning of evolving risk situations and of hazard events, public awareness and response capabilities, and event preparedness and management capacities (including flood and fire control).

Context management includes improved management of environmental and natural resources, improved social and livelihood circumstances, land-use zoning and management, 'land care' activity (conservation of soil, water, vegetation, etc.), the design and protection of structures and infrastructure, risk sharing mechanisms and financial instruments, and the assessment and control of the potential risk impacts of proposed development projects.

Underpinning activities include the development of information resources, networks and communication, public awareness raising, capacities for research on risks and risk reduction methods, education on risk reduction at all age levels, and the evaluation of the effectiveness of policies and practices.

Institutional mechanisms are essential for effective disaster risk reduction. Risk assessment and risk reduction practices need to be strongly supported by sound administration, law and political processes. This includes the following aspects, arranged in order from more tangible to less tangible items: organizational structures (departments, consultative bodies, etc.) and professional staff resources; incorporation of risk reduction into existing and new legislation; implementation of laws via codes, standards, documents, workshops, accountability, enforcement and evaluation; integration across sectors and government departments; and political recognition and financial commitment.

Disaster risk management thus spans a wide range of methods and activities – assessment and analysis, mapping and data analysis, public information, community participation, early warning systems, policy and regulation, project impact assessment, education programmes, conservation practices, and political processes. Political recognition is especially important.

To capitalize on the mutual benefits of the various activities, they are best pursued in a systematic integrated fashion. This can be termed *integrated disaster risk management*. The concept of *total disaster risk management*, developed in Asia, addresses the same goals.

While there are general approaches to risk reduction, the specific approaches must be tailored to local circumstances. Typically, the risk reduction activities will not be done as stand-alone projects, but will be implemented as integral components of other programmes, such as in specific development projects, water resources management, planning and land-use policies, environmental protection, and community development.

The global review of disaster risk reduction initiatives, *Living with Risk* (ISDR, 2002a), contains a rich resource of information with examples of effective risk reduction activities reported from countries around the world, some of which are applicable to Africa. The report of the ISDR Ad Hoc Discussion Group on Drought (ISDR, 2003a) is another useful resource. It provides a review of critical issues on drought, a set of recommendations for concrete actions, and extensive lists of relevant references and web sites. A review of the use of seasonal climate forecasts in disaster risk management is given in Dilley (2002).

25.6 INSTITUTIONAL DEVELOPMENT OF RISK REDUCTION IN AFRICA

The prevalence and persistence of climate-related disasters and their constraints on sustainable development have commanded the attention of many African leaders, and over the years have led to a number of national and regional initiatives.

Disaster reduction was identified as a priority area of action, and crucial for both economic and sustainable national development, at annual meetings of the African Ministerial Conference on the Environment held in conjunction with the preparation of the Johannesburg World Summit on Sustainable Development (WSSD) in 2002. The recommendations of the WSSD further amplified these concerns and the intimate relationship between the consequences of disasters and national development. Several of the Millennium Development Goals will require advances in disaster risk management if they are to succeed.

In the past, major regional-scale droughts have often motivated new institutional commitments to risk management. Extended droughts in the Sudan-Sahelian region in 1970–1974 led directly to the formation of the Permanent Interstate Committee for Drought Control in the Sahel (CILSS) in West Africa, in 1974. The 1984–1985 drought in the Horn of Africa that caused acute food insecurity and famine and affected more than eight million people directly influenced the creation of the Intergovernmental Authority on Drought and Development (IGADD) in 1986. Its successor, the Intergovernmental Authority on Development (IGAD), was formed mainly to address the persistent issues of drought, desertification and food insecurity. In southern Africa, political action for disaster management has also arisen from the protracted ravages of drought and disruptions of livelihoods, often in association with other emergencies. The Southern African Development Community (SADC) and its predecessor the Southern African Development Coordination Conference (SADCC) have identified food security as a priority area for strategic regional cooperation.

Regional specialized technical institutions also have been created to address the consequences of climatic hazards, often with the support of UN agencies such as the World Meteorological Organization (WMO), Food and Agriculture Organization (FAO), United Nations Environment Programme (UNEP), and United Nations Development Programme (UNDP). These include the Africa Centre for Meteorological Applications for Development (ACMAD) in Niamey, Niger; the Regional Centre for Training and Application in Agrometeorology and Operational Hydrology (AGRHYMET) also in Niger and linked to CILSS; the Drought Monitoring Centre at Harare, Zimbabwe (DMC-Harare), linked

to SADC; and the Drought Monitoring Centre at Nairobi, Kenya (DMC-Nairobi), linked to IGAD.

Progressive shifts are taking place to revise and expand the roles of IGAD and SADC to identify and manage risks on a regional basis, with initiatives to strengthen operational cooperation and the exchange of information among countries, and to adopt broader political and technical commitments to development and risk management policies. Challenges remain though to move national policies beyond earlier ad hoc, under-resourced and often un-coordinated emergency relief assistance functions.

Under this wider mandate of IGAD, disaster management has been highlighted and placed under a revised humanitarian and conflict resolution department. IGAD has explicitly reviewed the capabilities needed to ensure the availability of minimum needs for food, water, shelter, health and security, and to identify the steps that members and the region as a whole needed to take. These included frameworks of principles, policies, legislation and agreements, the roles of national, regional and international agencies, vulnerability assessments, early warning mechanisms, community awareness and capacities, and effective response mechanisms at times of stress. National officials have ranked hazards in the region into three categories based on a combination of past consequences and taking account of other factors, such as the pace of hazard onset, the potential magnitudes or severities that could be expected, the frequency of occurrence, and most importantly the possible impacts on society and the environment. Potential risks include pandemic and epidemic diseases, environmental hazards and industrial accidents.

IGAD's early warning systems encompass the social and economic factors in risk, including the provision of market and food prices on the Internet, reports about food production prospects and requirements, and assessments of land conditions and seasonal crop production using satellite and geographical information system (GIS) technologies. DMC-Nairobi has become an integral part of IGAD's subregional early warning system. International efforts are also important, particularly the Famine Early Warning System Network (FEWSNET) supported by the US Agency for International Development (USAID); the Global International Early Warning System (GIEWS) supported by FAO; the Vulnerability Assessment and Monitoring (VAM) programme of the UN World Food Programme (WFP); and the field assessments of food prospects and needs by the UN Office for the Coordination of Humanitarian Affairs (OCHA).

SADC's overall purpose is to foster the economic integration and the promotion of peace and security among its 14 member countries. It devotes considerable attention to issues of public vulnerability, covering potential disaster threats from climatic hazards, conditions of poverty, and disease, and has made disaster management a regional priority. An Ad-Hoc Working Group on Disaster Management was established in 1999, and an

Extraordinary Summit of SADC Heads of State and Government was convened in Maputo, Mozambique, in March 2000 to review the impacts caused by the floods that had just affected the region. At this summit, representatives of the SADC countries expressed the need for improved institutional arrangements for disaster preparedness and management of similar risks in the future. Some of the governments in the region have begun amending legislation to place greater emphasis on natural disaster risk.

In May 2000, the SADC Sub-Sectoral Committee on Meteorology, comprising the directors of the National Meteorological and Hydrological Services (NMHS), recommended that a regional project be formulated to address and strengthen the local capacities of NMHS for early warning and disaster preparedness. A month later, the SADC Committee of Ministers for Water recommended that a strategic and coordinated approach be developed to manage floods and droughts within the region. By the end of 2001, SADC had developed and approved a multi-sector disaster management strategy for the region and had drafted a Strategy for Floods and Drought Management. Spurred on by the severity of the floods earlier in the year and the inadequate disaster management linkages that the floods revealed among the nine countries affected, the SADC Council of Ministers in August 2002 approved an overarching SADC Disaster Management Framework. This incorporated an integrated regional approach and established a full Technical Steering Committee on Disaster Management.

SADC's technical units play critical roles in disaster risk reduction. The SADC Food, Agriculture and Natural Resources (FANR) directorate oversees regional food security issues and several other programmes related to the management of natural resources. Its Food Security Programme and related Regional Early Warning Programme provide member states and the international community with advance information on food security prospects in the region. These include food supply and demand assessments, projections on food imports and exports, the areas and populations threatened by food insecurity, and outlooks of threatening climate conditions that could trigger food insecurity. The activities are integrally supported by the SADC Regional Remote Sensing Unit, based in Harare, which monitors and maps land-use patterns, land degradation and desertification conditions, and trains agro-meteorologists in the use of satellite imagery products and GIS. Input is also provided by the SADC Drought Monitoring Centre (DMC-Harare), which collaborates with NMHS and international climate centres to provide seasonal climate outlooks for the region. As is the case in other parts of Africa, these activities are supported by many international organizations, such as FAO/GIEWS, FEWSNET, WMO, United States National Oceanic and Atmospheric Administration (NOAA), United States Geological Survey (USGS), and the International Research Institute for Climate Prediction (IRI).

The water sector in southern Africa has long given attention to the development of cooperative agreements on shared river basins. The need for interstate cooperation is particularly acute as there are more than ten shared watercourses in the region, with the largest, the Zambezi River, flowing through nine different countries. Initiatives on flood management now include regular regional consultations, a 50-station real-time hydrological monitoring system, international collaboration on modelling and forecasting river levels and flows and the production of more detailed flood risk maps. A Joint Operations Technical Committee for the Zambezi River oversees weekly exchanges of data and monthly meetings of partners during the critical rainy season.

Other SADC-led or influenced projects include a number related to land-use practices, the conservation of environmental factors that help reduce both flood and drought-prone conditions, and the risks posed by climate change. Collaboration with the World Health Organization (WHO) and climate-health researchers has led to the development of an experimental climate-based early warning scheme for malaria incidence in Botswana.

In West Africa, the 16-country Economic Community of West African States (ECOWAS) promotes cooperation and integration leading toward the goal of an economic union in West Africa, in addition to efforts to support conflict management in the region. ECOWAS programmes include a regional meteorological and water resource management programme, and subregional programmes for desertification control, but there is no designated subregional activity on natural disaster risk reduction or consolidated regional strategy on risk management.

However, the subregional programmes for desertification control do provide disaster risk reduction and risk management functions. These, together with the meteorological information programme, offer possibilities for the development of future subregional disaster risk reduction initiatives and for their integration into poverty reduction, environmental protection and sustainable development strategies. The subregional technical institutions ACMAD and AGRHYMET provide a basis for the engagement of scientific and technical inputs into the process. Their activities fulfil roles similar to those provided by the drought monitoring centres in East Africa and southern Africa. There remain further opportunities to link these and other institutional and technical capabilities for a more structured regional approach to monitoring hazards and pursuing disaster risk reduction in West Africa.

A number of African universities support research centres that specialize in disaster risk management and reduction, or in associated subject areas such as food security, human vulnerability, hydrology and climate. These centres provide a valuable resource of expertise and educational capacity for support of regional and national institutions.

A vital link in the dissemination of early warning information in Africa since 1997 has been the Regional Climate Outlook Forum

(RCOF). Typically of a few days duration, the forums are held once or twice a year, usually at times linked to the agricultural calendar, to review the latest climate information and to formulate a consensus seasonal forecast for the region. They also play an important role in facilitating dialogue among climate experts and representatives of climate-sensitive sectors and policy groups. Participants include water managers, health officers, farmers, food security experts and managers, journalists, seed company staff, agricultural extension workers, and disaster managers.

The forums are organized by ACMAD in West Africa, DMC-Nairobi in the Greater Horn, and DMC-Harare in southern Africa, with considerable technical and financial support from WMO, NOAA, IRI, World Bank, USAID and other international partners. They are normally hosted by a national meteorological service, and often are accompanied by workshops or training events for meteorologists and sectoral groups, including journalists.

In 2000, an international review of RCOF activities worldwide was jointly initiated by WMO, NOAA, IRI and the World Bank (Basher et al., 2001). The main aims were to identify how the forum outputs could be improved in both accuracy and relevance to users of information, and how forum processes could be made more sustainable overall. Regional forum participants and international experts contributed to a detailed peer-reviewed preparatory report and then met at an international workshop in Pretoria, hosted by the South African Weather Bureau. The review concluded that the forums had achieved a great deal and enjoyed strong user support. However, significant improvements were needed in forecasting methodology, linkages to user groups, and financial sustainability. It was agreed that there is a clear need to systematically develop and institutionalize the forum process. The meeting's Africa Working Group reinforced these points and set out a five-year plan of action for addressing the issues in Africa.

25.7 INTERNATIONAL STRATEGY FOR DISASTER REDUCTION

As already noted, disaster risk reduction increasingly is being recognized as of fundamental importance to sustainable development. The International Strategy for Disaster Reduction (ISDR) was established by United Nations General Assembly Resolutions 54/219 (December 1999) and 56/195 (January 2002) and is the centrepiece of the United Nations' efforts to address the challenge of disasters, principally through advocacy, coordination, and information dissemination (ISDR, 2003b; also see http://www.unisdr.org). The ISDR developed out of the earlier 1990–1999 International Decade for Natural Disaster Reduction (IDNDR) and is still in an early stage of evolution.

UN agencies and governments are increasingly using the ISDR as a primary international vehicle to develop and guide commitments and action. The strategy is implemented through the work of numerous international actors guided by the Inter-Agency Task Force on Disaster Reduction (IATF/DR), which meets twice-yearly, and the inter-agency ISDR Secretariat. The Secretariat is not supported by the regular UN budget and relies totally on voluntary donations from a number of governments.

A major achievement of the ISDR was to secure high visibility for disaster risk reduction at the WSSD. This was fostered through a background paper that was successively evolved in a participatory manner at the WSSD's regional preparatory meetings (ISDR, 2002b). The WSSD subsequently adopted much of the paper's agenda, thereby reinforcing international awareness of the need for risk and vulnerability reduction in order to secure sustainable development. However, the implication that development sectors should channel more investment into disaster risk reduction presents a new challenge for governments and donors, since at present most resources for disaster risk reduction are drawn predominantly from the humanitarian sector, which usually has minimal engagement in development decisions and poverty reduction strategies.

Within the UN system, numerous other initiatives are strengthening country capacities to reduce disaster risk and better manage risk, through programmes in UNDP, WMO, UNEP, WHO, FAO, UNESCO, World Bank and OCHA, for example. Civil society organizations are also very active, and include the International Federation of Red Cross and Red Crescent Societies (IFRC), and the ProVention Consortium. The latter is a project-oriented coalition of governments, international organizations, academic institutions, the private sector and civil society organizations working toward disaster risk reduction.

While the ISDR Secretariat coordinates international strategic directions, UNDP has the primary responsibility for national capacity-building in disaster risk reduction and for supporting humanitarian coordination at a national level through the UN Resident Coordinator system. An ISDR initiative is currently under way to develop national platforms (i.e. organizational coordination mechanisms) for disaster risk reduction. This project, which comprises advocacy, capacity-building, institutional strengthening and strategy development, is being undertaken mainly through the efforts of UNDP's Bureau for Crisis Prevention and Recovery and UNDP country teams and regional advisers on disaster reduction. Another UNDP group, the Nairobi-based Drylands Development Centre, is active in addressing the vulnerability of poor people to climatic shocks, particularly droughts and floods.

A major UNDP report concerning world vulnerability, due to be published late in 2003, highlights contemporary trends in the evolution of disaster risk and vulnerability patterns and advocates relevant policies and strategies for reducing disaster risk. This report is complementary to the ISDR's flagship publication *Living with Risk* (ISDR, 2002a). It is planned that further issues of these two reports will be combined in a single publication starting in 2004.

Partners of the ISDR have developed a number of specific weather and climate-related activities that are relevant to Africa. These include El Niño advisories issued by WMO and prepared as a collaborative effort between WMO and the IRI as a contribution to the ISDR (http://www.wmo.ch/index-en.html), and the ad hoc discussion group on drought referred to above. Another is the statement drafted by the ISDR Secretariat in response to the UN General Assembly's request (via Decision 57/547) to the UN Secretary General to report on 'the negative impacts of extreme weather events and associated natural disasters on vulnerable countries, in particular developing countries.'

In the climate change policy arena, the ISDR Secretariat and partners are working to promote the use of disaster risk reduction as a readily implemented component of climate change adaptation strategies. This includes the strategies of major donors as well as those associated with the UNFCCC, such as the National Adaptation Plans of Action (UNFCCC, 2002). At present only a few of the National Communications prepared under the Convention processes have any significant mention of disaster risk reduction activities. The ISDR Secretariat is advising on potential risk reduction elements to be included in a manual being prepared by the UNFCCC Secretariat to guide the preparation of national communications of Non-Annex 1 countries.

An initiative to prepare an authoritative, multi-stakeholder report on the topic of disaster risk reduction and climate change was launched in June 2003 at a side event to the meetings of the UNFCCC Subsidiary Bodies in Bonn. The side event was co-sponsored by the ISDR Secretariat, German Technical Cooperation (GTZ) and the International Centre for Climate Change and Disaster Preparedness, a centre established by the Netherlands Red Cross in 2002 in cooperation with the IFRC. An information product *DR + CC Infolink* has been initiated by a partnership between the ISDR Secretariat, the Red Cross Climate Centre and UNDP to stimulate cooperation and information exchange between the disaster risk reduction and climate change communities.

Steps are being taken to promote the involvement of disaster risk reduction experts in the next IPCC assessment process, which is about to start and will be completed in 2007. Work is also continuing under the ISDR on technical matters, such as the development of better databases on hazards, risks, vulnerabilities and disasters.

The ISDR Secretariat launched an African outreach programme in October 2002, hosted at UNEP offices in Nairobi. The initial goals are to develop enhanced awareness, networking and information flows. The office undertakes advocacy on the links between disaster risk reduction and other key issues, such as environmental protection, climate change, poverty alleviation and sustainable development, as well as promoting the development of practical strategies for disaster risk reduction. Information and knowledge exchange is achieved through ISDR Africa's website, http://www.unisdrafrica.org, a biannual publication (ISDR, 2003c), and through regional networking. An educational series for public awareness has started jointly with DMC-Nairobi.

ISDR Africa interacts with ECOWAS, IGAD and SADC and other key entities, such as the Economic Community of Central African States, the Common Market for Eastern and South Africa, the League of Arab States, the Southern African Research and Documentation Centre, the New Partnership for Africa's Development (NEPAD) and the African Union Secretariat. Initiatives also have been taken to work with women in Africa to study early warning practices from a gender perspective and to organize a regional conference on gender issues on disaster reduction and sustainable development.

Workshops to develop and enhance national platforms for disaster risk reduction were organized by ISDR and UNDP in Djibouti, Uganda and Madagascar in 2003. A regional consultation meeting on early warning was held in Nairobi in June 2003 to identify regional needs concerning the integration of warning in disaster risk policy and management and to prepare a collective African input for the Second International Conference on Early Warning (EWC-II) held in Bonn, October 2003.

The proposals of the ISDR ad hoc discussion group on drought (ISDR, 2003a) are very relevant to Africa. These concern the development of regional networks for drought risk reduction, the exchange of information and experience, strengthened collaboration on drought-related activities under the UNFCCC and the UN Convention to Combat Desertification (UNCCD), and improving the cooperation among existing drought early warning systems.

25.8 CONCLUSION

The climate is a lively, changing system and its extreme weather and climate events are natural features of the earth that human society must continue to adapt to, now and in the future. Africa has suffered badly from floods and droughts over the last few decades. Mind-numbing numbers of people have been killed or affected, largely because of the region's severe underlying vulnerabilities, arising from poverty, HIV/AIDS impacts, and not-infrequent failures of governance.

So far, only minimal changes have been observed in the climatic conditions that are most relevant to disasters, and the projected changes in these conditions from increased greenhouse gases for the near-term future are also relatively small compared to natural variability. At the same time, the uncertainties in the projections are large and climate change remains a very serious long-term threat. In responding to the climate change issue, it will be important to put high priority on steps that address the fundamental vulnerabilities and that build resilience to existing climatic fluctuations. Given the regional nature of climatic disasters, this will require a regionally coordinated approach.

There is a growing awareness in Africa of the linkages between development and disaster risk reduction. There are some well-established regional institutions that are playing crucial roles in early warning services and policy formation. Internationally, there is a rapid development of thinking and action on the issue of disaster risk reduction that Africa can not only tap into but also materially contribute to. The International Strategy for Disaster Reduction provides an important vehicle for facilitating this necessary exchange.

ACKNOWLEDGEMENTS

The authors wish to thank the staff of the ISDR Secretariat for their assistance in identifying and developing the many materials upon which this paper is based, in particular Feng Min Kan, Regional Coordinator of UNISDR Africa for information on African initiatives, Haris Sanahuja for preparing the disaster statistics, Terry Jeggle for expert review and inputs, and referees David Hastings, Ti Le-Huu and Stjepan Keckes for their valuable comments. The authors also wish to warmly acknowledge the many national, regional and international ISDR partners upon whose continued commitments and efforts the ISDR is built.

REFERENCES

Adger, W. N., Khan, S. R. and Brooks, N. (2003) *Measuring and enhancing adaptive capacity.* UNDP Adaptation Policy Framework Paper 7, UNDP, New York, USA.

AfDB and nine other organizations (2003) *Poverty and Climate Change: Reducing the Vulnerability of the Poor through Adaptation.* Report produced jointly by African Development Bank (AfDB); Asian Development Bank; Department for International Development, UK; Directorate-General for Development, European Commission; Federal Ministry for Economic Cooperation and Development, Germany; Ministry of Foreign Affairs, The Netherlands; Organization for Economic Cooperation and Development; United Nations Development Programme; the World Bank. 42pp.

Basher, R., Clark, C., Dilley, M. and Harrison, M. (2001) *Coping with the Climate: A Way Forward Summary and Proposals for Action. A multi-stakeholder review of Regional Climate Outlook Forums concluded at an international workshop, October 16–20, 2000, Pretoria, South Africa.* International Research Institute for Climate Prediction, Palisades, New York, USA. 28pp.

Bruce, J. P. (1999) Disaster loss mitigation as an adaptation. *Mitigation and Adaptation Strategies for Global Change,* **4**, 295–306.

Cornford, S. G. (2003) The socio-economic impacts of weather events in 2002. *WMO Bulletin,* **52** (3), 269–290.

Desanker, P., Magadza, C. and 28 other authors (2001) Africa. In IPCC (2001a), *Climate Change 2001: Impacts, Adaptation, and Vulnerability,* Chapter 10, pp. 487–531.

Devereux, S. (2002). *State of Disaster: Causes, Consequences and Policy Lessons from Malawi.* ActionAid report commissioned by ActionAid Malawi, 33pp.

Dilley, M. (2002) The use of climate information and seasonal prediction to prevent disasters, *WMO Bulletin,* **51** (1), 42–48.

Goddard, L., Mason, S. J., Zebiak, S. E., Ropelewski, C. F., Basher, R. and Cane, M. A. (2001) Current approaches to seasonal-to-interannual climate predictions. *Int. J. Climatol.,* **21**, 1111–1152.

Holloway, A. (2003) Disaster risk reduction in Southern Africa: hot rhetoric–cold reality. *African Security Rev.,* **12** (1), 29–38.

Hulme, M., Doherty, R., Ngara, T. and New, M. (2005) Global warming and African climate change: a re-assessment. This volume.

IDNDR (1992) Internationally agreed glossary of basic terms related to disaster management, in English, French, and Spanish. (Also in ISDR (2002a); and at website http://www.unisdr.org)

IFRC (2002) *World Disasters Report, 2002, Focus on Reducing Risk.* International Federation of Red Cross and Red Crescent Societies, Geneva, Switzerland, 240pp.

IPCC (2001a) *Climate Change 2001: The Scientific Basis.* Contributions of Working Group I to the Third Assessment Report of the Intergovernmental Panel on Climate Change. J. T. Houghton, Y. Ding, D. J. Griggs, *et al.* (Eds.). Cambridge and New York, Cambridge University Press, 881pp.

(2001b) *Climate Change 2001: Impacts, Adaptation, and Vulnerability.* Contributions of Working Group II to the Third Assessment Report of the Intergovernmental Panel on Climate Change. J. J. McCarthy, O. F. Canziani, N. A. Leary, D. J. Dokken, and K. S. White (Eds.). Cambridge and New York, Cambridge University Press, 1032pp.

ISDR (2002a) *Living with Risk, A Global Review of Disaster Reduction Initiatives (preliminary version, July 2002).* UN Secretariat of the International Strategy for Disaster Reduction, Geneva, 382pp.

(2002b) *Disaster Reduction and Sustainable Development: Understanding the Links Between Vulnerability and Risk to Disasters Related to Development and Environment.* Background paper developed for the World Summit on Sustainable Development, Johannesburg, 26 August–4 September 2002. UN Secretariat of the International Strategy for Disaster Reduction, Geneva, Switzerland, 24pp.

(2003a) *Drought: Living With Risk: An Integrated Approach to Reducing Societal Vulnerability to Drought.* Report of the ISDR Ad Hoc Discussion Group on Drought. UN Secretariat of the International Strategy for Disaster Reduction, Geneva, Switzerland, 41pp.

(2003b) *United Nations documents related to disaster reduction, Volume 2, 2000–2002.* UN Secretariat of the International Strategy for Disaster Reduction, Geneva, Switzerland, 452pp.

(2003c) *Disaster Reduction in Africa – ISDR Informs.* UN Secretariat of the International Strategy for Disaster Reduction, Nairobi, Kenya, 55pp.

Munich Re (2003) Annual Review: Natural Catastrophes 2002. In *Topics*, Jan. 2003, Munich Reinsurance, Munich, Germany, 48pp.

Swiss Re (2003) *Opportunities and Risks of Climate Change*. Swiss Reinsurance Company, Zurich, Switzerland, 28p.

UNAIDS (2003) *AIDS Epidemic Update 2002*. Joint United Nations Programme on HIV/AIDS (UNAIDS) and World Health Organization (WHO), 42pp.

UNDP (2003) *A Climate Risk Management Approach to Disaster Reduction and Adaptation to Climate Change*. Report of UNDP Expert Group Meeting on Integrating Disaster Reduction with Adaptation to Climate Change, Havana, 19–21 June, 2002.

UNFCCC (2002) *Decision 28/CP.7, Guidelines for the Preparation of National Adaptation Programmes of Action* (see http://unfccc.int/text/program/sd/ldc/documents/13a04p7.pdf)

PART IV

Capacity-building

26 Climate change mitigation analysis in southern African countries: capacity enhancement in Botswana, Tanzania and Zambia*

GORDON A. MACKENZIE

UNEP Risoe Centre on Energy, Climate and Sustainable Development, Roskilde, Denmark

Keywords

Climate change mitigation; greenhouse gas abatement; energy; forestry; capacity-building; institutional arrangements; Botswana; Tanzania; Zambia

Abstract

The climate change mitigation studies in Botswana, Tanzania and Zambia shared an institutional set-up in which a non-governmental centre, with close ties to the responsible government ministry, carried out the analytical work. The studies were carried out within the broader context of a methodological development project with a parallel worldwide country-study programme. Each study thus had a threefold purpose: to produce a country report; to establish or enhance national capacity for climate change mitigation analysis; and to contribute to the methodological development process.

This chapter summarizes and reviews the three country studies, discussing the institutional arrangements, the analytical methods used and the specific results obtained. The Botswana study concentrated on energy sector options and introduced a simple method to rank options on the basis of national development priorities. The Tanzania study widened the scope of that country's mitigation analysis to include forestry and land-use options. In Zambia, the analysis team focused on household energy and found a large potential for negative-cost mitigation options.

While the institutional set-up may not be immediately relevant for all country settings, the arrangement presents a number of advantages in terms of effectiveness, motivation and flexibility, which may be applicable elsewhere in Africa and indeed in other developing regions.

26.1 INTRODUCTION

Climate change mitigation country studies have been carried out since the beginning of the 1990s, in parallel with the development of methodology for mitigation analysis. One of the first 'rounds' of studies was that associated with the UNEP *Greenhouse Gas (GHG) Abatement Costing Studies*. This included a study of Zimbabwe, carried out in collaboration between a Zimbabwean government institution (the Department of Energy), a local non-governmental centre (NGC) (the Southern Centre for Energy and Environment – SCEE) and a 'Northern' counterpart institution (the UNEP Collaborating Centre on Energy and Environment – UCCEE).

This institutional structure, which proved successful in Zimbabwe, both in terms of producing a high-quality study report and in terms of capacity-building, has been emulated in other countries. Notable examples are the country studies of Botswana, Tanzania and Zambia, again associated with UNEP, carried out in parallel with country studies in other developing regions.

Capacity-building is a complex process that cannot be seen in isolation from the institutional, political and human-resource context in which it takes place (Cissé *et al.*, 1999). Moreover, each capacity-building activity often builds on earlier studies, already established institutions and trained professional staff, as was the case with the country studies described here. The challenge lies in engendering and encouraging institutional relationships, at the same time as assisting personnel to enhance their expertise in specific analytical techniques.

This article presents some of the results of the country studies of Botswana, Tanzania and Zambia, and discusses them with particular focus on the capacity-building process, the relationship with methodological development and the establishment and enhancement of a network of country teams within the Southern African region and beyond.

*This paper was finalized in 2001 when the author was seconded to Department of Energy, Ministry of Natural Resources, Maseru, Lesotho, working as Chief Technical Adviser for the DANCED project *Energy Management in Lesotho* under the employment of the Danish consulting firm RAMBØLL.

26.2 HISTORICAL PERSPECTIVE

The methodology for climate change mitigation analysis was developed through the 1990s and applied in numerous country study programmes. The first sector to be analyzed was energy, mainly because energy is the major contributor to GHG emissions in the form of carbon dioxide (CO_2) from fossil fuel combustion. The energy sector was also readily analyzed, well-developed analytical tools already existed, and moreover most of the analysts working on climate change mitigation analysis had a background in energy modelling and planning. The methodology used was thus based on well-developed analytical methods and tools from energy planning.

Methodological development at the UCCEE was initiated in 1991 following a decision of UNEP's Governing Council. A primary aim in the initial studies was to investigate and explain the differences in both emission-reduction potential and cost resulting from the two main types of analysis: the bottom-up or techno-economic approach and the top-down or macroeconomic approach. In recent years there has been less focus on the difference between the top-down and bottom-up approaches, with aspects of both approaches often being applied in studies whenever appropriate models are available. Methodological development has concentrated on clearly defining the cost concepts involved, and extending the analysis to sectors other than energy. The guidelines (Halsnæs et al., 1999) for mitigation analysis developed by the UCCEE discuss these issues in detail.

The methodology of mitigation analysis has been refined significantly since the first efforts, for example in the framework of the UNEP/GEF project *Economics of GHG Limitations*. That project extended the methodology beyond the energy sector, particularly to agriculture and forestry, and also developed further the concepts of cost (Christensen et al., 1998) as applied to mitigation. An essential ingredient of the UNEP studies from the outset was the objective of a comparable methodology for mitigation assessment. The concept of comparability evolved from an initial aim to compare results from different countries, into a more complex notion, involving common cost concepts and transparency of assumptions. The recognition of the importance of scenario definition, and in particular the baseline, combined with the national specificity of data and models, led to this modification of the idea of comparability (Halsnæs et al., 1994). While it can still be useful to compare results from different countries and indeed with the regional mitigation options (Mackenzie et al., 1998), this has to be done with caution.

In parallel with the methodological development and closely coordinated with it, there have been programmes of country studies, applying the methodology, producing country study reports, building institutional and personnel capacity, and feeding back into the methodological development. The UNEP programmes, coordinated by the UCCEE, have been instrumental in this development: the UNEP *GHG Abatement Costing Study (Phases One and Two)*, from 1992 to 1994, with 10 associated country studies, including Zimbabwe (Maya et al., 1992, 1993); the UNEP/GEF *Economics of GHG Limitations* and associated studies funded by Danida and UNDP/GEF, from 1996 to 1998, including Botswana (Zhou, 1999), Mauritius, Senegal, Tanzania (Mwandosya et al., 1999), Zambia (Yamba et al., 1999) and a regional study of the Southern African Development Community (SADC) region (Rowlands, 1998). Results for Africa were presented at a regional conference in May 1998 (Mackenzie et al., 1999).

Other country study programmes have, of course, been carried out within the same period, notably the German (GTZ) programme, the US Country Studies, and subsequently the GEF Enabling Activities aimed at assisting countries to prepare national communications to the United Nations Framework Convention on Climate Change (UNFCCC). There has been close coordination between the responsible institutions, and in many cases direct collaboration or follow-up on specific country studies. This coordination has helped to avoid overlaps and has exploited synergies. One of these programmes is discussed elsewhere in this book (see Dixon et al., this volume), and some have also been described earlier in the context of the broader capacity-building effort (Dixon, 1997; Liptow, 1997).

The institutional set-up, consisting of a local NGC closely associated with government through a Memorandum of Understanding (MOU), was first established in the Zimbabwe country study of the first UNEP series. The country study itself was funded by Danish International Development Assistance (Danida), which supported the Southern Centre for Energy and Environment in Harare in carrying out the analytical work on behalf of the Zimbabwe Department of Energy. This institutional model was subsequently applied in the country studies of Botswana, Tanzania and Zambia described in the following sections.

26.3 BOTSWANA – ACCOUNTING FOR NATIONAL PRIORITIES IN OPTION RANKING

The mitigation study of Botswana was carried out within the Danida project *Climate Change Mitigation in Southern Africa*, which also supported studies of Tanzania and Zambia, and which ran parallel with the UNEP/GEF project *Economics of Greenhouse Gas Limitations* coordinated by UCCEE. The Botswana study focused on the energy sector and analyzed the baseline economic, energy development and greenhouse gas emission scenarios, and the costing of plausible greenhouse gas abatement options in the energy sector of Botswana. EECG Consultants, led by Dr Peter Zhou, on behalf of the Ministry of Mineral Resources and Water Affairs, Botswana, carried out this first study of climate change mitigation in Botswana.

Table 26.1. CO_2 abatement options for Botswana in 2030, discount rate 10% (adapted from Zhou,1999).

Ranking number	Mitigation option	US$/tonne CO_2	Cumulative reduction CO_2 equivalent (million) tonnes)
1	Efficient lighting	−141.2	0.30
2	Paved roads	−113.7	0.60
3	Road freight to rail	−102.8	0.76
4	Zero tillage in agriculture	−93.7	0.77
5	Prepayment meters	−23.2	1.07
6	Geyser time switches	−17.6	1.50
7	Solar home systems	−14.2	1.50
8	Power factor correction	−9.9	1.54
9	Efficient boilers	−7.6	1.56
10	Petrol to diesel through differential pricing	0.0	1.56
11	Efficient motors	0.5	1.71
12	Vehicle inspection	0.7	2.02
13	Biogas from landfills	1.1	2.52
14	Solar water heaters	5.7	2.53
15	Biogas for rural households	11.5	2.67
16	Central PV electricity	17.8	2.88
17	Reforestation	71.3	2.96
18	Oil pipeline	199.7	3.05
19	Solar PV water pumps	202.9	3.11
20	Electrifying the railway line	663.1	3.17

The tools applied for the mitigation analysis were the Long-range Energy Alternatives Planning system (LEAP) (Heaps *et al.*, 1995) and the Greenhouse Gas Abatement Costing Model (GACMO) (Fenhann, 1999). LEAP was used to create an integrated model of the Botswana energy system and to produce the energy and GHG baseline scenarios for the time frame to 2005 and 2030.

Analysis of the baseline emissions showed that, in the short-term, the major contribution (59% of energy sector CO_2 emissions in 2005) is from power generation. In the longer term, emissions from the demand side (from fossil fuel combustion in domestic, agricultural, industrial and transport sectors) dominate. The distribution of CO_2 equivalent emissions in the baseline would suggest emphasizing GHG abatement in power generation in the short term and on the demand sectors in the long term. The selection of mitigation options, however, depends on the available opportunities in the sectors.

A portfolio of GHG abatement options was identified for the energy sector, and GACMO was used to calculate the abatement costs and the potential GHG reduction of these selected mitigation options. In the analysis it was ensured that the penetration rate of each option was a proportion of the total penetration capacity in the baseline produced by LEAP.

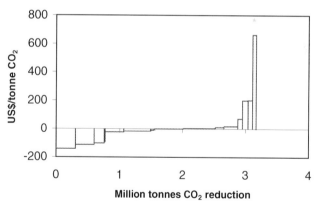

Figure 26.1. CO_2 abatement cost curve for Botswana in 2030, discount rate 10% (adapted from Zhou, 1999a).

The cost of GHG reduction and total reduction potential for each option were sequenced in a cost curve, shown in Figure 26.1. The cost curve is one way of presenting a GHG abatement strategy for a country. It is a graph showing a possible sequencing of mitigation actions based on the unit cost of reducing GHG emissions. Table 26.1 shows 20 mitigation options, nine of which are found to have negative cost. These relate to both electricity and transport-fuel savings. It should be remembered that these calculations

Table 26.2. Realignment of mitigation options based on national development priorities and macroeconomic impacts showing the ranking before and after taking into account these aspects (adapted from Zhou,1999).

Ranking (realigned)	Ranking (cost only)	Mitigation option	US$/tonne CO_2
1	12	Vehicle inspection	0.7
2	5	Prepayment meters	−23.2
	7	Solar home systems	−14.2
3	1	Efficient lighting	−141.2
	13	Landfill gas for power generation	1.1
4	2	Paved roads	−113.7
	15	Biogas for rural households	11.5
	19	Solar PV pumps	202.9
5	9	Efficient boilers	−7.6
	6	Geyser time switches	−17.6
	8	Power factor correction	−9.9
	11	Efficient motors	0.5
	17	Reforestation	71.3
6	14	Solar water heaters	5.7
7	4	Zero tillage in agriculture	−93.7
8	3	Road to rail freight	−102.8
	10	Petrol to diesel through differential	0.0
	16	Central PV plants	17.8
	18	Pipeline for petroleum products	199.7
	20	Electrifying the railway line	663.1

include only 'direct' costs of options. Costs involved in removing implementation barriers, etc. are not included. All these results thus require further study, especially with regard to implementation costs and other indirect costs involved in mitigation. For further discussion of negative costs, see the UNEP Mitigation Guidelines (Halsnæs et al., 1999). Indirect costs of mitigation are discussed in a study of Mauritius (Markandya et al., 1999).

On the basis of the cost curve, the mitigation strategy for Botswana would be expected to 'mop up' the opportunities in the household sector that have negative costs and low capital outlay, followed by the negative-cost transport and power-sector options, which require relatively higher capital outlay. The renewable energy options and the expensive transport options would probably be implemented in the long term when it becomes cost-effective to use them.

A meaningful GHG abatement strategy in developing countries, however, has to reflect national development policies and possible impacts on the economy. The Botswana study attempted to account for such aspects in a semi-quantitative way. Criteria evaluated included government policy regarding the specific mitigation measures, ease of implementability, impact on balance of payments, employment creation, social and health benefits, and non-climate environment improvement. Consideration was also

given where benefits accrue in another sector, such as deferred investment in additional power plants or enhancement of agricultural output. These issues were assessed using a scoring procedure that led to an alternative ranking of options, shown in Table 26.2. An important result of the study was thus the demonstration of a simple method for changing the ranking of options by accounting for implementation issues and estimated macroeconomic impacts.

26.4 TANZANIA – EXTENSION FROM ENERGY TO FORESTRY AND LAND USE

Tanzania provides an example of the coordinated interplay between different climate change studies, funded by different donors, avoiding overlapping and leading to capacity-building in the same institution. An initial mitigation study of Tanzania (CEEST, 1995) was carried out by the Centre for Energy, Environment, Science and Technology (CEEST) under the leadership of Professor Mark J. Mwandosya, with GTZ funding, and with substantive backstopping support from the UCCEE. This mitigation study built on the foundations of GHG inventory studies (CEEST, 1994) funded by UNEP and the US Country Studies Programme.

Land-use change and deforestation account for more than half of the anthropogenic GHG emissions from Tanzania. The scope of

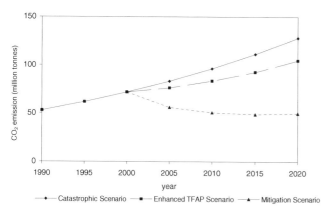

Figure 26.2. Comparison of CO_2 emissions from the three forestry scenarios for Tanzania (adapted from Mwandosya *et al.*, 1999).

the first mitigation study did not, however, allow sufficient analysis of forest and land-use sectors, the main concentration being on energy-sector mitigation options. The subsequent mitigation study, initiated in 1996 within the Danida project *Climate Change Mitigation in Southern Africa,* provided an opportunity to extend the analysis and focus on land-use and forestry options, especially the link between fuelwood and forestry. The study was carried out by CEEST on behalf of the Tanzanian Ministry of Energy and Minerals through a Memorandum of Understanding (MOU).

The analysis used the Comprehensive Mitigation Assessment Process (COMAP) model (Sathaye *et al.*, 1995) to study three scenarios relevant to forestry and land-use in Tanzania and to assess the climate change mitigation potential and cost of the various options. The scenarios comprised:

- A Catastrophic Scenario, in which severe deforestation takes place due to a hypothetical failure to follow forestry plans;
- An enhanced Tanzania Forest Action Plan (TFAP) Scenario, which embodies a number of measures within forestry;
- A Mitigation Scenario with additional measures, such as fuel substitution from biomass and more stringent protection of forest resources.

In the mitigation scenario, a combination of supply-side (sustainable forest management) and demand-side measures (use of forest resources including energy options) gradually reduces the rate of deterioration in the capacity to sequester carbon, leading to stabilization of the emission level. This analysis did not consider the impact of intensification and mechanization of agriculture, and therefore the status of clearing of forests for agriculture will remain as projected in the enhanced TFAP scenario. Figure 26.2 shows the comparison of the emissions from the three scenarios.

An important aspect of the Tanzania study has been the enhancement of the capabilities at CEEST to carry out forest sector mitigation analysis, linked to the energy-sector analysis, in

which expertise had already been developed. Through the mitigation studies and involvement in other climate-related activities, CEEST has networked with other Tanzanian institutions, both in academia and in government. Personnel were seconded from these institutions to contribute to the studies and to take part in workshops both nationally and internationally. This has enhanced national capacity-building significantly in the institutions that are directly involved in the implementation of the UNFCCC, the Kyoto Protocol and related negotiations (H. Meena, private communication).

26.5 ZAMBIA – NEGATIVE COST OPTIONS FOR HOUSEHOLDS?

The Zambia Country Study, also part of the Danida-funded project *Climate Change Mitigation in Southern Africa,* was carried out by the Centre for Energy, Environment and Engineering (Zambia) Ltd (CEEEZ) under the leadership of Professor Francis D. Yamba with participation from CEEEZ staff as well as staff seconded on a part-time basis from the Zambian Department of Energy (DOE). The focal point for climate change in Zambia was changed shortly before the start of the project, from the Ministry of Energy and Water Development – MEWD (parent ministry to the Department of Energy) to the Ministry of Environment and Natural Resources (MENR). The MOU was thus between CEEEZ and the MENR, while the substantive input to the study from the government side came primarily from the Department of Energy.

The study of Zambia followed a preliminary inventory and mitigation study carried out by CEEEZ (on behalf of MEWD) with support from GTZ. The study reported here was thus able to utilize and enhance this capacity at CEEEZ and DOE.

As in the Botswana study reported above, the energy system was modelled using LEAP in order to establish a baseline for energy demand and supply and for GHG emissions. The mitigation scenario was also constructed using LEAP. In addition, the GACMO spreadsheet tool was used to track mitigation options and to perform the detailed costing analysis.

The GHG inventory and projections of baseline emissions from energy, industry and land use indicate that anthropogenic emissions in Zambia are almost compensated by absorption of CO_2 in sinks. According to this estimate Zambia is, and is likely to remain, a relatively small net emitter compared with the very large carbon flows, mainly associated with land-use activities. Regrowth of natural forests after forest clearing, abandonment of managed cultivated land and reforestation plantations are the major carbon sinks in Zambia.

Although the overall picture indicates relatively stable net emissions, there is local deforestation in Zambia associated with land clearing for agriculture and charcoal production, and thus there is

Table 26.3. Total biogenic and non-biogenic CO_2 emissions in Zambia for 1995, 2010, 2030 in the baseline and mitigation scenarios as calculated by LEAP (million tonnes CO_2) (source: Yamba et al., 1999).

Source	1995 Baseline	2010			2030		
		Baseline	Mitigation	Reduction (%)	Baseline	Mitigation	Reduction (%)
Biogenic	14.9	22.3	20.3	9	37.9	33.8	11
Non-biogenic	2.1	4.2	3.4	19	10.6	7.1	32
Total	17.0	26.5	23.7	10	48.5	40.9	16

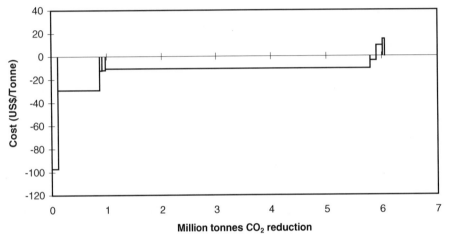

Figure 26.3. CO_2 abatement cost curve for Zambia in 2030, discount rate 10% (adapted from Yamba *et al.*, 1999).

an interest in exploring measures that can reduce this, and in so doing make Zambia a net sink of carbon in the future. The mitigation analysis focused particularly on options that could reduce the pressure on forest for fuel associated with charcoal production. Mitigation options include electrification, use of improved stoves and coal briquettes. The latter, although originating from coal, can be utilized at a higher efficiency for cooking than charcoal, resulting in proportionally less CO_2 emission. If competing with unsustainable charcoal, coal briquettes may therefore be considered a mitigation option.

In addition to household options, a few measures in the industrial and transportation sectors were considered in the analysis. In the industrial sector, the partial replacement of coal, diesel and fuel-oil fired boilers with electric boilers was considered. Under transport and government/service, use of ethanol–gasoline blend in petrol-propelled motor vehicles and improved maintenance of motor vehicles options, respectively, were considered.

The LEAP model allowed emissions of CO_2 originating from biomass and fossil fuel combustion to be accounted for separately. These are shown in Table 26.3 as biogenic and non-biogenic emissions, respectively; under baseline and mitigation scenarios for the year 1995 and projections to 2010 and 2030.

The individual mitigation options were subsequently analyzed using the GACMO spreadsheet tool, which allows a calculation of the levelized cost of GHG abatement for each option in a specific year. The results are plotted as an abatement-cost curve in Figure 26.3 and listed in Table 26.4. A noticeable feature of the Zambia results is the very large identified potential for negative cost options. As noted above in the discussion of the Botswana study, the mitigation costs shown do not include certain costs, such as those involved in overcoming implementation barriers and the indirect costs that may result from adopting the measures.

For example, much of Zambia's mitigation potential lies in replacing charcoal as the predominant urban household cooking fuel. (This assumes that charcoal is produced at least partly unsustainably.) The cost calculation does not include the cost that might be involved in overcoming popular resistance to a new fuel, or compensating for social costs, associated with the very large numbers of people involved in the charcoal 'fuel chain' from production to (informal) retail sale. This is among the interesting research questions that require further study. Also requiring further work in Zambia is an investigation of the connection between energy mitigation options and forestry. While a study of forest sector options was included in the Zambia country study referred

Table 26.4. CO_2 abatement options for Zambia in 2030, discount rate 10%
(adapted from Yamba et al., 1999)

	Mitigation option	US\$/tonne CO_2	CO_2 equivalent cumulative reduction (million tonnes)
1	Ethanol blend	−97.04	0.15
2	Use of improved stoves	−29.11	0.91
3	Efficient boilers in industry (1)	−12.16	0.95
4	Use of coal briquette stoves	−11.97	1.01
5	Increase use of electric stoves	−10.55	5.83
6	Efficient boilers in industry (3)	−3.87	5.93
7	Cement production	9.14	6.05
8	Efficient boilers in industry (2)	14.74	6.08

to here, the modelling of forestry and energy mitigation options were not fully integrated, again leaving scope for further work.

With regard to assessment of established capacity, there is no doubt that the study, and the previous GTZ-funded activity, contributed significantly to government capacity, almost entirely in the Department of Energy, since the study was carried out largely by seconded personnel who have now returned full time to the DOE. The study and its forerunner also helped to establish the credibility and acceptance of the centre – CEEEZ. Whether this capacity will be fully utilized in future climate related studies, connected with the National Communication to the UNFCCC, is not clear however, because of structural and institutional factors. This is not to say that the enhanced capacity at DOE is wasted in any way, or that at CEEEZ. Personnel at DOE have gained valuable experience in the use of analytical tools and have been exposed to the international network of planners and researchers engaged in the studies. This will significantly enhance the capability of DOE staff to carry out climate-related planning and implementation activities in the future.

26.6 INSTITUTIONAL ARRANGEMENTS AND CAPACITY-BUILDING

Responsibility for national climate change activities in all countries is vested in a government institution, most commonly the ministry of environment or the meteorological services department. Countries in southern Africa are no exception. This generally means that when a bilateral or multilateral agency interacts with a country regarding a mitigation study, or another climate-related study or activity, then that climate focal point takes on the task of conducting the study. This typically involves research, enhancement of capacity in modelling and analysis, coordination between different ministries and concentrated analytical work over a period of a year or two.

Such activities are not always compatible with the day-to-day routines of ministries or government departments. Civil servants are often geared to working on much shorter-term tasks, and frequently have to respond at very short notice to the demands of government. While the ministry or department has the ultimate responsibility for climate-related affairs, and there is an obvious interest in enhancing capacity among personnel to deal with the issues at a high professional level, the realities of government departments can inhibit the process of carrying out a study and building capacity among staff. Moreover, the multidisciplinary and multiministerial nature of the climate change mitigation problem is a factor that can hinder the effective execution of a climate change study. It is common, not only in developing countries, that different perceptions and priorities among ministries can make it difficult to collect data, perform analysis and build meaningful capacity across ministerial boundaries.

This potential incompatibility has been recognized both by ministries and donor organizations. The solution, as practised in Zimbabwe and subsequently Botswana, Tanzania and Zambia, has been to delegate responsibility for the study to an independent NGC with strong links to government. A necessary link, at contractual level, is a MOU that allows the centre to carry out the work on behalf of government. The centres also often comprise senior staff with high-level experience and contacts in government and they are often able to second staff from the relevant ministries to work on studies, providing suitable remuneration and motivation as well as the freedom to concentrate on the activity either full-time or part-time. Such secondment has the added advantage that seconded staff generally return to the ministry with enhanced capacity in the climate-related discipline, adding to the national capability at policy-making level.

There are nevertheless potential barriers against effective collaboration between government institutions and NGCs, which must be considered carefully and overcome. For example, ideological differences, institutional competition (especially for the allocation of external funds) and cases of mutual distrust may hamper more fruitful relationships. For NGCs to play a relevant role in capacity-building they have to work towards and maintain credibility, consistency and recognition or acceptance (Maya, 1997). This means that NGCs everywhere in developing countries experience significant challenges in consolidating themselves as stable institutions able to contribute to substantive analysis and policy formulation. They depend on financing from two sources: government contracts and donor funds. The former source has limited financial potential, and the latter creates the problem of increasing aid dependence. A possible solution is for the NGCs to develop into entities that can provide high-level research consultancy services to domestic clients, both government and private sector.

A factor that has contributed significantly to the enhancement of capacity in the countries mentioned here and other participating countries in the UNEP mitigation study programmes has been the close link between methodological development, training and production of the country study reports. Through parallel processes of developing, refining, applying and extending the methodology with regular workshops, a network of involved practitioners was established. Preparation of country presentations, discussion of central problem areas and related issues has without doubt strengthened the capability of teams, both in analytical terms and with respect to participation in the ongoing climate change negotiations. In this context every effort was made wherever possible to involve representatives both from the relevant NGCs and from the government institutions on whose behalf the studies were carried out.

The NGCs that were involved in the three country studies summarized here (EECG, CEEST and CEEEZ) were in existence before the respective studies were initiated, and indeed played a central role along with the UCCEE in formulating and refining the project proposal, finalizing funding and liaising with the appropriate government agencies. Nevertheless, involvement in the studies has significantly enhanced the capability of the centres to perform independent analyses of climate change mitigation. Each of the centres carried out its respective study in close collaboration with government and with direct participation of personnel from the relevant ministries, thus it is fair to assume that national capacity has been significantly enhanced.

The quality of the product, the mitigation study, may be judged by the respective country study reports (Zhou 1999; Mwandosya et al., 1999; Yamba et al., 1999). The capacity-building effect is more difficult to assess, as is the likelihood for continued involvement of the centres, or the seconded personnel, in climate-related activities. The latter is highly dependent on the institutional relationships that were built up before, during and after the study period, between NGCs, government departments and others. The importance of taking these relationships into account at

the very beginning of the process of capacity-building, when selecting the partners, has been emphasised elsewhere by Cissé et al. (1999).

Assessment of capacity-building activities in general is a difficult task, and one that has been receiving considerable attention in the development community, for example UNDP (1997), ECDPM (1998) and Horton (1999). Enquiries by the present author (private communications by H. Meena, 1999, F. D. Yamba, 1999, and P. Zhou, 1999) have confirmed that the studies conducted are being used, albeit to varying degrees in the three countries, as input to the National Communications to the UNFCCC. Involvement of the NGCs in the mitigation studies has moreover helped to establish or enhance networks across ministerial and academic boundaries. Through the studies, personnel from diverse organizations were able to attend, and contribute productively to, national and international workshops. This participation has contributed to the establishment of sustained capacity, since it is often these persons who are involved in climate negotiations at national level.

In view of the resources being devoted to capacity-building for climate change analysis, there is a clear need for more research into the various issues involved, such as institutional models, assessment of effectiveness, and the role of donor organizations and regional centres. Capacity-building for climate change analysis has much in common with capacity-building in other sectors and disciplines. Therefore, such research should take into account the considerable literature on the subject of capacity-building and its assessment, and activities such as that within the UN regime, for example UNDP (1997). The experience gained in the studies reported here, in the other studies in Africa (Senegal and Mauritius) under the UNEP/GEF project *Economics of GHG Limitations*, as well as those from other developing regions, is providing valuable input for such research already initiated at UCCEE (Villavicencio, 1999).

26.7 CONCLUSIONS

As stated by Cissé et al. (1999), ' . . . there is no definitive given model for the African context, the approach must be to take account of the particular circumstances of participating partners, and the nature and objectives of the planned activities.' Capacity-building for climate change analysis requires careful selection of partners from among government departments, research institutions and non-government centres or organizations. In many countries, the freedom of choice of institutions is limited by institutional, structural or political constraints. Donor agencies and governments would, however, do well to consider the benefits to national capacity, as well as the effectiveness in producing results, that can accrue from a judicious choice of institutional set-up, embodying flexibility, staff motivation and multisectoral access.

The series of mitigation studies in Botswana, Tanzania and Zambia has sought to enhance and consolidate the capacity for climate change mitigation analysis in these countries. The institutional set-up has proved successful as it did in earlier studies in Zimbabwe. There is reason to believe that this model for capacity-building will also be successful in other areas of study and planning, given the pressures of day-to-day routines in government departments, and given the tendency towards private sector participation. Independent centres provide flexibility, staff motivation and a multidisciplinary potential that is essential for climate change studies. At the same time, close links to government ensure national continuity and connection to the climate change negotiation process.

In summary, the words of Maya (1997) are still appropriate, referring to the group of independent centres in Africa, including EECG, CEEST and CEEEZ, as well as the SCEE: 'It can be said that a group of centres, working closely together, have been reasonably successful in building the necessary level of analytical capacity, participating in the global dialogue, and putting forward or defending the African position [in the global climate change agenda].'

ACKNOWLEDGEMENTS

This chapter builds extensively on the work of colleagues in Botswana, Tanzania and Zambia who carried out the studies: Peter Zhou of EECG, Botswana; Mark Mwandosya and Hubert Meena of CEEST, Tanzania; Francis Yamba, Dominic Mbewe, Charles Mulenga and Oscar Kalumiana of CEEEZ (the latter two seconded from the Department of Energy) in Zambia. The Tanzania study also owed much to the contribution from Willy Makundi of Lawrence Berkeley Laboratory, USA. A special acknowledgement is due to Shakespeare Maya and Norbert Nziramasanga at SCEE in Harare, who, working closely with colleagues at the Department of Energy, Zimbabwe, paved the way for the institutional collaboration that has proved successful in the other countries. I am also indebted to Shakespeare Maya for his insight into the problems and potential of capacity-building in Africa, and to Arturo Villavicencio for his analysis of the capacity-building process, which has contributed significantly to Section 26.6 of this chapter.

Finally, I would like to thank the reviewers for useful comments on the early draft of the manuscript.

REFERENCES

CEEST (1994) *Sources and Sinks of GHG in Tanzania for 1990.* Report No. 5/1994, CEEST, Dar es Salaam, Tanzania.
— (1995) *Technological and Other Options for the Mitigation of Greenhouse Gases in Tanzania.* CEEST, Dar es Salaam, Tanzania.

Christensen, J. M., Halsnæs, K. and Sathaye, J. (Eds.) (1998) *Mitigation and Adaptation Cost Assessment: Concepts, Methods and Appropriate Use.* UNEP Collaborating Centre on Energy and Environment, Risø National Laboratory, Roskilde, Denmark.

Cissé, M. K., Sokona, Y. and Thomas, J-P. (1999) *Capacity-building: Lessons from Sub-Saharan Africa.* ENDA-Tiers Monde, Dakar, Senegal. http://www.enda.sn/energie/cc/lessons.htm

Dixon, R. K. (1997) US Country Studies Programme. In c_2e_2 *news No. 9, May 1997,* UNEP Collaborating Centre on Energy and Environment, Risø National Laboratory, Denmark.

ECDPM (1998) *Approaches and Methods for National Capacity-building.* Report of a Workshop, September 1998, European Centre for Development Policy Management, Maastricht, the Netherlands. http://www.oneworld.org/ecdpm/pubs/cb.htm

Fenhann, J. (1999) Introduction to the GACMO Mitigation Model. In J. M. Callaway, J. Fenhann, R. Gorham, W. Makundi and J. Sathaye, (Eds.), *Sectoral Assessments, Handbook Reports: Economics of Greenhouse Gas Limitations.* UNEP Collaborating Centre on Energy and Environment, Risø National Laboratory, Roskilde, Denmark.

Halsnæs, K., Mackenzie, G. A., Swisher, J. N. and Villavicencio, A. (1994) Comparable Assessment of National GHG Abatement Costs. *Energy Policy,* **22** (11), 925–934.

Halsnæs, K., Callaway, J. M., Meyer, H. J. and Markandya, A. (1999) *Methodological Guidelines Main Reports: Economics of Greenhouse Gas Limitations.* UNEP, Risø National Laboratory, Roskilde, Denmark.

Heaps, C. and Lazarus, M. (1995) *LEAP User Manual.* Stockholm Environment Institute-Boston Centre, Boston, MA, USA.

Horton, D. (1999) *Evaluation of Capacity . . . Capacity for Evaluation.* Capacity.org (internet newsletter on capacity-building from ECPDM, Netherlands) issue 2, July 1999, European Centre for Development Policy Management, Maastricht, the Netherlands. http://www.capacity.org/Web_Capacity/Web/UK_Content/Content.nsf/(DocID)/6E0647964AB812B2C1256D8800509AD6?OpenDocument.

Instituto de Economía Energética (IDEE/FB) (1999) *Andean Region, Regional Studies: Economics of Greenhouse Gas Limitations.* UNEP Collaborating Centre on Energy and Environment, Risø National Laboratory, Roskilde, Denmark.

Liptow, H. (1997) German Support Programme to Implement the UNFCCC. In c_2e_2 *news No. 9, May 1997,* UNEP Collaborating Centre on Energy and Environment, Risø National Laboratory, Roskilde, Denmark.

Mackenzie, G. A. and Rowlands, I. H. (1998) Conclusions. In Rowlands, I. H. (Ed.), *Climate Change Cooperation in Southern Africa.* London, Earthscan.

Mackenzie, G. A., Turkson, J. K. and Davidson, O. R. (Eds.) (1998) *Climate Change Mitigation in Africa.* Proceedings, International Conference, Victoria Falls (Zimbabwe), 18–20 May 1998.

UNEP Collaborating Centre on Energy and Environment, Risø National Laboratory, Roskilde, Denmark.

Markandya, A. and Boyd, R. (1999) *The Indirect Costs and Benefits of Greenhouse Gas Limitations: Mauritius Case Study, Handbook Reports: Economics of Greenhouse Gas Limitations.* UNEP Collaborating Centre on Energy and Environment, Risø National Laboratory, Roskilde, Denmark.

Maya, R. S. (1997) Capacity-building under the UNFCCC: an African Perspective. In *c₂e₂ news No. 9, May 1997*, UNEP Collaborating Centre on Energy and Environment, Risø National Laboratory, Roskilde, Denmark.

Maya, R. S., Muguti, E., Fenhann, J. and Morthorst, P. E. (Eds.) (1992) *UNEP Greenhouse Gas Abatement Costing Studies.* Zimbabwe Country Study. Phase One. Risø National Laboratory, Roskilde, Denmark.

Maya, R. S., Nziramasanga, N. and Fenhann, J. (1993) *UNEP Greenhouse Gas Abatement Costing Studies.* Zimbabwe Country Study. Phase Two. Systems Analysis Department, Risø National Laboratory, Roskilde, Denmark.

Mwandosya, M. J. and Meena, H. E. (1999) *Climate Change Mitigation in Southern Africa.* Tanzania Country Study. UNEP Collaborating Centre on Energy and Environment, Risø National Laboratory, Roskilde, Denmark.

Rowlands, I. H. (Ed.) (1998) *Climate Change Cooperation in Southern Africa.* London, Earthscan.

Sathaye, J., Makundi, W. R. and Andrasko, K. (1995) *A Comprehensive Mitigation Assessment Process (COMAP) for Evaluation of Forestry Mitigation Options.* Lawrence Berkeley Laboratory, USA.

UCCEE (1995) *Climate Change Mitigation in Southern Africa.* Methodological Development, Regional Implementation Aspects, National Mitigation Analysis and Institutional Capacity-building in Botswana, Tanzania, Zambia and Zimbabwe. Phase 1, UNEP Collaborating Centre on Energy and Environment, Risø National Laboratory, Roskilde, Denmark.

UNDP (1997) *Capacity Assessment and Development in a Systems and Strategic Management Context.* Technical Advisory Paper No. 3, Management Development and Governance Division, Bureau for Development Policy, United Nations Development Programme, New York, USA.

Villavicencio, A. (1999) *Capacity-building for Climate Change Mitigation and Adaptation Analysis.* (unpublished working paper), UNEP Collaborating Centre on Energy and Environment, Risø National Laboratory, Roskilde, Denmark.

Yamba, F. D., Mbewe, D. J., Mulenga, C. and Kalumiana, O. S. (1999) *Climate Change Mitigation in Southern Africa.* Zambia Country Study. UNEP Collaborating Centre on Energy and Environment, Risø National Laboratory, Roskilde, Denmark.

Zhou, P. (1999) *Climate Change Mitigation in Southern Africa.* Botswana Country Study. UNEP Collaborating Centre on Energy and Environment, Risø National Laboratory, Roskilde, Denmark.

27 Capacity-building initiatives to implement the climate change convention in Africa

GEORGE MANFUL

UNFCCC Secretariat, Bonn, Germany

Keywords

Capacity-building; climate change; Conference of the Parties; greenhouse gas inventories; impacts; vulnerability and adaptation assessment; greenhouse gas mitigation in Africa; national communications; technology transfer; education, training and public awareness

Abstract

Capacity-building has long been recognized by African countries as one of the principal issues to be urgently addressed in order to support national and regional initiatives to effectively implement the provisions of the United Nations Framework Convention on Climate Change (UNFCCC) in ways that contribute to their efforts aimed at achieving sustainable development. This chapter briefly outlines the ecological and socio-economic impacts of projected climate change on Africa, and identifies the key capacity-building needs and concerns of African countries as described in their national communications, as well as in published literature. These needs include the capacity to (i) prepare national inventories of greenhouse gases; (ii) assess impacts, vulnerability and adaptation to climate change; (iii) undertake greenhouse gas mitigation analysis in major sectors of the economies; (iv) elaborate on education, training and public awareness programmes; and (v) cooperate in the development and transfer of climate friendly technologies. Other needs relate to capacities to mainstream the adaptation to climate change and thereby integrate the implementation of adaptation and mitigation measures within the framework of implementing national sustainable development plans and strategies.

Since the process for preparing national communications was not designed as the principal vehicle for identifying capacity-building needs and concerns, some of the needs identified within the initial national communications lack specificity and clarity of prioritization in the context of national development priorities. Additional financial support provided to African countries by the Global Environment Facility (GEF) since 2002 within the framework of the national capacity self-assessments may enable countries to further identify and prioritize their capacity-building needs and concerns for the implementation of the Convention, while exploring areas for synergies to support the implementation needs of other global environmental conventions from a development perspective.

There is an urgent need for adequate and predictable bilateral and multilateral assistance to help build the human and institutional capabilities of African countries to fully utilize the opportunities offered by the UNFCCC to advance the process of achieving sustainable development. African countries must therefore evaluate their options more closely and provide the needed political, financial and institutional support so that capacities built with donor support to address climate change would be sustainable.

27.1 INTRODUCTION

Climate change is perhaps the greatest environmental challenge confronting humankind in the twenty-first century. In the 1980s, scientific evidence about global climate change and its consequences led to growing concerns among scientists, policy and decision makers and the public.

Although Africa has contributed least to the anthropogenic emissions of greenhouse gases, it is most vulnerable to the ecological, economic and social impacts of projected climate change. Africa's share of global CO_2 emissions rose from 1.9% in 1973 to 3.1% in 2002, representing an emission of about 747 Mt of CO_2 in 2002. The emissions for Latin America and Asia (including China) are 844 Mt and 5,543 Mt respectively in 2002 (IEA, 2004). Africa is the only region of the world where poverty has been projected to rise within this century if adequate measures are not put

Climate Change and Africa, ed. Pak Sum Low. Published by Cambridge University Press. © Cambridge University Press 2005.

in place now to address the pervasive problems of land degradation and water scarcity (NEPAD, 2001). Nearly 40% of Africans
live below the poverty line, yet the basic resources for their existence are threatened by environmental degradation (NEPAD,
2001). The historical climate record for Africa, however, shows a
warming of approximately 0.7 C over most of the continent during
the twentieth century, a decrease in rainfall over large portions of
the Sahel, and an increase in rainfall in east central Africa (IPCC,
2001). The continent has the lowest conversion factor of precipitation to runoff, averaging 15%. Although the equatorial region
and coastal areas of eastern and southern Africa are humid, the
rest of the continent is dry sub-humid to arid. Land-use changes,
as a result of population and development pressures, continue to
be the major driver of land-cover change in Africa, with climate
change becoming an increasingly important contributing factor by
this mid-century. Resultant changes in ecosystems will affect the
distribution and productivity of plant and animal species, water
supply, fuelwood, and other services IPCC (2001). The grassland
areas of eastern and southern Africa, as well as vast areas currently under threat from land degradation and desertification, are
particularly vulnerable. The continent's vulnerability in the agricultural sector stems from the fact that many of its countries prac-
tise rain-fed agriculture (IPCC, 1998). The IPCC (2001) report
further revealed that the vulnerability of Africa is due in part to
the fact that the agriculture sector contributes 20–30% of the Gross
Domestic Product (GDP) in sub-Saharan Africa and 55% of the
total value of African exports. It identified the coastal nations of
west and central Africa, with low-lying lagoonal coasts, as being
susceptible to erosion and threatened by sea-level rise. The report
concluded that African populations may be faced with extreme climate events such as floods and droughts, and a significant number
of countries may be at risk primarily due to projected increases in
the incidence of vector borne diseases due to altered temperature
and rainfall patterns and reduced nutritional status. The tourism
industry, one of the fastest growing industries in Africa, will be
adversely affected because of the likely impacts on wildlife stem-
ming from projected droughts and/or a reduction in rainfall in the
Sahel and eastern and southern Africa.

Droughts and flooding in several countries on the continent
where sanitary infrastructure is inadequate will result in increased
frequency of epidemics, specifically enteric diseases. Increased
temperatures of coastal waters could aggravate cholera epidemics
in coastal areas. There is a wide consensus that climate change,
through increased extremes, will worsen food security in Africa.
The continent already experiences a major deficit in food production in many areas, and potential declines in soil moisture would be
an added burden. Food-importing African countries are at greater
risk of adverse climate change, and impacts could have as much
to do with changes in world markets as with changes in local and
regional resources and national agricultural economy. Because of

water stress, inland fisheries will be rendered more vulnerable as
a result of episodic drought and habitat destruction. The basic infrastructure for development – transport, housing, services – is
inadequate now, yet it represents substantial investments by governments. An increase in damaging floods, dust storms, and other
weather extremes would result in damage to human settlements
and infrastructure and affect human health.

Given the range and magnitude of the development constraints
and challenges facing most African nations, the overall capacity
for Africa to adapt to climate change is low. Although there is
uncertainty as to what the future holds, Africa must start planning
now to adapt to the adverse impacts of climate change. During
the 1990s, most African countries, with the assistance of multilateral and bilateral agencies, did prepare national environmental
action plans. Although these plans are at various stages of implementation, they do not address the issue of climate change in any
deliberate and systematic manner. It is also interesting to note that
in most African countries, these environmental action plans are
yet to be integrated into national sustainable development plans
where these do exist.

27.2 MULTILATERAL ACTION TO ADDRESS CLIMATE CHANGE

Over the past 20 years, significant progress has been made by
the international community in both understanding the complex
nature of the problem of climate change and in devising multilateral legal frameworks to tackle it. The United Nations Framework
Convention on Climate Change (UNFCCC,1992) (henceforth referred to as the Convention) is the first legally binding international
environmental instrument that deals exclusively with climate
change.[1]

The Convention sets out the overall framework for intergovernmental efforts to address climate change by establishing objectives, principles, and the commitments for both developed and
developing countries in accordance with their 'common but differentiated responsibilities'. It also defines the institutional framework for Parties to monitor its implementation, as well as to enable continued negotiations on the effective means of addressing the problem of climate change. Five years after its adoption,

[1] The Convention was adopted by acclamation on 9 May 1992 in Paris,
France, after 15 months of difficult but challenging negotiations by the Intergovernmental Negotiating Committee (INC) for a Framework Convention
on Climate Change established by the United Nations General Assembly
Resolution 45/212 of 21 December 1990. It was opened for signature at the
'Earth Summit' in Rio de Janeiro, Brazil, in June 1992, where it was signed
by the representatives of 154 states and the European Economic Commission
(now the European Union). On 19 June 1993, when the treaty was closed for
signatures, 165 states (including the European Economic Commission) had
signed. The Convention entered into force on 21 March 1994, subsequent
to the submission of the fiftieth instrument of ratification.

on 11 December 1997, Parties to the UNFCCC took another important step forward by adopting the Kyoto Protocol (UNFCCC, 1997a). New ground was therefore broken in the history of global environmental treaties with the establishment of legally binding limitations on the emissions of six greenhouse gases and innovative 'flexibility mechanism' aimed at finding cost-effective means of curbing such emissions. As of February 2005, 188 States and one regional economic integration organization have become Parties to the UNFCCC, of which 53 states are from Africa. At the same time, 30 out of the 53 African country Parties to the Convention have also ratified the Protocol, which entered into force on 16 February 2005. Since 1995, when the first session of the Conference of the Parties (COP 1) was held, Parties to the Convention have met annually to monitor progress in implementing the Convention and to continue negotiations on effective means of tackling various elements of the problem of climate change. The COP has taken several milestone decisions over the years at its annual sessions and these now constitute a detailed rulebook for the effective implementation of the Convention, some of which address the issue of capacity-building (UNFCCC, 1996, 1997, 1998, 1999, 2000, 2001, 2002, 2003).

The concept of capacity-building has acquired very wide and general usage and therefore has undoubtedly assumed different meanings in different contexts within literature. Although the term 'capacity-building' is mentioned in the Convention and in several decisions of the COP, only a limited effort has been made within the process to actually define the concept. The only attempt at defining capacity-building was in the context of the development and transfer of technologies, in the annex to the COP decision 4/CP.7. In this decision the term is defined in a much narrower sense, as a process which seeks to develop and strengthen existing scientific skills, capabilities, and institutions particularly for developing country Parties.

In this chapter, I have used the comprehensive definition proposed by the United Nations Conference on Environment and Development in 1992, in Agenda 21 (United Nations, 1992). Agenda 21 defined capacity-building as denoting the development of a country's human, scientific, technological, organizational, institutional and resources capabilities to address a problem. It suggested that the fundamental goal of capacity-building is to enhance the ability to evaluate and address the crucial questions related to policy and modes of implementation among development options, based on an understanding of environmental potential and limits and needs as perceived by the people of the country concerned.

Capacity-building to facilitate the effective implementation of the UNFCCC by developing countries has, over the years, become a key issue in intergovernmental discussions on climate change. Hence, as part of the Marrakesh Accords adopted in November 2001, the COP also adopted a comprehensive framework to guide the implementation of capacity-building activities in develop-

ing countries, which was annexed to decision 2/CP.7 (UNFCCC, 2001a). The purpose of the framework is to set out the scope of, and provide the basis of action on, capacity-building activities which, when implemented, should assist developing countries in developing and strengthening their capabilities to achieve the objective of the Convention. It was also meant to serve as a guide to the Global Environment Facility (GEF) as an operating entity of the financial mechanism of the Convention, and be considered by other multilateral and bilateral organizations in their capacity-building activities related to supporting developing countries implement the Convention and enable them to prepare for their effective participation in the Kyoto Protocol process. The principles underlying capacity-building efforts in developing countries include the need for these to be country-driven and implemented in a programmatic manner, which also takes into consideration the specific national circumstances of developing countries. The activities should also seek to maximize the synergies for implementing the UNFCCC and other global environmental agreements from a development perspective.

27.3 THE OPPORTUNITIES AND CHALLENGES OF ADDRESSING CLIMATE CHANGE IN AFRICA

While the phenomenon of climate change may be global in origin and thus require international cooperation to address its causes and impacts, there are several key actions to address climate change issues that could be taken at local and national levels. Not only do human activities contribute to the problem occurring at these levels, but also it is here where the impacts of climate change are also usually manifested. For example, the production and use of energy is one of the principal human activities that contributes most to the building up of greenhouse gases in the atmosphere. Energy is also one of the critical elements of the national development plans and programmes of all African countries, since they currently lack energy to run industrial establishments and transport. Although there are wide disparities in the national levels of commercial energy use across the continent, the majority of the African population does not have access to commercial energy. Lack of access to energy services constrains social development and economic growth (World Bank, 2004). The average electricity consumption in the Organization for Economic Cooperation and Development (OECD) countries in 2002 is about 16 times higher than in Africa. The consumption of electricity in 2002 was 514 kWh/capita in Africa, compared with 13,228 kWh/capita in the United States and the world average of 2,373 kWh/capita (IEA, 2004). Africa's commercial energy consumption would be expected to increase several fold in the coming decades as the continent strives to industrialize. The widespread application of energy efficiency practices, the increased use of renewable energy resources and the use of cleaner energy would be some of the major challenges African countries would face in their efforts to promote sustainable development, while trying to make their contribution to the global efforts aimed at addressing the causes and impacts of climate change. Tackling climate change offers unique opportunities and challenges to African countries, which are struggling to mobilize enough resources to address the overriding concerns of poverty alleviation, ensuring food and energy security, improving access to education and health, developing transport and communication infrastructure, as well as sanitation facilities, creating employment opportunities, and improving access to capital. To meet these challenges, sustained efforts should be made to develop and maintain scientific and analytical capability to understand the bio-geophysical impacts of climate change and their implications for national economies and societies. Discussing the capacity of African countries to address the complex issue of climate change, Cissé *et al.* (1999) noted that the continent does not have the requisite scientific and technical knowledge, basic infrastructure for public awareness creation, the financial resources, and institutional framework to tackle the problem. It is therefore imperative that adequate and sustained capacity is built to enable African countries to generate technological and other policy alternatives, as well as to analyse their applicability in various national contexts, their implications for national economies, and how the knowledge gained may be used to protect the national interest in the negotiations under the Convention. Efforts must also be made to effectively manage the capacity to facilitate the implementation of possible climate protection strategies (Sagar, 1999).

The Convention provides an excellent opportunity to contribute to global efforts at addressing the problems of climate change while at the same time pursuing legitimate socio-economic development goals. Article 4 paragraph 1 and Article 12 paragraph 1 of the Convention obligate all developing country Parties, including the African States that have ratified or acceded to the Convention, to provide the COP with information on national inventories of sources and sinks of anthropogenic greenhouse gases; and steps taken or envisaged to implement the Convention. These steps may include programmes containing measures to mitigate climate change and to facilitate adequate adaptation to climate change. They are also to implement, among others:

(i) The integration of climate change management considerations into national socio-economic, and environmental policies and programmes;

(ii) The promotion of education, training and public awareness of climate change issues;

(iii) The promotion and cooperation in the conduct of climate research and exchange of relevant scientific, technological, technical, socio-economic and legal information; and

(iv) The participation in the meetings of the Conference of the Parties to the Convention and its subsidiary bodies.

It is in the context of employing their limited financial, human and technical resources to address these commitments along with other national priorities that the scope and nature of Africa's capacity-building needs and priorities are to be established. Many African country Parties to the Convention have provided some information on their capacity-building needs and concerns in their initial national communications submitted to the COP via the UNFCCC secretariat.[2] These needs could be categorized in three broad levels (individual, institutional and systemic), which are, however, not mutually exclusive. The individual level needs refer mainly to human resource skills/expertise required to identify and solve specific problems. The institutional level capacity-building needs seek to address the needs of government agencies, corporate bodies, and community-based organizations, which create and utilize resources such as information, finance, social structures, etc. The systemic needs focus on the overall national economic and political, socio-cultural, general infrastructural, and inter-institutional/organizational configuration, which facilitate the development and implementation of policies, laws and administrative measures on the basis of agreed upon or set benchmarks.

27.4 IDENTIFICATION OF CAPACITY-BUILDING CONCERNS AND NEEDS FROM INITIAL NATIONAL COMMUNICATIONS AND OTHER SOURCES

The synthesis reports (UNFCCC, 2001b, 2002, 2003) prepared by the UNFCCC secretariat in response to the decisions of the COP have provided very useful information on the capacity-building needs and constraints as contained in the initial national communications of several African countries (UNFCCC, 2003).[3] The submissions made by Parties to the COP have also identified other needs. The GEF Capacity Development Initiative Regional Report for Africa (GEF, 2000) also contains important information on some capacity-building needs and constraints.

---

[2] The guidelines for the preparation of the initial national communications annexed to decision 10/CP.2 did not specifically request Parties to report on their capacity-building needs. They, however, required that reporting Parties may provide information on their financial and technological needs and constraints, and it is under this heading that several countries provided information on their capacity-building needs. The guidelines suggest that non-Annex I Parties 'may describe the financial and technological needs and constraints' associated with the following: (a) communication of information and further improvement of national communications, including reducing uncertainties about GHG emissions and sinks through appropriate institutions and capacity-building; (b) activities and measures envisaged under the Convention; (c) measures to facilitate adequate adaptation to climate change; and (d) assessment of national, regional and/or sub-regional vulnerability to climate change, which may include data-gathering systems to measure climate change effects or to strengthen such systems, and identification of a near-term research and development agenda to understand sensitivity to climate change.

[3] http://unfccc.int/program/imp/nai/natcsub.html

27.4.1 Information on capacity-building needs and concerns contained in initial national communications

Almost all African countries that are Parties to the Convention have received financial and technical support from the GEF and its implementing agencies to help prepare their initial national communications, of which 41 countries have made submissions to the COP via the UNFCCC secretariat. As indicated earlier, these communications identify to a varying degree of detail the capacity-building needs and concerns of the countries. In these reports, capacity-building needs were identified essentially under the heading of financial and technological needs and constraints. A number of countries outlined some of their needs under various sections of the national communications such as greenhouse gas (GHG) inventories, impacts, vulnerability and adaptation assessment, mitigation analysis, research and systematic observation and public awareness. They have also voluntarily proposed projects for further development within their national communications (UNFCCC,1996), which provide some indication of capacity-building needs (UNFCCC, 2004a).

Institutional capacity-building

Several African countries have expressed the need to strengthen their institutions, develop their human resources, and benefit from information sharing and networking, actions which would enable them to effectively implement their commitments under the Convention, and participate more fully in the intergovernmental processes relating to the UNFCCC and its Kyoto Protocol. More specifically, these needs include the establishment and strengthening of institutions such as national climate change committees and climate change secretariats, enabling them to formulate, implement and monitor national climate change projects and programmes. It is also necessary to strengthen key academic, scientific, technical and research institutions and non-governmental organizations, so as to enable them to undertake tasks relating to the implementation of the Convention. Other needs are to enhance capacity for policy-making and planning, linking science and policy, integrating climate change policies into national development strategies and plans; and establishing national or regional clearing houses for information sharing and networking on climate change issues.

Capacity-building related to the preparation of national greenhouse gases inventories

African countries have reported difficulties in preparing their national GHG inventories, primarily due to the lack of reliable data in key socio-economic sectors, such as land-use change, forestry and agriculture, and the lack of technical and institutional capacity relating to data collection, analysis and man-

agement, as well as methodological problems. The capacity-building needs identified by these countries in this area included the improvement of infrastructure, equipment and facilities; the development of local technical capacities and expertise; the establishment of reliable and effective database systems; improvement of data quality; the development of country-driven methodologies; and the enhancement of information-sharing and networking systems. To address some of these needs, the United Nations Development Programme (UNDP) is implementing a GEF-funded regional capacity building project for West and Francophone Central Africa, designed to improve the quality of GHG inventories of countries within the subregion. The project will strengthen the national institutional capacities of recipient countries to estimate national GHG emissions by employing good quality emission factors and other relevant data obtained through appropriate data collection procedures.

Capacity-building relating to assessments of vulnerability and adaptation to climate change

Assessments of vulnerability and adaptation by the African countries typically cover key socio-economic sectors, such as water resources, agriculture, forestry, coastal zones, fisheries, human settlements, human health and natural ecosystems (UNFCCC, 1996). Such assessments often require extensive collection and analysis of data, application of computer models and the interpretation of results. In this context, African countries identified the need to strengthen relevant national and regional institutions; build requisite national and regional expertise through adequate training to enhance research capacities, including data collection and analysis; and improve methodologies that take into account local conditions. Other needs identified include the improvement of systems for the effective exchange of information and sharing of experiences. Some specific capacity-building needs identified included the development of regional climate models for the Nile basin; linking of climate models to hydrological models and the application of predictive computer modelling and integrated assessment modelling; and improvement of integrated coastal zone management, including more advanced and sophisticated equipment for monitoring, especially coastal erosion.

Capacity-building relating to assessment and implementation of climate change mitigation options

Several African countries described in their national communications greenhouse gas mitigation options in key socio-economic sectors, such as transport, energy, agriculture, forestry and waste management, which reflect the specific needs and priorities of individual countries, some of which are common to most African countries. Capacity-building needs identified by these countries include: the analysis of socio-economic sectors for energy efficiency; renewable resources inventories and computerized databases; the installation and maintenance of renewable energy technologies, as well as the development of appropriate institutional and regulatory frameworks for the assimilation of these technologies.

Mackenzie (2005) discusses the capacity enhancement activities in mitigation analysis in three southern African countries: Botswana, Tanzania and Zambia. He also describes the framework and the outputs of the UNEP *GHG Abatement Costing Study (Phases One and Two)*, from 1992 to 1994, with 10 associated country studies, including Zimbabwe. Other studies include the UNEP/GEF *Economics of GHG Limitations* and associated studies funded by Danida and UNDP/GEF from 1995 to 1998, including Botswana, Ghana, Kenya, Mali, Mauritius, Senegal, Tanzania, Zambia and Zimbabwe, as well as a regional study of the Southern African Development Community (SADC) region. There are several other studies that African countries have participated in, which have resulted in significant capacity-building. These include the German Technical Cooperation (GTZ) programme, the United States Country Studies, and subsequently the GEF Enabling Activities aimed at assisting countries to prepare national communications to the Convention. There has been close coordination between the responsible institutions, and in many cases direct collaboration or follow-up on specific country studies.

Capacity-building relating to research and systematic observation

Justice *et al.* (2005) have highlighted the true partnership required between national and international scientists to address the knowledge gaps in Africa in the area of research and systematic observation and made reference to the agenda developed by the international research community for global change research through the International Geosphere–Biosphere Programme (IGBP), the System for Analysis, Research, and Training (START) and the World Climate Research Programme (WCRP), which focuses on those aspects of global change research that are directly relevant to developing countries and regional natural resource management. They further outlined the success that these collaborative partnerships have had in developing institutional capacity-building and thus minimizing the tendency of natural scientists to attempt ad hoc policy responses for areas and on topics beyond their expertise.

Most African countries that have submitted their national communications included information on their activities related to the implementation of Article 5 of the Convention (research and systematic observation). The initiatives mentioned above are serving to address the gaps and needs identified in the national communications in the area of research and systematic observation.

The Global Climate Observing System (GCOS)[4] secretariat, with funding from the GEF, is implementing a project that includes organizing regional workshops to develop project proposals to address the priority capacity-building needs of developing countries in relation to their participation in systematic observation networks and the follow-up regional action plans. It has prepared the second report on the adequacy of the global climate observation systems, based on an analysis of national reports and on the outcome of the above-mentioned regional workshops (UNFCCC, 2002).

The World Meteorological Organization (WMO), through its education and training programme, and a network of 23 Regional Meteorological Training Centres, is assisting developing countries by providing comprehensive instruction in the use and maintenance of modern equipment now available for observing and collecting data on weather and climate (WMO, 2002). In addition, WMO has a Climate Information and Prediction Services (CLIPS) project, which assists developing countries in improving their climate services through better use of contemporary and historical climate information. WMO provides capacity-building through seminars, regional climate outlook forums, specialist training programmes and user-oriented workshops.

Capacity-building relating to education, training and public awareness

Most African country Parties reported their initiatives, programmes and activities related to the implementation of Article 6 of the Convention (education, training and public awareness). The activities on public awareness reported by some African countries varied from broad information on climate change and its effects, to specific issues, such as the benefits of certain mitigation and adaptation options, energy conservation and conservation of natural resources. Although most of the activities reported were oriented towards the general public, some Parties also reported special awareness campaigns targeting specific groups, such as local communities, government officials, and decision makers within industry.

Capacity-building relating to the development and transfer of technology

Most African countries have expressed the need for the development and transfer of technologies, defined broadly as hardware, software and know-how, needed to assist their efforts in reducing

---

[4] GCOS was established in 1992 and is co-sponsored by the World Meteorological Organization, the Intergovernmental Oceanographic Commission of the United Nations Educational, Scientific and Cultural Organization, the United Nations Environment Programme and the International Council for Science.

the rate of increase of GHG emissions and enhancing their capacity to adapt to the adverse impacts of climate change. Many of the countries also expressed the importance of increasing access to new technologies, the need to adapt foreign technologies to local conditions, and the strengthening of local research institutions.

27.4.2 Information on capacity-building needs and concerns as identified and addressed through other sources

CC:TRAIN

CC:TRAIN was a joint training programme of the UNFCCC Secretariat and the United Nations Institute for Training and Research (UNITAR), in collaboration with the Secretariats of the GEF, IPCC, the UNDP, and with the assistance of the Information Unit on Climate Change (IUCC) of UNEP. The executing agency of the programme was UNITAR. The GEF, through the UNDP, provided the bulk of the project funds. The Swiss Government and the Government of New Zealand provided financial and technical support to the programme. CC:TRAIN evolved and became an important capacity-building initiative that supported implementation of the UNFCCC by strengthening the capacity of participating countries to undertake relevant technical and policy-oriented studies. The programme also worked to enhance the mechanisms that lead to increased political and public awareness of, and support for, the implementation of appropriate responses to climate change. This is crucial to the strategy of building local capacity and strengthening local institutions in both the production of training materials and the delivery of training and ongoing technical assistance by regional partners that have close and effective working relationships with countries in the three regions in which the project was implemented. The UNDP and participating governments provided in-kind support, as did the UNEP and several bilateral donors. Several African countries have benefited from the products developed by the programme.

The programme developed, tested and subsequently published on the Internet and on CD-ROM the following materials:

1. Workshop Package on Climate Change and the UNFCCC;
2. Preparing a National Greenhouse Gas Inventory;
3. Preparing a Climate Change Mitigation Analysis;
4. Preparing a Climate Change Vulnerability and Adaptation Assessment; and
5. Preparing a National Implementation Strategy; and, Workshop Package on the Preparation of Initial National Communications by Non-Annex I Parties.

Most packages are available in English, French and Spanish.

Hay (1999) concluded after evaluating the programme that CC:TRAIN identified and implemented a strategy that helped to address the growing demand for developing and applying indigenous capacity at national level in order to enhance the ability of non-Annex I Parties to undertake the diverse range of activities associated with implementing the Convention and making the required communications. These activities are related both to meeting international obligations with respect to reporting the results of national assessments and to harmonizing responses to climate change with national development goals.

Institutional capacity-building

Through a GEF-funded and UNDP-implemented project, UNITAR provided climate change focal points in Least Developed Countries (LDCs) in Africa with support to obtain, process and exchange data and improve electronic communication among each other and with the UNFCCC secretariat. This was achieved through equipping the focal points in the countries with computers and Internet connectivity. As a result of an assessment of capacity-building needs in several of these countries, and in close cooperation with a team of national centres of expertise in Africa, UNITAR has developed country-based training facilities. These programmes will propose, inter alia, training on scientific (regional and subregional climate modelling) and policy/decision-making tools, as well as provision of other relevant advisory sources of information related to climate change and sustainable development.

Capacity-building relating to the assessment of vulnerability and adaptation

A GEF-funded UNEP project, 'Assessments of Impacts and Adaptation to Climate Change (AIACC) in Multiple Regions and Sectors', has contributed substantially to climate change scientific capacity-building. The three-year global initiative was developed in collaboration with IPCC, and it is executed by the Global Change System for Analysis Research and Training (START) and the Third World Academy of Sciences (TWAS). Twenty-four individual research activities, including 11 projects in Africa, have been under implementation since 2002. Twelve of these are being undertaken in Africa to address gaps in existing assessments. This project has enhanced the scientific capacity of developing countries to assess climate change vulnerabilities and adaptations, and generate and communicate information useful for adaptation planning and action through training, technology transfer, and interaction with international assessment teams. The AIACC African regional studies cover important sectors which include agriculture and food security, water resources, biodiversity conservation, livelihood security, and human health (AIACC, 2004).

UNDP, with Swiss, Canadian and GEF funding, has developed an Adaptation Policy Framework (APF), which aims to help countries strengthen their capacity for preparing national plans and prioritizing adaptations to climate change. The work will also provide input for National Communications of non-Annex I Parties. A key innovation is that the APF will work from current climate variability and extremes, and assess recent climate experiences. It therefore relates both to the present and to the near and medium-term future. It will also assist countries in developing an adaptation baseline and mainstreaming adaptation into the sustainable development process. The APF promotes the involvement of stakeholders and public participation at the community level, as well as the integration of adaptation measures with natural hazard reduction and disaster prevention programmes (UNDP, 2004).

In 2001, the World Bank initiated a GEF-funded project which aims to improve national and regional assessments of the economic impact of climate change on the agricultural sector of eleven African countries, and to determine the economic value of various adaptation options (World Bank, 2001). The project deliverables are:

(a) Enhancement of the understanding of the impacts of climate change on the agricultural sector at the national and regional levels;

(b) Enhancement of the understanding of the performance of various adaptation measures in response to impacts of climate change on the agricultural sector at the national and regional levels;

(c) Development of databases and models to predict consequences of climate change impact and adaptation;

(d) Full country dialogue on climate change impact on and adaptation by the agricultural sector; and

(e) Establishment of capacity at both scientific and policy levels to address climate variability and climate change in the agricultural sector.

In addition, the World Bank is initiating a process to integrate climate variability and climate change adaptation concerns in projects in water resources management, agriculture, and land management through country-driven assessments.

The UNFCCC secretariat has created a compendium and a web page to facilitate access to information on methods for evaluating adaptation options.

Capacity-building for education, training and public awareness

A work programme on Article 6 of the Convention (education, training and public awareness) was approved by the Parties at the eighth session of the COP in 2002. It emphasized the need to enhance, strengthen or develop institutional and technical capacity. The objectives of the work programme are (i) to share experiences and practices; (ii) to enhance coordination and cooperation between activities at international and regional levels, including the identification of partners and networks; and (iii) to enhance

public participation in, and public access to, information on activities to address climate change and its effects, and to develop and implement adequate responses. Capacity-building needs in Africa for education, training and awareness include the development or the improvement of national programmes for formal or non-formal education, as well as raising awareness on climate change issues in academic and research institutions and among the public at large. There is a continued need for training of media practitioners, as well as a system for exchanging information among experts on the national, regional and international levels.

An African regional workshop was held in January 2004 in Banjul, the Gambia, to advance the effective implementation of the New Delhi work programme on Article 6 in Africa. The workshop explored opportunities and strategies for overcoming several barriers (e.g. lack of financial and human resources, competition from other problems and priorities, and linguistic differences) to strengthening climate change outreach, education and training, including the establishment of Article 6 information network clearing house to facilitate information exchange (UNFCCC, 2004c).

Capacity-building relating to the development and transfer of technology

The Expert Group on Technology Transfer, established by the COP in 2001, has initiated a set of actions to promote capacity-building activities, including the strengthening of capacity to conduct technology needs assessments and access to technology information.

UNDP and the Climate Technology Initiative (CTI) have developed handbooks to guide technology needs assessments and they are planning to conduct associated training. The International Energy Agency/Organization for Economic Cooperation and Development (IEA/OECD) countries and the European Commission launched the CTI at the first session of the COP in Berlin in 1995. The CTI promotes the objectives of the UNFCCC by fostering international cooperation for accelerated development and diffusion of climate-friendly technologies. About half of all African country Parties to the UNFCCC have received GEF funds to undertake 'phase II enabling activity' projects under the so-called 'top-up fund' arrangement, to conduct some capacity-building activities, which include the assessment of technology needs in some selected sectors of their national economies. Several African countries have also noted the potential of TT:CLEAR (a technology transfer information clearing-house mechanism) developed by the UNFCCC secretariat to provide useful information and services related to capacity-building for technology transfer. The technology transfer framework adopted by the COP in 2001 and annexed to its decision 4/CP.7, identified the following initial needs for capacity-building for technology transfer: the establishment or strengthening of relevant organizations and institutions;

training and expert exchange; and the development of scholarship and cooperative research programmes in relevant national and regional institutions. The framework further identified the need to build capacity for adapting to adverse effects of climate change, strengthen endogenous capacities and capabilities in research, development and technological innovation, and improve knowledge in the areas of energy efficiency and the utilization of renewable energy technologies (UNFCCC, 2001a).

Capacity-building relating to improved decision-making and participation in international negotiations

To facilitate the effective participation of Parties in the UNFCCC process, at its first session in 1995, the COP established the Special Trust Fund (also known as the 'participation fund') for this purpose (UNFCCC, 1995). The UNFCCC secretariat offers funding to one delegate per eligible country to enable them to participate in the sessions of the Convention's Subsidiary Bodies, with funding for a second delegate being provided to both least developed countries and small island developing states for the annual sessions of the COP. There is at least one instance in 2003 when delegates were unable to attend due to inadequate Trust Funds. The limited number of funded delegates from eligible Parties causes difficulties in their participation during sessions when meetings take place simultaneously. The UNFCCC secretariat also provides information support to Parties, enabling them to participate in the negotiation process more effectively. It maintains the secretariat's web site, produces a series of on-line and off-line information products, and provides for communications and media-related services.

As part of the capacity-building efforts, training workshops for African climate change negotiators were organized by UNEP Collaborating Centre on Energy and Environment (UCCEE) in 2001 and 2002, and by UNEP Risoe Centre on Energy, Climate and Sustainable Development (URC) in 2003. These workshops were aimed at improving the negotiating skills of the participants from the region.

Capacity-building relating to the clean development mechanism (CDM)

With the Kyoto Protocol's entry into force on 16 February 2005, two new and important sources of funding will become available to support the implementation of climate change projects, including those related to capacity-building in African countries: the Adaptation Fund and the Clean Development Mechanism (CDM). The Adaptation Fund will be operationalized by the GEF with proceeds from the CDM projects and with ontributions from donor countries. Its resources will assist African countries in meeting the costs of adaptation to the adverse effects of climate change. The CDM offers an opportunity to mobilize additional resources for the implementation of projects in Africa that help mitigate climate change while advancing sustainable development. It is expected that implementation of small-scale CDM projects, especially in the area of renewable energy, energy efficiency, fossil-fuel switching, agriculture, landfills, and low GHG emission vehicles, will channel additional resources to promote the transfer of environmentally safe and sound technology and know-how apart from those available under the GEF Trust Fund and the SCCF.

Several multilateral institutions and bilateral agencies are undertaking a variety of initiatives to enhance the capacities of African countries to implement CDM projects. For example, UNEP Risø Centre on Energy, Climate and Sustainable Development (URC) in Denmark, has implemented a four-year project on capacity-development for the CDM with funding from the Government of The Netherlands. The objective of this project is to generate in 12 developing countries a broad understanding of the opportunities offered by the CDM, and to develop the necessary institutional and human capabilities that will allow them to formulate and implement projects under the CDM. The United Nations Industrial Development Organization (UNIDO) is implementing a capacity-building project to assist host countries in preparing CDM project proposals relating to industry. Since 1999, UNIDO has supported a number of African countries in their capacity-building efforts focusing on issues related to the CDM, and subsequently it has assisted in the development of six national capacity-building programmes in Africa. UNIDO has also been implementing activities related to the identification, formulation and promotion of industrial CDM investment opportunities, including the development of methodologies for determining emissions reductions (UNFCCC, 2003).

The World Bank in 2003 completed its National Joint Implementation/Clean Development Mechanism (JI/CDM) Strategy Studies Programme (NSS Programme), which was launched in 1997 in collaboration with the government of Switzerland, with the objective of providing capacity-building assistance to the host countries relating to these two Kyoto Protocol mechanisms. The NSS programme was later expanded to include other donor countries (i.e. Germany, Australia, Finland, Austria and Canada), and it targeted nearly 30 of the Bank's client countries and promotes the integration of global climate change issues into their sustainable development programmes. In addition, the PCF plus programme, associated with the World Bank's Prototype Carbon Fund (PCF), provides capacity-building, especially project development training, as well as research and assistance with methodological issues related to the CDM.

The UNFCCC secretariat, through its CDM web site, is providing information on all aspects relating to the CDM, including the work of the Executive Board, as well as information and networking capacity needed by potential project participants and governments, including those in Africa. Key capacity-building needs continue to be the development of a national institutional framework to coordinate actions for the preparation, acceptance, revision, and implementation of CDM projects; the elaboration of studies about specific methodological and institutional aspects of the implementation of the CDM; and the enhancement or strengthening of technical capability to increase public awareness of the CDM.

Support for the least developed countries

The COP 7 in 2001 took several important steps to address the special needs of LDCs, including the establishment of a LDC fund and a LDC Experts Group (LEG) to facilitate their implementation of the Convention.

With a GEF-funded UNDP project, UNITAR has provided computers and information/technology support and training to 30 LDC climate change focal points in Africa, helping them to obtain, process and exchange data and improve electronic communication between them and the UNFCCC secretariat. Arrangements for the preparation of National Adaptation Programmes of Action (NAPAs) by 30 African LDC Parties are under way. NAPAs will articulate the urgent and immediate needs of LDCs relating to vulnerability and adaptation. The LEG is mandated to provide recommendations on capacity-building needs for the preparation of NAPAs. To further assist these LDCs in Africa with respect to their NAPA preparation, the LEG have held two workshops, the first in September 2002 in Dhaka, Bangladesh, and the second in September 2003 in Thimphu, Bhutan. The development of LDC NAPAs is seen as an effort that, in itself, contributes to capacity-building for identifying and prioritizing urgent adaptation-related activities at the national level. The work programme for LDCs includes activities relating to institutional capacity-building, in particular for national climate change focal points and for meteorological and hydrological services, as well as capacity-building in support of training in negotiating skills and language for LDC negotiators.

Capacity-building relating to observation systems, information and networking

The WMO is providing training in the establishment and use of climate data systems. In addition, it has prepared and implemented a strategy for capacity-building for supporting the establishment and maintenance of marine observation and information systems.

The UNFCCC secretariat, through its web site and its provision of CD-ROM, databases and media clipping services, provides the Parties with regular updates about the latest information on climate change. Information provided by some African countries in their national communications revealed that there is the need to enhance capacities in information and networking area and communication technologies (hardware and software).

27.5 GLOBAL ENVIRONMENT FACILITY SUPPORT FOR CAPACITY-BUILDING ACTIVITIES

The GEF has also been collaborating with African countries to address critical global environmental problems within the framework of New Partnership for Africa's Development (NEPAD). In this context, the GEF provided US$600,000 to support the elaboration of the environmental initiative described in NEPAD (GEF, 2003). This environmental initiative, which is firmly grounded in African realities, advocates a unique approach to addressing the disturbing environmental trends that exacerbate the growth of hunger, poverty, hopelessness and conflict. One of its priority goals is the building of the capacities of African countries to implement the global environmental conventions, such as the UNFCCC. The GEF's support for NEPAD is the latest in a series of action-oriented projects that have been administered by the GEF's implementing agencies UNDP, UNEP, and the World Bank and some of its executing agencies, including the African Development Bank and International Fund for Agricultural Development (IFAD). From 1991 to 2003, the GEF allocated US$904.5 million in grants for more than 380 projects in African countries (GEF, 2004).

The financial and technical assistance provided by the GEF to address capacity-building needs in African countries has been provided not only within the framework of its support for the category of 'enabling activity' projects, but also through the execution of medium-sized and full-sized mitigation projects it finances under its climate change focal area. The category of 'enabling activities' was first defined by COP1 to include 'measures such as planning and endogenous capacity-building, including institutional strengthening, training, research and education, that will facilitate implementation in accordance with the Convention, of effective response measures.'

27.5.1 Financing for capacity-building in priority areas

In implementing the COP decision 2/CP.4 (UNFCCC, 1998), the GEF Council in May 1999 approved additional funding limited to US$100,000 (also referred to as 'top up' funding) within the framework of enabling activity phase II projects (UNFCCC, 1998) under its expedited procedures to strengthen activities initiated by countries during the preparation of their initial national communications and may lead to project proposals for the second national communications. Forty-one African countries have submitted their national communications and most of them have received funding to undertake enabling activity phase II project activities in six areas. These are: (i) assessment of technology needs, the identification of sources and suppliers of these technologies, and the determination of modalities for the acquisition and absorption of those technologies; (ii) improvement of national activities for public awareness and education on climate change; (iii) effective participation in systematic observation networks; (iv) facilitation of national and/or regional access to the information provided by international centres and network; (v) design, formulation, management, implementation and evaluation of projects which would help Parties to fulfil their commitments under the Convention; and (vi) studies leading to the preparation of national programmes to address climate change, including the improvement of local emission factors.

27.5.2 Financing of national self-assessments of capacity-building needs

The specific mandate for the conduct of the country self-assessments arises from the provisions of the framework for capacity-building in developing countries annexed to decision 2/CP.7 (UNFCCC, 2001a). The framework makes reference to the special circumstances of small island developing states (SIDS) and LDCs. The GEF operational guidelines for the preparation of the National Capacity Needs Self-Assessments (NCSAs) proposals also addressed the need to maximize synergy between the UNFCCC and other global environmental agreements, as well as the special circumstances of SIDS and LDCs identified within the framework for capacity-building annexed to decision 2/CP.7. The guidelines further required that countries continue to identify specific needs, options, and priorities for capacity-building on a country-driven basis, accounting for prior capacities and past and current activities; promote participation of wide range of stakeholders; promote coordination vis-à-vis national coordinating mechanism focal points. The NCSA process seems to have been set in motion by the GEF Council in response to growing requirements from the UNFCCC and other global environmental agreements, particularly Convention on Biological Diversity (CBD) and United Nations Convention to Combat Desertification (UNCCD), that capacity-building be adequately addressed. The two relevant milestones in this process are the May 1999 GEF Council meeting, which approved the Capacity Development Initiative launched in January 2000, and the May 2001 GEF Council meeting, which reviewed the paper entitled 'Elements of strategic collaboration and a framework for GEF action for capacity-building for the global environment'. The latter set in motion the initiation of the country level self-assessments (GEF, 2001). As of February 2005, 34 African countries have launched NCSA at the cost of about US$6.8 million (GEF, 2005).

27.5.3 Future GEF support for capacity-building activities

The GEF has outlined the key elements of strategic collaboration and provided a framework for its action for capacity-building

for global environment (GEF, 2001). Among the activities which commenced in 2002 included the country self-assessment of capacity-building needs and priorities as mentioned above, pilot projects under the pathway of targeted capacity-building, and greater attention to the capacity-building components in regular GEF projects. Among the other elements which are included in the strategic framework are (a) critical capacity-building for least developed countries and small island developing states; (b) technical support and coordination; (c) consultation with other donors on capacity-building goals and efforts; and (d) the use of a strategic partnership between the GEF secretariat and the UNDP, with regular interagency consultation. The GEF Council would advise the GEF secretariat on the timing for action under the strategic elements and the framework above. After the completion of the country level assessments and subject to the Council's approval of the revised strategic elements and framework, it is anticipated that the resources required for the other modalities would be funded under GEF III (2002–2006). An average level of US$1–2 million per country is proposed during GEF III for initial country, crosscutting and foundational support. Targeted and streamlined capacity-building, which would be supported via the GEF's core activities, were also identified. Approximately US$200 million would be budgeted for targeted and streamlined capacity-building support and for enhanced capacity-building project components in all the focal areas of the GEF (GEF, 2001).

27.6 CONCLUSION

Africa is highly vulnerable to the adverse impacts of climate change due to its weak economic situation. Thus, building effective human, institutional and systemic capacities of the countries in the region, so as to enable them to address the challenges posed by climate change, is crucial to national and international efforts in improving their adaptive capacity to the adverse effects of climate change. Well-designed and coordinated capacity-building initiatives are required to support the efforts made by African countries to mainstream climate change into development programmes. Future support should assist to fully identify their capacity-building needs, to address the causes and impacts of climate change within the context of their national, and where appropriate, regional sustainable development plans and programmes, so as to enhance the sustainability of capacity-building initiatives and encourage replicating successful initiatives in other parts of Africa. The process of preparing national communications, and carrying out national capacity self-assessment offers excellent opportunities to identify and prioritize in a more strategic and targeted manner those areas for which capacities must be built or strengthened to implement the UNFCCC. This should, however, be seen as the first important step in the long and challenging process of building national capacities. The resources and opportunities available to them as

Parties to the UNFCCC, including those for the development of their institutional and systematic capacities and human resource capabilities, should be utilized in a synergistic manner and in ways that maximize national, regional and global benefits.

ACKNOWLEDGEMENTS

A special acknowledgement is due to Amrita N. Achanta for her invaluable contribution and insightful comments in the preparation of the manuscript. I would also like to thank Professor John E. Hay and Vivian Raksakulthai for their excellent review comments that have helped to improve the manuscript.

REFERENCES

AIACC (2004) Second AIACC Regional Workshop for Africa & Indian Ocean Islands. http://www.aiaccproject.org/meetings/meetings.html

Cissé, M. K., Sokona, Y. and Thomas, J. -P. (1999) *Capacity-Building: Lessons from Sub-Saharan Africa*. ENDA-Tiers Monde, Dakar, Senegal. http://www.enda.sn/energie/cc/lessons.htm

Desanker, P. V., Frost, P. G. H., Justice, C. O. and Scholes, R. J. (1997) *The Miombo Network: Framework for a Terrestrial Transect Study of Land-use and Land-cover Change in the Miombo Ecosystems of Central Africa*. IGBP Report No. 41. IGBP Secretariat, Stockholm.

GCOS (2002) *The Report of the Global Climate Observing System Secretariat to the Seventeenth Session of the Subsidiary Body for Scientific and Technological Advice of the COP*. http://unfccc.int/resource/docs/2002/sbsta/misc10.pdf

GEF (2000) *Capacity Development Initiative: Country Capacity Development Needs and Priorities: Report for Africa*. http://www.gefweb.org/Documents/Enabling_Activity_Projects/CDI/African_Report_revised_pdf

    (2001) *Elements of Strategic Collaboration and a Framework for GEF Action for Capacity-Building for the Global Environment*. http://www.gefweb.org/Documents/Council_Documents/GEF_C17/gef_c17.html

    (2003) *GEF Annual Report 2002: A Year of Renewed Commitment to Sustaining the Earth*. http://gefweb.org/Whats_New/Draft_Annual_Report_-_2-6-03.pdf

    (2004) *GEF Annual Report 2003: Making a difference for the environment and people*. http://www.gefweb.org/2003_Annual_Report.pdf

    (2005) The GEF project database. http://www.gefonline.org/home.cfm

Hay, J. E. (1999) Capacity-building for climate change. *Tiempo*, issue 31. http://www.cru.uea.ac.uk/tiempo/floor0/recent/issue31/t31a1.htm

IEA (2004) *Key World Energy Statistics from IEA. 2004 Edition.* http://library.iea.org/dbtwwpd/Textbase/nppdf/free/2004/keyworld2004.pdf

IPCC (1998) *The Regional Impacts of Climate Change: An Assessment of Vulnerability*. R. T. Watson, M. C. Zinyowera and R. H. Moss (Eds.). A Special Report of IPCC Working Group II. Cambridge, Cambridge University Press, 517pp.

    (2001) *Climate Change 2001: Working Group II: Impacts, Adaptation and Vulnerability*. J. J. McCarthy, O. F. Canziani, N. A. Leary, D. J. Dokken and K. S. White (Eds.). A Contribution of the Working Group II to the Third Assessment Report of the IPCC, Cambridge, Cambridge University Press, 1032pp.

Justice, C., Wilkie, D., Putz, F. and Brunner, J. (2005) Climate change in sub-Saharan Africa: assumptions, realities and future investments. This volume.

Mackenzie, G. A. (2005) Climate change mitigation analysis in southern African countries: capacity enhancement in Botswana, Tanzania and Zambia. This volume.

NEPAD (2001) *Right to Development: The New Partnership for Africa's Development*, Geneva, 54pp. http://www.avmedia.at/cgi-script/csNews/news_upload/.ENGL.pdf

NOAA (1997) *Workshop on Reducing Climate-Related Vulnerability in Southern Africa, Workshop Report, 1–4 October 1996, Victoria Falls, Zimbabwe*. NOAA OGP, Silver Spring, MD, USA.

Odada, E., Totolo, O., Stafford Smith, M. and Ingram, J. (1996) *Global Change and Subsistence Rangelands in Southern Africa: The Impact of Climatic Variability and Resource Access on Rural Livelihoods*. Report of a workshop on Southern African Rangelands today and tomorrow: social institutions for an ecologically sustainable future, Gaborone, 10–14 June 1996. GCTE Working Document No. 20. 99.

Sagar, A. (1999) Capacity-building and climate change: a review of some issues. In C. McCloskey (Ed.), *Policy Matters: Sustainable development and climate change*. http://www.iucn.org/themes/ceesp/Publications/newsletter/policy4.pdf

Scholes, R. J. and Parsons, D. A. B. (1997) *The Kalahari Transect: Research on Global Change and Sustainable Development in Southern Africa*. IGBP Report No. 42, IGBP Secretariat.

START (1995) *Regional Workshop on Global Changes in Southern, Central and Eastern Africa*. Workshop report, Gaborone, Botswana, July 4–7, 1994.

UNDP (2004) *The Adaptation Policy Framework*. http://www.undp.org/cc/apf.htm

UNFCCC (1992) *United Nations Framework Convention of Climate Change*, 30pp. http://unfccc.int/resource/docs/convkp/conveng.pdf

    (1995) *Report of the Conference of the Parties on its First Session, Berlin, 28–March 7 April 1995, addendum Part Two: Action Taken by the Conference of the Parties*. http://unfccc.int/resource/docs/cop1/07a01.pdf

    (1996) *Report of the Conference of the Parties on its Second Session, Geneva, 8–19 July 1996, Addendum Part Two: Action*

Taken by the Conference of the Parties. http://unfccc.int/resource/docs/cop2/15a01.pdf

(1997a) *Kyoto Protocol to the Convention on Climate Change.* 34pp. http://unfccc.int/resource/docs/convkp/kpeng.pdf

(1997b) *Report of the Conference of the Parties on its Third Session, Kyoto, 1–11 December 1997, Addendum Part Two: Action Taken by the Conference of the Parties.* http://unfccc.int/resource/docs/cop3/07a01.pdf

(1998) *Report of the Conference of the Parties on its Fourth Session, Buenos Aires, 2–14 December 1998, Addendum Part Two: Action taken by the Conference of the Parties.* http://unfccc.int/resource/docs/cop4/16a01.pdf

(1999) *Report of the Conference of the Parties on its Fifth Session, Bonn, 25 October–5 November 1999, Addendum Part Two: Action taken by the Conference of the Parties.* http://unfccc.int/resource/docs/cop5/06a01.pdf

(2001a) *Report of the Conference of the Parties on its Seventh Session, Bonn, 29 October-10 November, 2001 Addendum Part Two: Action taken by the Conference of the Parties.* http://unfccc.int/resource/docs/cop7/13a01.pdf

(2001b) *The Third Compilation and Synthesis of Initial Communications from Parties Not Included in Annex I to the Convention* http://unfccc.int/resource/docs/2001/sbi/14.pdf

(2002) *The Fourth Compilation and Synthesis of Initial Communications from Parties Not Included in Annex I to the Convention.* http://unfccc.int/resource/docs/2002/sbi/16.pdf

(2003) *Non-Annex I Implementation: Submitted National Communications from Non-Annex I Parties.* http://unfccc.int/program/imp/nai/natcsub.html

(2004a) *Report on the assessment of funding necessary to assist developing countries in fulfilling their commitments under the Convention prepared in the context of the Memorandum of Understanding between the Conference of the Parties and the Council of the Global Environment Facility.* http://unfccc.int/resource/docs/2004/sbi/18.pdf

(2004b) *List of projects submitted by Parties not included in Annex I to the Convention in accordance with Article 12, paragraph 4, of the Convention.* http://unfccc.int/resource/docs/2004/sbi/inf13.pdf

(2004c) Report on the African regional workshop on Article 6 of the Convention. http://unfccc.int/resource/docs/2004/sbi/07.pdf

United Nations (1992) *Report of the United Nations Conference on Environment and Development, Vol. 1 Agenda 21,* New York, 294pp.

WMO (2002) *Report of the World Meteorological Organization to Seventeenth Session of the Subsidiary Body for Scientific and Technological Advice of the COP on Capacity-Building and Technology Transfer Activities Related to "Adapting to Climate Change".*

World Bank (2001) *Impacts on Adaptation of Agro-Ecological Systems in Africa.* http://www.ceepa.co.za/Climate_Change/

World Bank (2004) A Brighter Future? Energy in Africa's Development. http://www.worldbank.org/html/fpd/energy/subenergy/energyinafrica.htm

28 Education and public awareness: foundations of energy efficiency

JOHN J. TODD[*]

University of Tasmania, Hobart, Australia

Keywords

Climate change mitigation; greenhouse gas abatement; education; community; teachers; teacher-education; Africa

Abstract

Increased population and modest economic growth are likely to result in a 10-fold increase in emissions of greenhouse gases for the African continent over the next two to three decades. If economic growth can be achieved with lower than expected fossil fuel use increases, then substantial global benefits would result. An important long-term opportunity for 'no regrets' action on climate change is education in both the formal (school) and informal (community) sectors. To be effective, the education programme must inform, enthuse and empower. This requires demonstration of the relevance of climate change mitigation and adaptation strategies to the individual. Within the school sector, it is conceivable that a well-planned education programme aimed at improving awareness of energy minimization actions and the importance of global climate could reach 100 million school children within three to four years. Community education must be flexible and support, but not attempt to manage, existing community organizations.

28.1 INTRODUCTION

The continent of Africa, with its 50 countries and vast differences in economic, cultural, and political systems, presents enormous challenges for those charged with the implementation of sustainable development and the design of responses to the threat of climate change. The magnitude of the problem, together with this diversity, immediately raises the question – is there anything that can be said that might be common to all these countries? In detail,

*Current address: Eco-Energy Options, Tasmania, Australia

probably not, but there are broad principles that can be applied. This article addresses one such principle relating to education, both formal school education and the less formal community education. The basic ideas presented here are not new; many environmental commentators, government and agency officials, and academics have stressed the ecological importance of formal and informal education. This discussion revisits these important topics and attempts to place a slightly different perspective on their role in climate change.

The article includes a brief introduction that reinforces some basic principles of anthropogenic driven climate change that are addressed in more detail in other articles in this book. Increases in per capita resource use (including, but not limited to, fossil fuels) and increasing population (with parallel alteration to land use) are considered by many scientists to be the key driving forces behind changes to atmospheric composition and climate change. African nations have high population growth and generally high (but erratic) growth rates in per capita resource use. Thus, there is potential for very significant increases in greenhouse gas emissions. Such a statement, particularly by a non-resident of Africa, might be seen by some as simply an abrogation of responsibility for one's own 'backyard'. After all, the per capita greenhouse gas emissions from industrialized countries, such as Australia, are an order of magnitude greater than those of developing nations. Obviously, much needs to be done in the developed world. But the potential problem is so great that every nation needs to play some part in the global challenge of climate change. In poorer nations, the cost of action might reasonably be borne by the developed world; however, even this is not possible without political acceptance that some action on climate change is warranted. At a pragmatic level, if individual countries are not able to demonstrate that reasonable efforts are being taken to mitigate emissions of greenhouse gases, then they will be in a weakened position if requesting international aid to help with adaptation to a modified climate.

Before focusing on the specific theme of this article, several more qualifications are warranted. This article is not suggesting

Climate Change and Africa, ed. Pak Sum Low. Published by Cambridge University Press. © Cambridge University Press 2005.

how things *should* be done; rather it aims at sowing ideas that others might be able to adapt to the particular circumstances that prevail within each country. So many poor practices, mistakes and contradictions abound in the industrialized nations that it is simply presumptuous for a commentator in an industrialized country to suggest how things should be done elsewhere.

The causes of, and responses to, climate change are exceedingly complex. They cannot be separated from all the other environmental, social and economic issues facing all countries. Thus, any discussion of education aimed at mitigating climate change can never be complete or all-inclusive. Also, the problems of poverty, disease, political instability, war, economic exploitation and famine are huge. All of them are destructive of the environment and directly or indirectly influence global climate. It is essential that decisions on environmental education discussed in this article are made with due consideration of these other problems.

Finally, it is acknowledged that academic debate continues on whether or not the observed global warming of the past decade is human induced or a natural fluctuation. Further, a few commentators argue that there may be benefits associated with climate change – carbon dioxide fertilization, increased rainfall in some arid regions, and so on. These views are not shared by the author of this article. The article is based on the premise that humans *are* influencing global climate and that the overwhelming outcome of climate change will be disadvantageous to humans and ecosystems.

28.2 WHERE TO FROM THE PRESENT?

All African nations are obviously striving for economic growth in the expectation of improved quality of life for their inhabitants and greater material wealth. Assistance of various kinds from international organizations and support from individual countries will reinforce this ongoing move towards greater development. International businesses will continue to seek to develop natural resources and access lower labour costs. All of this points to a development path for many African countries very similar to the development path of newly industrialized nations. These developments will, hopefully, improve the general well-being of millions of people and so they should not be obstructed. The biggest uncertainty is the time frame for this development. Problems of air, water and soil contamination will invariably increase. However, the extent to which ecological and social problems accompany the beneficial aspects of development will depend on the long-term planning and education.

The magnitude of the potential problem is illustrated by some background statistics. WRI (1992) reports that the average population growth rate (1985 to 1990) for the whole of Africa was about 3% per annum, compared to Europe's 0.25% and USA's 0.6%. Annual growth rates vary considerably from one country to

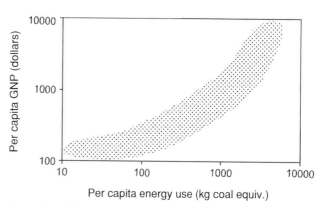

Figure 28.1. The observed relationship between per capita GNP and per capita energy use for various countries. Most countries fall within the shaded area of the graph. Adapted from Cutter *et al.* (1991).

another, ranging from 1.2% (Mauritius) to 3.8% (Côte d'Ivoire). The mid-1999 population is estimated at 771 million (PRB, 1999), notwithstanding the ravages of AIDS, famine and war, this number is likely to double in 25 to 35 years, with obvious implications for land use and energy use. Openshaw (2005) provides more detailed discussion of the population and resource issues facing Africa.

Over the decade 1979–1989 the annual Gross National Product (GNP) growth rates in USA and Europe were about 2.5%, yet 18 of the 50 African nations had annual GNP growth rates over the same period greater than 2.5% with the highest being Botswana with 11% (WRI, 1992). Even though economic growth is starting from very low levels in most African countries, there is clearly potential for large increases in fossil fuel use as industry and transport expand.

At present, Africa has very low per capita CO_2 emissions associated with fossil fuel use. Marland *et al.* (2003) estimates for 2000 were 1.06 t CO_2/person compared to the global average of 4.03 t CO_2/person and North America 19.25 t CO_2/person. On the other hand, climate change emissions associated with land clearing are high in Africa. Estimates for 1989 suggest a per capita emission of 2.3 t (CO_2 equivalent) for Africa compared to the global average of 1.2 t, and the USA 0.09 t (WRI, 1992). These estimates are undoubtedly only approximate, but they illustrate that the combined issues of fossil fuel use and land use must be addressed.

The link between average per capita GNP and energy use when national per capita GNP is plotted against per capita energy use for different countries shows very rapid growth in energy use as development begins (Figure 28.1). A poor country with double the per capita GNP of another even poorer country has roughly 10 times the per capita energy use. This suggests a very rapid increase in energy use as a country starts its industrial development. If, over the next 25 to 35 years, Africa's population doubles and per capita GNP increases by 50%, one might expect a five to ten fold increase in energy use. If this translates into a five to ten fold increase in total CO_2 release, then Africa would be emitting 50 to 100% as much CO_2 from fossil fuel use as North America is now. This is clearly undesirable. Perhaps the industrialized nations will halt, or even reverse, their use of fossil fuels within a few decades, but this is far from guaranteed, despite the Kyoto Protocol.

Per capita and total CO_2 emissions in the world's developing countries increased by 18.2% and 37.2% respectively between 1990 and 1999 (WRI, 2003), illustrating the overall growth in emissions even though there was large variability from one country to another, some even reducing emissions as economies deteriorated.

Thus, in simple terms, one must envisage a future for Africa that involves increased rates of land clearing due to increasing population. This will reduce the carbon sink associated with living biomass and increase the susceptibility of food production to climate change due to more intensive agriculture and greater use of marginal land. This will occur even if economic growth does not. If economic growth is achieved at a sufficient rate to increase per capita GNP, or even if GNP simply increases in absolute terms, then there will be substantially increased transport, electricity use and industrial fuel use leading to rapid increases in CO_2 emissions. It is naive to think that this can be avoided, but it should be possible to slow the growth of emissions and increase the resilience of the agricultural system without hampering economic growth. With economic growth should come reduced population growth, thus easing the land-clearing problem. So, the key seems to be to achieve economic growth, but to minimize growth in fossil fuel use as this occurs.

28.3 EDUCATION

Education is a very powerful tool for achieving social change. However, the capacity of an education programme to deliver information and training is not infinite. If climate change is to be included in a school curriculum or a community education programme, it will almost certainly be at the expense of some other topic. A judgment must be made on the merit of including or avoiding climate change as a topic for education. It is most likely that climate change issues will influence industrial development, transport and land use in most of the world over the next few decades. Some, of many, examples include accelerated development of more energy efficient transport and industry, carbon trading, increased carbon-based energy costs, and renewable energy supply. Failure to provide appropriate education will reduce the opportunities for the least developed countries to participate and benefit from these changes. There is also an ethical question here – is there a duty to inform all communities about the changes that human activities are causing to the atmosphere and the possible consequences? If people do have a right to such information, since it could profoundly affect their well-being in coming years,

it could be argued that those with the resources to disseminate the information through education programmes are morally obliged to do so.

Education has a crucial underlying role in dealing with the problem of climate change. Education is directed at the individual, the goal being to assist that individual in making an informed choice on matters influencing climate change. Education must provide appropriate information about the role of climate in the individual's well-being. Thus, three broad goals for a climate change education programme can be identified; it should:

(a) engender an understanding of how human activities might be influencing global climate;
(b) demonstrate that an individual's actions can contribute to slowing the rate of climate change; and
(c) prepare the individual for adoption of an appropriate response if local climate does change.

In order to achieve these three goals it is necessary to:

(a) decide what issues can/should be addressed at each level of an education programme (i.e. primary, secondary, tertiary, technical, community, in-service professionals);
(b) establish appropriate mechanism for implementing the education programme; and
(c) provide adequate financial and political support for the programme.

Starting with the question of financial and political support. Education programmes on specific topics such as climate change are not free of cost, even where the basic education infrastructure is in place. Teachers require training (new teachers and in-service teachers) and information must be provided (textbooks and demonstration equipment). These costs are modest compared to the full cost of education, but because they represent an additional cost on top of invariably stretched education budgets, they must have a strong commitment from senior administrative personnel and politicians if appropriate funds are to be provided. This might be a useful area for international organizations to play a major role, particularly in the least-developed countries.

This commitment by government is essential because of government's key role in formal education. But other major players in the provision of formal education, notably religious organizations, must also make suitable commitment to education for climate change.

The development of an appropriate mechanism for implementing education programmes must occur concurrently with the development of commitment and funding. If an implementation plan is not available, it is not possible to estimate costs, and so the possibility of 'hollow' government commitment arises. In other words, governments, or other educational organizations, can publicly announce their commitment to an education programme but then, when the cost becomes apparent, no action occurs. If costs are known early in the process, then governments can be asked from the start to not only make a moral commitment but a concrete financial commitment as well.

This approach lends itself to some preliminary work that could be applied to the whole of the African continent. An implementation model, together with costing, could be developed by an international organization. Any model should address the specific educational requirements for:

(a) teacher training, both new and in-service;
(b) training of community educators (i.e. the informal education providers);
(c) inclusion of climate change issues in a broad cross-section of tertiary courses (not just science, but administration, business, health and so on);
(d) provision of adequate teaching aids (text books and demonstration materials); and
(e) appropriate incentives to ensure that business and industry (local and multinational) include climate change education for all employees.

Having established a plan for implementation of a widespread education campaign, it is necessary to determine what information the educators should disseminate. The two broad issues that need to be addressed are:

(a) ways of slowing (or reversing) climate change, and
(b) preparing for climate change (i.e. adapting as climate changes).

The rate of climate change can be slowed by reducing emissions of greenhouse gases and by absorbing carbon out of the atmosphere through tree planting. Communities can be better prepared for climate change if they are aware it is likely to occur and if they maximize the diversity of their local economies.

The role of education in mitigation is to:

(a) dispel the myth that economic growth and material gains can only be achieved by using more and more energy;
(b) demonstrate that effective and efficient use of energy will allow the same tasks to be carried out but with less energy and at lower cost; and
(c) show that individuals and local communities can achieve a lot themselves without relying on governments or aid organizations.

The second crucial area of education is to increase diversity and flexibility so that adaptation to climate change is less disruptive. Education provides an opportunity for introducing people to different ways of doing things. It does not mean that traditional practices are abandoned, simply that people are aware that alternatives exist. For example, if long-term changes to rainfall occur, traditional crops may no longer be suitable for a region. If students have been exposed to different foods, then accepting a

change in the types of crops grown becomes easier. Similarly, all aspects of lifestyle and culture can, and do, change. Education is the mechanism for demonstrating that alternatives are available. Should circumstances require a change from traditional, proven approaches, then an informed community is much better placed to enter into a managed, and hopefully ordered, transition rather than a chaotic abandoning of the old ways.

28.3.1 Formal education sector

The formal education sector offers several advantages for spreading information about climate change. The physical infrastructure for education is already in place (at least in those regions where schools are provided). Teachers are trained in the provision of effective education and children are present with the express purpose of learning. Thus, provided information can be disseminated to the teachers, it can reach large numbers of school children. However, there are also disadvantages. School children are not today's decision makers, so the outcomes of the education will be delayed, possibly for a decade or more. Also, there are many social and environmental issues competing with basic skill development within the school curriculum. Introducing climate change may well mean displacing other important learning goals. For this reason, wherever possible, climate change issues should be spread across the teaching curriculum and used to illustrate and reinforce other practical skills.

The importance of appropriate training for teachers and other educators cannot be emphasized enough. Teachers must feel confident about dealing with the many issues surrounding climate change. They must understand the issues themselves. They must have the opportunity of updating their knowledge in this rapidly advancing area. It is not sufficient to just include training for new teachers; in-service teaching must also be provided. This is not a simple task. It is most efficient to bring groups of teachers together and provide them with fairly intensive training, possibly over one or two days. But this means removing them from the classroom during their training. In urban areas it might be possible to spread the training over several shorter periods outside normal school hours. But the teachers might reasonably expect some compensation for giving up their own time. Also, such an approach would probably be impossible for rural teachers.

An alternative approach might be to educate the teachers in their own teaching environment by sending trainers into classrooms to work with individual teachers. This is much less efficient and would require large numbers of trainers. The magnitude of the problem of training teachers is illustrated when one considers that there are roughly two million school teachers in Africa.[1] Simply

---

[1] This is a rough estimate assuming that there are about 200 million children in Africa between the ages of 6 and 14 (26% of the population) and that half of these children attend school. With school class sizes of 50, the number of teachers would be about 2 million.

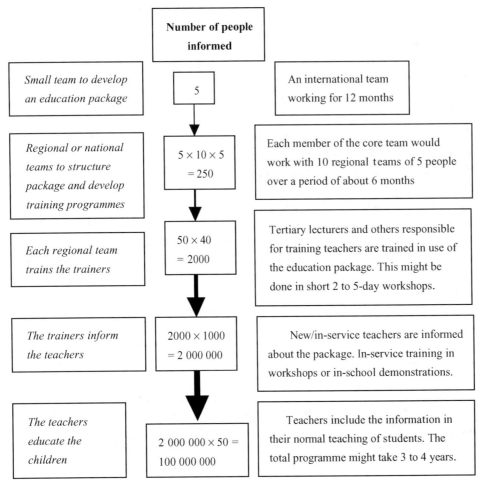

Figure 28.2. An example of how a cascade of teaching/learning could carry an understanding of climate change issues and mitigation opportunities from a small core group of experts to about 100 million children in a few years.

providing a basic textbook for each teacher would be a difficult task, but actually offering training to the majority of teachers becomes a huge undertaking.

However, one of the benefits of education is that it readily lends itself to a cascade of learning, an exponential growth in knowledge. Figure 28.2 provides an illustration of how, from a small team of dedicated educators and administrators, it might be possible to reach 100 million children in the space of three to four years. Perhaps this is a rather optimistic illustration, but the underlying principle is sound, and a realistic plan could be developed that would achieve the educational goal in a modest time frame.

School education material on climate change is plentiful in western countries. Books, Internet sites and entire school programmes are readily available. This material could be adapted to each African country and region relatively easily.

Tertiary education relating to climate change presents fewer logistic problems because the total number of universities is much less than the number of schools and tertiary teachers (lectur-

ers) might reasonably be expected to be able to develop relevant teaching materials themselves. Much good description of climate change science (from basic to quite technical) is available on the Internet, as are mitigation measures and adaptation suggestions. Individual lecturers could use material relevant to their region and subject area.

28.3.2 Community education

Working outside the formal education system provides a quite different set of advantages and disadvantages. Among the advantages, education can be directed at adults – in other words, the section of the community making decisions and participating in activities that contribute to emission of greenhouse gases. Since changes in climate will also affect the livelihood of many, it is more straightforward to demonstrate a personal relevance of the issues. There is also an active and enthusiastic network of community educators, many already dealing with climate change (e.g.

see Duke University, 2003). Many of these NGOs operate without direct funding from African governments, so the issue of redirecting scarce funds from other education requirements does not arise. However, the NGOs will have their own underlying agenda that may, or may not, complement other climate change education programmes in each country.

Another disadvantage of the informal education sector is that they do not have a 'captive' audience. They must convince the community that they are worth listening to and that their teachings will lead to improved quality of life. With so many messages being promulgated, communities can become sceptical about the worth of information that might not be immediately reinforced through practical demonstration. So, dealing with very long-term problems, such as climate change, requires a quite different approach compared to day-to-day environmental problems, such as clean water or improved agricultural output.

Some coordination of information collection and dissemination for community education across the whole African continent may have value. However, such coordination should not involve any management, direct or indirect, of the participating organizations, as this would be likely to discourage participation and reduce the flexibility of participating organizations through increased bureaucracy.

One example, of many, is used here to illustrate how education could achieve a significant reduction in greenhouse gas emission. The use of biomass for cooking causes land degradation in many regions (Hall and Scrase, 2005). Households using kerosene or gas for cooking may spend a significant proportion of their income on fuel. A simple, solar box-cooker can be constructed from local materials and waste products for the cost of a few litres of kerosene. Obviously not all meals can be cooked in such an appliance, but by cooking just one quarter of the meals, significant reductions in greenhouse gases can be achieved. Such cookers are well proven (SCI, 1999), simple designs are available, and good community responses have been achieved in many regions. In addition to climate change benefits, such cookers can have important health benefits through reducing mothers' and children's exposure to wood-smoke, and they offer a cheap and practical means of sterilizing water. Thus, if community educators are shown how to construct and use a solar cooker, and told of the many learning goals that can be illustrated, they can enthuse their local audience and the technology would gradually spread.

There are many other examples of domestic energy conservation or use of alternative energy sources for rural and urban households that lend themselves to education programmes. Energy conservation in transport, business, selection of materials, and industry all can provide many examples suitable for education programmes at all levels of learning. The energy reduction options can be used as examples across most disciplines, so it is not necessary to run extra lessons on energy conservation, it is simply a matter of in-

cluding the information in existing learning programmes, such as science, home economics, mathematics, art, reading and so on.

28.4 CONCLUSION

Global social equity requires that economic development take place across the 50 nations of Africa. But, even with modest economic growth, it is likely that Africa will be releasing 10 times as much CO_2 from fossil fuel use than it does now within a few decades. In absolute terms this is about the same as North and Central America are now emitting. Current scientific evidence suggests that such an imposition on the composition of the atmosphere is highly undesirable. Both developed and developing countries must seek an alternative path to future economic and social growth. The challenge, of course, is to gain the benefits of economic growth but without such a huge increase in CO_2 emissions. For example, if Africa's emissions of CO_2 increase by, say, a factor of three rather than a factor of 10, the global benefits would be substantial. It is known from experience in the industrialized nations that through improved efficiency and restructuring, it is possible to grow the economy while stabilizing or even reducing energy use. But, as yet, there is no model for economic growth starting from a very low economic base without resorting to huge increases in fossil fuel use.

One long-term action that might slow the emissions of greenhouse gases is education. The action will not solve the problem, but it will provide a sound base for more direct actions should they prove necessary in future. It also has direct social benefits irrespective of their impact on climate change. Thus, in the current jargon, it represents a 'no regrets policy'.

ACKNOWLEDGEMENT

The author wishes to thank his colleagues in Tasmania, Australia, for sharing their African teaching experiences and providing valuable insight into the enormous challenges faced by teachers struggling for adequate resources in many African countries.

REFERENCES

Cutter, S., Renwick, H. L. and Renwick, W. H. (1991) *Exploitation Conservation Preservation: A Geographic Perspective on Natural Resource Use*, 2nd Edition. New York, Wiley.

Duke University (2003) http://billie.lib.duke.edu/pubdocs/ngo/africa.asp

Hall, D. O. and Scrase, J. I. (2005) Biomass energy in sub-Saharan Africa. This volume.

Marland, G., Boden, T. A. and Andres, R. J. (2003) *Global, Regional, and National CO₂ Emissions. Trends; A Compendium of Data on Global Change*, Carbon Dioxide Information Analysis Center,

Oak Ridge National Laboratory, U.S. Department of Energy, Oak Ridge, TN.

Openshaw, K. (2005) Natural resources: population growth and sustainable development in Africa. This volume.

Population Reference Bureau (PRB) (1999) http://www.prb.org/prb/pubs/wpds99

SCI (1999) Solar Cookers International, http://www.accessone.com/~sbcn/index.htm

World Resources Institute (WRI) (1992) *World Resources 1992–93*; New York, Oxford University Press.

(2003) *Earth Trends the Environmental Information Portal*, http://earthtrends.wri.org/

PART V
_____

Lessons from the Montreal Protocol

29 Lessons for developing countries from the ozone agreements

K. MADHAVA SARMA

Environmental Consultant, Chennai, India

Keywords

Stratospheric ozone layer; ozone-depleting substances; Vienna Convention; Montreal Protocol; lessons; awareness; information; education and training; stakeholders; assessment panels; incremental costs; capacity development; country programme; new technologies; technology transfer; assessment of technologies; project preparation and implementation; policy instruments

Abstract

The international agreements to protect the ozone layer - the Vienna Convention of 1985 and the Montreal Protocol of 1987, have a near universal participation and have been successful. Developing countries have implemented these agreements for the past 14 years benefiting the global environment but without adverse effects on their economic development. The countries can apply the lessons from the ozone agreements to deal with the climate change agreements.

The principal lessons are: creation of public awareness; involvement of all the stakeholders in negotiation and implementation of the agreements; active participation in the scientific and technological panels related to the agreements; understanding the new concepts such as incremental costs; capacity-building for assessment of national status, impacts, and new technologies, technology transfer and project preparation and implementation and understanding, choosing and implementing appropriate policy instruments.

29.1 INTRODUCTION

The ozone agreements – the Vienna Convention for the Protection of the Ozone Layer (1985) and the Montreal Protocol on Substances that Deplete the Ozone Layer (1987) – have the objective of protection of the stratospheric ozone layer by phasing out the emissions of man-made ozone-depleting substances (ODS). A total of 185 governments, including almost all the developing countries, have ratified the agreements and have been implementing them successfully. These agreements pioneered the implementation of conditions necessary for ensuring that the economic growth of developing countries is not hampered by their implementation of global environmental agreements designed to solve specific global problems.

29.2 THE DEVELOPMENT OF THE OZONE AGREEMENTS

When scientists first raised the hypothesis linking chlorofluorocarbons (CFCs) to stratospheric ozone depletion in the early 1970s, it created a stir in a few countries, such as the USA. Most of the industries, manufacturing or using CFCs, scoffed at the theory and demanded more concrete proof on the link and on the adverse impacts of ozone depletion. The United Nations Environment Programme (UNEP), with the help of scientists, persuaded the world community that waiting for more certainty would increase the penalties from ozone depletion and that the precautionary principle should be followed to avoid serious consequences. The inter-governmental negotiations arranged by the UNEP in the early 1980s led to the Vienna Convention of 1985, through which the governments committed, in principle, to take action against a distant threat without waiting for a full proof of effect. An achievement of this Convention was to intensify the scientific research throughout the world, which kept providing information to the governments on the extent of ozone depletion and further evidence of the link between CFCs and ozone depletion.

It is this scientific information that persuaded the governments to take a first step, through the Montreal Protocol in 1987, to reduce the consumption of eight specified ODS gradually. This path-breaking Protocol recognised the inadequacy of such reductions in stopping ozone depletion and included a provision for adjustment and amendment of the Protocol based on a scientific assessment at least once in every four years. These assessments

Climate Change and Africa, ed. Pak Sum Low. Published by Cambridge University Press. © Cambridge University Press 2005.

studied the extent of ozone depletion, the adverse environmental effects of such depletion and the viability of alternative chemicals and processes to the ODS used in industries, and worked out policy options for control measures on the ODS. Scientists and industry throughout the world cooperated in contributing and preparing these assessments. Based on these assessments, the Protocol has been adjusted five times and amended four times in quick succession, in 1990, 1992, 1995, 1997 and 1999, to expand the list of controlled ODS and to prescribe early dates of total phase-out for each ODS.

29.3 KEY FEATURES OF THE OZONE AGREEMENTS PROCESS

According to Tolba and Rummel-Bulska (1998),[1] the process of the ozone agreements has five key features:

(a) Basing the negotiation process on scientific information, provided by international panels of experts.

(b) Moving forward one step at a time on the basis of science and availability of alternatives, recognizing that such incremental steps will not impose undue hardships on any country. The control measures of the Protocol at any stage are only interim steps, justified by the conclusions of the scientific and technological assessment at that stage. The measures are liable for change, if future assessments make it necessary.

(c) Creation of conditions for the participation of all countries of the world in the agreements, recognizing that any country keeping out could detract from the beneficial effect of the implementation of the agreements by other countries. Special attention was paid to the needs of the developing countries – *Shared but differential responsibility*. Developed countries provide the lead.

(d) Provision of financial assistance and technology transfer to developing countries.

(e) Trade restrictions on non-parties to discourage *free-riding* and encourage universal participation.

29.4 SIMILARITIES BETWEEN THE OZONE AND CLIMATE CHANGE ISSUES

Both the ozone and climate change issues deal with adverse impacts on the atmosphere of emissions of some gases in the process of economic and industrial development. These development activities have intensified in the past 300 years as a result of technological developments and have conferred immense benefits on humanity in raising their standards of living. The ozone-depleting chemicals were discovered and found useful in many applications,

such as refrigeration, air-conditioning, fire fighting and soil fumigation. The excessive emission of greenhouse gases, which cause climate change, arose because of the increasing consumption of energy in various socio-economic sectors, especially agricultural and industrial activities, and transportation, which modern humanity cannot do without. The challenge, in both cases, is how to avoid the adverse impacts on the atmosphere without interrupting the economic development to raise the standards of living of humanity.

In meeting this challenge for both issues, industry has to respond by innovating new and environment-friendly technologies. The pressures of the market mostly guide the response of the industry, and the market is influenced by consumers and by government regulations. Scientists, citizens, non-governmental organizations (NGOs) and the media play an important part in moulding the consumer preferences and promoting regulations. International organizations also play their part in catalysing the various stakeholders to arrive at international agreements to ensure a level playing field for all countries, industrialized or developing. These similarities between the two issues have resulted in the climate change agreements following closely the features of the ozone agreements as outlined in the previous section.

29.5 THE SUCCESS OF THE OZONE AGREEMENTS

The results of the ozone agreements in the last 10 years have been very impressive. The industry, with the signal given by the Protocol that alternatives to ODS will have a good market, cooperated in discovering the alternatives and facilitated the implementation of the phase-out of ODS. The production and consumption of ODS in the world have declined by more than 85% from 1986 to 2000 (UNEP, 2002). Scientists have predicted that the ozone layer will begin recovery in the next few years and will recover fully by the middle of this century if the Montreal Protocol is fully implemented (UNEP, 1998). The Protocol has been shown to be effective, and has resulted in significant benefits for human health, fisheries, agricultural production and building materials (Environment Canada, 1997). Technological innovation driven by the agreements has created economic and environmental benefits shared by all countries.

Almost all the countries of the world, including developing countries, have now ratified the ozone agreements. Fifty-one of the 53 African countries have ratified and are fully participating in the agreements. The Protocol provided for the developing countries a grace period of 10 years for implementing the control measures. A Multilateral Fund was established, contributed to by the developed countries, to meet the incremental costs of implementing the control measures by the developing countries. This fund has so far given out more than a billion US dollars to 130 developing countries, enabling them to take up more than

---

[1] See Chapter 5 in Tolba and Rummel-Bulska (1998).

2,000 projects to phase out ODS by introducing alternative technologies (Andersen and Sarma, 2002).[2]

29.6 LESSONS FROM THE OZONE AGREEMENTS

The United Nations Framework Convention on Climate Change (UNFCCC) of 1992 is being implemented. While the Kyoto Protocol of 1997 has, as of February 2003, not yet entered into force, many Governments are striving to expedite their ratification. Considering the many similarities between the ozone and climate change issues as outlined above, the experience of developing countries in successfully implementing the technological change promoted by the ozone agreements offers valuable lessons to them for participating effectively in the climate change agreements. Some of these lessons are discussed below.

29.6.1 Awareness, information, education and training

Environment-friendly technological change involves the efforts of governments, industries and other sectors, as well as every section of the population. Those who need to change their practices must be made aware of the need to change and they would require information before accepting alternatives. The adoption of alternatives would require education and training of all those involved in the change. One of the main challenges of the acceptance and implementation of the Montreal Protocol has been to create awareness and provide information to the thousands of industries that use the ODS on the need for, and the viability of, switching to ozone-friendly alternative chemicals and processes. In addition, the general public, which consumes the products of the industries, also needs to accept the changes. For example, one of the more difficult tasks of the implementation of the Protocol has been to persuade secretive military establishments to change over to ozone-friendly alternatives. The media and environmental NGOs in the USA, in collaboration with the Environmental Protection Agency (EPA), have played a large role in persuading the US military to take up the elimination of ODS consumption (Cook, 1996). Subsequently, the US military establishment played a positive role in helping the military establishments of other countries to phase out the ODS. The involvement of both industrial and environmental NGOs also enables governments to learn from the experience of industry and the public on the most cost-effective ways of bringing about the required change. A sufficiently aware citizenry will compel changes in attitude on the part of everyone concerned.

29.6.2 Involvement of stakeholders

The experience of countries in implementing the Montreal Protocol through the cooperation of every stakeholder – industry, scientists, technologists, the media, decision-makers, professionals and the public – can be replicated for the climate change treaties. Technological and attitudinal change is the bottom line in both the Montreal Protocol and the climate change agreements. One lesson of the Montreal Protocol is that given clear signals and incentives to invest in environmentally sound technologies, industry would enthusiastically participate in transforming itself. It is essential that industry be involved from the beginning by the governments in the process of negotiations, as well as in all the aspects of implementation of the environmental agreements. The success of the Protocol is also the result of involvement of thousands of individuals from an astonishingly diverse number of professions – engineers, fire fighters, standards makers, medical doctors, regulators, lawyers, agriculturists, officials, training specialists, customs officers and workshop technicians, to mention only some (Andersen and Sarma, 2002).[3] The developing countries must remember to involve all stakeholders when they tackle the climate change issue.

29.6.3 Global participation necessary and unavoidable

While any country, as a sovereign state, can keep out of the ozone agreements or the climate change agreements theoretically, in practice this is not possible for many reasons, such as technology dependence, political and moral pressures and trade measures. Further, the resources of global financial mechanisms of these agreements will be available only for those developing countries that commit themselves to the agreements.

Technology dependence

In the interdependent global economies of today, change of technologies in the industrialized countries has an impact on the developing countries. The developing countries, for example, could not have continued their reliance on products with ODS while other parts of the world move away from the ODS technologies. The climate change agreements impact on many areas and sectors, and the world is bound to move to new technologies in crucial areas such as energy, agriculture, etc. The developing countries need access to these technologies in order to keep up their economic progress. Such access can be facilitated by access to the resources of the financial mechanism of the climate change agreements.

Political and moral imperative

Ozone depletion would have become worse if even a few countries had kept on increasing their consumption and emissions of the

---

[2] See Chapter 6 of Andersen and Sarma (2002).

[3] See the Introduction in Andersen and Sarma (2002).

ODS. The adverse impacts of ozone depletion are almost universal. Similarly, climate change problems caused by unsustainable, environmentally negative patterns of production or consumption cannot be fully solved if even a few countries continue such patterns. The Parties to the agreements would, in order to ensure that the gains from their implementation of the agreements are not nullified by the actions of the non-parties, exert tremendous moral and political pressure on the non-parties to join in the agreements. The public opinion within the non-party countries too would exert similar pressure.

Trade measures

The trade measures of the ozone agreements ensured that non-parties could not obtain their requirements of ODS for maintenance of their existing equipment. They also would have been unable to export any equipment containing the ODS. While the climate change agreements have no similar trade provisions, the parties could observe trade discrimination against non-parties for environmental reasons. While controversial, Article XX of the General Agreement on Tariffs and Trade does permit such discrimination. The World Trade Organization, at its meeting in Seattle in December 1999, was under pressure from some countries to legitimize trade restrictions to safeguard the environment. The attempt did not succeed but the future could see a different result.

Global financial mechanisms

All the aid mechanisms – the Global Environment Facility (GEF), the World Bank, the regional development banks, the United Nations agencies and bilateral aid agencies – make their aid to promote new technologies relevant to the agreements conditional on the developing countries committing themselves to the objectives of the agreements.

Clean development mechanism

The UNFCCC and its Kyoto Protocol have no mandatory measures for developing countries. The Kyoto Protocol prescribes control measures for the greenhouse gas emissions of developed countries. However, the Clean Development Mechanism (CDM), envisaged in the protocol, allows the developed countries to fulfil a part of their obligations through assistance to developing countries to control their greenhouse gas emissions. Therefore, even without any legal obligations for control measures, the developing countries can be fully involved in the implementation of the control measures of the Kyoto Protocol. The resulting introduction of new technologies will benefit the developing countries, if the CDM projects can be implemented efficiently and effectively.

29.6.4 Participation in scientific and technological panels is a must

The international policy decisions on global environmental agreements are increasingly based on scientific and technological advice. The Montreal Protocol has set the trend in this regard. The successive amendments of the Protocol over the last 10 years have been almost wholly based on the recommendations of the three assessment panels of the Protocol – The Scientific, Environmental Effects, and Technology and Economic Assessment Panels. These Panels have over 500 experts from all regions of the world. Nearly a third are from developing countries, whose participation expenses in the Panels are met by the budget of the Protocol. The developing-country members have not only contributed tremendously to the work of the Panels but also gained greatly by being in regular touch with the innovators of cutting-edge technologies in the industrialized countries. The countries to which these members belonged gained too in getting expert advice from these members.

As pointed out by Canan and Reichman (2002),[4] 'Participants in the Montreal Protocol experience developed human and social capital through their networking activities, enhancing their environmental expertise and their internationalism.– This is especially beneficial among the developing country members of the ozone-layer community. The experience and contacts they have gained should reduce the developing-country dependence on outside consultants.'

Similarly, the evolution of control measures with regard to climate change will be greatly influenced by the reports and findings of the Intergovernmental Panel on Climate Change (IPCC), the UNFCCC's Subsidiary Body on Scientific and Technological Advice (SBSTA) and other expert bodies on greenhouse gases and on the viability of technologies to minimize greenhouse gas emissions. The findings of these bodies have direct impact on the international assistance to developing countries in particular areas such as energy. The developing countries need to participate meaningfully in these expert panels to ensure that their interests are represented. Such participation will help them to base their positions on the basis of science and to influence international decisions. It will also greatly enhance their scientific and technological capabilities to deal with the complexities of climate change (Benedick, 1998).[5]

29.6.5 Understanding the new concepts

The UNFCCC has introduced some new concepts that need to be understood in depth if the developing countries are to participate meaningfully in the implementation of the agreement.

---

[4] See Chapter 8 in Canan and Reichman (2002).
[5] See Chapter 9 in Benedick (1998).

Incremental costs

It was the Montreal Protocol that introduced, through political negotiations, the concept of incremental costs to be financed by the financial mechanism. The GEF, the operating entity of the financial mechanism of the UNFCCC, has also adopted this concept. While the concept may be clear intuitively, each particular project of the developing countries to be financed presents its own challenge in calculating the incremental costs. The GEF and the Executive Committee of the Multilateral Fund have, over time, clarified this concept through theoretical studies and their interpretive decisions. However, there is still some subjectivity in the calculations. The proceedings of the Executive Committee of the Multilateral Ozone Fund, which has disbursed more than a billion dollars as incremental costs for nearly 2000 projects for developing countries, show an astonishing resilience and ingenuity in interpreting incremental costs. These interpretations are results of political negotiations between the developed and developing countries in each particular case. To participate in such negotiations and to draw the full advantage of the financial mechanisms, the countries must be fully familiar with the intricacies of such calculations.

Clean development mechanism

The CDM of the Kyoto Protocol will provide the developing countries with an opportunity to improve their technology and environment. The elements of the CDM framework will promote sustainable development. However, the definitions of baseline, the emission reduction measurements and valuation of the benefits are extremely complex issues, which need to be understood by the developing countries to get full advantage of the mechanism.

29.6.6 Capacity development

An important lesson of the Montreal Protocol to the developing countries is the need for rapid capacity development in many areas if they are to deal competently with the climate change agreements.

Country programmes

Almost all the developing countries have developed, with the assistance of the Multilateral Fund, a country programme for the implementation of the Montreal Protocol. This programme has been arrived at by each country after assessing the status of ODS consumption in each of the sectors in that country and arriving at a strategy to phase out ODS without disrupting the country's economy. A similar assessment and delineation of strategy is necessary for each country with respect to the climate change agreements.

Each country has to assess its emissions of the greenhouse gases now and in the future, the impact of global emissions on itself, the adaptation measures needed to minimize the adverse impacts and the impact of implementation of the agreements by itself and others on its economy. It is true that assistance is given by the Global Financial Mechanisms to developing country Parties to prepare such assessments. However, the country will develop its capacity to deal with the problem only if the experts of the country acquire enough expertise to assess the situation for themselves.

Assessment of technologies

The developing countries will have to pay special attention to the technologies and sectors involved in, or affected by, the climate change agreements. It is important that the developing countries are aware of newly emerging technologies and trends and of the viability of these technologies and trends. Tall claims are often made on behalf of new technologies, which need to be verified. The new technologies, even when viable, are frequently replaced with improved versions in a short period. Some of them may have other adverse environmental impacts. Often it will be desirable for developing countries to wait for the technology picture to be clearer rather than rush to new technologies, which may be obsolete shortly.

Many international organizations, such as UNEP, UNDP, UNIDO and the World Bank, and national organizations, such as the EPA of the USA, maintain clearing houses for information on environmentally sound technologies. The OzonAction programme of UNEP maintains a data bank on such technologies. Developing countries would gain greatly by vigilance on this score.

In the implementation of the Montreal Protocol for example, the technologies to replace CFCs from 1985–1995 were predominantly using HCFCs, which also needed to be phased out under the Montreal Protocol, though after 30 years. Now many technologies have emerged where a double transition is not necessary and the CFCs can be replaced with ozone safe technologies. The technologies using HFCs (greenhouse gases) to replace CFCs are another example of the need to assess technologies for their suitability (UNEP, 1999).

The recipients of technologies must be extremely careful to ensure that they receive the right technologies. For example, some corporations in the industrialized countries transferred ODS technologies to developing countries in 1987 and 1988 knowing full well that the technologies would become obsolete shortly (Andersen and Sarma, 2002).[6] Also where alternative technologies have to be adopted to implement a Protocol within a stipulated time, many new untested technologies emerge to take advantage

---

[6] See Box 3.6 by Maneka Gandhi, Chapter 3, in Andersen and Sarma (2002).

of the demand. Better ones will soon replace these technologies. The developing countries cannot afford to make many transitions to new technologies, and hence they must get the full information and make wise choices on technologies. They must develop their capacity for assessment of technologies.

Keeping in touch with the latest technological trends is easier if the developing countries place their experts on the panels of organizations that assess new technologies. Even if the experts of the developing countries may not be as familiar with the latest technologies as the experts from the industrialized countries, the best experts in various areas in the developing countries can, through their membership of the expert panels, keep in touch with the latest trends in technology. They can also enrich the panels by giving their expertise on technology transfer to developing countries.

Technology transfer

The implementation of the Montreal Protocol has given clear lessons on the issue of technology transfer. In the abstract, technology transfer is discussed as though it is a gift by the industrialized countries to the developing countries. However, in practice, this is not so. Many environmentally sound technologies are in the public domain and can be adopted by the developing countries at no cost. In the case of many technologies, which are held by many companies, the companies are quite eager to sell their technologies at reasonable prices. It is only in the cases of a few technologies held by a few companies, that the holders of the technologies are reluctant to transfer or are willing to transfer only with tough conditions. Assistance will be available from the financial mechanisms and the CDM to facilitate technology transfer, but these mechanisms cannot assure transfer under conditions satisfactory to the developing countries if the companies holding the technologies do not agree to the conditions. The developing countries tried to incorporate guarantees of technology transfer into the Montreal Protocol but failed (Andersen and Sarma, 2002).[7]

Thus, while the Multilateral Fund for the implementation of the Montreal Protocol succeeded in getting most of the advanced technologies, some developing countries complained that they are unable to get the technology for manufacture of HFC 134A, a key substitute for CFCs (Andersen and Sarma, 2002).[8] In such cases, the developing countries will be well advised to initiate their own research and development to invent alternative technologies. China and South Korea, for instance, developed their own processes for manufacturing HFC 134A. Each country must develop its capacity for research and development in industrial processes crucial to its economic development to the best of its ability, using international assistance where available but without such assistance, if necessary.

Project preparation and implementation capacity

Most developing countries are only at the beginning of their development. They are characterized by inefficient paths to development. Energy inefficiency, obsolete technologies and outdated information networks characterize them. While the world community helps them to progress, all donors tend to maximize the returns on their investment and prefer the more organized and better-governed developing countries, rather than spending their money and time in countries that require a lot of effort to succeed. The global financial mechanisms offer the developing countries a unique opportunity to get the latest technologies at subsidized rates, in addition to investment to implement these technologies. The developing countries that are more prepared will get more resources from this exercise. They should, therefore, strive to improve their project preparation and implementation capabilities.

29.6.7 Policy instruments

One important lesson from the implementation of the Montreal Protocol is that the innovative policy instruments are essential for cost-effective implementation (Andersen and Sarma, 2002).[9] The instruments are of three broad categories, as discussed below.

Command and control

Once countries accede to a legal instrument and commit themselves to its objectives, an effective way of implementation is through a regulatory framework of laws, rules, and codes, etc., which mandate the actions required from all the stakeholders to achieve the objectives. Examples include the bans on the uses of CFCs in particular sectors; import quotas; changing the codes and manuals which earlier mandated the use of ODS; mandatory energy efficiency standards, and others. Such measures provide certainty for all stakeholders, who can take action to adjust their activities to suit the regulations. However, such regulations will be effective only through adequate monitoring and enforcement, which require resources, acquisition of skills, and training. Too rigid a regulation will reduce the flexibility to the stakeholders and the country in achieving the objectives. Thus, there is a need to carefully design regulations to incorporate sufficient flexibility. Companies find the best and most effective ways to implement control measures if they are given the freedom to innovate.

Market-based interventions

These are essentially pricing signals to affect the decisions taken by producers and consumers. These signals include taxes, targeted subsidies, tradable pollution permits or import quotas, etc. These promote cost effective implementation, give flexibility to

---

[7] See Chapter 3 in Andersen and Sarma (2002).
[8] See Chapter 4 in Andersen and Sarma (2002).

[9] See Chapter 6 in Andersen and Sarma (2002).

stakeholders and increase the revenues of the Government. Some examples are: (i) tax on CFCs levied by the USA to encourage alternatives; (ii) the emission trading system of the USA to reduce emissions of sulphur oxides, which cause acid rain; (iii) subsidies given to ozone-safe technologies by many countries; and (iv) auctionable permits used in Singapore.

Singapore's approach to the implementation of the Montreal Protocol has been very innovative (UNEP *et al.*, 1998).[10] Each quarter, the national quota of ODS allowed by the Protocol is allocated to the importers and users, half on the basis of historic consumption and half through open auction. Each firm must submit its sealed bid indicating the quantity it wants and the price it offers. Bids are then ranked by price and the lowest winning bid (the one that clears the market) serves as the quota price. This system gives an incentive to firms to adopt alternative technologies. The government, on its part, gains more revenues.

Voluntary agreements and public involvement

Another innovative method adopted in the implementation of the Montreal Protocol is voluntary agreements with industry in the implementation (Cook, 1996). Many countries, such as the USA, Germany, The Netherlands and Norway, have entered into such agreements with industry. *Eco labelling* helped consumers to choose products containing, or made with, ozone-safe chemicals and prompted the faster phase out of the ODS by the industries. Efforts to create awareness on the importance of protection of the ozone layer through the media, Ozone Day celebrations, publications and workshops ensured that both the producers and consumers got the message that fast phase out of CFCs by the producers is the only way to save the ozone layer. The clearing house of information of UNEP spearheaded these efforts. Its clearing house on alternative technologies was a great asset in educating the industries of developing countries to switch to ozone-safe alternatives. Similar methods will greatly assist in the implementation of the climate change agreements.

29.7 CONCLUSION

Climate change is a serious threat looming before humanity. All the countries will, sooner or later, have to take concrete actions

to face the threat. These actions require awareness, capacity-building, new environment-friendly technologies, innovative sustainable development paths and the resources to take such paths. The ozone agreements have shown the way the developing countries can cooperate with the efforts to solve global environmental problems without affecting their economic interests. They should utilise the lessons learnt to safeguard both their environmental and socio-economic interests, while implementing the climate change agreements.

ACKNOWLEDGEMENTS

The author would like to thank Robert Reinstein and an anonymous reviewer for their constructive comments on the original draft of the manuscript.

REFERENCES

Andersen, S.O. and Sarma, K.M. (2002) *Protecting the Ozone Layer: The United Nations History*. London, EarthScan.

Benedick, R. (1998) *Ozone Diplomacy: New Directions in Safeguarding the Planet, Enlarged Edition*, Cambridge, MA, USA, Harvard University Press.

Canan, P. and Reichman, N. (2002) *Ozone Connections: Expert Networks in Global Environmental Governance*. London, Greenleaf Publishing.

Cook, E. (1996) *Ozone Protection in the United States: Elements of Success*. World Resources Institute, Washington, DC.

Environment Canada (1997) *The Right Choice at the Right Time*.

Tolba, M. and Rummel-Bulska, I. (1998) *Global Environmental Diplomacy: Negotiating Environmental Agreements for the World, 1973–1992*. Cambridge, MA, USA, MIT Press.

UNEP (1998) *Report of the Scientific Assessment Panel of the Montreal Protocol*. Ozone Secretariat, UNEP, Nairobi, Kenya.

(1999) *Report of the Technology and Economic Assessment Panel of the Montreal Protocol on Alternatives to HFCs HCFCs*. Ozone Secretariat, UNEP, Nairobi, Kenya.

(2002), *Production and Consumption of ODS, 1986–2000*. Ozone Secretariat, UNEP, Nairobi, Kenya.

UNEP, US NASA and The World Bank (1998) *Protecting our Planet, Securing our Future*. The World Bank, Washington, DC.

---

[10] See Box P, Part IV in UNEP *et al.* (1998).

30 Opportunities for Africa to integrate climate protection in economic development policy*

STEPHEN O. ANDERSEN

United States Environmental Protection Agency, USA

Keywords

Climate change; ozone protection; environment; business opportunities; economic development; Montreal Protocol; Kyoto Protocol; sustainable development; corporate leadership; technology cooperation; sustained financing

Abstract

Protection of the environment is increasingly recognized as a prerequisite to economic development because a healthy workforce and sustainable access to natural resources are the foundation of any economy. Protection of the stratospheric ozone layer is being accomplished by diplomacy, corporate leadership, technology cooperation and financing. African and other developing countries can apply lessons from ozone layer protection to address climate change. Economically successful strategies can be developed for international diplomacy, domestic regulation, corporate leadership, consultancy, financing and technology development and adaptation.

30.1 INTRODUCTION

Africa and other developing countries face daunting challenges to simultaneously increase the standard of living and to protect the global and local environment. It is increasingly recognized that protection of the environment is a prerequisite to economic development because a healthy workforce and sustainable access to natural resources are the foundation of any economy. Many of the multinational companies most important to national economic development have environmental policies that give priority to investments in countries most able to support sustainable development.

Climate protection is particularly important to African countries because they will be among the first to experience the consequences of climate change (droughts, floods and sea-level rise), and will be among the least financially able to mitigate the damage.

This chapter explains how the phase-out of ozone-depleting substances (ODS) under the Montreal Protocol on Substances that Deplete the Ozone Layer (1987) has complemented economic development and may serve as a model for international cooperation on other environmental protection efforts, particularly climate change (see Benedick, 1991 and 1998; Andersen, 2000; Andersen and Sarma, 2002; Canan and Reichman, 2002; Parson, 2002).

30.2 THE MONTREAL PROTOCOL MOBILIZED TECHNICAL INNOVATION

When the Montreal Protocol was signed in 1987, few alternatives to ozone-depleting substances were available or even imagined. Industry predicted that technical progress would be slow, that costs would be exorbitant, that valuable equipment would be abandoned, and that safety might be compromised.[1] As it turned out, these predictions were never realized. In fact, in many cases the new technologies are now more energy efficient, more affordable and more reliable than those based on chlorofluorocarbons (CFCs) and other ozone-depleting substances (Cook, 1996; Le Prestre *et al.*, 1998; Andersen and Sarma, 2002).

30.3 THE IMPORTANCE OF INDUSTRY LEADERSHIP TO THE MONTREAL PROTOCOL

The Montreal Protocol is a remarkable success in a variety of ways. Foremost, it is environmentally successful because the

---

*The views presented here are the views of the author and do not necessarily represent the views of the U.S. EPA.

[1] To chronologically review perspectives on the history of the Montreal Protocol see: Molina and Rowland (1974); Dotto and Schiff (1978); Brodeur (1986); Miller and Mintzer (1986); Cogan (1988); Gribbin (1988); Nantze (1991); Benedick (1991 and 1998); Cagin and Dray (1993); Tolba and Rummel-Bulska (1998); Canan and Reichman (2002); Grundmann (2001); Parson (2002) and Andersen and Sarma (2002).

Climate Change and Africa, ed. Pak Sum Low. Published by Cambridge University Press. © Cambridge University Press 2005.

controls placed on the production of ozone-depleting substances are working: atmospheric concentrations of chlorine and bromine are coming down; and ozone depletion will soon be reversing. It is diplomatically successful because it is proving that the global commons can be wisely managed with stringent obligations, adequate financing, global cooperation and flexible procedures. It is economically successful because it was affordable to business and virtually unnoticed in consumer prices. It is technically successful because engineers used the necessity of change to thoroughly redesign systems, resulting in better efficiency, better reliability and better overall value for consumers.

One reason for this success is that industry and governments worldwide learned to cooperate to accomplish common objectives. Looking back, it is possible to trace many unanticipated ways that organizations with little previous experience working together, ultimately helped each other to overcome technical challenges (Andersen and Morehouse, 1997; Andersen, *et al.*, 1997; Andersen 1998).

Another reason for success is that companies in developing countries and their governments realized that their economies would be most affected by environmental despoilment but least able to respond to its destructive effects.[2] Lost productivity and public spending for health programmes slow economic development. Funds needed to counter the harmful effects of ozone depletion compete with other needs.

Worldwide, environmentally concerned children influenced their families and friends to support boycotts against CFC products. Businesses worked hard to shift from obsolete technology that damaged the ozone layer to modern CFC-free technology that was often more profitable.

International organizations modernized their operations to more effectively protect the global environment (see UNEP, 1996, 2000). Until the Montreal Protocol, it was a tradition of United Nations organizations to seek technical advice primarily from government and academic experts who were recommended through national diplomatic channels. Former UNEP Executive Director, Dr Mustafa Tolba, broke with this tradition and engaged business leaders and local technical experts (see DeCanio, 1991 and 1994; Brack, 1996; Tolba and Rummel-Bulska, 1998). These experts had the practical experience necessary to identify technology that would be appropriate to developing country circumstances and be better poised to integrate new technology in existing manufacturing facilities.

---

[2] Depletion of the stratospheric ozone layer allows more ultraviolet radiation in sunlight to reach the earth's surface. This increases skin cancer and cataracts and suppresses the human immune system; it damages agricultural crops and natural ecosystems and it accelerates deterioration of paints and plastics.

30.4 CLIMATE PROTECTION IS MOVING TOWARD TECHNICAL OPTIMISM AND BUSINESS OPPORTUNITY

When the Kyoto Protocol to the UNFCCC was adopted in December 1997, a vocal minority of global industry predicted dire consequences from possible regulations to reduce the emissions of greenhouse gases. Among these predictions were an end to economic growth, drastic changes in life-style, mass starvation and political chaos.

Now, only a few years after the Kyoto Protocol was signed, industry leaders are challenging the minority view and seeking ways to protect the climate in ways that stimulate technical innovation and save money. For example, John Browne (CEO of British Petroleum), Stephen M. Wolf (Chairman and CEO of US Airways), and an increasingly large number of executives from almost every kind of business have publicly agreed that action is needed now to avert the consequences of climate change (see www.pewclimate.org). British Petroleum already has an action plan to increase investment in photovoltaic electricity generation, particularly in countries like Africa where solar energy is strong and reliable and where remote areas need reliable electric service. Companies worldwide are reporting impressive financial savings from energy conservation programmes that reduce greenhouse gas emissions.

As an important part of the global community, industry must accept its responsibility to help protect the climate. For our part, BP Amoco plans to serve our global customers by supplying traditional and solar energy in a way that is environmentally sustainable.

John Browne, Group Chief Executive, BP Amoco PLC

Dozens of multinational companies are making Corporate Climate Leadership Pledges. Dow Chemical is pledging a 2% energy reduction per year. IBM has a goal to reduce PFC emissions by 40% by 2003, and Motorola has a goal of a 50% PFC reduction by 2010. Mitsubishi Electric, Ontario Hydro, Philips, Sanyo and United Technology have goals that will reduce emissions 25% by 2007–2010. Shell plans to reduce energy consumption 10% by 2002 and British Petroleum 10% from 1990 to 2010. Japanese and German steel producers are working to reduce energy intensity by 20–30%. Other companies, including Dow Chemical, Mitsubishi Electric, Philips, Sanyo and Seiko-Epson, have pledged to dramatically increase energy efficiency. Ford, General Motors, Honda, Mitsubishi, Nissan and Toyota are leading the way in electric, fuel cell, hybrid and other ultra-high efficiency technologies. Semiconductor manufacturers are planning expanded production as customers use computers to save energy and have pledges to reduce their own manufacturing emissions of potent greenhouse gases (see Andersen and Zaelke, 2003).

'Toyota has the means, and therefore the responsibility, to design and market cars that help protect the climate and perform as well as conventional technology.'

See http://www.toyota.com/about/environment/

Many industry associations are advocating climate protection. Some examples are the Alliance for Responsible Atmospheric Policy, the Business Council for Sustainable Energy, the International Climate Change Partnership, the International Cooperative for Environmental Leadership, the Pew Centre on Global Climate Change, and the World Semiconductor Council.

30.5 DEVELOPING COUNTRIES ARE USING GLOBAL PARTNERSHIPS AND STRATEGIES

Developing countries innovate in remarkable ways to protect the global environment, rivalling the efforts of even the most strident developed countries. For example, the government of Thailand announced the end of 1996 as its target date for eliminating CFCs used as refrigerants and as blowing agents for insulating foam in domestic refrigerators. This challenging target date was only one year later than the phase-out date for developed countries and was the first phase-out of the refrigerator sector in any developing country. The ambitious phase-out goals were ultimately achieved through corporate leadership and unprecedented technical cooperation. But the crown jewel of the project was that the Thai Government complemented this technical achievement by prohibiting the manufacture or import of CFC refrigerators after 1996: the first trade barrier enacted by any developing country in support of global environmental protection.

Vietnam is another developing country that has been progressive in environmental protection. In 1995, more than 40 multinational companies from seven countries pledged to help the Government of Vietnam protect the ozone layer by investing only in modern, environmentally acceptable technology in their Vietnam projects. In September 2002, leading multinational companies operating in Vietnam agreed to apply the same climate protection policy in Vietnam as used in their home countries. These companies also pledged to only manufacture and import products that achieve high energy efficiency, so as to reduce the total cost of ownership for their Vietnamese customers (see Andersen and Sarma, 2002).

30.6 DEVELOPING COUNTRIES AND COUNTRIES WITH ECONOMIES IN TRANSITION ARE SOURCES OF TECHNOLOGY

Countries with economies in transition and developing countries made remarkable contributions to the development of new

Table 30.1. Ozone protection technologies developed in Countries with Economies in Transition (CEIT) and developing countries.

Countries	Ozone protection technologies
Brazil	Low solids flux without solvent
Brazil	Caster oil for spray foam blowing
China and India	'EcoFrig'
Former East Germany	Hydrocarbon refrigerator
India	Neem replacement for methyl bromide
Republic of Korea	New HFC-134a production pathway
Mexico	Open-air HC aerosol filling
Poland	Military halon bank management

technology to protect the ozone layer. Some technology such as Neem oil, a natural pesticide to replace methyl bromide as a soil fumigant, is grown primarily in India and Africa. Other technology could have been developed anywhere but was first perfected in a developing country. For example, low solids flux used to solder electrical circuits without the need for solvents was perfected in Brazil. More recently, a company from the former East Germany was more willing to commercialize hydrocarbon refrigerators because they were more willing to take business risks than their competitors (see Andersen and Sarma, 2002).

30.7 STRATEGIES FOR DEVELOPING COUNTRIES TO INTEGRATE ECONOMIC DEVELOPMENT AND CLIMATE PROTECTION

African and other developing countries can develop national and regional strategies to help protect the climate that can attract increased investment. National experts can participate in technology cooperation and assessments as a way to learn first-hand how to invest wisely for the future. Joint projects can result in all sides winning. Strategies for achieving the integration of economic development with climate protection are presented below.

30.7.1 Diplomatic strategies

Work with national scientists and their networks to understand the importance of climate stability to economic development. Refocus diplomatic interventions to get beyond science debates. Encourage economic development projects that protect the climate while saving money. Identify and encourage national business leaders to participate in international technical committees.

30.7.2 Domestic regulatory innovation strategies

International finance agencies know that most investments are futile unless they complement effective domestic environmental policy. A country's demonstrated ability to link policy to environmental goals makes it a preferred candidate for future funding when regulations are necessary to guarantee investment success. National government can attract increased funding by developing a reputation for regulatory innovation and good governance.

30.7.3 Domestic industry strategies

Go on record as an advocate for energy efficiency whenever cost effective. Ask business partners, suppliers, customers and professional associations for new ideas on saving energy. Use the best ideas to develop investment proposals for upgrading business operations. Share your investment plans with government environmental diplomats and economic development authorities so they can watch for opportunities to fund your projects.

30.7.4 Consulting strategies

Developing countries that place their most effective technical experts on international technical committees will gain far more than countries that make appointments as political rewards.

30.7.5 Recognition strategies that encourage sustained funding

Future funding depends on both project and institutional success. Project success is achieved when work is completed on time and on budget, and the expected environmental benefits are forthcoming. Institutional success is achieved when donors, intermediaries and recipients are satisfied with the successful results and are fully credited for their contribution.

Institutional success can be rewarded with thoughtful personal letters or with public recognition such as project completion ceremonies, certificates of appreciation or with publications documenting the role of the most important individuals and organizations. It is natural that donors will want to work again with partners that make these extra efforts. Donor organizations will also be more successful in renewing their funding if they have evidence that the work is appreciated.

30.7.6 Technology development strategies

Developing countries can innovate independently or can be important partners to multinational technology commercialization teams. Developing countries can take advantage of world-class experts, agile and inventive business organizations, flexible regulations and unique market situations. For example, an African location may be better able to create adequate infrastructure for alternative fuel vehicles than locations where vehicles will travel further from base. Developing countries should catalogue national capabilities and offer to be demonstration projects.

30.7.7 Corporate leadership strategies

Ask national and multinational companies to demonstrate technical and environmental leadership. Every country in Africa has multinational enterprises and joint ventures. Multinational companies have easy access to technical information and may have policies to help protect the climate. Use the Internet to study environmental policy of corporate leadership pledges like those described above for Thailand and Vietnam. Host a meeting of industry representatives to talk about how international environmental polices can be used locally to support economic development. Explain the importance of environmental protection and ask how governments can help implement policies.

30.7.8 Streamline government for cooperation on climate protection

It is very frustrating to begin technology cooperation projects and then find out that outdated national laws inhibit success. Make sure that donors can efficiently import necessary equipment without onerous fees or long delays at Customs. Work with partners to improve the operation of financial intermediaries so projects can be completed on time and within budget. Remember that the long-term gains of technology cooperation are far more valuable than the short-term profits or satisfaction from demanding that outdated bureaucratic procedures are exactly followed.

30.8 LESSONS FOR CLIMATE PROTECTION-CONCLUSIONS AND HOPE FOR THE FUTURE

Although climate protection is more daunting than ozone layer protection, corporate and military leadership is already showing the world that technology can reduce greenhouse gas emissions. Technology cooperation under the Montreal Protocol mobilized experts from fiercely competing companies to share information and to speed technology development and application. Military organizations broke with precedents and made ozone layer protection a priority for environmental security. Leading companies announced ambitious goals that stimulated suppliers to do it better without ODS. Public gratitude and confidence rewarded innovation.

The history of stratospheric ozone protection is repeating itself in climate protection. Multinational companies are declaring climate protection goals. Industry associations are supporting an international protocol. National governments are choosing emission reduction strategies. Perhaps most remarkably, industry is working with government to protect the climate through voluntary measures complemented with regulatory incentives.

ACKNOWLEDGEMENTS

The author is indebted to the valuable comments of Dr Suely Carvalho, Dr Pak Sum Low, Dr Alan Miller, and the anonymous reviewers. The encouragement of Dr Pak Sum Low on completing this paper is greatly appreciated.

REFERENCES

Andersen, S. O. (1998) *Newest Champions of the World: Winners of the 1997 Stratospheric Ozone Protection Awards.* US EPA, 430-K-98-003, Washington, DC.

(2000) Industrial responses to stratospheric ozone depletion and lessons for global climate change. In M. Tolba (Ed.), *Responding to Global Environmental Change, Encyclopedia of Global Environmental Change.* New York, John Wiley and Sons.

Andersen, S. O. and Morehouse, E. T. (1997) The ozone challenge – industry and government learned to work together to protect environment. *ASHRAE® Journal*, September, 33–36.

Andersen, S. O. and Zaelke, D. (2003) *Industry Genius, Inventions and People Protecting the Climate and Fragile Ozone Layer.* Sheffield, UK, Greenleaf Publishing.

Andersen, S. O. and Sarma, M. K. (2002) *Protecting the Ozone Layer, The United Nations History.* London, Earthscan Publishing.

Andersen, S. O., Frick, C. and Morehouse, E. T. (1997) *Champions of the World: Stratospheric Ozone Protection Awards.* USEPA, 430-R-97-023, Washington, DC.

Benedick, R. (1991) *Ozone Diplomacy: New Directions in Safeguarding the Planet,* Cambridge, MA, USA, Harvard University Press, and the updated edition, Benedick, R. (1998) *Ozone Diplomacy: New Directions in Safeguarding the Planet, Enlarged Edition,* Cambridge, MA, USA, Harvard University Press.

Brack, D. (1996) *International Trade and the Montreal Protocol.* London, Royal Institute of International Affairs/Earthscan.

Brodeur, P. (1986) Annals of Chemistry in the Face of Doubt, *The New Yorker,* June 9.

Cagin, S. and Dray, P. (1993) *Between Earth and Sky: How CFCs Changed Our World and Endangered the Ozone Layer.* New York, Pantheon.

Canan, P. and Reichman, N. (2002) *Ozone Connections: Expert Networks in Global Environmental Governance.* London, Greenleaf Publishing.

Cogan, D. G. (1988) *Stones in a Glass House: CFCs and Ozone Depletion.* Washington, DC, Investor Responsibility Research Center.

Cook, E. (1996) *Ozone Protection in the United States: Elements of Success.* Washington, DC, World Resources Institute.

Dotto, L. and Schiff, H. (1978) *The Ozone War.* Garden City, NY, USA, Doubleday & Co.

DeCanio, S. J. (1991) Managing the transition: lessons from experience. In *Report of the Economic Options Committee.* UNEP, Nairobi, Kenya.

(1994) The dynamics of the phaseout process under the Montreal Protocol. In *Report of the Economic Options Committee.* UNEP, Nairobi, Kenya.

Gribbin, J. (1988) *The Hole in the Sky: Man's Threat to the Ozone Layer.* New York, Bantam.

Grundmann, R. (2001) *Transnational Environmental Policy: Reconstructing Ozone.* London, Routledge.

Le Prestre, P. G., Reid, J. D. and Morehouse, E. T. (Eds.) (1998) *Protecting the Ozone Layer: Lessons, Models and Prospects.* Boston, MA, USA, Kluwer Academic Publishers.

Miller, A. and Mintzer, I. (1986) *The Sky Is the Limit: Strategies for Protecting the Ozone Layer.* Washington, DC, World Resources Institute.

Molina, M. and Rowland, F. S. (1974) Stratospheric sink for chlorofluoromethanes: chlorine atom-catalyzed destruction of ozone. *Nature,* **249**, June 28.

Nantze, J. (1991) *What Goes Up: The Global Assault on Our Atmosphere.* New York, William Morrow Publishing.

Parson, E. A. (2002) *Protecting the Ozone Layer: Science and Strategy.* New York, Oxford University Press,

Tolba, M. and Rummel-Bulska, I. (1998) *Global Environmental Diplomacy: Negotiating Environmental Agreements for the World, 1973–1992.* Cambridge, MA, USA, MIT Press.

UNEP (1996) *Handbook for the International Treaties for the Protection of the Ozone Layer.* Nairobi, Kenya, United Nations Environment Programme.

(2000) *Action on Ozone.* Nairobi, Kenya, United Nations Environment Programme.

31 Ozone depletion and global climate change: is the Montreal Protocol a good model for responding to climate change?

ROBERT A. REINSTEIN

Reinstein & Associates International, Inc., Rockville, MD, USA

Keywords

Climate change; ozone depletion; Vienna Convention; Montreal Protocol; UNFCCC; Kyoto Protocol

Abstract

This chapter, originally written in 1996 during the negotiations on the Kyoto Protocol to the United Nations Framework Convention on Climate Change (UNFCCC) and recently updated, compares the situation of global climate change as addressed by the Kyoto Protocol with the situation of stratospheric ozone depletion as addressed by the Montreal Protocol on Substances that Deplete the Ozone Layer in 1987. It concludes that, while the two situations have many similarities, they are different in a number of regards. The manner in which the legally binding emission targets included in the Kyoto Protocol were determined has created a treaty that cannot serve as an effective longer-term approach to climate change. It is argued that the future approach should not be modelled strictly along the lines of the Montreal Protocol. Alternative means of encouraging all countries to reduce greenhouse-gas emissions need to be developed in a broad spirit of international cooperation.

31.1 INTRODUCTION

In 1987 the nations of the world agreed on a treaty, the Montreal Protocol, to reduce emissions of gases that deplete the stratospheric ozone layer. This agreement has been amended several times since 1987 to reflect increased knowledge and experience regarding the impact of different chemicals on the ozone layer and the options for reducing their emissions. Most observers point to the Montreal Protocol as an example of an extremely effective agreement for responding to the global environmental threat that it was negotiated to address. Some of those directly involved in negotiating the Montreal Protocol, as well as many environmental advocates, also argue that it is a very good model that should be adapted as an approach to respond to global climate change.

Representatives from most of the countries of the world engaged in a process from 1995 to 1997 of negotiating additional commitments under the United Nations Framework Convention on Climate Change (UNFCCC). The result of those negotiations was the Kyoto Protocol. Many essential details of the Kyoto Protocol were not resolved in Kyoto in 1997 and countries have only recently reached political agreement on these. The basic structure of the Kyoto Protocol is, in many ways, quite parallel to the approach taken under the Montreal Protocol.

As one of those involved directly in the negotiation of the Montreal Protocol and also of the UNFCCC, I do not believe the two situations are sufficiently comparable for the climate-change regime to benefit from such a direct adaptation of the Montreal Protocol approach, as will be explained later in this paper.

The fact that the Kyoto Protocol evolved in the way that it did is one reason a number of countries are having difficulties as they proceed with ratification and implementation. Moreover, the one feature that should have been adapted, namely emission targets negotiated from the 'bottom up' was not used for the Kyoto Protocol, and this results in a number of problems.

31.2 THE NATURE OF THE ENVIRONMENTAL PROBLEM

With regard to the nature of the environmental concern addressed, on the one hand, by the Vienna Convention[1] and Montreal Protocol and, on the other hand, the UNFCCC and the Kyoto Protocol, the two situations are both similar and also quite different.

When the Montreal Protocol was first negotiated, and earlier, concern was centred on the depletion of ozone (O_3) in the earth's upper atmosphere – the stratosphere. The depletion was believed to be due to chemical reactions that were assumed to be occurring

---

[1] Vienna Convention on the Protection of the Ozone Layer (1985)

Climate Change and Africa, ed. Pak Sum Low. Published by Cambridge University Press. © Cambridge University Press 2005.

there over long periods as a result of emissions of certain chemicals produced intentionally to meet various human needs. The earth's stratospheric ozone layer is critically important to human health and the well-being of many ecosystems because it shields the earth from ultraviolet radiation from the sun that would otherwise be damaging to life on earth. Excessive doses of ultraviolet radiation (UV-B in particular) can cause skin cancer and other diseases in humans and animals and can disrupt the normal functioning of ecosystems (UNEP, 2003).

The stratospheric ozone layer is in a sense a 'global commons' belonging to all the earth, not just to nations of human beings organized politically into 'sovereign' entities but also to all forms of life, which depend on its existence for their well-being. All suffer if it is damaged or destroyed, and all who are involved in producing emissions of ozone-depleting substances (ODSs) must in theory participate in the response in order for the ozone layer to be protected effectively.

In this sense, the concern regarding potentially dangerous interference in the global climate system resulting from human emissions of heat-trapping gases (greenhouse gases, or GHGs) is comparable to the ozone layer problem. The climate system, like the ozone layer, lies outside the jurisdiction of any nation, yet all nations are potentially affected by changes that may occur as a result of human activities, and virtually all must participate for the response to be effective.

However, with regard to the types and distribution of impacts, the two issues are rather dissimilar. Ozone depletion and its impacts are greatest at high latitudes, i.e. toward the North and South Poles. Moreover, the threat of skin cancer is greatest among light-skinned people, who also tend to live farther from the equator and toward the Poles. Thus, people in the Scandinavian countries, Canada, Australia and New Zealand are at the greatest risk from ozone depletion, while dark-skinned people living near the equator are at the least risk.

Impacts of global climate change, on the other hand, are quite different from those resulting from the depletion of the stratospheric ozone layer. These impacts are projected to be distributed in a rather uneven manner across the entire earth. Further, not all impacts are negative, as in the case of ozone depletion. Some countries will benefit from a more temperate climate or increased agricultural productivity, or both, although the IPCC has noted that the majority of impacts, including indirect impacts, are likely to be negative (IPCC, 2001a).

Sea-level rise, one of the more serious consequences of climate change, will be experienced by all maritime countries, but not with uniform impacts. The gradual shifting of the tectonic plates that make up the earth's surface causes some areas to rise over long periods and others to sink. Thus, the impact of GHG-induced sea-level rise may be offset in some locations, but in others it (among

other factors) may add to the natural rise in sea level due to the sinking of the land.

Some countries, especially those that are more economically and technologically advanced, will be better able to adapt to sea-level rise and other adverse impacts of climate change. Others, especially the least developed countries and small islands and low-lying coastal regions, will have much more difficulty adapting to the possible consequences of global climate change.

31.3 THE STATE OF SCIENTIFIC UNDERSTANDING

With regard to degree of scientific understanding of the problem itself, the environmental and health impacts of the problem, and the technical and economic aspects of possible responses, there are again differences between the situations of ozone depletion and global climate change.

The role of the stratospheric ozone layer in protecting the earth and its ecosystems from harmful UV radiation was well understood for many years prior to the negotiation of the Montreal Protocol. The damaging effects of excess UV-B were recognized and acknowledged. Doctors had long been urging their patients to avoid excess exposure to sunlight, or at the least to use effective sun screens.

In addition, ozone depletion had been verified by observations and by special aircraft equipped with sensors to monitor the state of the ozone layer. The causes of this depletion of ozone concentration had not yet been proven as of 1987, but certain chemicals were demonstrated to destroy ozone in laboratory tests under conditions comparable to those in the stratosphere, most notably the chlorofluorocarbons (CFCs) and bromofluorocarbons (halons). Chemical analysis showed that these chemicals, being relatively inert (not prone to enter into chemical reactions) under conditions near the earth's surface, would mostly rise through the lower atmosphere after being emitted and reach the stratosphere. Finally, increased concentrations of these chemicals had been observed in the stratosphere (WMO, 1994, 1998, 2002).

There was no 'smoking gun' in the sense of direct observation of the CFCs and halons actually destroying ozone in the stratosphere, but the theory and indirect evidence of causality were quite persuasive to most scientists. The discovery of the Antarctic ozone hole in the mid-1980s[2] was a watershed in terms of public perceptions of a global environmental threat caused by ozone depletion.

---

[2] Although the first decreases in Antarctic ozone hole were observed in the early 1980s, it was not until Farman et al. (1985) published their paper in Nature that the results became more widely known in the international community.

By comparison, the science concerning global climate change is extremely complex. Many different layers of the atmosphere must be analysed, with many different physical and chemical reactions. The roles of clouds, the oceans, land masses, vegetation, etc., must all be taken into account, as well as complex interactions with radiation from the sun and other parts of space. And then multiple and interrelated effects on temperature, precipitation, storm frequency and severity, sea-level rise and so forth should be projected not merely on a global basis but on a regional basis, and sufficiently disaggregated to allow individual countries to estimate the impacts of climate change in their national context.

The giant general circulation models that are used in powerful computers to project the possible effects of human GHG emissions have produced results that are generally consistent with observed climate patterns. The Summary for Policymakers of the Intergovernmental Panel on Climate Change (IPCC) working group on the science of climate change states that most of the recently observed increases in surface temperatures are probably due to human causes (IPCC, 2001b). However, it also acknowledges that there is other evidence that is inconsistent with the warming predicted by the models, such as the much smaller warming observed by satellites and weather balloons in the lower troposphere, and that neither clouds nor radiation are well represented in the current models.

A further unresolved question is what natural mechanisms caused significant warming of the planet in earlier geological periods. While ozone depletion appears to be only a recent phenomenon and distinctly related to human activity, the earth's climate system has undergone many major changes over millions of years, some taking place in the space of only a few decades, as a result of natural causes that are still not understood.

One argument that has been raised by environmentalists and some others is that the science regarding global climate change is likely to be confirmed, just as the science of ozone depletion was confirmed, and therefore nations must act immediately. This is a logical fallacy. One cannot assume that because scientific questions regarding ozone depletion were resolved in favour of confirming the environmental threat, the uncertainties regarding global climate change will also be resolved in the same way. The scientific aspects of the two issues – the physical and chemical processes – are quite distinct. It may be that the theory of human-induced global climate change will be confirmed in future years, but this has nothing to do with the earlier confirmation of ozone depletion.

While much work has been done on estimating the impacts of climate change, such as studies indicating possible increases in the incidence of vector-borne diseases like malaria, a great deal less is known about the potential impacts than is known about the impacts of ozone depletion.

The state of understanding of the economic consequences of various response options also differs significantly between ozone depletion and climate change. As described below, panels of experts (including industry experts) have analysed in detail possible alternatives to the use of ODSs, including technical aspects and costs of each. Much less is known about the potential costs of some of the major options for limiting GHG emissions, for reasons that will also partly be explained below.

31.4 THE NATURE OF THE RESPONSE OPTIONS

The initial international response to concerns about ozone depletion focused on eight chemicals that had been demonstrated in the laboratory to destroy ozone under conditions similar to those in the stratosphere and that had also been detected in increasing concentrations at stratospheric altitudes. The chemicals were five CFCs and three halons. These chemicals were artificially manufactured by a small number of companies and used for a limited range of specific applications, such as refrigerants, foam-blowing agents, fire-retarding agents and solvents. Most of the production and consumption of the chemicals took place at that time in the developed economies of the Organization for Economic Cooperation and Development (OECD) countries and those of central and eastern Europe.

Because there were very limited sources of supply of the ODSs, the negotiators of the Montreal Protocol focused on reducing this supply as a means of limiting the eventual emissions of the chemicals into the atmosphere. The simplest way to do this was by controlling production directly. However, to assure equitable distribution of this restricted supply among historical users dependent on these chemicals, supply available for internal consumption (defined as domestic production plus imports minus exports) was also controlled.

Getting industry's initial agreement to reduce and ultimately eliminate the use of CFCs was not easy, since billions of dollars would need to be spent on research on substitutes and capital investment in production and consumption-related equipment. However, it was facilitated by the fact that in most cases the same companies producing the CFCs were also the producers of many of the chemicals which ultimately replaced them in the majority of applications – the hydrochlorofluorocarbons (HCFCs) and hydrofluorocarbons (HFCs). Some users of these chemicals, particularly small firms, had problems. And not all uses of the CFCs and halons were substituted through the use of alternative chemicals. But in general the transition away from the eight ODSs controlled by Montreal Protocol has been fairly orderly, and at a cost considered acceptable by most of industry.

Since 1987 additional chemicals have been added to the list of ODSs controlled by Montreal Protocol, including carbon tetrachloride, methyl chloroform (1,1,1-trichloroethane) and methyl bromide. The schedule for reducing the use of the controlled substances has been advanced several times, as noted, based on increased knowledge and experience. The original 1987 Montreal Protocol required a 50% reduction in CFC emissions (production and consumption) by 1998. The 1990 London Amendment moved to a total phase-out of CFC emissions by the year 2000, and the 1992 Copenhagen Amendment advanced the phase-out date to 1 January 1996. The assessments carried out under Article 6 of the Montreal Protocol (described below) were critical in getting nations' (and companies') agreement to move rapidly in response to increased scientific confirmation of the severity of the environmental problem.

The nature of the response options for global climate change differs considerably from the situation with ozone depletion. Most GHGs occur naturally and are emitted in very large volumes through natural processes. Without the presence of these GHGs in the atmosphere, the natural greenhouse effect could not occur and the earth would be too cold to support human and other forms of life as we know them today. The most important GHG by far is water vapour, which is constantly being evaporated from the water and land surfaces of the earth, forming clouds and being removed from the atmosphere as rain, snow and other forms of precipitation.

Thus, unlike the ODSs controlled by the Montreal Protocol, anthropogenic (human-produced) GHG emissions are mostly a by-product of various human activities rather than resulting from intentional production in chemical plants. The primary anthropogenic GHGs are carbon dioxide (CO_2), methane (CH_4) and nitrous oxide (N_2O). All of these are produced naturally as well as through human activities. Many of the human activities leading to GHG emissions are basic to human survival, such as energy use for heating, cooking and basic manufacturing, agriculture and forestry. These activities are carried out in every part of the world, not just primarily in the developed nations of the OECD. With CFCs and halons, there were only a handful of producers throughout the world. On the other hand, with CO_2, for example, every human being who burns coal, oil or natural gas for whatever reason is a 'producer' of this key GHG. Any farmer raising cattle or sheep anywhere in the world is a 'producer' of methane.

It is clear that 'production' of CO_2, CH_4 and N_2O is not going to be 'phased out' as was the case for CFCs. Nor are the activities that lead to the by-product release of these gases going to be phased out, since they are mostly associated with meeting basic human needs. Providing food is essential for human survival. Consuming energy is also often essential to survival (cooking and space heating) and is central to any developed economy. Thus, simply applying the Montreal Protocol approach to limiting human GHG emissions is not appropriate.

31.5 THE ROLE OF ASSESSMENTS OF SCIENCE, TECHNOLOGY AND ECONOMICS

One respect in which the international responses to ozone depletion and global climate change are alike is the need for an integrated approach to understanding the problem itself and the various options for responding to it. Such an approach must take into account the evolution both of the scientific understanding of the problem and its impacts and also of the complex technological and economic aspects of different response options. Both treaties recognize this need in establishing a dynamic process for assessing the state of understanding of all these factors as an essential input to the implementation of the agreement.

Article 6 of the Montreal Protocol requires an assessment of the control measures (Article 2) at least every four years 'on the basis of available scientific, environmental, technical and economic information', and calls for the establishment of panels of experts in each of these areas to prepare conclusions and recommendations. In practice, because the Parties to the Protocol have held meetings every year, the assessment panels have been producing technical and economic assessments almost every year. Provisions are contained in Article 2 for adding new ODSs to the annexes of the Protocol and for making a number of adjustments to the emission reduction schedules or to the 'ozone depletion potentials' of the ODSs on the basis of the Article 6 assessments. These assessments have thus provided the basis for the additional control measures added in the London Amendment, the Copenhagen Amendment and the Vienna Amendment.

This assessment process has been a key element in the effective implementation of the Montreal Protocol. The science assessment has been critical in confirming the original hypothesis of ozone depletion by CFCs and other chemicals. But more importantly, the assessment of technical and economic feasibility with full participation by industry experts has been the main vehicle for advancing the reduction schedule of emissions of ODSs. Because those who are most directly affected by the decisions of the Protocol Parties are a part of the decision-making process, they are much more willing to accept and carry out the decisions of the Parties. And because they are involved, they also can anticipate the direction of the future development of control measures and factor this understanding into their investment decisions regarding both current commercial products and also research and development of future products.

An assessment mechanism is also included in the UNFCCC, and is contained, in effect, in Articles 4.2(d), 9 and 12. However, the actual work of the assessments is divided between the Conference of the Parties (COP) to the UNFCCC, the Subsidiary Body for Scientific and Technological Advice (SBSTA) and the independent IPCC. This assessment process is also used by the Kyoto Protocol. The IPCC has so far carried out both the science assessment and the technical and economic assessment work in some ways similar to that of the expert panels under Article 6 of the Montreal Protocol.

The assessment process for global climate change is, as yet, not really comparable to the work done under the Montreal Protocol. The situation for ODSs involves the reduction and eventual elimination over time of the production of the problem chemicals and their replacement in various uses by other chemicals or by techniques that do not require chemicals. The basic approach was clear from the beginning; the only issue was how to set the interim reduction targets over time so that they corresponded to the availability of economically and environmentally acceptable alternatives.

From the point of view of assessing alternative technologies and techniques that can reduce GHG emissions, the situation is very different from that of the ODSs. As noted above, most anthropogenic GHG emissions are not the result of direct production of the GHGs but the indirect consequence of other actions. Because they are also produced naturally in very large quantities, they will not be 'phased out' in the same way as ODSs. Even under the assumption of stringent emission reductions to stabilize atmospheric concentrations of anthropogenic GHGs at current levels, global emissions from human activities would still be left at some non-zero level significantly below today's emissions. Thus, the question posed to experts involved in the assessment process is not how quickly can the GHGs be phased out at acceptable costs, but simply what are the economically and environmentally acceptable ways of reducing their emissions.

Although there are fewer major GHGs than ODSs, the number of situations giving rise to their emission to the atmosphere is orders of magnitude larger than those resulting in ODS emissions. Almost every sector of the economy is affected. And, unlike the situation for most ODSs, the mix of human activities and the associated technologies giving rise to GHG emissions differ greatly from country to country. For example, an alternative to fossil-fuel use for transportation that may work very well for one country or group of countries may not work for others because of such factors as population density and the mix and associated costs of alternative sources of energy.

The experts gathered under the Montreal Protocol assessment process tend to come up with specific judgements that, for example, emissions of a certain ODS could be reduced by 50% by a certain year through the use of economically and environmentally acceptable alternatives. These recommendations then serve as the basis for decisions by the Parties to the Protocol to set reduction targets that are applied to all countries. No such judgements are possible in the case of the GHGs because of the large differences in national circumstances that affect both technical options and associated costs, as acknowledged by the IPCC Third Assessment Report (IPCC, 2001c). Thus, the role of the technical and economic assessment bodies under the UNFCCC must be quite different from that of the assessment panels under the Montreal Protocol.

31.6 THE ROLE OF DEVELOPING COUNTRIES

It was noted earlier that the impacts of global climate change on most developing countries are likely to be much more serious than the impacts of ozone depletion as compared with the impacts on industrialized countries. The role of developing countries in responding to climate change in the near term is also relatively more important than their near-term role in responding to ozone depletion.

Under the Montreal Protocol developing countries were given a 10-year grace period before the control measures undertaken by other countries would be binding on them. This reflected acknowledgment that most of the emissions (production and consumption) of CFCs and halons were in the industrialized countries and that this would continue to be the case for many years to come. The industrialized countries in turn pledged to continue to supply these chemicals for basic domestic needs of developing countries during the transition phase and to make available alternatives under favourable terms as soon as they were ready for introduction into the market. It was expected that the near-term availability of CFCs and halons and the longer-term availability of alternatives would act to limit major investments in CFCs and halon production and consumption technologies in developing countries. In any case, the magnitude of emissions from developing countries was not expected to grow significantly until several decades in the future.

It is quite a different case with global climate change. By 2020, CO_2 emissions from developing countries are projected to exceed those from industrialized countries (IEA, 2000). The investments made in developing countries over the next few years in power plants and other facilities that will release GHGs during their operation will be in place for many years to come, perhaps even half a century or more. For this reason (and because developing countries are potentially at greater risk from climate change), it is essential that these countries be involved in the global response as early as possible.

For this reason, any approach for limiting GHG emissions should be designed with inclusion of developing countries in the relatively near-to-medium term clearly in mind. Reduction targets of the kind adopted under the Kyoto Protocol do not lend themselves to acceptance by any but a few developing countries. They have been advocated mostly by and for industrialized countries with relatively stable populations and mature economies. They are not at all appropriate for developing countries with growing populations and human needs and aspirations for significant economic growth.

31.7 THE NATURE AND APPROPRIATENESS OF EMISSIONS TARGETS

The critical argument regarding application of the Montreal Protocol model to global climate change is how targets and timetables for reductions of emissions might be a key element of the response strategy for climate change. In view of the preceding discussion of a number of aspects of the two different environmental issues, the answer to this question is that, if such targets are included, they should not be in the form contained in the Montreal Protocol.

Under the Montreal Protocol, emissions of ODSs, or in a few cases, small groups of similar ODSs, are to be reduced according a precise, legally binding schedule. Each Party 'shall ensure' that the calculated levels of production and consumption of each ODS or group of ODSs does not exceed a fixed percentage of a certain established base level as of a specified future time period. This is essentially a gas-by-gas approach. It focuses on the unique specific uses of each ODS and establishes reduction schedules based on the technology and economic assessment of the feasibility of switching to alternatives within a specified time frame. Differences in the mix of ODSs used from one country to another are therefore mostly accommodated by considering each ODS separately.

Production limits are enforced through direct controls on each producer, which is administratively feasible due to the small number of producers. Consumption limits are enforced by controlling the supply of each ODS available for consumption, i.e. through controls on production, imports and exports. Again, this is feasible because the original number of controlled substances was small. Since then a large number of additional ODSs have been added to the annexes of the Montreal Protocol, but most of these are closely related and produced in the same facilities in relatively small volumes. This approach was also possible because the controls were applied directly to the substances themselves.

The Montreal Protocol approach does not take into account differences in the mix of uses of individual ODSs. Thus, countries that still used a large quantity of CFCs for aerosol sprays in 1986 (the base period for CFCs) had a relatively easy time achieving reductions as compared with countries that had earlier phased out such uses. A type of adjustment for certain uses, however, is provided through essential-use exemptions, which partially take into account certain sector-specific aspects.

In the Montreal Protocol, the targets at each stage (Montreal, London, Copenhagen, etc.) were established through a 'bottom-up' process, taking into account the results of the assessment process. They attempted at each stage to secure commitment to those emission-limitation actions that were agreed to be both technically and economically feasible. Moreover, they were negotiated simultaneously among governments and also with the affected

industries. The result at each stage was an agreement that had a high probability of ratification and successful implementation.

Given the broad range of human activities that result in GHG emissions, it is difficult to see how such an approach could be applied to reducing such emissions. As already noted, most GHGs are not 'produced' by commercial firms established for such purposes, but rather are the by-product of many activities carried out almost daily by almost every household and every consumer in the world. The situation is not improved greatly by considering the sources or materials that may be used in a way that causes GHG emissions. Every combustible organic material is a potential source of GHG emissions. Every cow and every sheep is a potential source.

Under such circumstances, it is extremely difficult for any government to 'ensure' that emissions would not exceed a certain level in a certain year. Regulating the behaviour of every individual consumer or farmer is quite impossible. Broad measures such as taxes and efficiency standards contribute to reducing emissions, but they cannot 'ensure' the achievement of specific reductions.

The major factors influencing emissions include rates of population growth, economic growth, world oil prices and agricultural production. The ability of experts to calculate the potential for reductions is much more limited than in the case of most ODSs. Although broad ranges or estimates of emission limitation potential can be provided, as in the IPCC Working Group III Summary (IPCC, 2001c), these generally do not take into account significant differences in circumstances between individual countries or even regional differences within a single country. It would be quite impossible for these experts to establish a proposed schedule for reductions, as is done under the Montreal Protocol assessment process, that would be appropriate for and acceptable to even a moderate-sized subset of countries, such as the OECD countries.

Yet the Kyoto Protocol imposes such specific legally binding limits on GHG emissions. During the negotiations, most countries had favoured a simple approach of a single 'flat-rate' emissions target for all industrialized countries. A few countries, however, argued that the large differences in national circumstances required that any targets be 'differentiated' to take into account these large differences. This was agreed in the end, but only at the final negotiating session in Kyoto. Thus, the negotiators had less than two weeks to decide how such differentiation should be determined.

The result was a set of emission targets, set out in Annex B to the Kyoto Protocol that was based primarily on politics, that is, 'top-down' rather than 'bottom-up'. There appears to be no evidence of analysis of the type done under the Montreal Protocol assessment process of what specific measures were available to each country for limiting its emissions, how effective these measures would be within the time frame of the Kyoto targets (2008–2012) and what the economic, social and other environmental impacts of the measures might be.

Further, there was little consultation with industry regarding the technical, economic and political feasibility of the measures that might need to be adopted to meet the targets. Such consultation was an intrinsic part of the negotiation process under the Montreal Protocol, and helped to ensure that the emission limitation commitments agreed internationally would be ratified in capitals and implemented within each country.

One point worth mentioning is the supposed benefit of emissions targets and timetables as a means of encouraging the development of new and more 'environmentally friendly' technologies. This aspect of targets has sometimes been referred to as 'technology forcing' or 'induced technology development and dissemination'. The Montreal Protocol approach is often cited as proof that targets can have this positive result. Advocates claim the targets in the Montreal Protocol sent a strong signal to industry, which then developed the technologies necessary to meet the targets.

This is not quite an accurate description of what actually happened under the Montreal Protocol process. The targets were in fact developed, as described above, with the full participation of industry experts and had already been assessed to be technologically and economically feasible. In this sense, they are 'bottom-up' targets based on knowledge of existing technology and changes that are possible in the real world.

In contrast, the 'technology-forcing' approach using targets as the driving factor would be a 'top-down' approach. Target advocates for GHG emission reductions in effect admit the difficulty in developing a 'bottom-up' approach due to the complexity of all the many different sources and variations in GHG emissions from country to country, and they wish to overcome this difficulty through a target-driven 'top-down' approach. But such an approach is unlikely to be acceptable to a number of governments once they have understood the consequences for their national situations.

A final consideration related to a target-based approach to limiting GHG emissions is the role of developing countries. As was explained above, the situation of developing countries with regard to climate change is almost the opposite of the situation with regard to ozone depletion. Developing countries are under relatively greater risk from climate change as compared with industrialized countries, and will be the largest source of problem emissions in a few decades (IEA, 2000; IPCC, 2001c). It is thus essential to involve developing countries in the global response (both mitigation and adaptation) as soon as possible.

If targets and timetables for GHG emission reductions by industrialized countries become the basic global response to global climate change, as implied by the present text of the Kyoto Protocol, a dilemma would be created. Most developing countries would not accept such targets, and the industrialized countries alone cannot reduce emissions sufficiently to offset growing emissions from developing countries in order to keep atmospheric concentrations of GHGs at levels likely to be judged acceptable. The world would thus be split into two camps: one whose efforts are inadequate to deal with the problem and the other who cannot join in the commitments of the first group and still meet the socio-economic needs of their own present and future populations.

31.8 SUMMARY AND CONCLUSIONS

The situations of ozone depletion and global climate change, while similar in a few regards, are sufficiently different that an approach other than legally binding targets and timetables for emission reductions, of the type used in the Montreal Protocol for addressing ozone depletion, must be developed for responding to climate change.

Ozone depletion and climate change are similar in that:

- Both issues involve a 'global commons' outside the jurisdiction of any country;
- Both originally involved a relatively small number of gases that are emitted as a result of human activities and that either destroy ozone or trap heat in the earth's atmosphere;
- Both require the participation of all major countries for an effective response;
- Both involve a gradual increase of scientific understanding over time, while requiring some actions in the near term in the face of some scientific uncertainty;
- Both responses should include and be guided by an integrated assessment of the state of evolving scientific understanding of the problem itself and its potential impacts, as well as the technical and economic aspects of possible response options.

The two situations differ, however, in several important respects:

- The science regarding global climate change is more complex and less convincing to major affected industries in 2001 than the science regarding ozone depletion was at the time of completion of the Montreal Protocol in late 1987;
- The science of the impacts of global climate change is far more complex than the science of ozone depletion impacts;
- The impacts of climate change may be either negative or positive for a particular country or region, while the impacts of ozone depletion are essentially negative for all countries;
- The impact of ozone depletion on humans is greatest on light-skinned populations living at high latitudes (i.e. in relatively advanced industrialized countries), while the impacts of climate

change are more complex and distributed in quite different ways, with the largest negative impacts probably falling on developing country inhabitants;

- Most production and use of the original ozone-depleting substances was concentrated in the industrialized countries, and their projected uses (now replaced by substitutes) are also predominantly in these countries. On the other hand, emissions of greenhouse gases from developing countries will exceed those from industrialized countries in a little more than a decade;

- Ozone-depleting substances are basically commercial chemicals produced intentionally by a relatively small number of producers, while most human-related greenhouse gas emissions are a by-product of many different activities carried out by almost all elements of societies;

- Because of their nature, production and use of ozone-depleting substances can gradually be phased out over time (in some cases less than 10 years), while greenhouse-gas emissions can never be entirely phased out;

- The possible response options for reducing greenhouse-gas emissions are far more complex and diverse than those for limiting ozone-depleting substances;

- The costs of major greenhouse-gas reductions are orders of magnitude higher than the costs of phasing out ozone-depleting substances, and affect virtually every sector of the economy;

- There are significant differences in the mix of and relative costs of climate change response options available to different countries or even to regions within countries, while most alternatives to ozone-depleting substances are equally applicable in most countries.

Because of these significant differences, the approach of the Montreal Protocol in establishing targets and timetables for reducing and eventually eliminating emissions of ozone-depleting substances is not directly applicable to the situation of global climate change. Alternative means of encouraging all countries to reduce greenhouse-gas emissions need to be developed in a broad spirit of international cooperation.

Continued pursuit of emission reduction targets applicable only to industrialized countries, and often to only some of these countries, risks creating a response that would have only minimal effect on global GHG emissions because most major emitters of GHGs are likely to remain outside of any agreement that would impose such targets. Moreover, it is also likely that the debate surrounding this kind of 'agreement' would be highly confrontational. This risks leaving a number of key players feeling disenchanted with the international process and less likely to cooperate in further efforts to respond to the global environmental challenge. In such a situation, everyone would lose, because there would be

no effective action to deal with global climate change, and the overall process and institutions within which such actions should be developed and agreed would have been damaged by the failed attempt.

ACKNOWLEDGEMENTS

The author would like to acknowledge the constructive comments of K. M. Sarma, former Executive Secretary of the Ozone Secretariat, UNEP, and the critical comments of two anonymous reviewers, which have helped to improve the manuscript.

REFERENCES

Enders, A. and Porges, A. (1992) Successful conventions and conventional success: saving the ozone layer. In K. Andersen and R. Blackhurst (Eds.), *The Greening of World Trade Issues*, chapter 7. New York, Harvester Wheatsheaf.

Farman, J. C., Gardiner, B. G. and Shanklin, J. D. (1985) Large losses of total ozone in Antarctica reveal seasonal ClOx/NOx interaction. *Nature* **315**, 207.

International Energy Agency (IEA) (2000) *World Energy Outlook 2000*. Paris, International Energy Agency.

IPCC (2001a) *Climate Change 2001: Impacts, Adaptation and Vulnerability. Contribution of Working Group II to the Third Assessment Report of the Intergovernmental Panel on Climate Change*. J. J. McCarthy, O. F. Canziani, N. A. Leary, D. J. Dokken and K. S. White (Eds.), Cambridge, Cambridge University Press.

(2001b) *Climate Change 2001: Synthesis Report. Third Assessment Report of the Intergovernmental Panel on Climate Change*. R. T. Watson (Ed.), Cambridge, Cambridge University Press.

(2001c) *Climate Change 2001: Mitigation. Contribution of Working Group III to the Third Assessment Report of the Intergovernmental Panel on Climate Change*. B. Metz, O. Davidson, R. Swart and J. Pan (Eds.), Cambridge, Cambridge University Press.

UNEP (2003) *Environmental Effects of Ozone Depletion and its Interactions with Climate Change: 2002 Assessment*. Nairobi, Kenya, United Nations Environment Programme.

WMO (1994) *Scientific Assessment of Ozone Depletion: 1994*. WMO Global Ozone Research and Monitoring Project – Report No. 37, Geneva, Switzerland, World Meteorological Organization.

(1998) *Scientific Assessment of Ozone Depletion: 1998*. WMO Global Ozone Research and Monitoring Project – Report No. 44, Geneva, Switzerland, World Meteorological Organization.

(2002) *Scientific Assessment of Ozone Depletion: 2002*; WMO Global Ozone Research and Monitoring Project – Report No. 47, Geneva, Switzerland, World Meteorological Organization.

Index

Note: page numbers in *italics* refer to figures and tables